GROUND PENETRATING RADAR THEORY AND APPLICATIONS

Ground Penetrating Radar Theory and Applications

Edited by

HARRY M. JOL

ELSEVIER

Amsterdam • Boston • Heidelberg • London • New York • Oxford
Paris • San Diego • San Francisco • Singapore • Sydney • Tokyo

Elsevier Science
Radarweg 29, PO Box 211, 1000 AE Amsterdam, The Netherlands
The Boulevard, Langford Lane, Kidlington, Oxford OX5 1GB, UK

First edition 2009

Copyright © 2009 Elsevier B.V. All rights reserved

No part of this publication may be reproduced, stored in a retrieval system, or transmitted in any form or by any means, electronic, mechanical, photocopying, recording, or otherwise, without the prior written permission of the publisher

Permissions may be sought directly from Elsevier's Science & Technology Rights Department in Oxford, UK: phone (+44) (0) 1865 843830; fax (+44) (0) 1865 853333; email: permissions@elsevier.com. Alternatively you can submit your request online by visiting the Elsevier web site at http://elsevier.com/locate/permissions, and selecting *Obtaining permission to use Elsevier material*

Notice
No responsibility is assumed by the publisher for any injury and/or damage to persons or property as a matter of products liability, negligence or otherwise, or from any use or operation of any methods, products, instructions or ideas contained in the material herein. Because of rapid advances in the medical sciences, in particular, independent verification of diagnoses and drug dosages should be made

British Library Cataloguing-in-Publication Data
A catalogue record for this book is available from the British Library

Library of Congress Cataloging-in-Publication Data
A catalog record for this book is available from the Library of Congress

ISBN: 978-0-444-53348-7

For information on all Elsevier publications
visit our website at elsevierdirect.com

Printed in Slovenia
08 09 10 11   10 9 8 7 6 5 4 3 2 1

Working together to grow
libraries in developing countries

www.elsevier.com | www.bookaid.org | www.sabre.org

ELSEVIER   BOOK AID International   Sabre Foundation

Contents

Preface   *xiii*
Contributors   *xv*

Part I   Ground Penetrating Radar (GPR) Principles   1

1   Electromagnetic Principles of Ground Penetrating Radar   3
A.P. Annan

1.1	Introduction		4
1.2	Ground Penetrating Radar Basic Principles		5
	1.2.1	Overview	5
	1.2.2	Maxwell's equations	6
	1.2.3	Constitutive equations	6
	1.2.4	Material properties	7
1.3	Wave Nature of Electromagnetic Fields		8
	1.3.1	Wave properties	10
	1.3.2	Ground penetrating radar source near an interface	11
	1.3.3	Reflection, refraction, and transmission at interfaces	13
	1.3.4	Resolution and zone of influence	14
	1.3.5	Scattering attenuation	16
1.4	Signal Measurement		17
	1.4.1	Time ranges and bandwidth	18
	1.4.2	Center frequency	19
	1.4.3	Ground penetrating radar signal acquisition	20
	1.4.4	Characterizing system response	20
	1.4.5	Recording dynamic range	22
	1.4.6	Antennas	23
	1.4.7	Antenna directivity	24
	1.4.8	Antenna shielding	27
1.5	Survey Methodology		29
	1.5.1	Sampling criteria	29
	1.5.2	Ground penetrating radar surveys	30
	1.5.3	Common-offset reflection survey	30
	1.5.4	Multioffset common midpoint/wide-angle reflection and refraction velocity sounding design	31
	1.5.5	Transillumination surveys	31
1.6	Data Analysis and Interpretation		33
	1.6.1	Dewow	34
	1.6.2	Time gain	34
	1.6.3	Deconvolution	35
	1.6.4	Migration	36

	1.6.5	Topographic correction	36
	1.6.6	Two-dimensional and three-dimensional data visualization	37
1.7	Summary		37

2 Electrical and Magnetic Properties of Rocks, Soils and Fluids — 41
Nigel J. Cassidy

2.1	Introduction		41
2.2	Electromagnetic Material Properties: Basic Theory		43
2.3	Permittivity and Conductivity – The Electrical Parameters of Dielectrics		44
	2.3.1	Permittivity – ε	45
	2.3.2	Conductivity – σ	54
	2.3.3	Permeability μ – the magnetic parameters of dielectrics	55
2.4	Material Properties – Relationship to Electromagnetic Wave Characteristics		57
	2.4.1	Loss factor and skin depth	59
2.5	The Properties of Real Materials – Practical Evaluations		60
2.6	Characterising the Response of Real Materials		62
	2.6.1	Basic mixing models	63
	2.6.2	Volumetric and inclusion-based mixing models	64
2.7	Summary		66
Acknowledgements			67

3 Ground Penetrating Radar Systems and Design — 73
Steven Koppenjan

3.1	Introduction and Background		73
3.2	Methodology – Types of Ground Penetrating Radar		74
	3.2.1	Impulse	75
	3.2.2	Swept frequency-modulated continuous wave	75
	3.2.3	Stepped frequency-modulated continuous wave	76
	3.2.4	Gated, stepped frequency-modulated continuous wave	76
3.3	Radio Frequency Specifications and Definitions		77
	3.3.1	Dynamic range	77
	3.3.2	Bandwidth	78
	3.3.3	Range resolution	78
	3.3.4	Lateral resolution	79
	3.3.5	Unambiguous range	79
3.4	General Design Criteria for Ground Penetrating Radar		80
	3.4.1	System performance	81
3.5	Impulse Ground Penetrating Radar		81
	3.5.1	Theory of operation: Impulse radar	81
	3.5.2	System design parameters: impulse radar	84
	3.5.3	Implementation of an impulse ground penetrating radar	85
3.6	Continuous-Wave Ground Penetrating Radar		86
	3.6.1	Theory of operation – stepped-frequency, continuous-wave radar	86
	3.6.2	System design parameters: stepped-frequency radar	92
	3.6.3	Implementation of a gated, stepped-frequency, ground penetrating radar	93

4 Antennas
David J. Daniels
99

4.1	Introduction		99
4.2	Basic Antenna Parameters		102
	4.2.1	Energy transfer from antennas	102
	4.2.2	Gain	104
	4.2.3	Directivity	105
	4.2.4	Coupling energy into the ground	105
	4.2.5	Antenna efficiency	106
	4.2.6	Sidelobes and back lobes	106
	4.2.7	Bandwidth	106
	4.2.8	Polarisation – linear, elliptical, circular	107
	4.2.9	Antenna phase centre	108
	4.2.10	Antenna patterns	108
	4.2.11	Time sidelobes and ring-down	109
	4.2.12	Antenna footprint	110
4.3	Antennas for Ground Penetrating Radar		112
	4.3.1	Introduction	112
	4.3.2	Coupling into a dielectric	113
	4.3.3	Time domain antennas	115
	4.3.4	Frequency domain antennas	124
	4.3.5	Array antennas	128
4.4	Summary		133
4.5	Definitions		133

5 Ground Penetrating Radar Data Processing, Modelling and Analysis
Nigel J. Cassidy
141

5.1	Introduction		141
5.2	Background and Practical Principles of Ground Penetrating Radar Data Processing		143
5.3	Ground Penetrating Radar Data Processing: Developing Good Practice		145
5.4	Basic Ground Penetrating Radar Data Processing Steps		148
	5.4.1	Data/trace editing and 'rubber-band' interpolation	148
	5.4.2	Dewow filtering	150
	5.4.3	Time-zero correction	150
	5.4.4	Filtering	152
	5.4.5	Deconvolution	158
	5.4.6	Velocity analysis and depth conversion	158
	5.4.7	Elevation or topographic corrections	159
	5.4.8	Gain functions	161
	5.4.9	Migration	164
	5.4.10	Advanced imaging and analysis tools	166
	5.4.11	Attribute analysis	167
	5.4.12	Numerical modelling	168
5.5	Processing, Imaging and Visualisation: Concluding Remarks		171
Acknowledgements			172

Part II Environmental Applications — 177

6 Soils, Peatlands, and Biomonitoring — 179
James A. Doolittle and John R. Butnor

- 6.1 Introduction — 179
- 6.2 Soils — 180
    - 6.2.1 Soil properties that affect the performance of ground penetrating radar — 180
    - 6.2.2 Soil suitability maps for ground penetrating radar — 181
    - 6.2.3 Ground penetrating data and soil surveys — 185
    - 6.2.4 Uses of ground penetrating radar in organic soils and peatlands — 190
- 6.3 Biomonitoring — 192

7 The Contribution of Ground Penetrating Radar to Water Resource Research — 203
Lee Slater and Xavier Comas

- 7.1 Introduction — 203
- 7.2 Petrophysics — 206
- 7.3 Hydrostratigraphic Characterization — 209
- 7.4 Distribution/Zonation of Flow and Transport Parameters — 214
- 7.5 Moisture Content Estimation — 217
- 7.6 Monitoring Dynamic Hydrological Processes — 224
    - 7.6.1 Recharge/moisture content in the vadose zone — 225
    - 7.6.2 Water table detection/monitoring — 228
    - 7.6.3 Solute transport in fractures — 229
    - 7.6.4 Studies of the hyporheic corridor — 231
    - 7.6.5 Studies of the rhizosphere — 232
    - 7.6.6 Carbon gas emissions from soils — 232
- 7.7 Conclusions — 237

8 Contaminant Mapping — 247
J.D. Redman

- 8.1 Introduction — 247
- 8.2 Contaminant Types — 248
- 8.3 Electrical Properties of Contaminated Rock and Soil — 249
    - 8.3.1 Electrical properties of NAPLs — 249
    - 8.3.2 Electrical properties of soil and rock with NAPL contamination — 250
    - 8.3.3 Biodegradation effects — 253
    - 8.3.4 Inorganics — 253
- 8.4 Typical Distribution of Contaminants — 254
    - 8.4.1 DNAPL — 254
    - 8.4.2 LNAPL — 255
    - 8.4.3 Inorganics — 255
    - 8.4.4 Saturated and unsaturated zone — 256
- 8.5 GPR Methodology — 256

8.6	Data Processing and Interpretation			257
	8.6.1	Visualization		257
	8.6.2	Trace attributes		257
	8.6.3	Data differencing		257
	8.6.4	AVO analysis		258
	8.6.5	Detection based on frequency-dependent properties		258
	8.6.6	Quantitative estimates of NAPL		258
8.7	Case Studies			259
	8.7.1	Controlled DNAPL injection		260
	8.7.2	Controlled LNAPL injection		262
	8.7.3	Accidental spill sites		262
	8.7.4	Leachate and waste disposal site characterization		264
8.8	Summary			265
	Terms for Glossary			269

Part III Earth Science Applications — 271

9 Ground Penetrating Radar in Aeolian Dune Sands — 273
Charlie Bristow

9.1	Introduction		274
9.2	Sand Dunes		274
9.3	Survey Design		277
	9.3.1	Line spacing	277
	9.3.2	Step size	277
	9.3.3	Orientation	278
	9.3.4	Survey direction	278
	9.3.5	Vertical resolution	278
9.4	Topography		279
	9.4.1	Topographic surveys	280
	9.4.2	Topographic correction	281
	9.4.3	Apparent dip	281
9.5	Imaging Sedimentary Structures and Dune Stratigraphy		281
9.6	Radar Facies		282
9.7	Radar Stratigraphy and Bounding Surfaces		283
9.8	Aeolian Bounding Surfaces		285
	9.8.1	Reactivation surfaces	285
	9.8.2	Superposition surfaces	285
	9.8.3	Interdune surfaces	286
9.9	Dune Age and Migration		288
9.10	Stratigraphic Analysis		288
9.11	Ancient Aeolian Sandstones		290
9.12	Three-Dimensional Images		290
9.13	Pedogenic Alteration and Early Diagenesis		291
	9.13.1	Evaporites	291
	9.13.2	Environmental noise	291
	9.13.3	Diffractions	293

		9.13.4	The water table	293
		9.13.5	Multiples	293
	9.14	Conclusions		294
	Acknowledgments			294

10 Coastal Environments — 299
Ilya V. Buynevich, Harry M. Jol and Duncan M. FitzGerald

	10.1	Introduction	299
	10.2	Methodology	301
	10.3	Ground Penetrating Radar Strengths in Coastal Environments	303
	10.4	Ground Penetrating Radar Limitations in Coastal Environments	304
	10.5	Ground Penetrating Radar Studies in Coastal Environments	305
	10.6	Examples of Ground Penetrating Radar Images from Coastal Environments	305
		10.6.1 Record of coastal progradation	306
		10.6.2 Signatures of coastal erosion	307
		10.6.3 Coastal Paleochannels	308
		10.6.4 Ground penetrating radar signal response to lithological anomalies in coastal dunes	310
		10.6.5 Deltas	312
		10.6.6 Reservoir characterization – hydrocarbon and hydrogeology	313
	10.7	Summary	314
	Acknowledgments		315

11 Advances in Fluvial Sedimentology using GPR — 323
John Bridge

	11.1	Introduction	323
	11.2	Scales of Fluvial Deposits and GPR Resolution	324
	11.3	Examples of Use of GPR in Fluvial Sedimentology	327
		11.3.1 South Esk, Scotland	327
		11.3.2 Calamus, Nebraska	329
		11.3.3 Brahmaputra (Jamuna), Bangladesh	331
		11.3.4 Niobrara, Nebraska	336
		11.3.5 South Saskatchewan, Canada	340
		11.3.6 Sagavanirktok, northern Alaska	343
		11.3.7 Fraser and Squamish Rivers, Canada	349
		11.3.8 Pleistocene outwash deposits in Europe	350
		11.3.9 Mesozoic deposits of SW USA	353
	11.4	Concluding Discussion	354
	Acknowledgments		355

12 Glaciers and Ice Sheets — 361
Steven A. Arcone

	12.1	Introduction	361
	12.2	Antarctica	363
		12.2.1 Alpine glaciers: Dry valleys	365

		12.2.2	Polar firn: West Antarctica	367
		12.2.3	Englacial stratigraphy: West Antarctica	371
		12.2.4	Ice shelf: McMurdo Sound	373
		12.2.5	Crevasses: Ross Ice Shelf	376
	12.3	Alaska		379
		12.3.1	Temperate valley glacier: Matanuska Glacier	380
		12.3.2	Temperate valley glacier: Gulkana Glacier	382
		12.3.3	Temperate firn: Bagley Ice Field, Alaska	384
		12.3.4	Temperate hydrology: Black Rapids Glacier	385
	12.4	Summary		388

Part IV Engineering and Societal Applications — 393

13 NDT Transportation — 395

Timo Saarenketo

	13.1	Introduction		396
	13.2	GPR Hardware and Accessories		397
		13.2.1	General	397
		13.2.2	Air-coupled systems	398
		13.2.3	Ground-coupled systems	398
		13.2.4	Antenna configurations	399
		13.2.5	Antenna and GPR system testing	399
		13.2.6	Accessory equipment	400
	13.3	Data Collection		401
		13.3.1	General	401
		13.3.2	Data collection setups and files	403
		13.3.3	Positioning	404
		13.3.4	Reference sampling	405
	13.4	Data Processing and Interpretation		405
		13.4.1	General	405
		13.4.2	GPR data preprocessing	406
		13.4.3	Air-coupled antenna data processing	407
		13.4.4	Ground-coupled data processing	408
		13.4.5	Determining dielectric values or signal velocities	410
		13.4.6	Interpretation – automated vs. user controlled systems	411
		13.4.7	Interpretation of structures and other objects	411
	13.5	Integrated GPR Data Analysis with Other Road Survey Data		413
		13.5.1	General	413
		13.5.2	GPR and FWD	413
		13.5.3	Profilometer data	414
		13.5.4	GPS, digital video and photos	415
		13.5.5	Other data	416
	13.6	GPR Applications on Roads and Streets		416
		13.6.1	General	416
		13.6.2	Subgrade surveys, site investigations	416
		13.6.3	Unbound pavement structures	419
		13.6.4	Bound pavement structures and wearing courses	420

		13.6.5	GPR in QC/QA	423
		13.6.6	Special applications	425
	13.7	Bridges		425
		13.7.1	General	425
		13.7.2	Bridge deck surveys	426
		13.7.3	Other bridge applications	428
	13.8	Railways		429
		13.8.1	General	429
		13.8.2	Data collection from railway structures	430
		13.8.3	Ballast surveys	431
		13.8.4	Subgrade surveys, site investigations	432
	13.9	Airfields		433
	13.10	Summary and Recommendations		435

14 Landmine and Unexploded Ordnance Detection and Classification with Ground Penetrating Radar 445
Alexander Yarovoy

	14.1	Introduction	445
	14.2	Electromagnetic Analysis	446
	14.3	System Design	455
	14.4	GPR Data Processing for Landmine/UXO Detection and Classification	462
	14.5	Fusion with Other Sensors	469
	14.6	Overall Performance of GPR as an UXO/Landmine Sensor	472
	14.7	Conclusion	473

15 GPR Archaeometry 479
Dean Goodman, Salvatore Piro, Yasushi Nishimura, Kent Schneider, Hiromichi Hongo, Noriaki Higashi, John Steinberg and Brian Damiata

	15.1	Introduction		479
	15.2	Field Methods for Archaeological Acquisition		481
	15.3	Imaging Techniques for Archaeology		482
	15.4	Depth Determination		484
	15.5	Case Histories		485
		15.5.1	Case History No. 1: The Forum Novum, Tiber Valley, Italy	486
		15.5.2	Case History No. 2: The Villa of Emperor Trajanus of Rome, Italy	488
		15.5.3	Case History No. 3: Wroxeter Roman Town, England	494
		15.5.4	Case History No. 4: Saitobaru Burial Mound No. 100, Japan	495
		15.5.5	Case History No. 5: Saitobaru Burial Mound No. 111, Japan	498
		15.5.6	Case History No. 6: Monks Mound, Cahokia, Illinois	501
		15.5.7	Case History No. 7: Jena Choctaw Tribal Cemetery, Louisiana	502
		15.5.8	Case History No. 8: Glaumbaer Viking Age, Iceland	505
	Acknowledgments			507

Index 509

Preface

Ground penetrating radar (GPR) is a rapidly growing field that has seen tremendous progress in the development of theory, technique, technology, and range of applications over the past 15–20 years. GPR has also become a valuable method utilized by a variety of scientists, researchers, engineers, consultants, and university students from many disciplines. The diversity of GPR applications includes a variety of areas such as the study of groundwater contamination, geotechnical engineering, sedimentology, glaciology, and archaeology. This breath of usage has lead to GPR's rapid development and pre-eminence in geophysical consulting and geotechnical engineering, as well as inspiring new areas of interdisciplinary research in academia and industry. The topic of GPR has gone from not even being mentioned in geophysical texts a little over a decade ago to being the focus of hundreds of research papers and special issues of journals dedicated to the subject. The explosion of literature devoted to GPR theory, technology, and applications has led to this book which provides an overview and up-to-date synthesis of select areas in this swiftly evolving field. The book also provides sufficient background and case studies to allow both practitioners and newcomers to the area of GPR to use the volume as an accessible handbook and primary research reference.

This publication begins with a part that focuses on the fundamental aspects of GPR including electromagnetic principles of GPR (Annan), electrical and magnetic properties of rocks, soils and fluids (Cassidy), systems and design (Koppenjan), antennas (Daniels), and data processing, modeling and analysis (Cassidy). The next part covers environmental applications of GPR and includes topics relating to soils, peatlands and biomonitoring (Doolittle and Butnor), water resources (Slater and Comas), and contaminant mapping (Redman). The third part looks at applications relevant to the field of earth science and includes topics on aeolian dune sands (Bristow), coastal environments (Buynevich et al.), fluvial sedimentology (Bridge), and glaciers and ice sheets (Arcone). The volume is rounded out with a part on engineering and societal applications of GPR that cover NDT transportation (Saarenketo), landmine and UXO detection and classification (Yarovoy), and Archaeometry (Goodman et al.).

Individual chapters provide a review of the current state of GPR development as well as contemporary issues which the author(s) feels are most appropriate. The authors are leaders in their respective fields and are employed in a variety of settings including industry, consulting, government agencies, and academic institutes. By the very nature of this approach, chapters will reflect the author's strengths and will not be uniform in format. The reader will find some chapters have a theoretical focus, while others are more mathematical, and yet others take a case study approach. Each chapter also includes numerous references to direct the interested individual to further information and a more detailed examination of each topic. Many of the references are scientific journals and reports, but where authors felt it advantageous, various other sources have been cited. Ultimately, this book reflects a wide range of disciplines and perspectives that show how the field of GPR has a sound theoretical and practical base from which to grow in the future. In addition, the publication should bring students up to date on the latest subsurface GPR-imaging techniques. It will provide guidance to geophysical consultants, researchers, and engineers who want to move into new applications and/or expand their capability for efficient and effective subsurface investigation. It will also allow individuals from outside the field to gain ample information on select topics relating to GPR.

The following people are thanked for their technical peer reviews of one or more chapters in this book: S. Arcone, M. Bano, W. Barnhardt, J. Bridge, I. Buynevich, N. Cassidy, M. Collins, X. Comas, I. Craddock, D. Daniels, J. Doolittle, A. Enders, D. Goodman, S.-E. Hamran, S. Koppenjan, I. Lunt, C. Peterson, D. Redman, M. Sato, T. Savelyev, T. Scullion, R. Versteeg, B. Welch, R. Young, and A. Yarovoy. In addition, the individuals who reviewed chapters at the request of the authors are also thanked. The time and expertise of these individuals has significantly improved the quality of the book, but the responsibility for the content of each chapter ultimately rests with the authors.

Brian Moorman is thanked for his insight and vision in getting the project off the ground. Linda Versteeg's guidance and support greatly aided in getting the book completed. I thank my family (Carleen, Brianna, and Connor) and University of Wisconsin-Eau Claire for their support throughout this endeavor. Finally, I acknowledge, due to various reasons, the volume has been delayed at several stages and it has taken the patience of many of the authors, as well as the publisher, to complete this publication – their perseverance is much appreciated.

Harry M. Jol
University of Wisconsin-Eau Claire

Contributors

Peter Annan
Sensors and Software Inc.
1040 Stacey Court
Mississauga, ON L4W 2X8
Canada
E-mail: apa@sensoft.ca

Steven A. Arcone
US Army ERDC-CRREL
72 Lyme Road
Hanover, NH 03755-1290
USA
E-mail: Steven.a.arcone@usace.army.mil

John Bridge
Binghamton University
Department of Geological Sciences and Environmental Studies
PO BOX 6000, Binghamton, NY 13902-6000
USA
E-mail: jbridge@binghamton.edu

Charlie Bristow
School of Earth Sciences
Birkbeck College University of London
Malet Street
London WC1E 7HX
United Kingdom
E-mail: c.bristow@ucl.ac.uk

John Butnor
USDA Forest Service
Southern Research Station
705 Spear Street
South Burlington, VT 05403
USA
E-mail: jbutnor@fs.fed.us

Ilya V. Buynevich
Coastal Systems Group
Geology & Geophysics Department
MS #22
Woods Hole Oceanographic Institution

Woods Hole, MA 02543
USA
E-mail: ibuynevich@whoi.edu

Nigel Cassidy
School of Physical and Geographical Sciences
William Smith Building
Keele University
Staffordshire, ST5 5BG
United Kingdom
E-mail: n.j.cassidy@esci.keele.ac.uk

Xavier Comas
Department of Geosciences
Florida Atlantic University
777 Glades Road
Boca Raton, FL 33432
USA
E-mail: xcomas@fau.edu

Brian Damiata
Costen Institute of Archaeology
University of California Los Angeles
Box 95159, A210 Fowler Building
Los Angeles, CA 90095-1510
USA
E-mail: damiata@ucla.edu

David Daniels
Antennas and Electronics Division,
ERA Technology
Cleeve Road
Leatherhead, Surrey, KT22 7SA
United Kingdom
E-mail: david.daniels@era.co.uk

James Doolittle
USDA – Natural Resources Conservation Service – National Soil Survey Center
11 Campus Blvd., Suite 200
Newtown Square, PA 19073
USA
E-mail: jim.doolittle@lin.usda.gov

Duncan FitzGerald
Department of Earth Sciences
Boston University
675 Commonwealth Avenue
Boston, MA 02215
USA
E-mail: dunc@bu.edu

Dean Goodman
20014 Gypsy Land
Woodland Hills, CA 91364
USA
E-mail: gal_usa_goodman@msn.com

Noriaki Higashi
Saitobaru Archaeological Museum
Saito City, Miyazaki Prefecture
Japan
E-mail: higashi-noriaki@pref.miyazaki.lg.jp

Hiromichi Hongo
Saitobaru Archaeological Museum
Saito City, Miyazaki Prefecture
Japan
E-mail: hongo-hiromichi@pref.miyazaki.lg.jp

Harry Jol
Department of Geography and Anthropology
University of Wisconsin-Eau Claire
105 Garfield Avenue
Eau Claire, WI 54702-4004
USA
E-mail: jolhm@uwedc.edu

Steven Koppenjan
Special Technologies Laboratory
5520 Ekwill Street
Santa Barbara, CA 93111
USA
E-mail: koppensk@nv.doe.gov

Yasushi Nishimura
UNESCO-ACCU
757 Horen-cho
Nara 630-8113
E-mail: yasushi@nabunken.go.jp

Salvatore Piro
Consiglio Nazionale Delle Ricerche istituto per le Technologie
Applicate ai Beni Culturali
00016 Monterondo
SCALO, Rome c.post.10
Italy
E-mail: salvatore.piro@itabc.cnr.it

David Redman
Sensors and Software Inc.
1040 Stacey Court
Mississauga, ON L4W 2X8
Canada
E-mail: dr@sensoft.ca

Timo Saarenketo
Roadscanners Oy
Urheilukatu 5-7
P.O.Box 2219
FIN-96101 Rovaniemi
Finland
Email: timo.saarenketo@roadscanners.com

Kent Schneider
Underground Imaging Solutions, Inc.,
9790 Misty Cove Lane
Gainesville, GA
USA
E-mail: krschne@bellsouth.net

Lee Slater
Earth & Environmental Sciences
Rutgers-Newark
101 Warren St.
Newark, NJ 07102
USA
E-mail: lslater@andromeda.rutgers.edu

John Steinberg
Fiske Center for Archaeological Research
University of Massachusetts Boston
100 Morrissey Boulevard
Boston, MA 02125
USA
E-mail: john.steinberg@umb.edu

Alexander Yarovoy
Int. Research Centre for Telecom and Radar
Delft University of Technology
Mekelweg 4
2628 CD Delft
The Netherlands
E-mail: a.yarovoy@ewi.tudelft.nl

PART I

GROUND PENETRATING RADAR (GPR) PRINCIPLES

CHAPTER 1

Electromagnetic Principles of Ground Penetrating Radar

A.P. Annan

Contents

1.1. Introduction	4
1.2. Ground penetrating Radar Basic Principles	5
1.2.1. Overview	5
1.2.2. Maxwell's equations	6
1.2.3. Constitutive equations	6
1.2.4. Material properties	7
1.3. Wave Nature of Electromagnetic Fields	8
1.3.1. Wave properties	10
1.3.2. Ground penetrating radar source near an interface	11
1.3.3. Reflection, refraction, and transmission at interfaces	13
1.3.4. Resolution and zone of influence	14
1.3.5. Scattering attenuation	16
1.4. Signal Measurement	17
1.4.1. Time ranges and bandwidth	18
1.4.2. Center frequency	19
1.4.3. Ground penetrating radar signal acquisition	20
1.4.4. Characterizing system response	20
1.4.5. Recording dynamic range	22
1.4.6. Antennas	23
1.4.7. Antenna directivity	24
1.4.8. Antenna shielding	27
1.5. Survey Methodology	29
1.5.1. Sampling criteria	29
1.5.2. Ground penetrating radar surveys	30
1.5.3. Common-offset reflection survey	30
1.5.4. Multioffset common midpoint/wide-angle reflection and refraction velocity sounding design	31
1.5.5. Transillumination surveys	31
1.6. Data Analysis and Interpretation	33
1.6.1. Dewow	34
1.6.2. Time gain	34
1.6.3. Deconvolution	35
1.6.4. Migration	36
1.6.5. Topographic correction	36
1.6.6. Two-dimensional and three-dimensional data visualization	37
1.7. Summary	37
References	38

1.1. INTRODUCTION

Ground penetrating radar (GPR) is now a well-accepted geophysical technique. The method uses radio waves to probe "the ground" which means any low loss dielectric material. In its earliest inception, GPR was primarily applied to natural geologic materials. Now GPR is equally well applied to a host of other media such as wood, concrete, and asphalt.

The existence of numerous lossy dielectric material environments combined with the broad radio frequency spectrum leads to a wide range of GPR applications. The same methodology can be applied to glaciology and to nondestructive testing of concrete structures; the spatial scale of applications varies from kilometers to centimeters.

The most common form of GPR measurements deploys a transmitter and a receiver in a fixed geometry, which are moved over the surface to detect reflections from subsurface features. In some applications, transillumination of the volume under investigation is more useful. Both concepts are depicted in Figure 1.1. An example of GPR response is shown in Figure 1.2.

Use of radio waves to sound the earth was contemplated for decades before results were obtained in the 1950s (El Said, 1956; Waite and Schmidt, 1961). Waite's demonstration of ice sheet sounding with aircraft radar altimeters leads to radio echo sounding in many locations around the world. From this start, there was a gradual transition of the concepts to sounding soils and rocks, which began in the 1960s, and has continued ever since.

From the early days, applications have mushroomed, our knowledge of the basic physics has grown in leaps and bounds, and the nature of material understanding has blossomed (Davis and Annan, 1989). A succinct historical summary is given by Annan (2002). Excellent discussions of the use of GPR for geologic stratigraphy can be found in Bristow and Jol (2003) and for hydrogeology in Rubin and Hubbard (2005).

Ground penetrating radar has evolved its own natural set of terminology, common understandings, and practical application procedures. The objective of

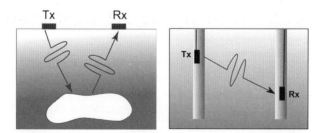

Figure 1.1 Ground penetrating radar (GPR) uses radio waves to probe the subsurface of lossy dielectric materials. Two modes of measurement are common. In the first, reflected or scattered energy is detected. In the second, effects on energy transmitted through the material are observed.

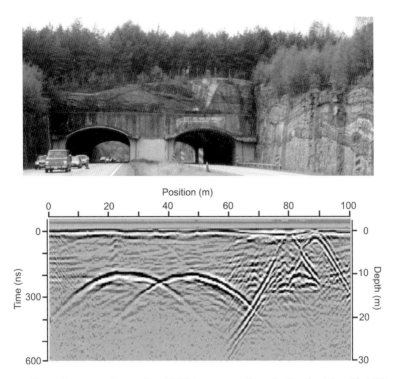

Figure 1.2 Ground penetrating radar (GPR) cross section obtained with a 50-MHz system traversing over two road tunnels. Ground penetrating radar signal amplitude is displayed as a function of position (horizontal axis) and travel time (vertical axis).

this chapter is to provide a succinct overview of the key concepts, physical issues, and practical experiences that underpin the language, practice, and interpretation of GPR today.

1.2. GROUND PENETRATING RADAR BASIC PRINCIPLES

1.2.1. Overview

The foundations of GPR lie in electromagnetic (EM) theory. The history of this field spans more than two centuries and is the subject of numerous texts such as Jackson (1962) and Smythe (1989). This overview outlines the basic building blocks needed to work quantitatively with GPR.

Maxwell's equations mathematically describe the physics of EM fields, while constitutive relationships quantify material properties. Combining the two provides the foundations for quantitatively describing GPR signals.

1.2.2. Maxwell's equations

In mathematical terms, EM fields and relationships are expressed as follows:

$$\overline{\nabla} \times \overline{E} = -\frac{\partial \overline{B}}{\partial t} \qquad (1.1)$$

$$\overline{\nabla} \times \overline{H} = \overline{J} + \frac{\partial \overline{D}}{\partial t} \qquad (1.2)$$

$$\overline{\nabla} \cdot \overline{D} = q \qquad (1.3)$$

$$\overline{\nabla} \cdot \overline{B} = 0 \qquad (1.4)$$

where $\overline{E}$ is the electric field strength vector (V/m); q is the electric charge density (C/m^3); $\overline{B}$ is the magnetic flux density vector (T); $\overline{J}$ is the electric current density vector (A/m^2); $\overline{D}$ is the electric displacement vector (C/m^2); t is time (s); and $\overline{H}$ is the magnetic field intensity (A/m).

Maxwell succinctly summarized the work of numerous researchers in this compact form. From these relationships, all classic EMs (induction, radio waves, resistivity, circuit theory, etc.) can be derived when combined with formalism to characterize material electrical properties.

1.2.3. Constitutive equations

Constitutive relationships are the means of describing a material's response to EM fields. For GPR, the electrical and magnetic properties are of importance. Constitutive equations (Equations (1.5), (1.6) and (1.7)) provide a macroscopic (or average behavior) description of how electrons, atoms, and molecules respond *en masse* to the application of an EM field.

$$\overline{J} = \tilde{\sigma}\overline{E} \qquad (1.5)$$

$$\overline{D} = \tilde{\varepsilon}\overline{E} \qquad (1.6)$$

$$\overline{B} = \tilde{\mu}\overline{H} \qquad (1.7)$$

Electrical conductivity $\tilde{\sigma}$ characterizes free charge movement (creating electric current) when an electric field is present. Resistance to charge flow leads to energy dissipation. Dielectric permittivity $\tilde{\varepsilon}$ characterizes displacement of charge constrained in a material structure to the presence of an electric field. Charge displacement results in energy storage in the material. Magnetic permeability $\tilde{\mu}$ describes how intrinsic atomic and molecular magnetic moments respond to a magnetic field. For simple materials, distorting intrinsic magnetic moments store energy in the material.

$\tilde{\sigma}$, $\tilde{\varepsilon}$, and $\tilde{\mu}$ are tensor quantities and can also be nonlinear (i.e. $\tilde{\sigma} = \tilde{\sigma}(E)$). For virtually all practical GPR issues, these quantities are treated as field-independent scalar qualities. (In other words, the response is in the same direction as the exciting field and is independent of field strength.) Although these assumptions are seldom fully valid, to date, investigators working on practical applications have seldom been able to discern such complexity.

Material properties can also depend on the history of the incident field. Time history dependence manifests itself when the electrical charges in a structure have a finite response time, making them appear as fixed for slow rates of field change and free to move for faster rates of field change. To be fully correct, Equations (1.5), (1.6) and (1.7) should be expressed in the following form (only Equation (1.5) is written for brevity):

$$\bar{J}(t) = \int_0^\infty \tilde{\sigma}(\beta) \cdot \overline{E}(t - \beta) d\beta \tag{1.8}$$

This more complex form of the constitutive equations must be used when physical properties are dispersive. For most GPR applications, assuming the scalar constant form for ε, μ, σ suffices with ε and σ being the most important.

For GPR, the dielectric permittivity is an important quantity. Most often, the terms relative permittivity or "dielectric constant" are used and defined as follows:

$$\kappa = \frac{\varepsilon}{\varepsilon_0} \tag{1.9}$$

where ε_0 is the permittivity of vacuum, 8.89×10^{-12} F/m.

1.2.4. Material properties

The subject of electrical properties (ε, μ, σ) of materials is a wide-ranging topic. Background can be found in Olhoeft (1981, 1987) and Santamarina et al. (2001). Discussion here is limited to the common basic issues. In most GPR applications, variations in ε and σ are most important while variations in μ are seldom of concern.

Ground penetrating radar is most useful in low-electrical-loss materials. If $\sigma = 0$, GPR would see very broad use since signals would penetrate to great depth. In practice, low-electrical-loss conditions are not prevalent. Clay-rich environments or areas of saline groundwater can create conditions where GPR signal penetration is very limited.

Earth materials are invariably composites of many other materials or components. Water and ice represent the few cases where a single component is primarily present. A simple beach sand is a mixture of soil grains, air, water, and ions dissolved in water. Soil grains will typically occupy 60-80% of the available volume. Understanding the physical properties of mixtures is thus a key factor in the interpretation of a GPR response.

Mixtures of materials seldom exhibit properties directly in proportion to the volume fraction of the constituent components. In many respects, this complexity can make quantitative analysis of GPR data impossible without ancillary information.

Although the subject of mixtures is complex, the big picture of GPR perspective is simpler. In the 10-1000 MHz frequency range, the presence or absence of water in the material dominates behavior with the general picture being as follows:

- Bulk minerals and aggregates in mixtures generally are good dielectric insulators. They typically have a permittivity in the range of 3-8 (depending on mineralogy and compaction) and are usually insulating with virtually zero conductivity.
- Soils, rocks, and construction materials have empty space between the grains (pore space) available to be filled with air, water, or other material.
- Water is by far the most polarizable, naturally occurring material (in other words, it has a high permittivity with $\kappa \approx 80$).
- Water in the pore space normally contains ions, and the water electrical conductivity associated with ion mobility is often the dominant factor in determining bulk material electrical conductivity. Resulting soil and rock conductivities are typically in the 1–1000 mS/m range.
- Since water is invariably present in the pore space of natural (geologic) materials, except in such unique situations where vacuum drying or some other mechanism assures the total absence of water, it has a dominant effect on electrical properties.

Empirically derived forms such as the Topp relationship (Topp et al., 1980) and variations of Archie's law (Archie, 1942) have long demonstrated the relationship between permittivity, electrical conductivity, and volumetric water content for soils. More advanced relationships, such as the BHS model (Sen et al., 1981), use effective media theory models to derive a composite material property from constituents. Referring to the reference materials and other chapters of this text will provide a more substantive view of this subject.

1.3. WAVE NATURE OF ELECTROMAGNETIC FIELDS

Ground penetrating radar exploits the wave character of EM fields. Maxwell's equations (Equations (1.1)–(1.4)) describe a coupled set of electric and magnetic fields when the fields vary with time. Depending on the relative magnitude of energy loss (associated with conductivity) to energy storage (associated with permittivity and permeability), the fields may diffuse or propagate as waves. Ground penetrating radar is viable when conditions yield a wave-like response.

The wave character becomes evident when Maxwell's equations are rewritten to eliminate either the electric or the magnetic field. Using the electric field, rewriting yields the transverse vector wave equation

$$\overline{\nabla} \times \overline{\nabla} \times \overline{E} + \mu\sigma \cdot \frac{\partial \overline{E}}{\partial t} + \mu\varepsilon \cdot \frac{\partial^2 \overline{E}}{\partial t^2} = 0 \quad (1.10)$$

$$\uparrow \qquad \uparrow \qquad \uparrow$$
$$\;\, A \qquad\;\; B \qquad\;\; C$$

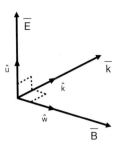

Figure 1.3 The electric field, E, magnetic field, B, and the propagation directions, k, are orthogonal. $\hat{u}$, $\hat{w}$ and $\hat{k}$ are orthogonal unit vectors.

Ground penetrating radar is effective in low-loss materials where energy dissipation (term B) is small compared to energy storage (term C).

Solutions to the transverse wave equation (1.10) take the form depicted in Figure 1.3. The electric and magnetic fields are orthogonal to each other and also to the spatial direction of the field movement, $\hat{K}$.

Such solutions are referred to as plane wave solutions to Maxwell's equations. With GPR, the electric field is the field normally measured and it has the following form:

$$\overline{E} = f(\overline{r} \cdot \overline{k}, t)\,\hat{u} \qquad (1.11)$$

where $\overline{r}$ is a vector describing spatial position and $f(\overline{r} \cdot \hat{k}, t)$ satisfies the scalar equation

$$\frac{\partial^2}{\partial \beta^2} f(\beta, t) - \mu\sigma \frac{\partial}{\partial t} f(\beta, t) - \mu\varepsilon \frac{\partial^2}{\partial t^2} f(\beta, t) \equiv 0 \qquad (1.12)$$

where $\beta = \overline{r} \cdot \hat{k}$ is the distance in the propagation direction.

In low-loss conditions

$$f(\beta, t) \approx f(\beta \pm vt)e^{\mp \alpha \beta} \qquad (1.13)$$

where

$$v = \frac{1}{\sqrt{\varepsilon\mu}}, \quad \alpha = \frac{1}{2}\sigma\sqrt{\frac{\mu}{\varepsilon}} \qquad (1.14)$$

are velocity and attenuation, respectively. The wave nature is indicated by the fact that the spatial distribution of the fields translates in the β direction between observation times as depicted in Figure 1.4.

In many formulations, discussions are given in terms of sinusoidal excitation with angular frequency ω. In this form

$$f(\beta, t) = A \exp\left(-\left(i\left(\frac{\beta}{v} - \omega t\right)\right)\right) e^{-\alpha \beta} \qquad (1.15)$$

where A is the peak signal amplitude.

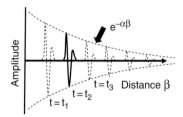

Figure 1.4 In low-loss environments, EM propagate at a finite velocity and decay in amplitude with minimal pulse shape change.

Sinusoidal signals are characterized by both excitation ω and spatial wavelength λ, where $\lambda = 2\pi v/\omega$.

1.3.1. Wave properties

Key wave field properties are velocity, v, attenuation, α, and EM impedance, Z (Annan, 2003). Wave properties for a simple medium with fixed permittivity, conductivity, and permeability are most easily expressed if a sinusoidal time variation is assumed. The variation of v and α versus sinusoidal frequency, f, are shown in Figure 1.5 (note $\omega = 2\pi f$).

All the wave properties exhibit similar behavior. At low frequencies, wave properties depend on $\sqrt{\omega}$, which is indicative of diffusive field behavior. At high frequencies, the properties become frequency-independent (if ε, μ, and σ are frequency-independent). The high-frequency behavior is the character of most importance to GPR.

The transition from diffusion to propagation behavior occurs when the electric currents change from conduction (free charge)-dominant to displacement (constrained charge) current-dominant behavior. For a simple material, the transition frequency is defined as follows:

$$f_t = \frac{\sigma}{2\pi\varepsilon} \qquad (1.16)$$

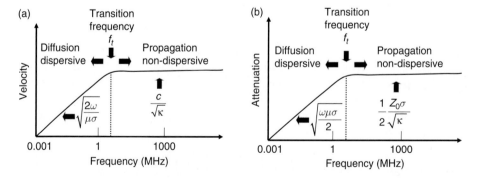

Figure 1.5 Variation in velocity and attenuation in a simple medium with nondispersive physical properties. c and Z_0 are the velocity and impedance of free space (i.e., a vacuum).

In the high-frequency plateau above f_t in Figure 1.4, all frequency components travel at the same velocity and suffer the same attenuation. An impulsive signal will travel with its shape intact, which is propagation without dispersion (Annan, 1996). In this case, the velocity, attenuation, and impedance can be expressed as follows:

$$v = \frac{1}{\sqrt{\varepsilon \cdot \mu}} = \frac{c}{\sqrt{\kappa}} \qquad (1.17)$$

$$\alpha = \sqrt{\frac{\mu}{\varepsilon}} \cdot \frac{\sigma}{2} = Z_0 \cdot \frac{\sigma}{2 \cdot \sqrt{\kappa}} \qquad (1.18)$$

$$Z = \sqrt{\frac{\mu}{\varepsilon}} = \frac{Z_0}{\sqrt{\kappa}} \qquad (1.19)$$

with the right most expression being valid when magnetic property variations are assumed negligible, making $\mu = \mu_0$, where $\mu_0 = 1.25 \times 10^{-6}$ H/m is the free-space magnetic permeability. In the above, c is the speed of light (3×10^8 m/s) and Z_0 is the impedance of free space.

$$Z_0(\Omega) = \sqrt{\frac{\mu_0}{\varepsilon_0}} = 377 \qquad (1.20)$$

The "GPR plateau" normally exhibits a gradual increase in velocity and attenuation with frequency. Two primary factors cause this increase. First, water starts to absorb energy more and more strongly as frequency increases toward the water relaxation frequency in the 10-20 GHz range (Hasted, 1972). Even at 500 MHz, water losses can start to be seen in otherwise low-loss materials. Second, scattering losses are extremely frequency-dependent and become important at high frequencies as discussed later.

To put the wave properties in perspective, typical values of v are in the range of 0.07–0.15 m/ns (or 0.2–0.5 when normalized to the velocity in air). Typical values of α are 1 dB/m with high loss of 10–100 dB/m and very low loss setting being 0.01–0.1 dB/m. Typical impedance values are 100–150 Ω.

1.3.2. Ground penetrating radar source near an interface

Only very simple forms of EM fields have been discussed to this point. In practice, fields are generated by finite-sized transmitters and must be detected by measurement sensors. Further, GPR sources are normally deployed close to the ground.

Figure 1.6 (Annan, 2003) depicts how the wavefront from a finite source impinges on the ground. The field at any point along the ground interface can be visualized locally as a planar wave impinging on the boundary at a specific incidence angle defined by geometry (source height and lateral distance). Locally the signal is reflected and refracted according to Snell's law and the Fresnel coefficients (see Section 1.3.3).

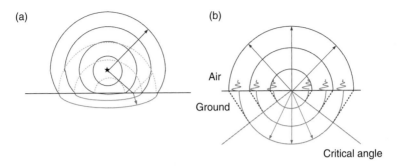

Figure 1.6 Wavefronts spreading out from a localized source. In (a), the source is located above the ground. The dotted lines indicate the reflected signal. In (b), the source is located on the air–ground interface. The dashed lines indicate refracted waves. The oscillating lines indicate evanescent waves.

The formal mathematical analysis has been the subject of much research. References such as Sommerfeld (1949), Wait (1962), Brekhovskikh (1960), Annan (1973), and Ward and Hohmann (1987), provide more detailed discussions.

Most GPR is conducted with the source on the ground. The limiting case of the source right at the interface is depicted in Figure 1.6b. The incident and reflected waves in the air coalesce into an upgoing spherical wave. In the ground, the transmitted signal divides into two parts, a spherical wave and a planar wavefront traveling at the critical angle, which links the direct spherical air wave and the spherical ground wave. Near the interface, the spherical ground wave extends into the air as an evanescent field.

The various wave fields are clearly separate in space and time when distances from the source are large compared to the wavelength or the pulse spatial length. For short distances from the source, the separation of the events becomes blurred but the essential concepts are valid. Signal paths between a transmitter and a receiver on the surface can be treated as rays following the paths depicted in Figure 1.7.

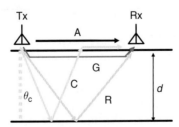

Figure 1.7 Signal paths between a transmitter and a receiver on the surface treated as rays following the paths. A is the direct airwave, G is the direct ground wave, R is the reflected wave, and C is the critically refracted wave.

1.3.3. Reflection, refraction, and transmission at interfaces

Ground penetrating radar methods normally depend on detection of reflected or scattered signal. Planar boundaries provide the simplest model for qualifying behavior. The Fresnel reflection (and transmission) coefficients (Jackson, 1962; Born and Wolf, 1980) quantify how the amplitudes of the EM fields vary across an interface between two materials, as depicted in Figure 1.8.

The direction of travel also changes (i.e., the wavefront is refracted) in accordance with Snell's law

$$\frac{\sin \theta_1}{v_1} = \frac{\sin \theta_2}{v_2} \qquad (1.21)$$

When $v_1 > v_2$, medium 2 has a critical angle beyond which energy cannot propagate from medium 1 to 2. The critical angle is determined by setting $\theta_1 = 90°$. The critical angle plays a role in many GPR responses.

Vector-field EM waves separate into two independent components defined by field orientation with respect to the boundary. Components are referred to as the TE (transverse electric field) and TM (transverse magnetic field).

The incident, reflected, and transmitted field strengths are related by the following equation:

$$I + R \cdot I = T \cdot I \qquad (1.22)$$

R and I are determined by requiring Snell's law to be satisfied, the electric and magnetic fields in the plane of the interface to be continuous, and the electric current and magnetic flux density crossing the interface must be equal on both sides.

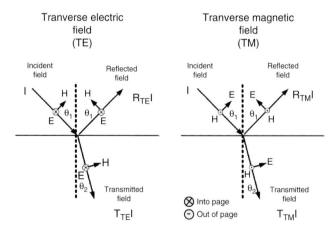

Figure 1.8 Electromagnetic (EM) waves are transverse vector wave fields. For any given propagation direction, two independent fields exist. For planar interfaces, it is tradition to discuss the two waves, one with the electric field in the interface plane called transverse electric (TE) and one with the magnetic field vector in the interface plane called transverse magnetic (TM).

The result is

$$R_{TE} = \frac{Y_1 \cdot \cos\theta_1 - Y_2 \cdot \cos\theta_2}{Y_1 \cdot \cos\theta_1 + Y_2 \cdot \cos\theta_2} \quad (1.23)$$

$$R_{TM} = \frac{Z_1 \cdot \cos\theta_1 - Z_2 \cdot \cos\theta_2}{Z_1 \cdot \cos\theta_1 + Z_2 \cdot \cos\theta_2} \quad (1.24)$$

and

$$T_{TE} = 1 + R_{TE} \quad (1.25)$$

$$T_{TM} = 1 + R_{TM} \quad (1.26)$$

where Z_i and Y_i are the impedances and admittances ($Y_i = 1/Z_i$) of the ith material. The critical factor is that an EM impedance contrast must exist for there to be a response.

When the EM wave is vertically incident on the interface ($\theta_1 = \theta_2 = 0°$), there is no distinction between a TE and a TM wave, and the TE and TM reflection coefficients become identical (for the field components).

1.3.4. Resolution and zone of influence

Given that GPR detects objects at a distance, how accurately can the object be located and what degree of information can be extracted about the geometry of the object? Resolution indicates the limit of certainty in determining the position and the geometrical attributes of a target (such as the size, shape, and thickness) and is controlled by the observation process.

Ground penetrating radar resolution consists of two components, namely the longitudinal (range or depth) resolution length and the lateral (angular or sideways displacement) resolution length as depicted in Figure 1.9.

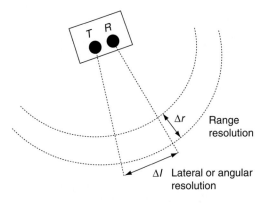

Figure 1.9 Resolution for ground penetrating radar (GPR) divides into two parts, namely range resolution and lateral (or angular) resolution.

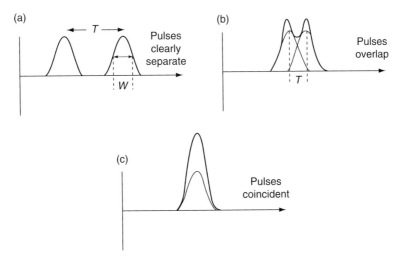

Figure 1.10 Temporal pulses with half width W. (a) Pulses are clearly separable when $T >> W$. (b) Two pulses are said to be distinguishable until $T \approx W$. (c) When $T << W$ then two events are not distinguishable.

Resolution is a fundamental concept common to most wave phenomenon-based detection methods. Explanation in terms of systems that generate a pulse and detect the echoes demonstrates the concept most readily. These echoes may arrive simultaneously, overlap or be separated in time as depicted in Figure 1.10.

When two responses are present, how closely spaced in time can they be and still be distinguished from one another? If two pulses are coincident in time, the result is one event with a larger amplitude.

Figure 1.10 depicts the extremes for two pulses. By characterizing a pulse by its width at half amplitude, W, the widely accepted opinion is that two pulses are distinguishable if separated by half their "half width". If they are separated in time by less than this amount, then they will most likely be interpreted as a single event.

The temporal pulse separation concept translates into spatial resolution (see Annan, 2005). The radial resolution length is expressed as follows:

$$\Delta r \geq \frac{Wv}{4} \qquad (1.27)$$

The pulse width and the velocity in the material dictate resolution length. The radial resolution length is independent of distance from the source in an ideal world. In practice, at larger distances, pulse dispersion and attenuation will affect radial resolution.

The lateral resolution length is as follows:

$$\Delta l \geq \sqrt{\frac{vrW}{2}} \qquad (1.28)$$

where r is the distance to the target.

The lateral resolution depends on the velocity, the pulse width, as well as the distance from the system. The larger the distance, the greater the lateral resolution length.

The lateral resolution is closely related to the Fresnel zone concept, which expresses the concept in terms of interference of monochromatic (sinusoidal) signals. With GPR, the pulse width, W, in time is directly related to the bandwidth, B, which is also directly related to the center frequency, f_c. If one uses the relationship

$$W = \frac{1}{B} = \frac{1}{f_c} \qquad (1.29)$$

and notes that the center frequency wavelength is

$$\lambda_c = f_c/v \qquad (1.30)$$

then lateral resolution length can be expressed as

$$\Delta l = \sqrt{\frac{d\lambda_c}{2}} \qquad (1.31)$$

which is identical to the Fresnel zone radius for monochromatic signals of frequency f_c.

Discussion of these concepts from the seismic point of view is given by Berkhout (1984), Knapp (1991), and Burhl et al. (1996). An interesting analysis of broadband Fresnel zones is presented by Pearce and Mittleman (2002).

1.3.5. Scattering attenuation

Ground penetrating radar signals are invariably transmitted through complicated media. The signals encounter heterogeneous electrical and magnetic properties on many scales. Smaller-scale heterogeneities generate weak or undetectable responses but their presence has an impact on the signals as they pass by. The heterogeneities extract energy as the EM field passes and scatter it in all directions.

Figure 1.11 illustrates how scattering can be viewed from an energy viewpoint. At any point on a wavefront, the incident signal with power per unit area impinges on local, small-scale scatters, which are characterized by spatial size, a, and number per unit volume, N.

The electric or magnetic field will attenuate with a scattering attenuation coefficient α_s (Annan, 2005). In other words, the electric field will decrease with distance r as

$$E = E_0 \, e^{-\alpha_s r} \qquad (1.32)$$

where

$$\alpha_s = \frac{NA}{2} \qquad (1.33)$$

and A is the scattering cross section of scatters.

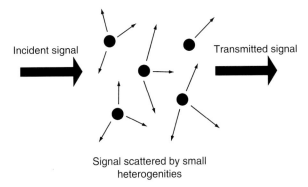

Figure 1.11 Ground penetrating radar (GPR) signals are scattered by heterogeneities in material properties, which reduce the transmitted signals.

Scattering attenuation is very frequency-dependent; examining the response of small scatters referred to as Rayleigh scattering is informative. The Rayleigh scattering cross section of a pulse is expressed as follows:

$$A = Ca^6 f^4 \tag{1.34}$$

where C is a constant with units of $1/m^4\ Hz^4$, a is the sphere radius, and f is the frequency.

Scattering attenuation must be added to the ohmic or material loss attenuation to determine the full attenuation the GPR signal will see as it travels through a heterogeneous lossy dielectric medium.

$$\alpha_{total} = \alpha_{ohmic} + \alpha_{scattering} \tag{1.35}$$

The effect of volume scattering was recognized very early by the radio echo sounding community (see Davis, 1973, Watts and England, 1976) as a limiting factor in temperate ice sounding. Volume scattering is more important in ice because the ohmic attention is much smaller than in most soil and rock materials.

1.4. SIGNAL MEASUREMENT

Ground penetrating radar systems are conceptually simple; the objective is to measure field amplitude versus time after excitation. The heart of a GPR system (Figure 1.12) is the timing unit, which controls the generation and detection of signals. Most GPRs operate in the time domain; however, frequency domain measurements are now being used to synthesize the time domain response. Since time of flight and seismic-like records are most easily understood and most commonly used in GPR, discussion here will be in the time domain, which applies whether signals are measured directly in time or synthesized.

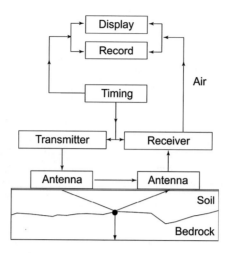

Figure 1.12 Block diagram depicting main components of a ground penetrating radar (GPR) system.

Characterization of a radar system is a complex task because there are many issues that impact the operation and use of the system. Electronic instrumentation factors governing GPR characterization are signal generation, signal capture method, signal processing, performance factor, dynamic range, center frequency and bandwidth, reliability, and portability. Antennas transform electrical signals to and from vector EM fields.

1.4.1. Time ranges and bandwidth

Ground penetrating radar systems typically need to record data with timing accuracies of less than 10 ps over time durations of 10 000 ns. Measurement bandwidth is application-dependent and relates directly to resolution. Resolution involves the two related topics of "transmitter blanking" and "target separation" as illustrated in Figure 1.13.

Transmitter blanking is caused by the inability of a receiver to detect signals until after the transmitter has finished transmitting. It is both a bandwidth and a dynamic range issue. The transmitting source usually emits a very large signal, and if the receiver is in close proximity to the transmitter as is usually the case in GPR, then the receiver will see the very large direct transmitted signal. If this signal is sufficiently large, the receiver electronics will be overloaded and will not detect reflected signals. The time duration of the transmit pulse varies inversely with the bandwidth.

Resolution length Δr dictates necessary bandwidth to be

$$B \geq \frac{v}{4\Delta r} \qquad (1.36)$$

In this discussion, bandwidth is defined by the -3-dB spectral level.

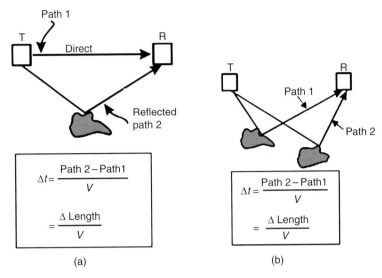

Figure 1.13 (a) Transmitter blanking occurs when the direct signal traveling from the transmitter to the receiver overlaps in time with the reflected signals. (b) If two targets yield similar path lengths, the differences in travel time can be small causing reflected pulses to overlap. To resolve two events, the path length difference must exceed half the pulse width multiplied by the velocity.

1.4.2. Center frequency

Bandwidth does not define GPR frequency. A frequency band anywhere in the spectrum will satisfy resolution needs. For GPR to be effective, attenuation dictates keeping frequency as low as possible above the transition frequency. The attenuation in natural materials is a combination of electrical losses and scattering loss, both of which increase with increasing frequency. The lower the frequency, the greater the likelihood of obtaining signal penetration into the medium.

Ground penetrating radar signals are characterized by the bandwidth to center frequency ratio

$$R = \frac{B}{f_c} \qquad (1.37)$$

and every effort is made to make R as large as possible. The instrumentation goal is always to maximize B and minimize f_c with the practical limiting value of R being above unity. Most practical GPRs achieve a monopulse wavelet character (i.e., $R = 1$) and are most appropriately referred to as ultra wideband radars, impulse, or baseband radars.

Since $R \approx 1$, GPRs are normally characterized by their center frequency, f_c. A 100-MHz GPR is interpreted as one with a 100 MHz bandwidth centered at 100 MHz with a corresponding temporal width of 10 ns.

1.4.3. Ground penetrating radar signal acquisition

The ideal GPR system is shown in Figure 1.14. In this system, the transmitter electronics deliver an electrical signal to a transmitting antenna, which energizes the surroundings. The receiving antenna detects the transmitted fields and translates them into an electrical signal.

In this perfect system, the output signal from the transmitter electronics, $p(t)$, is fed to the transmitting antenna, and the signal returned from the receiving antenna, $r(t)$, would be digitized at a rapid rate through high-dynamic-range analog-to-digital (A/D) converters and recorded, processed, and displayed.

Analog-to-digital converters having sufficiently high sampling rates with sufficient dynamic range have not been available although this is changing. Alternate signal capture techniques have been used for many years and are still the mainstay of signal acquisition in GPR systems. These techniques measure the response without having to directly digitize at the high rate required to capture the radio frequency signal. Synchronized operation of the transmitter and the receiver is exploited. Time domain systems use equivalent time sampling, while frequency domain systems use mixing or heterodyning. More discussions can be found in Annan (2005) and Koppenjan (this volume).

1.4.4. Characterizing system response

Although GPR data can be captured in different ways, data are normally reduced to digital amplitude versus time. Numerous factors in the electronics, antenna structure, and support systems invariably cause measured signals to be distorted from the ideal. Major issues are how "clean" is the system excitation impulse, how do the dynamic range limitations of the receiver distort the signal, and what contribution do antenna and other structure parasitic responses make.

Systems respond as shown in Figure 1.15. The transmit excitation signal has an onset at 0 time and lasts for a duration of W. In a perfect world, all system signals would vanish after W.

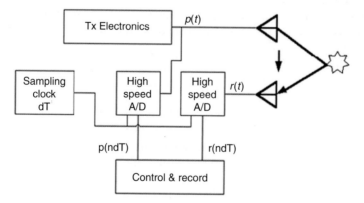

Figure 1.14 Ideal ground penetrating radar (GPR) system with full capability and raw wideband digital recording.

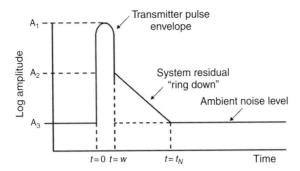

Figure 1.15 Idealized ground penetrating radar (GPR) impulse response envelope amplitude (log) versus time. The key parts of the response are the ambient noise level A_3 (before pulse and at late time), the peak signal A_1, and initial residual system response A_2.

Real-system electronics have transient (nonideal) responses, which persist in time after the ideal signal would normally terminate. In addition, the signals are launched and detected using antennas. Antennas create finite time delays in the currents that move around the antenna structure. This finite delay is critical to efficient launch and detection of signals.

In addition, antennas have to be carried, mounted, and interconnected using cables and other mechanical structures. These ancillary structures can also carry induced electric currents, which in turn generate reradiated fields with finite time delays.

Ground penetrating radar systems always exhibit self-generated signals after the transmit pulse has terminated. This residual system response should decrease quickly with time after the transmit pulse and ideally be below the ambient noise level. The residual response is often referred to as system "ring-down."

Echoes from the ground will return during this "ring-down" period. If the residual signal from the system is larger than the ground response, then the ground response will be masked and not visible. The result can be considered to be extended transmitter blanking.

Ground penetrating radar system ring-down response (no ground present) can be characterized by a simple model expressed as follows.

$$A = A_3, \quad t < 0 \tag{1.38}$$

$$A = A_1, \quad 0 < t < W \tag{1.39}$$

$$A = A_2 - \delta(t - W), \quad W < t < t_n \tag{1.40}$$

$$A = A_3, \quad t > t_n \tag{1.41}$$

where

$$t_n = \frac{A_2 - A_3}{\delta} + W \tag{1.42}$$

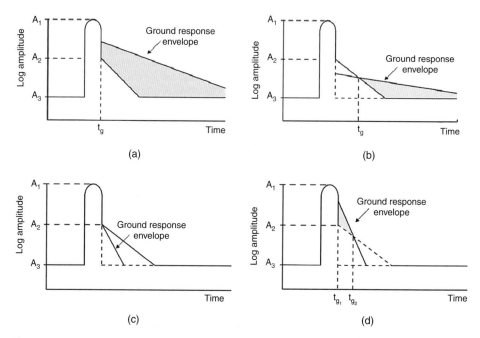

Figure 1.16 (a) Case where system "ring-down" is faster than the ground response fall off but initial system residual is larger than the ground response. Ground response becomes detectable after t_g. (b) Case where ground response is smaller than system residual response resulting in no detection of ground response. (c) Case where ground response initially exceeds system residual response but ground response decays more quickly than system "ring-down" rate, leaving ground response detectable only between time t_{g1} and t_{g2}. (d) In some cases, the system response might be static and subtracted from the combined signal to make ground response visible at all times.

and A_1 is the peak signal during the transmit pulse, A_2 is the residual signal amplitude at time W, A_3 is the ambient noise level of the measurement, δ is the residual "ring-down" rate (normally in dB/ns), and W is the pulse width

A wide variety of conditions can occur depending on ground conditions and system designs. Some of these behaviors are depicted in Figure 1.16a–d. The system residual response may mask some or all of the ground responses over some time periods.

In some situations, the system residual response can be estimated and subtracted. Processing called background subtraction is often attempted to improve system performance. Unfortunately, the residual response is not easily determined and is not invariant since it is often associated with the antenna system and its support structure and cabling. The "ring-down" or residual electric currents in the structure are affected by the local ground conditions, which in turn makes the "ring-down" character vary with measurement location.

1.4.5. Recording dynamic range

Figure 1.17 shows typical GPR response together with a binary dynamic range scale typifying the digitization levels of a 16-bit A/D converter. In this particular

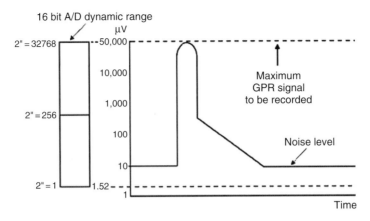

Figure 1.17 Relationship of detected ground penetrating radar (GPR) signals to record system dynamic range illustrated by using 16-bit binary sampling.

example, A_1 is chosen to be 50 000 µV and A_3 to be 10 µV. Maximum and minimum recording of the receiver should bracket A_1 and A_3.

The choice shown gives the smallest signal resolved as 1.5 µV and the maximum recording value of the A/D is $A_1 = 50\,000$ µV. For this example, the noise level A_3 is greater than the least significant bit of the A/D converter.

For rigid-system geometries, where there is no variation in the position of the components, the value of A_1 can be controlled and the design objectives met. For bistatic antennas with arbitrary antenna geometries and orientations, as well as possible changes in frequencies of the antennas, the value of A_1 can be highly variable, and it is virtually impossible to guarantee that A_1 will always remain within the dynamic range of the recording system. As a result, it is common to see the peak signal during the transmit pulse limited (clipped) by the recording system.

This is not a major problem for many GPR applications as the information in the transmit pulse may not be needed. Loss of signal during this time is common in time domain radars (transmitter blanking). In the situations where signals from shallow depths are critical, systematic attention must be given to the system dynamic range and the antenna geometry to assure signal fidelity.

1.4.6. Antennas

Ground penetrating radar antennas create and detect key EM fields. The transmit antenna must translate the excitation voltage into a predictable temporal and spatial distributed field. The receiving antenna must detect the temporal variation of a vector component of the EM field and translate it into a recordable signal.

The following are desired antenna characteristics.

a. The exact source and detection locations must be definable.
b. The transmitter and receiver responses (transfer functions converting electric field to/from voltage current) must be time and space invariant.

c. The vector character of the field linking the source voltage and received voltage must be quantifiable.
d. The bandwidth of antennas must match the system application needs.

These requirements are difficult to achieve. Efficient field generation and detection requires finite size of antennas. Current travel time across the antenna dimension must be comparable to the temporal rate of change of the exciting voltage or field. In frequency domain terminology, the antenna dimension must be similar to the wavelength of the signals.

For efficient operation, finite-size antennas must be used. They have the following characteristics:

a. Field creation and detection occurs over a spatially (and temporally) distributed region. (In other words, source and detection points are imprecise.)
b. Field transit time (or wavelength) in GPR applications depends on the host environment and is not invariant. (In other words, antenna response cannot be perfectly invariant.)
c. A spatially distributed antenna means less-precise vector characterization of the response since isolation of response to a single vector component becomes geometrically difficult.
d. Finite-dimension antennas have a preference to emit energy at frequencies that resonate on the structure. Bandwidth is best maximized by damping an antenna, which makes it less sensitive to its surroundings and less efficient.

Antennas that have been proven most effective for GPR are short electric dipoles. Resistively loaded small dipoles yield a fair degree of faithfulness to desired predictable and invariant behavior while retaining some efficiency.

1.4.7. Antenna directivity

The directional characteristics of a short electric dipole antenna are controlled by the ground. Although the analysis of this problem is complex, the basic characteristics can be explained. Background for this can be found in Annan (1973), Annan et al. (1975), Engheta et al. (1982), and Smith (1984).

A short, center-fed electric dipole is depicted in Figure 1.18. The relative electric field amplitude at a large distance is donut-shaped as depicted in Figure 1.19. No energy is radiated from the ends of the antenna, while energy is radiated uniformly in the plane perpendicular to the dipole axis. Figure 1.20 shows orthogonal cross sections through the donut commonly referred to as the TE and TM patterns.

When the dipole antenna pattern is placed on the ground surface, the pattern changes to that shown in Figure 1.21. The change in directivity is caused by the refractive focusing associated with the air-ground interface. This pattern represents the far-field radiated component of the fields. Near the antennas the fields are more complex and require numerical simulation.

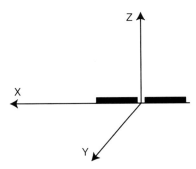

Figure 1.18 Geometry and coordinate system for an electric dipole antenna. When placed on the ground, the X–Y plane would represent the ground surface.

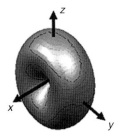

Figure 1.19 Three-dimensional presentation of pattern of a small electric dipole is donut-like in a uniform material. There is no radiation off the ends of the dipole.

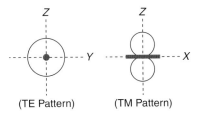

(TE Pattern)   (TM Pattern)

Figure 1.20 Orthogonal cross sections through the three-dimensional (3D) dipole pattern in Figure 1.19 referred to as the transverse electric TE (H plane) or transverse magnetic (TM) (E plane) directivity patterns.

The peaks in the TE (H plane) pattern occur at the critical angle of the air–ground interface.

$$\theta_c = \sin^{-1}\left(\frac{v_g}{c}\right) = \sin^{-1}\left(\frac{1}{\sqrt{K_g}}\right) \tag{1.43}$$

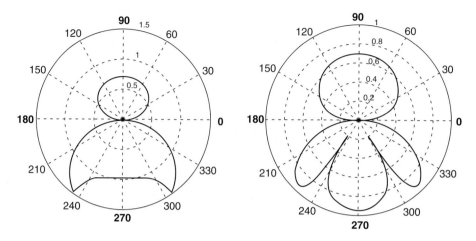

Figure 1.21 When the dipole is on the ground surface, directivity is drastically altered and depends on ground permittivity. The TE and TM patterns shown here are for ground permittivity of 3.2.

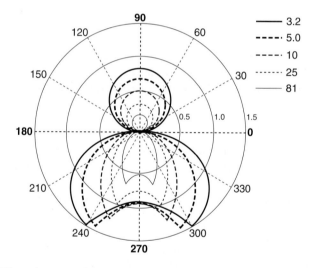

Figure 1.22 When the ground permittivity changes, the patterns change. The transverse electric (TE) pattern is shown for permittivities ranging from ice (low) to water (high).

Subsurface nulls occur in the TM (E plane) pattern in the critical-angle direction.

Antenna directivity is ground-dependent. As the ground properties change, the antenna directivity changes. Figure 1.22 shows a sequence of patterns as K_g is carried from 3.2 to 80.

The effect of antenna elevation off the ground surface is also important. In real-field situations, surface roughness and the need to transport antennas over the surface can limit close ground contact. Antenna elevation modifies antenna

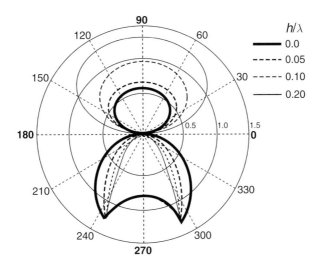

Figure 1.23 Antenna elevation also impacts on directivity. The change in transverse electric (TE) directivity is shown as a function of height normalized against center frequency wavelength.

directivity as shown in Figure 1.23. More signals are transmitted upward into the air and antenna efficiency is reduced. Further, ground reflection proximity generates time-delayed reverberation on the antenna affecting ring-down.

1.4.8. Antenna shielding

Ground penetrating radar antennas are normally placed close to the air–ground interface; a shield, when present, is a "container," which encloses the antenna as depicted in Figure 1.24.

As shown in Figure 1.25, signals can travel from a transmitter to a receiver along a number of paths. The purpose of shielding is to selectively enhance some signals and suppress others. Key shield design objectives are

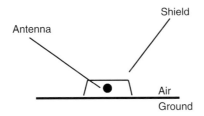

Figure 1.24 A ground penetrating radar (GPR) antenna shield encloses the antenna to minimize coupling with signals in the air. These signals may be generated by the GPR system itself or from an external source.

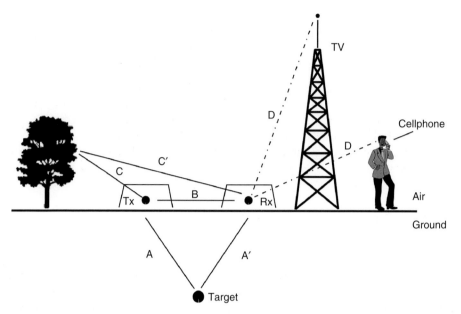

Figure 1.25 A ground penetrating radar (GPR) system emits and detects radio wave signals. There are many possible signals and paths and the objective is to maximize the target response and minimize others.

a. maximize the energy on the path AA' to and from the subsurface target (i.e., focus or direct signal downward);
b. minimize the direct transmitter to receiver energy on path B;
c. minimize the energy that escapes into the air as on path CC'; and
d. minimize external EM noise as indicated by signals D.

Given these laudable benefits, what are the drawbacks? Antenna shielding requires a structure that has an EM response. Energy travels from the transmitting antenna to both the transmitter shield and the receiving antenna shield and then to the receiving antenna as indicated in Figure 1.26. Shielding-generated signals can be large and reverberate for a long period of time, greatly increasing the system ring-down.

Besides the EM response of the shield itself, an effective shield leads to larger transducer size, greater weight, and increased manufacturing cost penalties. Antenna shielding can create more problems than it solves; as a result, many unshielded antenna configurations are employed, particularly at low frequencies where size and weight are critical considerations.

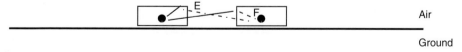

Figure 1.26 Antenna shields must interact with radio waves to be effective. The shield can generate additional response, which may be detrimental and interfere with the desired measurement unless extreme care is taken in the shield design.

Shielded antennas are most common for higher-frequency GPR systems (typically above 100 MHz) where antennas are smaller. Tight ground coupling minimizes signal leakage into the air and is often a part of the shielding design.

Practical antenna shielding considerations are as follows:

a. Shielding is never perfect no matter what claims are made.
b. Not all applications require shielding. The highest fidelity and maximum depth of penetration at open sites may be obtained without shielded antennas.
c. Even with the most ideal shield, spurious signal leakage can and does occur. The most experienced GPR users can be fooled occasionally and misinterpret these responses.

1.5. SURVEY METHODOLOGY

1.5.1. Sampling criteria

The objective of GPR surveys is to obtain information about the subsurface structures indirectly by using radio waves. The EM field as a function of space and time must be sampled and recorded. Survey design must adhere to fundamental sampling principles.

For a given sinusoidal frequency f, time and spatial sampling intervals, Δt and Δx, must obey

$$\Delta t \leq \frac{1}{2f} \tag{1.44}$$

$$\Delta x \leq \frac{v}{2f} \tag{1.45}$$

to satisfy Nyquist sampling criteria. For transient GPR signals with a bandwidth to center frequency ratio of unity, this translates to

$$\Delta t \leq \frac{1}{3f_c} \tag{1.46}$$

$$\Delta x \leq \frac{v}{3f_c} \tag{1.47}$$

The preceding criteria are ideal and the use of values that are half as large is more appropriate; thus

$$\Delta t \leq \frac{1}{6f_c} \tag{1.48}$$

$$\Delta x \leq \frac{v}{6f_c} \tag{1.49}$$

are recommended, and these criteria should be implicit in all survey design.

1.5.2. Ground penetrating radar surveys

Ground penetrating radar measurements fall into two categories, reflection and transillumination, as originally depicted in Figure 1.1. Reflection surveys using a single transmitter and a single receiver are the most common, although multiple source and receiver configurations are used occasionally for some specialized applications. Survey design discussion can be found in Annan and Cosway (1992) and Annan (2005) while frequency selection issues are addressed by Annan and Cosway (1994).

1.5.3. Common-offset reflection survey

Common-offset surveys deploy a single transmitter and receiver with a fixed offset or spacing between the units at each measurement location. The terminology for such a survey is single-fold common offset. The transmitting and receiving antennas have specific polarization character for the field generated and detected. The antennas are deployed in a fixed geometry (i.e., separation, s, and orientation) and measurements made at regular station intervals (Δx) as depicted in Figure 1.27. Data at uniform spacing are normally desired if advanced data processing and visualization techniques are to be applied.

The objective of reflection surveys is to map subsurface reflectivity versus spatial position. Variations in reflection amplitude and time delay indicate variations in v, α, and Z. Ground penetrating radar reflection surveys are traditionally conducted on "straight" survey lines and systems are designed to operate in this fashion. Area coverage most often entails data acquisition on a rectilinear grid of lines, which cover the area, such as that depicted in Figure 1.28.

The parameters defining a common-offset survey are GPR center frequency, the recording time window, the time sampling interval, the station spacing, the antenna spacing, the line separation spacing, and the antenna orientation.

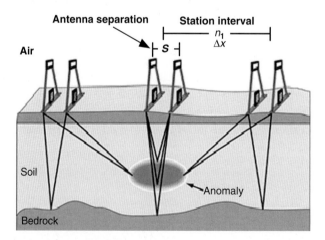

Figure 1.27 Schematic illustration of common-offset, single-fold profiling along a line showing major survey specification parameters.

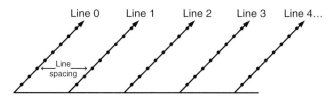

Figure 1.28 A survey area spanned by a number of survey lines. The ground response is measured at discrete points along the survey line. Although field practice may be more erratic, data of this format are key to most systematic data processing and visualization.

The GPR frequency selection is synonymous with defining both GPR pulse width and bandwidth. Application exploration depths and resolution requirements determine bandwidth and also dictate temporal and spatial sampling intervals. Most often, attenuation is an issue; so frequencies are kept as low as possible to maximize penetration even if resolution is compromised.

1.5.4. Multioffset common midpoint/wide-angle reflection and refraction velocity sounding design

The common midpoint (CMP) or wide-angle reflection and refraction (WARR) sounding mode of operation is the EM equivalent to seismic refraction and wide-angle reflection. Common midpoint soundings are primarily used to obtain an estimate of the radar signal velocity versus depth in the ground by varying the antenna spacing and measuring the change in the two-way travel time as illustrated in Figure 1.29.

Multioffset measurements can be performed at every station resulting in a multifold reflection survey. Two benefits are that CMP stacking can improve signal-to-noise ratio (Fisher et al., 1992) and that a full velocity cross section can be derived (Greaves et al., 1996). Multifold GPR surveys are seldom performed because they are time consuming, more complex to analyze, and most of the cost-effective benefit is obtained with well-designed, single-fold surveys.

1.5.5. Transillumination surveys

Transillumination GPR measurements (Annan and Davis, 1978) are less common. Most uses involve GPR measurements in boreholes for engineering and

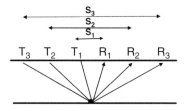

Figure 1.29 Procedure for conducting a common midpoint (CMP) velocity sounding. Systematically varying antenna separation varies the signal path in the ground while keeping the point of reflection fixed enabling wave properties to be estimated.

environmental studies (Olhoeft, 1988, Olsson et al., 1992). Applications to image pillars and walls are increasing. In both cases, estimates of v and α are derived from signal travel time and amplitude measurements.

The survey parameters are GPR frequency, station interval, time window, temporal sampling interval, and borehole spacing or wall/pillar thickness. Antenna orientation is seldom an issue since boreholes are usually slim, and the traditional electric dipole axis is aligned with the borehole. Some special borehole systems for larger diameter holes have been designed to have source and receiver directionality. For walls and pillars, maximizing vector-field coupling needs consideration.

Transillumination surveys require geometry information as auxiliary data to the GPR measurement to ensure correct interpretation. Deviations in geometry will introduce errors into values of v and α derived from GPR time and amplitude observations.

Zero-offset profiling (ZOP) is a quick and simple survey method to locate velocity anomalies or attenuation (shadow) zones (Gilson et al., 1996). This technique is a quick way for defining anomalous zones. The borehole methodology is depicted in Figure 1.30(a). The transmitter and the receiver are moved from station-to-station in synchronization. In a uniform environment, the received signal should be the same at each location.

Multioffset gather (MOG) surveying provides the basis of tomographic imaging. The objective is to measure a large number of angles passing through the volume between the boreholes as depicted in Figure 1.30(b). As with ZOP measurements, borehole geometry plays a critical role in data analysis.

A critical factor in borehole survey design is borehole depth, D, to borehole separation, S, ratio. Ideally, $D/S > 2$ is desirable for several reasons. First, keeping D/S large maximizes the range of view angles through the ground. Second, refracted wave events can mask direct arrivals (Figure 1.31). The same criteria apply for walls and pillars.

While maximizing D/S, S should be kept sufficiently large to be in the far-field of the antenna. In other words

$$S > \frac{v_g}{f_c} \qquad (1.50)$$

where v_g is the ground velocity and f_c is the GPR center frequency.

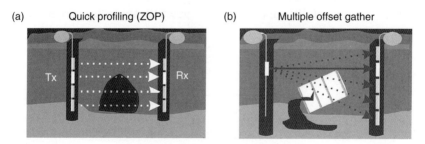

Figure 1.30 (a) Illustration of a transillumination zero-offset profiling (ZOP). (b) Illustration of a transillumination multioffset gather (MOG). Combining MOGs for each transmitter station provides the data for tomographic imaging.

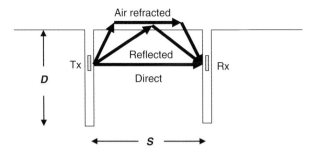

Figure 1.31 For borehole transillumination surveys, D/S should be kept as large as practical. Direct signals may be masked by faster refracted airwave arrivals.

1.6. DATA ANALYSIS AND INTERPRETATION

Transforming GPR data into application-specific information can follow two paths. The first is common to all geophysical methods, namely the GPR response measured is presented in a section, plan, or volume form to indicate anomalous target location. The second is to extract quantifiable wave property variables such as velocity, attenuation, or impedance and then translate the wave properties into application-specific quantities (also see other chapters in this volume).

Typical processing flow for GPR data is depicted in Figure 1.32. Data processing focuses on the highlighted areas: data editing, basic processing, advanced processing, and visualization/interpretation processing. Processing is usually an iterative activity; data will flow through the processing loop several times. Batch

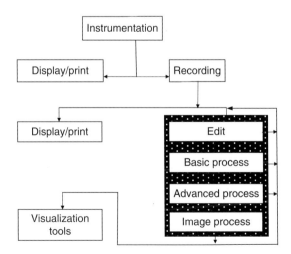

Figure 1.32 Overview of ground penetrating radar (GPR) data processing flow. Processing can vary from simple editing to total transformation of GPR information into different forms such as velocity versus depth.

processing with limited interactive control may be applied on large datasets after initial iterative testing on selected data samples has been performed.

Advanced data processing methods require varying degrees of interpreter bias to be applied and result in data that are significantly different from the raw input information. Such processes include well-known seismic processing operations (Yilmaz, 2000) such as spatial and temporal filtering, selective muting, dip filtering, deconvolution, and velocity semblance analysis as well as more GPR-specific operations such as background removal, multiple-frequency antenna mixing, and polarization mixing (Tillard and Dubois, 1992; Roberts and Daniels, 1996). The succession of GPR Conference Proceedings document GPR evolution (see reference list) and are highly informative on GPR data processing evolution.

Processing discussion here is limited to key GPR issues, which differ substantially from the plethora of seismic processing documents.

1.6.1. Dewow

A unique aspect of GPR data arises from the close proximity of receiver to transmitter. The fields near the transmitter contain low-frequency energy associated with electrostatic and inductive fields, which decay rapidly with distance. This low-frequency energy often yields a slowly time-varying component to the measured field data. This energy causes the base level of the received signal to bow up or down. This effect has become known as baseline "wow" in the GPR lexicon.

The "wow" signal process can be suppressed by applying a high-loss temporal filter to the detected signal. This process is referred to as "dewow." In older analog systems, dewow process was applied in an analog circuit at the time of acquisition. Modern systems apply digital filtering in real time or in post acquisition.

1.6.2. Time gain

Radar signals are very rapidly attenuated as they propagate into the ground. Signals from greater depths are very small when compared to signals from shallower depth. Simultaneous display of these signals requires conditioning before visual display. Equalizing amplitudes by applying a time-dependent gain function compensates for the rapid fall off in radar signals from deeper depths. Figure 1.33 indicates the general nature of the amplitude of radar signals versus time. Such time-varying amplification is referred to time gain and range gain when manipulating GPR data.

Attenuation in the ground can be highly variable. A low-attenuation environment may permit exploration to depths of tens of meters. In high-attenuation conditions, depth of penetration can be less than 1 m. Display of GPR data versus time must accommodate the low and high attenuation extremes as depicted in Figure 1.33. The concept of time-varying gain is depicted in Figure 1.34, which applies a spherical and exponential compensation gain.

There are various ways of applying time gain to radar data. Gain should be selected based on an a priori physical model, not user whim, with the objective of minimizing artifacts created by the process.

Electromagnetic Principles of Ground Penetrating Radar

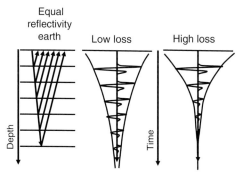

Figure 1.33 Layered earth model of equal reflectivity horizons and impulse response with envelope of reflection amplitude depicted. In reality, GPR for signal attenuation can be very high, making signal amplitudes small in a short time.

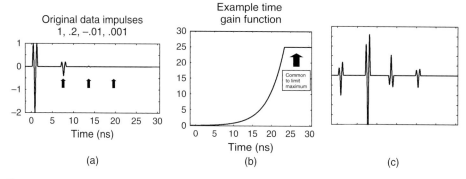

Figure 1.34 Concept of time-varying gain where signal amplification varies with time to compensate for attenuation. (a) shows a radar trace with four signals of decreasing amplitude with time, (b) shows a time gain function while (c) shows the result of multiplying (a) by (b). All four events are visible in (c).

Time gain is a nonlinear operation. Filtering operations before and after time gain will not be equivalent. The example in Figure 1.34 clearly shows a change in pulse shape after the time gain process.

1.6.3. Deconvolution

The purpose of deconvolution is normally to maximize bandwidth and reduce pulse dispersion to ultimately maximize resolution. Examples of deconvolution and other types of filtering are given by Todoeschuck et al. (1992) and Turner (1992). Deconvolution of GPR data has seldom yielded a great deal of benefit. Part of the reason for this is that the normal GPR pulse is the shortest and the most compressed that can be achieved for the given bandwidth and signal-to-noise conditions. Instances where deconvolution has proven beneficial occur when extraneous reverberation or system reverberation is present.

A closely related topic is inverse Q filtering. The higher GPR frequencies tend to be more rapidly attenuated resulting in lower resolution with increasing depth. Inverse Q filtering attempts to compensate for this effect. Irving and Knight (2003) present an excellent discussion of this topic.

1.6.4. Migration

Migration is spatial deconvolution (Fisher et al., 1992), which attempts to remove source and receiver directionality from reflection data. The goal is to reconstruct the geometrically correct radar reflectivity distribution of the subsurface. Migration requires knowledge of the velocity structure, which often makes it an interactive process as background velocity is iteratively adjusted to optimize the image.

Figure 1.35 shows how migration modifies the tunnel images shown in Figure 1.2. Several types of migration (Kirchoff, Stolt, reverse time, and finite difference) are possible. Proper GPR migration attempts to compensate for antenna directivity. Most seismic discussions of migration are applicable to the GPR, and Yilmaz (2000) discusses the subject extensively. A unique aspect of GPR is the magnitude of topography compared to depth of exploration. Migration processing that includes topography has been described by Lehmann and Green (2000).

1.6.5. Topographic correction

Because of the shallow exploration depth of GPR, compensating for topography is often important. For minor surface variations, time-shifting data traces can largely compensate for topographic variations. The reflection-profiling example in Figure 1.36 shows stratigraphy and a reflection from the water table. The profile begins in a flat area and climbs up a slope. The data have been time shift compensated for elevation along the profile line, which renders the water table reflection essentially flat.

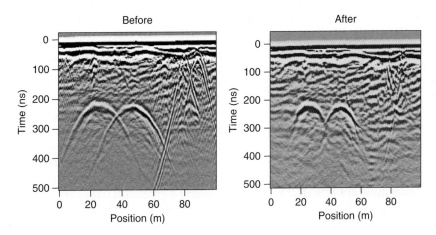

Figure 1.35 Ground penetrating radar (GPR) cross sections obtained with a 50-MHz system traverse over two road tunnels (shown in Figure 1.2) before and after migration.

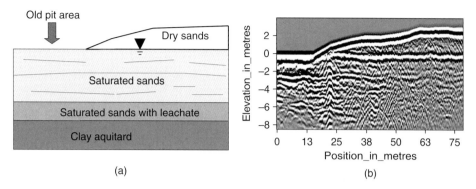

Figure 1.36 Illustration using trace time shifting to compensate for topographic variation. Note the water table is essentially flat in the ground penetrating radar (GPR) section.

1.6.6. Two-dimensional and three-dimensional data visualization

Initially, GPR data were always displayed as reflection cross sections (essentially vertical slices through the ground along the transect surveyed). As processing and computer processing power have advanced, more and more data presentation is in the form of 3D volume (voxel) rendering and time/depth slices (plan maps). The best results are obtained when migration and some form of signal rectification or trace Hilbert transform envelope (White, 1991) is applied. Excellent examples are given by Sigurdsson & Sigurdsson and Overgaard (1996), Grasmueck (1996), Lehmann and Green (1999), and McMechan et al. (1997). Today, 3D presentations are common as computer power and visualization tools have advanced rapidly.

1.7. SUMMARY

Over the last four decades, there has been a continuous evolution of the GPR method. Reliable equipment is commercially available from numerous sources. The basic physics of the problem are well-understood and quantitative analysis of data is expanding rapidly.

Although a majority of GPR use remains in the qualitative realm and many problems can be addressed this way, the future lies in more quantitative use of the information in GPR data. The ability to translate EM observations into other useful engineering or scientific observations is critically important to the widespread use of the technique in a quantitative way. Considerable research is occurring in this area; the inversion-type work described in van der Kruk (2001) is a good example.

The ability of GPR to image quickly is a powerful aspect of the method. Today, instruments are beginning to appear with arrays of sensors, which provide for wide-area coverage quickly. Acquisition of wide-area coverage data lends itself to fast 3D subsurface imaging.

At this point, the method is on the eve of its next major evolution into these other realms. Much scientific work is being done and instrument development occurring, which will open up new vistas for GPR!

REFERENCES

Annan, A.P., 1973, Radio interferometery depth sounding: Part I – Theoretical discussion. Geophysics, Vol. 38, pp. 557–580.

Annan, A.P., Waller, W.M., Strangway, D.W., Rossiter, J.R., Redman, J.D. and Watts, R.D., 1975, The electromagnetic response of a low-loss, 2-layer, dielectric earth for horizontal electric dipole excitation. Geophysics, Vol. 40, No. 2, pp. 285–298.

Annan, A.P., 1996, Transmission dispersion and GPR. JEEG, Vol. 0, January 1996, pp. 125–136.

Annan, A.P., 2002, The history of ground penetrating radar. Subsurface Sensing Technologies and Applications, Vol. 3, No. 4, October 2002, pp. 303–320.

Annan, A.P., 2003, Ground penetrating radar: Principles, procedures & applications. Sensors & Software Inc. Technical Paper.

Annan, A.P., 2005, Ground penetrating radar, in near surface geophysics, in D.K. Butler (eds), Society of Exploration Geophysicists, Tulsa, OK, USA, Investigations in Geophysics No. 13, pp. 357–438.

Annan, A.P. and Cosway, S.W., 1992, Ground penetrating radar survey design. Proceedings of the Symposium on the Application of Geophysics to Engineering and Environmental Problems, SAGEEP'92, Oakbrook, IL, April 26–29, pp. 329–351.

Annan, A.P. and Cosway, S.W., 1994, GPR frequency selection. Proceedings of the Fifth International Conference on Ground Penetrating Radar, Kitchener, Ontario, Canada, June 12–16, pp. 747–760.

Annan, A.P. and Davis, J.L., 1978, Methodology for radar transillumination experiments: Report of activities, Geological Survey of Canada, Paper, 78–1B, pp. 107–110.

Archie, G.E., 1942, The electrical resistivity log as an aid in determining some reservoir characteristics, Trans. AIME, Vol. 146, pp. 54–62.

Born, M. and Wolf, E., 1980, Principles of Optics, 6th edition, Pergamon Press, Oxford, UK.

Brekhovskikh, L.M., 1960, Waves in Layered Media, Academic Press, New York.

Berkhout, A.J., 1984, Seismic Resolution: Resolving Power of Acoustical Echo Techniques. Geophysical Press, London and Amsterdam.

Bristow, C.S. and Jol, H.M., 2003, Ground Penetrating Radar in Sediments, Geological Society, London, Special Publications, 211.

Burhl, M., Vermeer, G.J.O. and Kiehn, M., 1996, Fresnel zones for broadband data. Geophysics, Vol. 61, pp. 600–604.

Davis, J.L., 1973, The Problem of Depth Sounding Temperate Glaciers. M.Sc. Thesis, University of Cambridge, 106 p.

Davis, J.L. and Annan, A.P., 1989, Ground penetrating radar for high-resolution mapping of soil and rock stratigraphy. Geophysical Prospecting, Vol. 37, pp. 531–551.

El Said, M.A.H., 1956, Geophysical prospection of underground water in the desert by means of electromagnetic interference fringes, Proc. I.R.E., Vol. 44, pp. 24–30 and 940.

Engheta, N., Papas, C.H. and Elachi, C., 1982, Radiation patterns of interfacial dipole antennas. Radio Science, Vol. 17, pp. 1557–1566.

Fisher, E., McMechan, G.A. and Annan, A.P., 1992, Acquisition and processing of wide-aperture ground penetrating radar data. Geophysics, Vol. 57, p. 495.

Gilson, E.W., Redman, J.D., Pilon, J.A. and Annan, A.P., 1996, Near surface applications of borehole radar. Proceedings of the Symposium on the Application of Geophysics to Engineering and Environmental Problems, Keystone, Colorado, April 28–May 2, pp. 545–553.

Grasmueck, M., 1996, 3-D ground-penetrating radar applied to fracture imaging in gneiss. Geophysics, Vol. 61, No. 4, pp. 1050–1064.

Greaves, R.J., Lesmes, D.P., Lee, J.M. and Toksoz, M.N., 1996, Velocity variation and water content estimated from multi-offset, ground penetrating radar. Geophysics, Vol. 61, pp. 683–695.

Hasted, J.B., 1972, Liquid water -dielectric properties, in F. Franks (ed.), Water a Comprehensive Treatise, Vol. 1, The Physics and Physical Chemistry of Water, Plenum Press, New York, pp. 255–310.

Irving, J.D. and Knight, R.J., 2003, Removal of wavelet dispersion from ground-penetrating radar data. Geophysics, Vol. 68, No. 3, pp. 960–970.

Jackson, J.D., 1962, Classical Electrodynamics, John Wiley and Sons, New York.
Knapp, R.W., 1991, Fresnel zones in the light of broadband data. Geophysics, Vol. 56, pp. 354–359.
Lehmann, F. and Green, A.G., 1999, Semiautomated georadar data acquisition in three dimensional. Geophysics, Vol. 64, No. 3, pp. 719–731.
Lehmann, F. and Green, A.G., 2000, Topographic migration of georadar data: Implications for acquisition and processing. Geophysics, Vol. 65, No. 3, pp. 836–848.
McMechan, G.A., Gaynor, G.C. and Szerbiak, R.B., 1997, Use of ground-penetrating radar for 3-D sedimentological characterization of clastic reservoir analogs. Geophysics, Vol. 62, pp. 786–796.
Olhoeft, G.R., 1981, Electrical properties of rocks, in Touloukian, Y.S., Judd, W.R. and Roy, R.F. (eds), Physical Properties of Rocks and Minerals, Vol. II, McGraw-Hill, New York, 548 p.
Olhoeft, G.R., 1987, Electrical properties from 10^{-3} to 10^{+9} Hz – Physics and chemistry. Proceedings of the 2nd International Symposium on the Physics and Chemistry of Porous Media, American Institute of Physics Conference Proceedings, Vol. 154, pp. 281–298.
Olhoeft, G.R., 1988, Interpretation of hole-to-hole radar measurements., Proceedings of the Third Technical Symposium on Tunnel Detection, January 12–15, 1988, Golden, CO, pp. 616–629.
Olsson, O., Falk, L., Forslund, O., Lundmark, L. and Sandberg, E., 1992, Borehole radar applied to the characterization of hydraulically conductive fracture zones in crystalline rock. Geophysical Prospecting, Vol. 40, pp. 109–142.
Pearce, J. and Mittleman, D., 2002, Defining the Fresnel zone for broadband radiation. Physical Review E, Vol. 66, p. 056602.
Roberts, R.L. and Daniels, J.J., 1996, Analysis of GPR polarization phenomena. JEEG, Vol. 1, No. 2, pp. 139–157.
Rubin, Y and Hubbard, S.S. (eds), 2005, Hydrogeophysics, Springer, Dordrecht, The Tetherlands. Klewer.
Santamarina, J.C., Klein, K. and Fam, A., 2001, Soils and Waves: Particulate Materials Behavior, Characterization and Process Monitoring, John Wiley & Sons, Sussex, England, 508 pp.
Sen, P.N., Scala, C. and Cohen, M.H., 1981. A self-similar model for sedimentary rocks with application to the dielectric constant of fused glass beads. Geophysics, Vol. 46, pp. 781–795.
Sigurdsson, T. and Overgaard, T., 1996, Application of GPR for 3D visualization of geological and structural variation in a limestone formation. Proceedings of the Sixth International Conference on Ground Penetrating Radar (GPR'96), September 30–October 3, 1996, Sendai, Japan.
Smith, G.S., 1984, Directive properties of antennas for transmission into a material half-space: IEEE Trans. Antennas propagate, Vol. AP-32, pp. 232–246.
Smythe, W.R., 1989, Static & Dynamic Electricity, Taylor & Francis, A SUMMA book.
Sommerfeld, A., 1949, Partial Differential Equations in Physics, Academic Press, New York.
Tillard, T. and Dubois, J-C., 1992, Influence and lithology on radar echoes: Analysis with respect to electromagnetic parameters and rock anisotropy. Fourth International Conference on Ground Penetrating Radar, June 8–13, 1992, Rovaniemi, Finland. Geological Survey of Finland, Special Paper 16, 365 pp.
Todoeschuck, J.P., Lafleche, P.T., Jensen, O.G., Judge, A.S. and Pilon, J.A., 1992, Deconvolution of ground probing radar data, in ground penetrating radar, ed. J. Pilon. Geological Survey of Canada, Paper 90-4, pp. 227–230.
Turner, G., 1992, Propagation deconvolution. Fourth International Conference on Ground Penetrating Radar, June 8–13, 1992, Rovaniemi, Finland. Geological Survey of Finland, Special Paper 16, 365 pp.
Topp, G.C., Davis, J.L. and Annan, A.P., 1980, Electromagnetic determination of soil water content: Measurements in Coaxial Transmission Lines. Water Resources Research, Vol. 16, No. 3, pp. 574–582.
van der Kruk, J., 2001, Three Dimensional Imaging of Multi-Component Ground Penetrating Radar. Ph.D. Thesis, Delft University of Technology.
Wait, J.R., 1962, Electromagnetic Waves in Stratified Media, Pergamon Press, Oxford, pp 85–95.
Waite, A.H. and Schmidt, S.J., 1961, Gross errors in height indication from pulsed radar altimeters operating over thick ice or snow. IRE International Convention Record, Part 5, pp. 38–54.
Ward, S.H. and Hohmann, G.W., 1987, Electromagnetic theory for geophysical applications in electromagnetic methods in applied geophysics – Theory: 1, in Nabighian, M.N. Society of Exploration Geophysicists, Tulsa, OK, pp. 131–311.

Watts, R.D. and England, A.W., 1976, Radio-echo sounding of temperate glaciers: Ice properties and sounder design criteria. Journal of Glaciology, Vol. 21, pp. 39–48.
White, R.E., 1991, Properties of instantaneous seismic attributes, The Leading Edge. Society of Exploration Geophysicists, Vol. 10, No. 7, pp. 26–32.
Yilmaz, O., 2000, Seismic Data Analysis – Processing, Inversion, and Interpretation of Seismic Data, Society of Exploration Geophysicists, p. 2027.

GPR CONFERENCE REFERENCES

Proceedings of the International Workshop on the Remote Estimation of Sea Ice Thickness Centre for Cold Ocean Resources Engineering (C-CORE), St. John's Newfoundland, September 25–26, 1979.
Proceedings of the Ground Penetrating Radar Workshop, Geological Survey of Canada, May 24–26, 1988, Ottawa, Ontario, Canada, GSC paper 90-4.
Abstracts of the Third International Conference on Ground Penetrating Radar, United States Geological Survey, May 24–26, 1990, Lakewood, Colorado, USA.
Proceedings of the Fourth International Conference on Ground Penetrating Radar, Geological Survey of Finland, June 8–13, 1992, Rovaniemi, Finland.
Proceedings of the Fifth International Conference on Ground Penetrating Radar (GPR'94), June 12–16, 1994, Kitchener, Ontario, Canada.
Proceedings of the Sixth International Conference on Ground Penetrating Radar (GPR'96), September 30–October 3, 1996, Sendai, Japan.
Proceedings of the Seventh International Conference on Ground Penetrating Radar (GPR '98), May 27–30, 1998, Lawrence, Kansas, USA.
Proceedings of the Eighth International Conference on Ground Penetrating Radar (GPR 2000), May 23–26, 2000, Goldcoast, Australia, SPIE Vol. 4084.
Proceedings of the Ninth International Conference on Ground Penetrating Radar (GPR 2002), Apr 29–May 2, 2002, Santa Barbara, California.
Proceedings of the Tenth International Conference on Ground Penetrating Radar (GPR 2004), June 21–24, 2004, Delft, The Netherlands, Vol. I and Vol. II.
Proceedings of the Eleventh International Conference on Ground Penetrating Radar (GPR 2006), June 19–22, 2006, Columbus, Ohio.

CHAPTER 2

ELECTRICAL AND MAGNETIC PROPERTIES OF ROCKS, SOILS AND FLUIDS

Nigel J. Cassidy

Contents

2.1. Introduction	41
2.2. Electromagnetic Material Properties: Basic Theory	43
2.3. Permittivity and Conductivity – The Electrical Parameters of Dielectrics	44
2.3.1. Permittivity – ε	45
2.3.2. Conductivity – σ	54
2.3.3. Permeability μ – the magnetic parameters of dielectrics	55
2.4. Material Properties – Relationship to Electromagnetic Wave Characteristics	57
2.4.1. Loss factor and skin depth	59
2.5. The Properties of Real Materials – Practical Evaluations	60
2.6. Characterising the Response of Real Materials	62
2.6.1. Basic mixing models	63
2.6.2. Volumetric and inclusion-based mixing models	64
2.7. Summary	66
Acknowledgements	67
References	67

2.1. INTRODUCTION

The ability of ground penetrating radar (GPR to provide 'real-time', high-resolution, stratigraphically related, cross-sectional images of the subsurface is the technique's unique selling point and even novice users can quickly start to interpret GPR sections with some degree of confidence. As such, it is unrivalled amongst all the near-surface geophysical techniques and, arguably, is one of the most popular non-invasive subsurface characterisation tools for engineers, archaeologists and geologists. Unfortunately, the ease of both use and data interpretation is also GPR's downfall, as many inexperienced practitioners fail to fully appreciate the true nature of GPR wave propagation and its interaction with the subsurface materials. Key to this is the understanding that a GPR section is not a picture or an image of the subsurface *per se* but is, instead, the time-dependent, recorded response of the subsurface materials to the propagation of electromagnetic (EM) energy across a relatively narrow range of radio wave frequencies, typically

10 MHz–2 GHz. Consequently, it is important that users understand the physical meaning of a material's electrical and magnetic properties and how these relate to GPR signal attenuation and wave propagation velocities. As such, the following chapter covers topics such as permittivity, conductivity, lossy dielectrics, polarisation and relaxation mechanisms, etc., all in the context of practical GPR.

As a topic in its own right, the characterisation and analysis of materials at radio frequencies (or dielectric spectroscopy as it is often referred to) is much older than GPR, with some of the leading scientists of the past 100 years being involved in ground-breaking discoveries on the nature of matter and its interaction with EM energy. Much of our practical understanding, through either quantum or classical molecular approaches, can be traced back to the middle of the nineteenth century and the first few decades of the twentieth century, with the pioneering work of Debye (1929) still being relevant to GPR research today. The development of airborne radar during the Second World War led to a dramatic increase in radio frequency (RF), dielectric-related research, as did the development of microwave ovens in the late 1970s. Nowadays, thanks to our insatiable desire for new and improved mobile telecommunications and computing devices, the subject is as popular as ever. A quick review of the text and journal literature will reveal the true extent and depth of the subject with specialist disciplines in dielectric spectroscopy, telecommunication electronics, antenna design, material science, theoretical physics and colloidal science, all with their own terminology and approaches to describing the same phenomena: the interaction of propagating EM energy with materials. From a GPR users' perspective, this vast pool of knowledge and information can be bewildering, and it is often difficult to extract any practical understanding of the subject, even from the most basic of texts. This chapter conveys the topic from a practical perspective without resorting to detailed mathematical concepts, derivations, or proofs and is, therefore, less theoretically rigorous than some readers may like. I provide appropriate background and further reading for those with a particular theoretical bent and have broken down the subject into theoretical aspects (the microscopic and molecular/atomic nature of materials and their behaviour in EM fields) and more practical aspects (the GPR-related properties of materials, mixtures and methods for their measurement). Simplifications and assumptions have to be made and as we are primarily interested in GPR frequencies, the discussion and the scope of the subject are restricted to a relatively narrow frequency range of approximately 10 MHz–2 GHz. Important aspects, such as the polarisation and relaxation phenomena of water molecules, include higher-frequency effects, but, in general, optical (very high frequency) or low–frequency, EM-related phenomena will not be covered. Readers unfamiliar with the physical and mathematical concept of complex numbers with real and imaginary components may need to undertake a little revision, as these are key to the understanding of frequency-dependent material behaviour.

In basic terms, our understanding of EM waves, materials and their mutual interactions can be classified into either electrical phenomena (electric fields, permittivity and conductivity, etc.) or magnetic phenomena (magnetic fields, permeability, magnetic susceptibility, etc.). To some extent, this traditional classification still exists in many fields of EM study, and geophysicists, RF engineers, material scientists and physicists all seem to have their own favoured descriptive approaches and nomenclature, which can be contradictory and confusing to the uninitiated. In the following discussions, only common GPR-related terminology will be used (e.g., the symbol epsilon, ε, will be used for the complex-valued permittivity and the relative, real component, ε_r, used instead of the 'dielectric constant', κ), and mathematical formula and symbology will follow geophysical conventions rather than engineering texts. Ultimately, my brief account does not do the subject full justice, either theoretically or mathematically. For a more in-depth account of EM material behaviour, the seminal work of Von Hippel (1954) is an ideal place to start, along with the texts of Daniel (1967), Hill et al. (1969) and Hasted (1973). More recent texts of Gladkov (2003), Kremer and Schonhals (2002) and Von Hippel (1995) are recommended for the specialist reader, and for those with a very healthy library budget, the excellent, if expensive, two-volume set *Handbook of Low and High Dielectric Constant Materials and Their Applications* by Nalwa (1999) is a must.

2.2. ELECTROMAGNETIC MATERIAL PROPERTIES: BASIC THEORY

The starting point for any discussion on the nature of materials under the excitation of a propagating RF, EM wave is Maxwell's EM field equations [Equations (2.1)–(2.4)] and more importantly the constituent relations [Equations (2.5)–(2.7)]. These vector equations quantitatively describe the spatially and temporally varying coupled electric and magnetic fields and their interdependence. They are valid for the whole of the frequency spectrum and describe the EM energy storage and dissipation process for all materials. In their classical, time domain, differential form they are given by Equations (2.1)–(2.7) for heterogeneous, isotropic, linear and stationary media (Balanis, 1989):

Faraday's Law of Induction

$$\nabla \times \boldsymbol{E} = -\partial \boldsymbol{B}/\partial t \qquad (2.1)$$

Maxwell's modified circuit Law

$$\nabla \times \boldsymbol{H} = \partial \boldsymbol{D}/\partial t + \boldsymbol{J} \qquad (2.2)$$

Gauss' theorem in electrostatics

$$\nabla \cdot \boldsymbol{D} = \rho \qquad (2.3)$$

Gauss' theorem in magnetostatics

$$\nabla \cdot \boldsymbol{B} = 0 \qquad (2.4)$$

where

$\boldsymbol{E}$ – electric field strength vector (in volts per metre – V/m)
$\boldsymbol{H}$ – magnetic field strength vector (in amperes per metre – A/m)
$\boldsymbol{D}$ – electric flux density vector (in coulombs per metre squared – C/m^2)

B – magnetic flux density vector (in Tesla – T)
J – current density vector (in amperes per metre squared – A/m^2)
ρ – charge density (in coulombs per metre cubed – C/m^3)
∇ – del vector operator
× – vector cross product
• – vector dot product

The associated constitutive relations introduce the relevant material property parameters of permittivity, magnetic permeability and conductivity. These are the following:

$$\boldsymbol{D} = \varepsilon \boldsymbol{E} \qquad (2.5)$$

where ε – permittivity of the material (in Farads per metre – F/m)

$$\boldsymbol{J} = \sigma \boldsymbol{E} \qquad (2.6)$$

where σ – conductivity of the material (in Siemens per metre – S/m)

$$\boldsymbol{B} = \mu \boldsymbol{H} \qquad (2.7)$$

where μ – permeability of the material (in Henrys per metre – H/m)
In Equations (2.5)–(2.7), the properties are shown as simple constants. This is for the ideal case of a uniform, homogeneous material with no losses, anisotropy, or frequency dependence. In reality, most natural and man-made materials exhibit some degree of loss, frequency dependence and an element of anisotropy in one, if not all, of the parameters.

2.3. PERMITTIVITY AND CONDUCTIVITY – THE ELECTRICAL PARAMETERS OF DIELECTRICS

Subsurface materials are often described as dielectrics, with the parameters permittivity and conductivity loosely termed as their 'dielectric properties'. Strictly speaking, the term 'dielectric' describes a class of non-conducting materials that can accommodate a propagating, alternating EM field and, as such, only materials containing bound electrical charges alone can be called true dielectrics (e.g., crystalline solids). If there are any free charges available, then under the influence of the applied EM field, these will flow through the material (e.g., as in the pore fluids of saturated sands) producing attenuation and loss of energy. In reality, all subsurface materials possess some form of free charge, such as conduction electrons and ions, and these materials are best described as 'lossy dielectrics', which show some degree of EM attenuation. In extreme cases, a material containing a high degree of free charges is effectively a conductor where the majority of the EM energy will be lost in the conduction process as heat. This is the reason why GPR is ineffective in higher-conductivity environments (e.g., saline conditions and high clay contents).

2.3.1. Permittivity – ε

Put simply, permittivity describes the ability of a material to store and release EM energy in the form of electric charge and classically relates to the storage ability of capacitors. Alternatively, it can be described as the ability to restrict the flow of free charges or the degree of polarisation (in F/m) exhibited by a material under the influence of an applied electric field. It is usually quoted in terms of a non-dimensional, relative permittivity term (ε_r) where

$$\varepsilon_r = \text{permittivity of the material } (\varepsilon)/\text{permittivity of free space or vacuum } (\varepsilon_0) \tag{2.8}$$

The permittivity of free space (or permittivity constant) is given as 8.8542×10^{-12} F/m and differs negligibly from the permittivity of air. In some older texts, the relative permittivity of a material is sometimes referred to as the 'dielectric constant' and given the symbol κ. This practice is less common now, but unfortunately, the phrase has stuck and many users still refer to the real component of the complex permittivity as the dielectric constant.

The permittivity of subsurface materials can vary dramatically, especially in the presence of free and bound water, and is usually a complex, frequency-dependent quantity with real (storage) and imaginary (loss) components. The permittivity value of a material is often simplified to its constant, low-frequency (or static) real component with the loss term ignored. This is convenient for the approximate calculation of radar wave velocities and wavelengths, but it is too general for a detailed analysis. Table 2.1 lists the relative permittivity and conductivity of some common subsurface materials at 100 MHz and their typical range under natural conditions. They are 'typical' values derived from experiment and illustrate the influence of free and bound water, i.e., wetter higher, drier lower (Table adapted from Conyers and Goodman, 1997; Reynolds, 1997; Daniels, 2004).

To illustrate how RF EM energy interacts with a material, it is best to take a simplified version of the classical atomic approach, rather than the quantum one, and imagine the material to be composed of atoms (with positively charged nucleus and negatively charged electron cloud) or more simply, as a collection of bound particles (Figure 2.1). Let us consider the material as being uniform and excited by an incident pulse of RF EM energy that propagates through the material. In the absence of the applied EM field, the charges are unpolarised and there is a net zero charge across the material. As the propagating, incident EM pulse travels through the material, the charges become physically displaced or *polarised* in relation to their original position. Charge concentration occurs at both the local atomic scale (e.g., the negative centre of the electron cloud becomes slightly offset from the positive centre of the nucleus) and at the edges of the material where there are no neighbouring charges to balance the effect. On the rising edge of the incident pulse, energy is 'transferred' to the particles in the form of charge separation (energy storage) and released in the trailing edge. A *dipole moment* is induced in the materials, and a net *dipole moment density* is generated across the polarised charges. In simple materials, the dipole moment density is proportional to the strength of the applied

Table 2.1 Typical values of relative permittivity (real component) and static conductivity for common subsurface materials at an antenna frequency of 100 MHz

Material	Static conductivity, σ_s (mS/m)	Relative permittivity, ε_{ave}
Air	0	1
Clay – dry	1–100	2–20
Clay – wet	100–1000	15–40
Concrete – dry	1–10	4–10
Concrete – wet	10–100	10–20
Freshwater	0.1–10	78 (25 °C)–88
Freshwater ice	1–0.000001	3
Seawater	4000	81–88
Seawater ice	10–100	4–8
Permafrost	0.1–10	2–8
Granite – dry	0.001–0.00001	5–8
Granite – fractured and wet	1–10	5–15
Limestone – dry	0.001–0.0000001	4–8
Limestone – wet	10–100	6–15
Sandstone – dry	0.001–0.0000001	4–7
Sandstone – wet	0.01–0.001	5–15
Shale – saturated	10–100	6–9
Sand – dry	0.0001–1	3–6
Sand – wet	0.1–10	10–30
Sand – coastal, dry	0.01–1	5–10
Soil – sandy, dry	0.1–100	4–6
Soil – sandy, wet	10–100	15–30
Soil – loamy, dry	0.1–1	4–6
Soil – loamy, wet	10–100	10–20
Soil – clayey, dry	0.1–100	4–6
Soil – clayey, wet	100–1000	10–15
Soil – average	5	16

electric field, E, with the constant of proportionality being referred to as the *permittivity*, ε. The leading and trailing edges of the pulse supply energy to the separating charges in the form of acceleration that, in turn, generates a small *displacement current* that produces radiating EM energy. This localised energy is slightly out of phase with the incident pulse, and the net result is to 'slow down' the main body of the propagating wave. As such, the permittivity parameter is directly linked to the propagation velocity of the EM wave in the material (see later).

If the separating charges are free to move and can physically interact (as in the dipolar molecules of free water, for example), then the displacement and polarisation process converts some of the EM energy into heat during the particle interactions. As such, a component of energy loss is introduced into the polarisation process that acts out of phase with the energy storage and release mechanism. This phenomenon occurs in most materials, and therefore, the permittivity is usually described as a complex quantity, with the real component representing the 'instantaneous' energy storage–release mechanism and the imaginary component representing the energy dissipation. Both components are typically frequency-dependent, with the

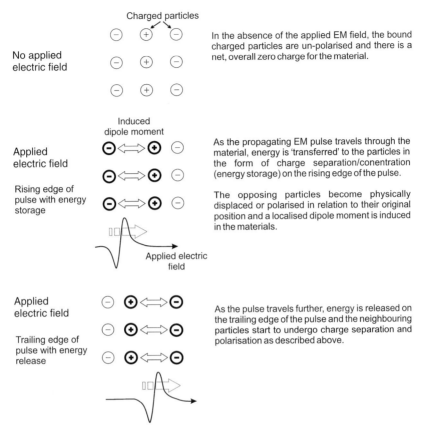

Figure 2.1 Conceptual diagram illustrating the process of energy storage/release, charge polarisation and the development of a dipole moment occurring when an electromagnetic (EM) wave propagates through a material.

imaginary component combining with that caused by the conductivity to give a total loss for the material (see later). The frequency dependence of the polarisation process is a manifestation of the permittivity *relaxation* phenomena where the time-dependent displacement mechanism is acting at different rates to the alternating, applied electric field. Below the relaxation frequency, the particles are able to 'react quickly' to the applied field and stay in phase with its changes. At, and above, the relaxation frequency, they cannot keep up with the rapidly changing field and spend most of their time in motion, therefore, producing significant loss of energy as heat to the surrounding matrix. This process is best illustrated by the relaxation and polarisation effects of free water molecules (see later) and is very important for GPR and, of course, microwave ovens! In general, most materials display a range of permittivity relaxation mechanisms that are represented, in each case, by an overall decrease in the value of the real component of the permittivity and a peaking in the imaginary component as frequency increases. The peak value of this distribution is called the *relaxation frequency*. The different individual relaxation mechanisms (e.g., electronic,

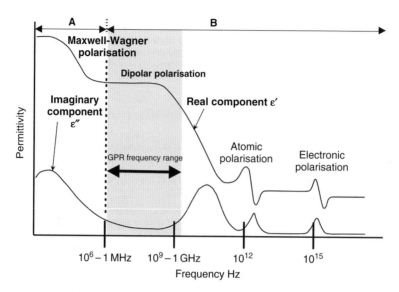

Figure 2.2 Permittivity relaxation phenomena associated with an idealised, lossy dielectric material. Region A represents the polarisations associated with free-charge and material interactions whilst region B represents molecular and atomic polarisations associated with bound charges. (Adapted from Daniels, 1996).

atomic, dipolar and Maxwell–Wagner) combine to give an overall response for the material. These can be divided into two separate groups:

- *Bound charge effects* – related to the relaxation response of individual atoms or molecules and includes electronic, atomic and dipolar polarisation.
- *Free charge effects* – describes the relaxation phenomena of 'trapped' free ionic charges in water and on grain surfaces and includes the so-called Maxwell–Wagner polarisation effect.

Figure 2.2 illustrates the permittivity response of an idealised, damp, lossy dielectric material across the whole EM spectrum and includes the common relaxation phenomena described above. For GPR, the most important is the dipolar polarisation of bound and/or free charges (usually water), whilst the others tend to fall outside the GPR frequency range. However, this is a simplified representation of idealised material and, typically, material mixtures exhibit a range of overlapping relaxation mechanisms with varying relaxation frequencies.

2.3.1.1. Electronic and atomic polarisation

For most dry solid materials, these are the only polarisation mechanisms that occur and give rise to a nearly constant permittivity over the GPR frequency range. They operate at atomic or sub-atomic scales and relate to the displacement of electron clouds or individual bonded atoms (Von Hippel, 1954). Although less significant for GPR applications in temperate regions, electronic and atomic polarisation effects are important in arid and glacial environments and provide a practical, high-frequency limit for the value of permittivity (often referred to as the optical permittivity).

2.3.1.2. Dipolar or orientational polarisation

This relaxation mechanism occurs in materials in which the molecules have permanent electric dipole moments. These may be covalently bonded gases, liquids, or ionically bonded crystalline solids with lattice defects. Whatever form they take, there are various theories used to describe their behaviour. The fundamental principles are discussed in detail in Von Hippell (1954) and Hasted (1973), whilst more generalised, yet application-specific, descriptions can be found in King and Smith (1981) and Daniels (1996, 2004). Although it is unnecessary to cover the subject in detail, it is important to understand how the polarisation mechanisms behave and, more significantly, how the relaxation of dipolar water molecules affects the complex permittivity response of multi-phase materials such as soils, sands and porous rocks.

2.3.1.2.1. Simple polar materials – 'free' water

Pure water (H_2O) is arguably the most important material exhibiting dipolar relaxation behaviour and is the classical example of a polar liquid. Strictly speaking, this behaviour relates only to molecules that are free to rotate (i.e., 'free' water) and not to those for which rotation is restricted by the action of other charges (e.g., the 'bound' water that adheres electrochemically to grain surfaces). Polar liquids can be well-modelled as a collection of isolated molecules with individual dipole moments. In the absence of an electric field, these dipoles randomly change orientation through the action of thermal agitation by neighbouring molecules, resulting in an equilibrium state with zero net volume density of polarisation. If an electric field is applied, each of these moments experiences a torque acting in a uniform direction that attempts to orientate the dipole moments parallel to the applied field. Thermal agitation, molecular inertia and the breaking resistance of the weak, intermolecular hydrogen bonds oppose this orientational torque, resulting in a time delay before maximum net polarisation is reached. In this higher energy state, the net volume density of polarisation is directly related to the DC or static permittivity, ε_s. If the applied field is removed, then the reverse situation occurs and the dipole moments relax with random realignments until they reach their original equilibrium state. The temporal response of this process is described by a relaxation time τ and is related to the relaxation frequency by $f_{\text{relax}} = 1/2\pi\tau$.

If the same material is subjected to an alternating electric field whose frequency is much less than the relaxation frequency (i.e., $f \ll f_{\text{relax}}$), then the net dipolar orientations do not lag the field variations. As a result, energy is passed to the molecules as the field increases and is released as the field value decreases. This produces the energy storage and release mechanism, which determines the value of the real component of the complex permittivity. As the process operates, the molecules undergo charge rotation and transport but the rate of the reaction is effectively too quick for the displacement current to have any appreciable effect on the imaginary component of the permittivity. As the frequency increases into the relaxation range, the net orientation starts to lag behind the field variations and the polarisation mechanism is underdeveloped. Consequently, less energy is passed to the molecule and the value of the real component of the permittivity is reduced. However, the molecules are now spending an increasing amount of time in rotational motion and the degree of charge transport is increased, generating a substantial out-of-phase displacement current. The value of the imaginary component of the permittivity increases and the action of molecular collisions produces energy loss in the form of heat. At the critical or peak relaxation frequency, the molecules are almost in permanent resonant motion, with the loss and the imaginary component of the permittivity having achieved their maximum values. As the frequency increases further, the field variations become too fast for the molecular orientations to respond and the net orientation remains in a 'non-polarised' state. Both the real and imaginary components of the permittivity decrease and stabilise at their high frequency or 'optical' molecular values. This phenomenon was first described by Debye (1929) for the permittivity relaxation of a simple, dilute solution of dipolar molecules in a non-polar liquid. The model produces a broad relaxation in the real component and a peaked distribution in the imaginary component. The real and imaginary components are easily separated to produce mathematical expressions for the complex permittivity of water (Von Hippel, 1954), which is as follows:

Permittivity, real component

$$\varepsilon'(\omega) = \varepsilon_\infty + \frac{\varepsilon_s - \varepsilon_\infty}{1 + \omega^2 \tau^2} \tag{2.9}$$

Permittivity, imaginary component

$$\varepsilon''(\omega) = (\varepsilon_s - \varepsilon_\infty) \frac{\omega \tau}{1 + \omega^2 \tau^2} \tag{2.10}$$

where

ε_s – static, DC or very low-frequency value of the permittivity
ε_∞ – optical or very high-frequency value of the permittivity
τ – permittivity relaxation time
ω – angular frequency

For pure, free water at room temperature (approximately 25 °C), the permittivity response is governed by a relaxation time of $\tau = 8.28$ ps (8.28×10^{-12} s or a critical frequency of approximately 19 GHz), a static relative permittivity of $\varepsilon_s = 81$ and a high-frequency relative permittivity of $\varepsilon_\infty = 5.6$ (Kaatze, 2000; Arkhipov, 2002). The response is shown in Figure 2.3, where it is clear that free water losses will only start to have a significant affect with the higher frequency surveys (i.e., above 500 MHz).

As thermal effects are component part of the polarisation process, there is a strong dependence of the permittivity upon temperature. As the temperature reduces, the permittivity spectrum of water shifts down-frequency with increasing values (Daniels, 2004). At 0 °C, but before freezing, the static permittivity rises to $\varepsilon_s \cong 88$ and the critical frequency reduces to about 9 GHz (King and Smith, 1981).

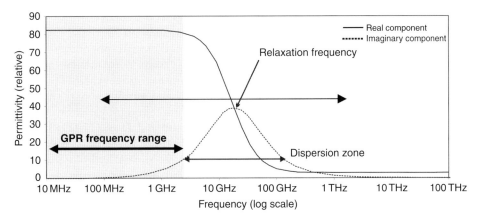

Figure 2.3 Permittivity spectrum of free, pure water at room temperature illustrating the single Debye relaxation mechanism at approximately 19 GHz.

Consequently, the GPR losses are substantially increased and there can be a significant effect on GPR signal attenuation, dispersion and propagation velocity above 100 MHz (Arcone et al., 1998, 2002). Upon freezing, the water molecules form ionic bonds, which prevent molecular rotation. Other process then cause relaxation in the kilohertz range and render the GPR losses insignificant (Arcone and Delaney, 1984; Delaney and Arcone, 1984). This is why GPR works very well in ice with very deep penetration (Arcone, 2002). Although bodies of ice can take many different physical forms (i.e., individual crystals, polycrystalline masses, congelation lake ice, frazil river ice and glacial ice), each still has relaxation frequencies below 100 kHz and relative permittivity of $\varepsilon = 3.15$–3.18. Clearly, the presence of frozen water in soils and other materials will have a bearing on the bulk permittivity of the materials as a whole. Fortunately, for most practical near-surface GPR applications, it can be assumed that the permittivity of frozen soils will be a constant, frequency-independent and lossless value of approximately $\varepsilon_s = 3$–5 (King and Smith, 1981).

Similar temperature-dependent permittivity characteristics would normally be expected in common subsurface materials but this is not always the case. Anomalous responses have been reported in soils, sands and rocks (e.g., Hoekstra and Delaney, 1974; Campbell, 1990; Pepin et al., 1995; Persson and Berndtsson, 1998; Or and Wraith, 1999; Wraith and Or, 1999; Chen and Or, 2006; Escorihuela et al., 2007). In these instances, the thermal characteristics are very much dependent on the water content and the nature/form of the component materials. In general, however, permittivity tends to increase with decreasing temperature, and for most practical GPR applications, the temperature effect can be considered unimportant, particularly if the surveys are collected over a relatively short period of time (i.e., days).

2.3.1.2.2. Bonded water

The classical, dipolar relaxation response of water is appropriate only for volumes of 'free' water (e.g., the main pore water component of a porous material). In soils and rocks, a proportion of the intergranular water bonds to the surface of the mineral grains generating a microscopic layer of absorbed water with restricted molecular rotation (Saarenketo, 1998). This is a well-known dielectric phenomenon that results in a pore fluid component whose critical relaxation frequency, and loss mechanism, is shifted to a frequency much lower than that of free water. This is due to the increased inertia of the bonded molecules and produces a permittivity spectrum that is highly sensitive to low-level water content variations. It has been observed in soils, clays and rocks (e.g., Hoekstra and Doyle, 1971; Dobson et al., 1985; Hallikainen et al., 1985; Fam and Dusseault, 1998; Friedman, 1998; Escorihuela et al., 2007) and is the subject of a lot of current hydrological GPR research. Unfortunately, the effect is non-trivial and depends on the degree of saturation, the form and distribution of the mineral phases (i.e., percentage of clay and rock particles), the amount of compaction, the percentage pore space and the ionic conductivity. At low saturation levels (typically 2–20% with volume), highly variable permittivity spectrums can be seen from different types of soil mixtures, particularly if phyllosilicate clay minerals are present. In general, however, the following simplifying assumptions can be made for GPR frequencies when dealing with wet soils with low clay contents:

- The bonded relaxation response is more pronounced at lower frequencies <200 MHz.
- In 'clay-deficient' materials, such as fine silts and sands, the percentage of bonded water is low (typically 1–5 vol%) and the pore fluid permittivity is equivalent to that of free water above this upper threshold value. At saturation levels less than 1–2 vol%, the pore fluid consists of bonded water with the effective permittivity properties of the bulk material.
- For saturation levels between 2 and 20 vol%, the pore fluid permittivity properties of common wet soils can be highly variable and primarily depends on the thickness of the water layer. This is governed by the grain size of the host medium and is most significant in very fine-grained material. Clay particles in the soil produce greater permittivity variations than rock particles at low saturation levels due to the increased electrochemical action of the surface charges.

2.3.1.2.3. Complex polar materials

Although the polarisation theory of Debye is a good approximation for pure and dilute polar liquids, it is often inappropriate for more complex polar materials (e.g., hydrocarbon contaminants, solvents and molecular mixtures). These have added orientational restrictions and, in the case of solids, more complex polarisation arrangements. Improved models by Onsager, Kirkwood and Fröhlich (Von Hippel, 1954) have provided better constraints on the permittivity and illustrate that the true relaxation response is usually broader than the Debye formulation. Cole and Cole (1941) recognised this and proposed an empirical formula that often successfully describes the permittivity behaviour of natural liquids, solids and heterogeneous mixtures. It is regularly used in GPR studies and includes an additional 'broadening' factor (α), based on either a distribution of individual relaxation times or the result of molecular collisions. From Cole and Cole (1941), the complex permittivity is given as

$$\varepsilon^*(\omega) = \varepsilon_\infty + \frac{\varepsilon_s - \varepsilon_\infty}{1 + (j\omega\tau)^{1-\alpha}} \quad (2.11)$$

where the complex permittivity ε^* is related to its real ε' and imaginary ε'' components by the following equation:

$$\varepsilon^*(\omega) = \varepsilon'(\omega) + j\varepsilon''(\omega) \quad (2.12)$$

For most subsurface materials, the Cole–Cole formulation is a good representation of the observed permittivity behaviour (e.g., Hoekstra and Delaney, 1974; Taherian et al., 1990; Olhoeft and Capron, 1993; Wensink, 1993; Friel and Or, 1999), with α varying between 0 and 0.7 ($\alpha = 0$ is equivalent to the Debye formulation).

2.3.1.3. Free charge and interfacial polarisation

These mechanisms describe the relaxation effects produced in porous, inhomogeneous materials and mixtures (e.g., the Maxwell–Wagner effect) that, in general, relate to the accumulation of free charges at material barriers or the preferential distribution of charge along material faces (Figure 2.4). In general, the

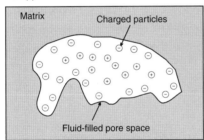

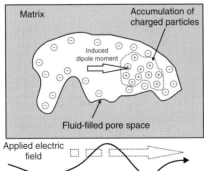

Figure 2.4 Conceptual diagram illustrating the process of interfacial polarisation in a porous material saturated with weakly conducting fluid.

relaxation response can be described by a Cole–Cole model (with a high α value) and accounts for the very high material permittivities observed below 1 MHz. In practice, interfacial polarisation effects are unimportant for most GPR frequency ranges but must be considered at the very low-frequency end of the spectrum, particularly in fine-grained, saturated, porous materials. Further information can be found in Chen and Or (2006), Chelidze and Gueguen (1999), Chelidze et al. (1999), Lima and Sharma (1992), Knight and Nur (1987) and Sen (1981).

2.3.2. Conductivity – σ

In simple terms, conductivity describes the ability of a material to pass free electric charges under the influence of an applied field. In metals, these charges relate to the free electrons of the metal atoms, whilst in fluids they are represented by the charges of dissolved anions and cations (e.g., Na^+, Ca^{2+}, Cl^-, CO_3^{2-}). These charge carriers rapidly accelerate to a terminal velocity generating internal conduction currents. As they propagate, they randomly collide against other atoms, ions or electrons, which produces energy loss in the form of heat (Figure 2.5). At low GPR frequencies, the charge response is effectively instantaneous, and the conduction current is in phase with the electric field (Turner and Siggins, 1994). In this case, the conductivity σ can be represented by a real, static or DC value, σ_s, in S/m and is usually the quantity quoted in published texts.

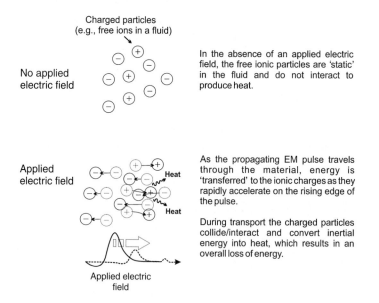

Figure 2.5 Conceptual diagram illustrating the process of conduction in a material containing free charges such as anions and cations.

At higher frequencies, the inertial effect of the accelerating charges produces a lag in the physical response and a conduction current that is out of phase with the electric field variations. The conductivity must now be described by a complex, valued quantity where the imaginary component represents the out-of-phase component of the current. This typically increases with frequency and adds to the energy storage effect of the permittivity. Often considered small or insignificant for radar frequencies (e.g., Turner and Siggins, 1994), the imaginary component is commonly ignored and the conductivity can be simplified to its real component only. However, natural materials can be very complicated and various free charges may be present within a small subsurface volume (e.g., electrons, mobile cations/anions in fluids, lattice and dislocation ions in solids, surface charges and free polymer charges). Each has different inertial properties (King and Smith, 1981) and, potentially, dissimilar effects on the imaginary component of the conductivity. This is particularly true in electrolyte solutions where the 'heavy' anions and cations are slow to respond to the applied field variations. Nevertheless, the degree of energy storage associated with the imaginary component is likely to be substantially less than the degree of energy loss represented by the real component (King and Smith, 1981). Consequently, the effect of conductivity relaxations will be small and, therefore, the simplifying assumption of a static conductivity value is appropriate in many cases.

2.3.3. Permeability μ – the magnetic parameters of dielectrics

In most circumstances, the magnetic effect of materials (i.e., diamagnetic, paramagnetic and superparamagnetic phenomena) has little effect on the propagating GPR wave (Olhoeft, 1998) and their magnetic permeability (μ) is often simplified to the free-space value of 1.26×10^{-6} H/m. However, ferromagnetic minerals can have a considerable effect on GPR wave velocity and signal attenuation. Iron, nickel and their sulphides/oxides have substantial ferromagnetic relaxation phenomena with mechanisms that relate to the development and reorientation of electron spin magnetic moments and the redistribution of magnetic domain wall boundaries (Von Hippel, 1954; Olhoeft, 1998). Magnetic relaxation frequencies primarily depend on both domain/grain size and the physical structure/size of the magnetic material (e.g., single-domain, nanometre-sized particles or larger, multi-domain, micro-to-millimetre-sized polycrystalline gains). In most common subsurface geological materials, the amount of ferromagnetic material is considered unimportant (typically <2%). However, appreciable amounts of magnetite, maghemite and hematite (the key ferromagnetic minerals in natural materials) can be found in some igneous rocks, iron-rich sands and soils, generating relaxation and loss effects comparable to those produced by the permittivity (Olhoeft and Capron, 1993; Cassidy, 2008).

From a practical GPR aspect, magnetic properties are important only if their effects are a significant proportion of the electrical response. The relative permittivity of most common subsurface materials ranges from approximately $\varepsilon_{ave} \cong 3$–$30$, the lowest value of which would equate to a volume magnetic susceptibility of 0.3 mks units (Mulay, 1963) and, therefore, an equivalent magnetite volume of some 7%. This is much higher than that naturally found in many rocks, sediments and soils and rare in man-made materials. Consequently, the assumption that magnetic effects are unimportant is justified as long as 'iron-free' subsurface materials are present. If this is not the case (such as magnetite-rich igneous rocks, hematitic sands and made-made smelting wastes), a non-unity value of relative magnetic permeability must be assumed and the effect of the magnetic components can be greater than that of the electrical components. As with the permittivity, the magnetic permeability of these materials is complex and frequency-dependent, with the real component representing the energy storage/release mechanisms and the imaginary component representing the loss effect (Cassidy, 2008).

The development of the Mars radar exploration programmes, such as MARSIS and SHARAD, has led to a renewed interest in the influence of magnetic minerals on GPR signals because significant quantities of magnetically lossy, iron-oxide materials are believed to exist on the Martian surface (Bertheliera et al., 2000; Stillman and Olhoeft, 2004, 2006; Pettinelli et al., 2005). The work has found applicability in terrestrial applications where similar mineralogical conditions exist (e.g., Chassagne et al., 2002; Van Dam et al., 2002; Robinson, 2004), with recent research illustrating that relatively low percentages of magnetite (10%) can have a significant effect on the velocity and attenuation (Figure 2.6) of the GPR signals (e.g., Mattei et al., 2007; Cassidy, 2007a, 2008).

The complex magnetic permeability of magnetically lossy materials varies significantly with percentage composition, grain size, mineral type and, of course, porosity. Figure 2.7 gives some idea of what to expect from a natural sample of dry,

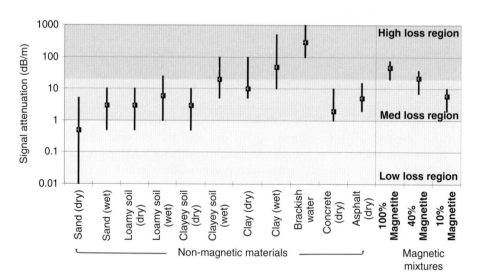

Figure 2.6 Typical material attenuation values, in dB/m, for a range of common near-surface materials and nano-to-microscale magnetite mixtures across the frequency range of 200–1200 MHz/1.2 GHz). Solid bars illustrate the full attenuation range, whilst the square markers represent 'typical' (materials) or average (magnetite) values. Data compiled, in part, from Daniels (1996, 2004) and Cassidy (2008).

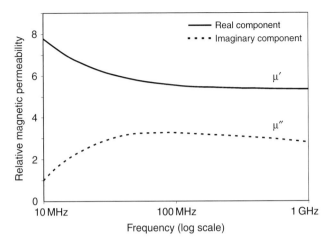

Figure 2.7 Typical measured values of the complex, frequency-dependent magnetic permeability for a solid, dry, igneous rock containing approximately 20%, millimetre-sized magnetite gains by volume.

solid igneous rock that contains ~20 vol% magnetite as millimetre-sized grains. The permeability spectrum shows a broad relaxation across the whole frequency range but with more pronounced effects at lower frequencies (<100 MHz). It is likely that multiple magnetic domains were present in the relatively large magnetite grains and that the domain walls mutually interacted under the influence of the applied EM field. This leads to a degree of latency in the overall relaxation response of the individual grains, resulting in a broad, bulk permittivity spectrum and a macroscopic relaxation response that is shifted towards the lower frequencies. Similar behaviour has been observed in biomedical ferrofluids (Pankhurst et al., 2003), and in terms of signal velocity and attenuation, the net result is a relative permeability that has the same degree of influence on the propagating GPR wave as does the permittivity. When compared to equivalent non-magnetic materials, it is likely that the user will see GPR wave velocities reduce and signal attenuation increase with the effect most pronounced in dry crystalline materials.

2.4. Material Properties – Relationship to Electromagnetic Wave Characteristics

So far, material properties have been considered in terms of their individual electrical and magnetic responses at the 'microscopic' scale without direct reference to their combined influence on the velocity and attenuation of GPR waves in 'real' materials. Because of their complexity, nearly all subsurface materials can be considered as mixtures of some form with a complex, frequency-dependent electric response and, potentially, a corresponding magnetic response. For a harmonic plane wave, with $e^{j\omega t}$ dependence, propagating in an electrically conducting dielectric medium of uniform magnetic properties, the propagation, attenuation, phase coefficient, intrinsic impedance and velocity in the direction of propagation can be derived directly from the EM wave equations. For frequency-independent, dielectrically lossless (but not conductively lossless) materials, these are given as follows (Lorrain et al., 1988; Balanis, 1989; Daniels, 1996; Reynolds, 1997):

Complex propagation constant

$$\gamma = \alpha + j\beta = \sqrt{(\sigma + j\omega\underline{\varepsilon})j\omega\underline{\mu}} \qquad (2.13)$$

Attenuation coefficient (Np/m)

$$\alpha = \omega\sqrt{\underline{\mu\varepsilon}}\left(\frac{1}{2}\left[\sqrt{1+\left(\frac{\sigma}{\omega\underline{\varepsilon}}\right)^2}-1\right]\right)^{1/2} \qquad (2.14)$$

Phase coefficient (rad/m)

$$\beta = \omega\sqrt{\underline{\mu\varepsilon}}\left(\frac{1}{2}\left[\sqrt{1+\left(\frac{\sigma}{\omega\underline{\varepsilon}}\right)^2}+1\right]\right)^{1/2} \qquad (2.15)$$

Intrinsic impedance of the medium (ohms)

$$\eta = \sqrt{\frac{j\omega\underline{\mu}}{\sigma + j\omega\underline{\varepsilon}}} \qquad (2.16)$$

Velocity (m/s)

$$v = \frac{c}{\left(\frac{\mu\varepsilon}{2}\left[\sqrt{1+\left(\frac{\sigma}{\omega}\right)^2}+1\right]\right)^{1/2}} \qquad (2.17)$$

where

c – velocity of EM waves in free space ($\sim 3 \times 10^8$ m/s)
ω – angular frequency
$\underline{\varepsilon}$ – absolute permittivity (real component)
ε – relative permittivity (real component)
$\underline{\mu}$ – absolute permeability
μ – relative permeability
σ – conductivity (real component)

For dielectrically lossy materials, with a complex conductivity and permittivity (but non-complex magnetic permeability), a real effective conductivity and real effective permittivity can be used in place of the specific constituent parameters above (King and Smith, 1981), where

Complex permittivity

$$\underline{\varepsilon}^* = \underline{\varepsilon}' - j\underline{\varepsilon}'' \qquad (2.18)$$

Complex conductivity

$$\sigma^* = \sigma' - j\sigma'' \qquad (2.19)$$

Real effective permittivity

$$\underline{\varepsilon}^e = \underline{\varepsilon}' - \frac{\sigma''}{\omega} \qquad (2.20)$$

Real effective conductivity

$$\sigma^e = \sigma' + \underline{\varepsilon}''\omega \qquad (2.21)$$

These effective parameters describe the combined EM energy loss and storage mechanisms of conductivity/permittivity relaxation and represent currents that are either in phase (σ^e) or out of phase (ε^e) with the electric field during the polarisation and relaxation processes. These are the parameters that are usually determined during experimental measurements (Turner and Siggins, 1994; Cassidy, 2007b) and can be inserted in the velocity and attenuation equations above (i.e., use σ^e and ε^e instead of σ and ε in 2.17 and 2.14) to determine the true velocity and attenuation characteristics of the lossy materials at any given frequency. In most cases, the conductivity component is considered independent of frequency, real-valued and, in general, relates only to the ionic conductivity of internal fluids and/or the surface charge conductivity of clay minerals. As such, the two components can be combined to give a *complex effective permittivity* expression that describes the total loss and storage effects of the material as a whole (Reynolds, 1997)

$$\underline{\varepsilon}_e^* = \underline{\varepsilon}' - j\left(\underline{\varepsilon}'' + \frac{\sigma_s}{\omega\underline{\varepsilon}_0}\right) \qquad (2.22)$$

where

$\underline{\varepsilon}_0$ – permittivity of free space (8.854×10^{-12} F/m)
$\underline{\sigma}_s$ – bulk static conductivity of the material (S/m).

2.4.1. Loss factor and skin depth

Other useful parameters are the loss tangent or more appropriately the *loss factor*, P, and the *skin depth*, δ, which are used to describe the loss component of a material and are related to the conductivity, permittivity and attenuation coefficient by the following equation:

$$\frac{1}{\alpha} = \delta \quad \text{and} \quad P = \left(\frac{\sigma' + \omega\underline{\varepsilon}''}{\omega\underline{\varepsilon}' - \sigma''}\right) \qquad (2.23)$$

These practically useful parameters help us to assess how 'lossy' a lossy dielectric is and, therefore, can provide a guide to the physical effects of attenuation on the GPR wave. The skin depth (δ) is the distance (m) that a plane wave has to travel before its amplitude has reduced by factor of $1/e$, or approximately 37% (Reynolds, 1997). It is helpful in evaluating the penetration distance of GPR waves and the likely amplitude of any reflections (neglecting spreading losses). The loss factor can be used as a limiting expression for the appropriateness of low-loss assumptions and describes the ratio of EM energy loss factors ($\sigma' + \omega \underline{\varepsilon}''$) to energy storage ($\omega \underline{\varepsilon}' - \sigma''$). For relatively dry, low conductivity materials, this will be much less than 1 and the loss factor can then be approximated to (Daniels, 1996)

$$P \cong \left(\frac{\sigma'}{\omega \underline{\varepsilon}'} \right) \quad (2.24)$$

This is considered to be the 'low-loss' condition and allows the velocity and wavelength to be approximated to

$$\text{Velocity, } v(m/s) = \frac{c}{\sqrt{\varepsilon \mu}} \quad (2.25)$$

$$\text{Wavelength, } \lambda(m) = \frac{v}{f} \quad (2.26)$$

where f is the frequency of the propagating wave in the material (in Hertz or cycles per second).

From this, initial estimates of target depth and resolution can be determined for a particular survey and provides the first step in GPR section interpretation. What is important to realise is that, in general, the attenuation of a GPR signal is proportional to, and strongly controlled by, the frequency (Equation 2.14), whilst the velocity is not so dependent on frequency in relative terms (Equation 2.17). As such, higher frequencies attenuate significantly more than low frequencies, and there is a relative loss of the high-frequency signal component as the wave propagates. This is particularly common in damp/wet subsurface materials. The frequency dependence of the attenuation also explains why, relatively speaking, lower-frequency GPR signals penetrate further than higher frequencies.

2.5. THE PROPERTIES OF REAL MATERIALS – PRACTICAL EVALUATIONS

For most GPR users, the low-loss condition is a useful and convenient approximation for common materials, and when used in combination with Table 2.1 (typical relative permittivity and static conductivity values), it provides an appropriate method for the calculation of GPR wave velocities, wavelengths and attenuation rates in a material. For a more detailed characterisation of porous, natural and man-made materials (such as soils and rocks), it is necessary to consider the media as a macroscopic assemblage, or mixture, of individual components, which include the matrix material (grains), pore space (air) and pore fluids (water and/or other fluids). There has been a wealth of research into the RF dielectric properties of materials, with several of empirical and theoretical models being developed and/or refined for all kinds of materials ranging from ice to concrete. In conjunction with this, our technical ability to measure the frequency-dependent permittivity, conductivity and permeability of real materials has improved dramatically over the past 20 years, and it is not unusual to see complex permittivity measurements routinely being undertaken on site with portable measurement equipment such as the Adek Percometer^{TM} (Adek Ltd., 2008). In the laboratory, both time domain reflectometry (TDR) and vector network analyser methods are commonplace, and there is a plethora of information available on measurement techniques and applications (refer to Bryant, 1988; Flemming et al., 1990; Baker-Jarvis and Grosvenor, 1996; Nalwa, 1999; Kremer and Schonhals, 2002; Evett and Parkin, 2005 for more detailed accounts on the measurement techniques). Unfortunately, much of the experimental work is valid only for a narrow range of materials (e.g., fine quartz sandstones or loamy soils), and it is difficult to say what the 'typical' value of complex effective permittivity is for a given material in the GPR bandwidth. For the strong-willed and generally curious, the following additional references (along with those already cited) will provide a comprehensive account of the experimental results, techniques and analysis methods for most common subsurface materials:

Soils – Arcone et al. (2008), Logsdon (2005), Lebron et al. (2004), Regalado et al. (2003), Francisca and Rinaldi (2003), Curtis (2001), Shang et al. (1999b), Starr et al. (1999), Darayan et al. (1998), Klein and Santamarina (1997), Heimovaara et al. (1996), Heimovaara (1994) and Topp et al. (1980, 1982).

Rocks – Sweeney et al. (2007), West et al. (2003), Robinson and Freidmen (2003), Kyritsis et al. (2000), Nichol et al. (2003) and Ulaby et al. (1990).

Concrete – Ekblad and Isacsson (2007), Adous et al. (2006); Filali et al. (2006), Wang et al. (2006), Van Damme et al. (2004), Soutsos et al. (2001), Shang et al. (1999a), Robert (1998), Rhim and Buyukozturk (1998) and Alqadi et al. (1995).

Although there is no typical complex effective permittivity spectrum that is valid for all materials, the measured characteristics of a damp, lossy soil across the frequency range of 10 MHz–1 GHz is a good indicator of what to expect in a granular porous material that contains a natural moisture content level (Figure 2.8).

The effect of the static conductivity on the imaginary component of the effective permittivity can be seen in the low-frequency part of the spectrum (<20 MHz) with relatively high values of ε'' that drop to a minimum at about 12 MHz. The imaginary component will never be zero at this minimum point as there is always the loss effect of weakly bound water relaxations and interfacial polarisations (i.e., the Maxwell–Wagner effect). As the frequency increases (>50 MHz), the dipolar relaxation response of free water starts to affect its spectrum; the real part of the effective permittivity drops smoothly with the corresponding imaginary part rising. However, the permittivity spectrum remains relatively broad as the more extreme effects of the water-based dipolar loss do not occur at these frequencies. Although

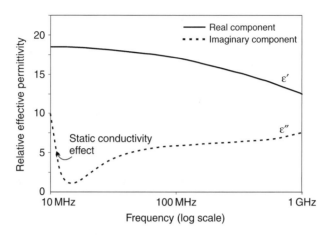

Figure 2.8 Typical (measured) representative values of the complex, frequency-dependent effective permittivity of a damp, lossy soil.

this is a simplified example, it does illustrate the effect of water on the effective permittivity spectrum, in the form of either dipolar losses or static conductivity effects. The effect of water is a constant theme within GPR, and as a basic rule of thumb; the relative permittivity of material increases with increasing water content (and therefore its velocity decreases), whilst the attenuation (in dB/m) is proportional to frequency and conductivity combined.

2.6. CHARACTERISING THE RESPONSE OF REAL MATERIALS

The Cole–Cole model mentioned earlier simulates the effective permittivity characteristics of water and simple materials well but is less suited to more complex mixtures such as soils and rocks. Other empirical models include the Cole–Davidson model (Davidson and Cole, 1951), the Power–Law model (Knight and Nur, 1987; Taherian et al., 1990) and the equivalent circuit methods (e.g., Rinaldi and Francisca, 1999; Sternberg and Levitskaya, 2001), which have all been used to describe natural materials. Olhoeft (1979) developed an interesting empirical model for dry mixtures that related the permittivity to the bulk density of the materials. Unfortunately, it is applicable only to truly 'dry' materials and is less suited for practical GPR applications. The Cole–Davidson model is an extension of the Cole–Cole formulation for conductive pore fluids, whilst the Power–Law model provides a generic formulation in which the frequency dependence of the permittivity is of the following form:

$$\varepsilon^*(\omega) \propto \omega^{n-1}, \quad \text{where } n<1 \qquad (2.27)$$

This is the basis of the 'universal dielectric response' function of Jonscher (1977) and the 'constant Q' behaviours reported by other researchers (e.g., Turner and Siggins, 1994; Bano, 1996; Hollender and Tillard, 1998; Bano, 2004).

In general, all of these models exhibit permittivity relaxation characteristics similar to the Cole–Cole formulation and tend to be used to describe experimental results by iteratively fitting the modelled complex effective permittivity spectrum to its measured counterpart. As such, they are limited in their practical application as often a user wishes to determine the effective permittivity spectrum of material without measurement or, alternatively, extract real material property information (e.g., water content) from a basic estimate/analysis of the permittivity. To accommodate this, a number of 'mixing models' have been developed that attempt to simulate the bulk effective permittivity of a material from a knowledge of its component parts (i.e., the individual permittivities of each part and their relative volumetric percentage). Unfortunately, the complexity of material mixtures makes it impossible to develop a single model that accurately describes the permittivity response of all materials, although a range of basic and advanced models exist. Most of the more advanced models relate to granular scale effects and require a knowledge of the matrix material, texture, effective surface area, grain shape, porosity, density, etc. Relevant examples are the composite sphere models of Friedman (1998), the self-similar models of Ghosh and Fuchs (1991, 1994), the geometrical models of Thevanayagam (1995) and the four-component, semi-empirical model of Dobson et al. (1985). In general, they are too complex for practical GPR applications, and simpler models are more useful.

2.6.1. Basic mixing models

There are a number of empirical formulations that can be considered as 'basic' mixing models, with, arguably, the most popular being the formula of Topp et al. (1980). This model fits a third-order polynomial function to the observed permittivity response of sandy/loamy soils as determined from TDR experiments. Appropriate for frequencies in the 10 MHz–1 GHz range, in general, it agrees reasonably well with the observed values across a wide range of water contents (~5–50%). For accurate results, it requires the selection of appropriate polynomial coefficients from an evaluation of the experimental data. However, a widely accepted, generalised formula is given by Annan (1999) as

$$\varepsilon' = 3.03 + 9.3\theta_v + 146(\theta_v)^2 - 76.6(\theta_v)^3 \quad (2.28)$$

where θ_v is the volumetric water content and the material is assumed to be low-loss with a 'dry-state' permittivity of $\varepsilon_s \cong 3-4$. This is not the only empirical, polynomial model and is often considered inappropriate for clays and organic-rich soils (Friedman, 1998). Other models include those of Curtis (2001), Sabburg et al. (1997), Dasberg and Hopmans (1992), Roth et al. (1990) and Hallikainen et al. (1985), plus a complementary model for the conductivity based on Archie's law (Annan, 1999). Their use as true mixing models is limited, primarily because they do not account for the imaginary component of the permittivity or the true relaxation response. However, they are extremely useful for determining natural water contents (Weiler et al., 1998) and can be used with other mixing models to provide a more realistic evaluation of the true permittivity.

2.6.2. Volumetric and inclusion-based mixing models

These are of more use, practically, and determine the effective permittivity of a mixture from a knowledge of its component parts. In general, they assume that the material is a multi-phase mixture of geometrically simple shapes or inclusions in a matrix (e.g., solid spheres in a fluid) or a composite of uniform layers. There are a range of applicable formulations – complex refractive index model (CRIM), Maxwell–Garnet theory (MGT), effective medium theory (EMT), Looyenga model, Hanai–Bruggeman and Bruggeman–Hanai–Sen (BHS) models, etc., all with slightly different approaches to determining the macroscopic permittivity. Detailed applications and comparisons can be found in Fiori et al. (2005), Johnson and Poeter (2005), Mironov et al. (2004), Carcione et al. (2003), Cosenza et al. (2003), Hu and Liu (2000), Tsui and Matthews (1997), Sihvola (1989, 2000), Sihvola and Alanen (1991) and Sen et al. (1981). These authors demonstrated that for relatively simple porous materials (sands, rocks, low-clay content soils, etc), the models correlate well with the experimental data. Of these, the CRIM, plus its derivatives, and the BHS model (BHS) have become the most popular for GPR-based hydrological/contaminant applications because they are simple to apply, robust and accurate over the GPR frequency range (e.g., Endres and Knight, 1992; Darayan et al., 1998; Persson and Berndtsson, 2002; Ajo-Franklin et al., 2004; Loeffler and Bano, 2004; Cassidy, 2007b).

The complex refractive index model is strictly a one-dimensional, layered medium model and has been shown to be effective for medium-to-coarse grained, multi-phase mixtures involving simple granular materials (e.g., semi-spherical sand grains) and moderate-to-low viscosity fluids. It has the advantage of being a volumetric model that requires only knowledge of a material's permittivities and their fractional volume percentages and can be used on both the real and imaginary components of the complex permittivity (Tsui and Matthews, 1997). In its general form, the CRIM formula is written as follows:

$$\varepsilon^e_{mix} = \left(\sum_{i=1}^{N} f_i \sqrt{\varepsilon_i} \right)^2 \qquad (2.29)$$

where

ε^e_{mix} – complex bulk effective permittivity of the mixture
f_i – volume fraction of the ith component
ε_i – complex permittivity of the ith component.

Any number of phases can be included but, in most cases, a three-phase model is appropriate with ε_w, ε_g, and ε_m representing the measured complex effective permittivities of water, gas (air) and matrix, respectively. As such, the CRIM formula becomes

$$\varepsilon^e_{mix} = \left[(\phi S_w \sqrt{\varepsilon_w}) + \left((1-\phi)\sqrt{\varepsilon_m}\right) + \left(\phi(1-S_w)\sqrt{\varepsilon_g}\right) \right]^2 \qquad (2.30)$$

where

ϕ – porosity
S_w – water saturation (i.e., percentage of pore space filled with fluid)
ε^e_{mix} – effective permittivity of the mixture
$\varepsilon_w, \varepsilon_g, \varepsilon_m$ – permittivities of the water, gas and matrix phases, respectively.

The practical use of CRIM is illustrated in Figure 2.9 in which the modelled complex effective permittivity spectrum of a damp sandy soil with 20% water content and <2% clay content is compared to its measured values. The porosity and the moisture contents of the soil were determined using standard laboratory methods (Avery and Bascomb, 1982), a CRIM-based model developed from the typical permittivity value of a dry sandy soil (matrix component, $\varepsilon_m = 4$) and a Cole–Davison relaxation characterisation of water that included a static conductivity of 10 mS/m. In general, the mixing model performs well over the GPR frequency range (10 MHz–1 GHz), with an excellent fit to the real component of the permittivity and a less accurate but "within-error" fit to the imaginary component. The effects of low-frequency losses due to bonded water or interfacial polarisations are not included in the Cole–Davidson model and it is likely that this affected the accuracy results near the bottom end of the frequency range. In addition, the minor clay fraction is likely to have had an effect on the static conductivity value with a slight, but non-trivial, increase in the conductivity at the lowest frequencies. As such, the CRIM mixing model is slightly underestimating the loss component below ~100 MHz, but for most practical applications, it provides a good analogy to the response of real materials.

The BHS model (sometimes referred to as the Hanai–Bruggeman–Sen model) is similar to CRIM in that it is a volumetric-based mixing model that incorporates the material's complex permittivities and porosity but includes a factor for the shape of

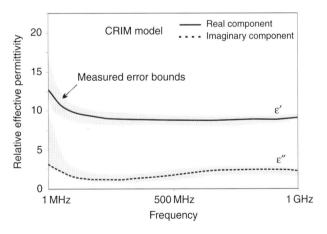

Figure 2.9 Comparison between the measured and complex refractive index model (CRIM)-modelled complex, frequency-dependent effective permittivity of a damp, lossy, sandy soil containing approximately 20% water content and <2% clay content.

the pores/grains (Hanai, 1968; Sen et al., 1981). In its general implicit form, the two-phase BHS model is given as follows (Sen et al., 1981):

$$\phi = \left(\frac{\varepsilon_{mix}^e - \varepsilon_m}{\varepsilon_w - \varepsilon_m}\right)\left(\frac{\varepsilon_w}{\varepsilon_{mix}^e}\right)^c \qquad (2.31)$$

where

ϕ – porosity
ε_{mix}^e – complex effective permittivity of the mixture
ε_w, ε_m – the complex permittivities of water and matrix phases, respectively.

In this formulation, c, is a factor related to the geometrical shape of the grains and ranges from $c = 1/3$ for spherical grain/inclusions, $c = 0$ for needles and $c = 1$ for plates (Sen et al., 1981). The formulation is based on a self-similar model of 'spheres within spheres', and therefore, the user can build multi-phase mixing models from combinations of two-phase mixtures (e.g., the bulk pore air/water permittivity can be made up from a two-phase model of air, water and the saturation index). Using this multi-phase approach, Loeffler and Bano (2004) compared the relative performance of the BHS, Topp and CRIM mixture models to the evaluated moisture contents of saturated and unsaturated sands in a controlled "sand-box" experiment. In-situ permittivities were determined from GPR wave velocities and water contents derived from the volume of water injected. All three formulations compared reasonably well against the measured values with the CRIM model producing the best results.

From a practical aspect, the choice of mixing model is dependent on individual circumstances (i.e., what is known about the subsurface materials in the first place), but in general, it is fair to expect that any of these three approaches (Topp, BHS or CRIM) will provide reasonably accurate estimates of effective bulk permittivity for most common subsurface materials.

2.7. SUMMARY

In this chapter, the physical meaning of material properties (permittivity, conductivity and permeability) has been discussed from both a theoretical and a practical standpoint. As stated in the introduction, key to our understanding of GPR is the fact that the image we see on the screen is not a cross-section of the subsurface but is, instead, the time-dependent response of the subsurface materials to the propagation of EM energy, as recorded at the receiving antenna. In the relatively narrow range of radio wave frequencies associated with GPR, typically 10 MHz–2 GHz, it is the macroscopic effect of the subsurface materials that is pertinent to our final interpretations, and therefore, it is important that users fully understand how these relate to GPR signal attenuation and wave propagation velocities.

The physical nature of materials has been discussed in detail (particularly water, as it has the strongest influence on our GPR signals), and the reader should now be in a position to assess the effect of moisture content and material type on the propagation velocity and attenuation of GPR signals to a high degree of confidence. Similarly, the reader should have at least a basic understanding of how to extract meaningful material property information from their GPR data. Ultimately, what has been presented here is only a brief review of a subject that has been at the forefront of science research for over a century. Hopefully the cited texts and papers, both historical and recent, should provide the dedicated and curious reader with the knowledge to expand into specialist subject areas that are not only fascinating but have applicability way beyond the field of GPR.

ACKNOWLEDGEMENTS

I would like to thank the reviewers, Maksim Bano and Steven Arcone for their constructive and thoughtful comments, which have helped improve this chapter considerably.

REFERENCES

Adek Ltd. 2008. http://adek.ee.

Adous, M., Queffelec, P. and Laguerre, L., 2006, Coaxial/cylindrical transition line for broadband permittivity measurement of civil engineering materials. Measurement Science and Technology, Vol. 17, No. 8, pp. 2241–2246.

Ajo-Franklin, J.B., Geller, J.T. and Harris, J.M., 2004, The dielectric properties of granular media saturated with DNAPL/water mixtures. Geophysical Research Letters, Vol. 31, p. L17501, doi:10.1092/2004GL020672.

Alqadi, I.L., Hazim, O.A., Su, W. and Riad, S.M., 1995, Dielectric-properties of portland-cementconcrete at low radio frequencies. Journal of Materials in Civil Engineering, Vol. 7, No. 3, pp. 192–198.

Annan, A.P., 1999, Ground Penetrating Radar: Workshop Notes, June 1999 edition, Sensors and Software Inc., Ontario, Canada.

Arcone, S.A., 2002, Airbourne-radar stratigraphy and electrical structure of temperate firn: Bagley Ice Fiel, Alaska. Journal of Glaciology, Vol. 48, No. 161, pp. 317–334.

Arcone, S.A. and Delaney, A.J., 1984, Dielectric measurements of frozen silt using time domain reflectometry. Cold Regions Science and Technology, Vol. 9, pp. 39–46.

Arcone, S.A., Grant, S.A., Boitnott, G.E. and Bostick, B., 2008, Complex permittivity and minerology of grain size fractions in a wet silt soil. Geophysics, (in press). Vol. 73, no. 3., DOI:10.1190/1.2890776.

Arcone, S.A., Lawson, D.E., Delaney, A.J., Strasser, J.C. and Strasser, J.D., 1998, Ground-penetrating radar reflection profiling of groundwater and bedrock in an area of discontinuous permafrost. Geophysics, Vol. 63, No. 5, pp. 1573–1584.

Arcone, S.A., Prentice, M.L. and Delaney, A.J., 2002, Stratigraphic profiling with ground penetrating radar in permafrost: A review of possible analogues for Mars. Journal of Geophysical Research, Vol. 107, p. E11, Article no. 5108.

Arkhipov, V.I., 2002, Hierarchy of relaxation times in water. Journal of Non-Crystalline Solids, Vol. 305, No. 1–3, pp. 127–135.

Avery, B.W. and Bascomb, C.L. (eds), 1982, Soil Survey Laboratory Methods, Soil Survey Technical Monograph No. 6., Soil Survey of England and Wales, Bartholomew Press, Dorking, UK.

Baker-Jarvis, J. and Grosvenor, J. H., 1996, Dielectric and Magnetic Measurements from −50°C to 200°C and in the Frequency band 50 MHz to 2 GHz, United States Department of Commerce, National Institute of Standards and Technology, NIST Technical Note 1355 (revised), Bolder Colorado, USA.

Balanis, C.A., 1989, Advanced Engineering Electromagnetics, John Wiley & Sons, New York.

Bano, M., 1996, Constant dielectric losses of ground-penetrating radar waves. Geophysical Journal International, Vol. 124, pp. 279–288.

Bano, M., 2004, Modelling of GPR waves for lossy media obeying a complex power law of frequency for dielectric permittivity. Geophysical Prospecting, Vol. 52, pp. 11–26.

Bertheliera, J.J., Neya, R., Costard, B.F., Hamelina, M., Meyera, A., Martinatc, B., Reineixc, A., Hansend, T.H., Banoe, M., Kofmanf, W, Lefeuvreg, F. and Paillouh, P., 2000, The GPR experiment on NETLANDER. Planetary and Space Science, Vol. 48, pp. 1161–1180.

Bryant, G.H., 1988, Principles of Microwave Measurement, IEE Electrical Measurement Series 5. Peter Peregrinus Ltd, London, UK.

Campbell, J.E., 1990, Dielectric properties and influence of conductivity in soils at one to fifty megahertz. Soil Science Society of America Journal, Vol. 54, pp. 332–341.

Carcione, J.M., Seriani, G. and Gei, D., 2003, Acoustic and electromagnetic properties of soils saturated with salt water and NAPL. Journal of Applied Geophysics, Vol. 52, pp. 177–191.

Cassidy, N.J., 2007a, Frequency-dependent attenuation and velocity characteristics of magnetically lossy materials. IEEE Proceedings of the 4th International Workshop on Advanced Ground Penetrating Radar, Naples, pp. 142–146.

Cassidy, N.J., 2007b, Evaluating LNAPL contamination using GPR and dielectric analysis: Practical implications for signal attenuation and attribute analysis studies. Journal of Contaminant Hydrology, Vol. 94, No. 1–2, pp. 49–75.

Cassidy, N.J., 2008, GPR frequency-dependent attenuation and velocity characteristics of nano-to-micro scale, lossy, magnetite-rich materials. Near Surface Geophysics, (in press).

Chassagne, C., Bedeaux, D. and Koper, G.J.M., 2002, The interpretation of dielectric spectroscopy measurements on silica and hematite sols. Journal of Colloid and Interface Science, Vol. 255, pp. 129–137.

Chelidze, T.L. and Gueguen, Y., 1999, Electrical spectroscopy of porous rocks: A review – I. Theoretical models. Geophysical Journal International, Vol. 137, pp. 1–15.

Chelidze, T.L., Gueguen, Y. and Ruffet, C., 1999, Electrical spectroscopy of porous rocks: A review – II. Experimental results and interpretation. Geophysical Journal International, Vol. 137, pp. 16–34.

Chen, Y.P. and Or, D., 2006, Effects of Maxwell-Wagner polarization on soil complex dielectric permittivity under variable temperature and electrical conductivity. Water Resources Research, Vol. 42, No. 6. Article no. W06424 doi:10.1029/2005WR004744.

Cole, K.S. and Cole, R.H., 1941, Dispersion and absorption in dielectrics. Journal of Chemical Physics, Vol. 9, pp. 341–351.

Conyers, L.B. and Goodman, D., 1997, Ground-penetrating radar: An introduction for archaeologists, AltaMira Press, Sage Publications, California, USA.

Cosenza, P., Camerlynck, C. and Tabbagh, A., 2003, Differential effective medium schemes for investigating the relationship between high-frequency relative dielectric permittivity and water content of soils. Water Resources Research, Vol. 39, No. 9, Article no. 1230 doi:10.1029/2002WR001774.

Curtis, J.O., 2001, Moisture effects on the dielectric properties of soils. IEEE Transactions on Geoscience and Remote Sensing, Vol. 39, No. 1, pp. 125–128.

Daniel, V.V., 1967, Dielectric Relaxation, Academic Press, London, UK.

Daniels, D.J., 1996, Surface Penetrating Radar, Radar, Sonar, Navigation and Avionics Series 6, The Institute of Electrical Engineers, London, UK.

Daniels, D.J. (ed.), 2004, Ground Penetrating Radar, Radar, Sonar, Navigation and Avionics Series 15, The Institute of Electrical Engineers, London, UK.

Darayan, S., Liu, C., Shen, L.C. and Shattuck, D., 1998, Measurement of electrical properties of soil. Geophysical Prospecting, Vol. 46, pp. 477–488.

Dasberg, S. and Hopmans, J.W., 1992, Time domain reflectometry calibration for uniformly and nonuniformly wetted sandy and clayey loam soils. Soil Science Society of America Journal, Vol. 56, pp. 1341–1345.

Davidson, D.W. and Cole, R.H., 1951, Dielectric relaxation in glycerol, propylene glycol and n-propanol. Journal of Chemical Physics, Vol. 29, pp. 1484–1490.

Debye, P., 1929, Polar Molecules, Chemical Catalog Co., New York, USA.

Delaney, A.J. and Arcone, S.A., 1984, Field dielectric measurements of frozen silt using VHF pulses. Cold Regions Science and Technology, Vol. 9, pp. 29–37.

Dobson, M.C., Ulaby, F., Hallikainen, M.T. and El-Rayes, A., 1985, Microwave dielectric behaviour of wet soil-part II: Dielectric mixing models. IEEE Transactions on Geoscience and Remote Sensing, Vol. 23, No. 1, pp. 35–46.

Ekblad, J. and Isacsson, U., 2007, Time-domain reflectometry measurements and soil-water characteristic curves of coarse granular materials used in road pavements. Canadian Geotechnical Journal, Vol. 44, No. 7, pp. 858–872.

Endres, A. and Knight, R., 1992, A theoretical treatment of the of microscopic fluid distribution on the dielectric properties of partially saturated rocks. Geophysical Prospecting, Vol. 37, pp. 531–551.

Escorihuela, M.J., De Rosnay, P., Kerr, Y.H. and Calvert, J.C., 2007, Influence of bound-water relaxation frequency on soil moisture measurements. IEEE Transactions on Geoscience and Remote Sensing, Vol. 45, No. 12, pp. 4067–4076.

Evett, S.R. and Parkin, G.W., 2005, Advances in soil water content sensing: The continuing maturation of technology and theory. Vadose Zone Journal, Vol. 4, No. 4, pp. 986–991.

Fam, M.A. and Dusseault, M.B., 1998, High-frequency complex permittivity of shales (0.02–1.30 GHz). Canadian Geotechnical Journal, Vol. 35, pp. 524–531.

Filali, B., Rhazi, J.E. and Ballivy, G., 2006, Measuring dielectric properties of concrete by a wide coaxial probe with an open end. Canadian Journal of Physics, Vol. 84, No. 5, pp. 365–379.

Fiori, A., Benedetto, A. and Romanelli, M., 2005, Application of the effective medium approximation for determining water contents through GPR in coarse-grained soil materials. Geophysical Research Letters, Vol. 32, p. L09404, doi:10.1092/2005GL022555.

Flemming, M.A., Plested, G.A. and Urquhart, A.J., 1990, Dielectric Measurement Techniques at Microwave Frequencies, AEA Technology, USA.

Francisca, F.M. and Rinaldi, V.A., 2003, Complex dielectric permittivity of soil-organic mixtures (20 MHz–1.3 GHz). Journal Of Environmental Engineering-ASCE, Vol. 129, No. 4, pp. 347–357.

Friedman, S.P., 1998, A saturation degree-dependent composite spheres model for describing the effective dielectric constant of unsaturated porous media. Water Resources Research, Vol. 34, No. 11, pp. 2949–2961.

Friel, R. and Or, D. 1999, Frequency analysis of time domain reflectometry (TDR) with application to dielectric spectroscopy of soil constituents. Geophysics, Vol. 64, No. 3, pp. 707–718.

Ghosh, K. and Fuchs, R., 1991, Critical behavior in the dielectric properties of random self-similar composites. Physical Review B, Vol. 44, No. 14, pp. 7330–7343.

Ghosh, K. and Fuchs, R., 1994, Optical and dielectric properties of self-similar composites. Physica A, Vol. 207, No. 14, pp. 185–196.

Gladkov, S.O., 2003, Dielectric Properties of Porous Media, Springer-Verlag, Berlin, Germany.

Hallikainen, M.T., Ulaby, F.T., Dobson, M.C., El-Rayes, M.A. and Wu, L., 1985, Microwave dielectric behaviour of wet soil-part 1: Empirical models and experimental observations. IEEE Transactions on Geoscience and Remote Sensing, Vol. 23, No. 1, pp. 25–35.

Hanai, T., 1968, Electrical properties of emulsions, in Sherman, P. (ed.), Emulsion Science, Academic Press, San Diego, CA, pp. 354–478.

Hasted, J.B., 1973, Aqueous Dielectrics, Chapman and Hall, London, UK.

Heimovaara, T.J., 1994, Frequency domain analysis of time domain reflectometry waveforms 1. Measurement of the complex dielectric permittivity of soils. Water Resources Research, Vol. 30, No. 2, pp. 189–199.

Heimovaara, T.L., de Winter, E.J.G., van Loon, W.K.P. and Esveld, D.C., 1996, Frequency-dependent dielectric permittivity measurements from 0 to 1GHz: Time-Domain reflectometry measurements compared with frequency domain network analyser measurements. Water Resources Research, Vol. 32, No. 12, pp. 3603–3610.

Hill, N.E., Vaughan, W.E., Price, A.H. and Davis, M., 1969, Dielectric Properties and Molecular Behaviour, Van Nostrand Reinhold Company, London.

Hoekstra, P. and Delaney, A., 1974, Dielectric properties of soils at UHF and Microwave frequencies. Journal of Geophysical Research, Vol. 79, No. 11, pp. 1699–1708.

Hoekstra, P. and Doyle, W.T., 1971, Dielectric relaxation of surface absorbed water. Journal of Colloidal and Interface Science, Vol. 36, No. 4, pp. 513–521.

Hollender, F. and Tillard, S., 1998, Modelling ground-penetrating radar wave propagation and reflection with the Jonscher parameterization. Geophysics, Vol. 63, No. 6, pp. 1933–1942.

Hu, K. and Liu, C.R., 2000, Theoretical study of the dielectric constant in porous sandstone saturated with hydrocarbon and water. IEEE Transactions on Geoscience and Remote Sensing, Vol. 38, No. 3, pp. 1328–1336.

Johnson, R.H. and Poeter, E., 2005, Iterative use of the Bruggeman–Hanai–Sen mixing model to determine water saturations in sand. Geophysics, Vol. 70, No. 5, pp. K33–K38.

Jonscher, A.K., 1977, The ?universal? dielectric response. Nature, Vol. 267, pp. 673–679.

Kaatze, U., 2000, Hydrogen network fluctuations and the microwave dielectric properties of liquid water. Subsurface Sensing Technologies and Applications, Vol. 1, No. 4, pp. 377–391.

King, R.W.P. and Smith, G.S., 1981, Antennas in Matter, MIY Press, Cambridge, MA, USA.

Klein, K. and Santamarina, J.C., 1997, Methods of broad-band dielectric permittivity measurements (soil-water mixtures, 5 Hz to 1.3GHz). Geotechnical Testing Journal, Vol. 20, No. 2, pp. 168–178.

Knight, R.J. and Nur, A., 1987, The dielectric constant of sandstones, 60 kHz to 4 MHz. Geophysics, Vol. 52, No. 5, pp. 644–654.

Kremer, F. and Schonhals, A. (eds), 2002, Broadband Dielectric Spectroscopy, Springer-Verlag, Berlin, Germany.

Kyritsis, A., Siakantari, M., Vassilikou-Dova, A., Pissis, P. and Varotsos, P., 2000, Dielectric and electrical properties of polycrystalline rocks at various hydration levels. IEEE Transactions on Dielectrics and Electrical Insulation, Vol. 7, No. 4, pp. 493–497.

Lebron, I., Robinson, D.A., Goldberg, S. and Lesch, S.M., 2004, The dielectric permittivity of calcite and arid zone soils with carbonate minerals. Soil Science Society of America Journal, Vol. 68, No. 5, pp. 1549–1559.

Lima, O.L.A. and Sharma, M.M., 1992, A generalised Maxwell-Wagner theory for membrane polarization in shaly sands. Geophysics, Vol. 57, No. 3, pp. 431–440.

Loeffler, O. and Bano, M., 2004, GPR measurements in a controlled vadose zone: Influence of the water content. Vadose Zone Journal, Vol. 3, pp. 1082–1092.

Logsdon, S.D., 2005, Soil dielectric spectra from vector network analyzer data. Soil Science Society of America Journal, Vol. 69, No. 4, pp. 983–989.

Lorrain, P., Corson, D.R. and Lorrain, F., 1988, Electromagnetic Fields and Waves, W. H. Freemann and Company, New York, USA.

Mattei, E., De Santis, A., Pettinelli, E. and Vannaroni, G., 2007, Effective frequency and attenuation measurements of glass beads/magnetite mixtures by time-domain reflectometry. Near Surface Geophysics, Vol. 5, pp. 77–82.

Mironov, V.L., Dobson, M.C., Kaupp, V.H., Komarov, S.A. and Kleshchenko, V.N., 2004, Generalized refractive mixing dielectric model for moist soils. IEEE Transactions on Geoscience and Remote Sensing, Vol. 42, No. 4, pp. 773–785.

Mulay, L.N., 1963, Magnetic Susceptibility, Interscience Publishers, London, UK.

Nalwa, H.S. (ed.), 1999, Handbook of Low and High Dielectric Constant Materials and Their Applications, (Two-Volume Set), Academic Press, London, UK.

Nichol, C., Smith, L. and Beckie, R., 2003, Time domain reflectometry measurements of water content in coarse waste rock. Canadian Geotechnical Journal, Vol. 40, No. 1, pp. 137–148.

Olhoeft, G.R., 1979, Tables of Room Temperature Electrical Properties of Selected Rocks and Minerals with Dielectric Permittivity Statistics. U.S. Department of Interior USGS Open File Report, 79–993, pp. 1–24.

Olhoeft, G.R., 1998, Electrical, magnetic and geometric properties that determine ground penetrating radar performance. Proceedings of the 7th International Conference on Ground Penetrating Radar (GPR'98), USA, pp. 477–483.

Olhoeft, G.R. and Capron, D.E., 1993, Laboratory Measurements of the Radio Frequency Electrical and Magnetic Properties of Soils Near Yuma, Arizona. U.S. Department of Interior USGS Open file Report, pp. 93–701.

Or, D. and Wraith, J.M., 1999, Temperature effects on soil bulk dielectric permittivity measured by time domain reflectometry: A physical model. Water Resources Research, Vol. 35, No. 2, pp. 371–383.

Pankhurst, Q.A., Connolly, J., Jones, S.K. and Dobson, J., 2003, Applications of magnetic nanoparticles in biomedicine. Journal of Physics D: Applied Physics, Vol. 36, pp. R167–R181.

Pepin, S., Livingston, N.J. and Honk, W.R., 1995, Temperature dependent measurement errors in time domain reflectometry determinations of soil water. Soil Science Society of America Journal, Vol. 59, pp. 38–43.

Persson, M. and Berndtsson, R., 1998, Texture and electrical conductivity effects on temperature dependency in time domain reflectometry. Soil Science Society of America Journal, Vol. 62, pp. 887–893.

Persson, M. and Berndtsson, R., 2002, Measuring nonaqueous phase liquid saturation in soil using time domain reflectometry. Water Resources Research, Vol. 38, No. 5, p. 1064, doi: 10.1092/2001WR000523.

Pettinelli, E., Vannaroni, G., Cereti, A., Pisani, A.R., Paolucci, F., Del Vento, D., Dolfi, D., Riccioli, S. and Bella, F., 2005, Laboratory investigations into the electromagnetic properties of magnetite/silica mixtures as Martian soil simulants. Journal of Geophysical Research, Vol. 110, p. E04013, doi:10.1029/2004je002375.

Regalado, C.M., Carpena, R.M., Socorro, A.R. and Moreno, J.M.H., 2003, Time domain reflectometry models as a tool to understand the dielectric response of volcanic soils. Geoderma, Vol. 117, No. 3–4, pp. 313–330.

Reynolds, J.M., 1997, An Introduction to Applied and Environmental Geophysics, John Wiley & Sons, Chichester, England.

Rhim, H.C. and Buyukozturk, O., 1998, Electromagnetic properties of concrete at microwave frequency range. ACI Materials Journal, Vol. 95, No. 3, pp. 262–271.

Rinaldi, V.A. and Francisca, F.M., 1999, Impedance analysis of soil dielectric dispersion (1MHz–1GHz). Journal of Geoenvironmental Engineering, Vol. 25, No. 2, pp. 111–121.

Robert, A., 1998, Dielectric permittivity of concrete between 50 MHz and 1 GHz and GPR measurements for building materials evaluation. Journal of Applied Geophysics, Vol. 40, No. 1–3, pp. 89–94.

Robinson, D.A., 2004, Calculation of the dielectric properties of temperate and tropical soil minerals from ion polarizabilities using the Clausius–Mosotti equation. Science Society of America Journal, Vol. 68, pp. 1780–1785.

Robinson, D.A. and Friedman, S.P., 2003, A method for measuring the solid particle permittivity or electrical conductivity of rocks, sediments, and granular materials. Journal of Geophysical Research-Solid Earth, Vol. 108, No. B2, Article no. 2076 doi no.10.1029/2001JB000691.

Roth, K., Schulin, R., Flhler, H. and Attinger, W., 1990, Calibration of time domain reflectometry for water content measurement using a composite dielectric approach. Water Resources Research, Vol. 26, No. 10, pp. 2267–2273.

Saarenketo, T., 1998, Electrical properties of water in clay and silty soils. Journal of Applied Geophysics, Vol. 40, pp. 73–88.

Sabburg, J., Ball, J.A.R. and Hancock, N.H., 1997, Dielectric behaviour of moist swelling clay soils at microwave frequencies. IEEE Transactions on Geoscience and Remote Sensing, Vol. 35, No. 3, pp. 785–787.

Sen, P.N., 1981, Relation of certain geometrical features to the dielectric anomaly of rocks. Geophysics, Vol. 46, No. 12, pp. 1714–1720.

Sen, P.N., Scala, C. and Cohen, M.H., 1981, A self-similar model for sedimentary rocks with application to the dielectric constant of fused glass beads. Geophysics, Vol. 46, No. 5, pp. 781–795.

Shang, J.Q., Umana, J.A., Bartlett, F.M. and Rossiter, J.R., 1999a, Measurement of complex permittivity of asphalt pavement materials. Journal of Transportation Engineering-ASCE, Vol. 125, No. 4, pp. 347–356.

Shang, J.Q., Rowe, R.K., Umana, J.A. and Scholte, J.W., 1999b, A complex permittivity measurement system for undisturbed/compacted soils. Geotechnical Testing Journal, Vol. 22, No. 2, pp. 168–178.

Sihvola, A.H., 1989, Self-consistency aspects of dielectric mixing theories. IEEE Transactions on Geoscience and Remote Sensing, Vol. 27, No. 4, pp. 403–415.

Sihvola, A.H., 2000, Mixing rules with complex dielectric coefficients. Subsurface Sensing Technologies and Applications, Vol. 1, No. 4, pp. 393–415.

Sihvola, A.H. and Alanen, E., 1991, Studies of mixing formulae in the complex plane. IEEE Transactions on Geoscience and Remote Sensing, Vol. 29, No. 4, pp. 679–687.

Soutsos, M.N., Bungey, J.H., Millard, S.G., Shaw, M.R. and Patterson, A., 2001, Dielectric properties of concrete and their influence on radar testing. NDT & E International, Vol. 34, No. 6, pp. 419–425.

Starr, G.C., Lowery, B. Cooley, E.T. and Hart, G.L., 1999, Soil water content determination using network analyser reflectometry methods. Soil Science Society of America Journal, Vol. 63, pp. 285–289.

Sternberg, B.K. and Levitskaya, T.M., 2001, Electrical parameters of soils in the frequency range from 1 kHz to 1 GHz, using lumped-circuit methods. Radio Science, Vol. 36, No. 4, pp. 709–719.

Stillman, D.E. and Olhoeft, G.R., 2004, GPR and magnetic minerals at mars temperatures. Proceedings of the Tenth International Conference on Ground Penetrating Radar, Vol. 2, pp. 735–738.

Stillman, D.E. and Olhoeft, G.R., 2006, Electromagnetic properties of Martian analog minerals at radar frequencies and Martian temperatures. Proceedings of the 37th Lunar and Planetary Science Conference, 2002.pdf, pp. 1–2.

Sweeney, J.J., Roberts, J.J. and Harbert, P.E., 2007, Study of dielectric properties of dry and saturated green river oil shale. Energy and Fuels, Vol. 21, No. 5, pp. 2769–2777.

Taherian, M.R., Keyon, W.E. and Safinya, K.A., 1990, Measurement of dielectric response of water-saturated rocks. Geophysics, Vol. 55, No. 12, pp. 1530–1541.

Thevanayagam, S., 1995, Frequency-domain analysis of electrical dispersion of soils. Journal of Geotechnical Engineering, Vol. 121, No. 8, pp. 619–628.

Topp, G.C., Davis, J.L. and Annan, A.P., 1980, Electromagnetic determination of soil water content: I. Measurements in coaxial transmission lines. Water Resources Research, Vol. 16, No. 3, pp. 574–582.

Topp, G.C., Davis, J.L. and Annan, A.P., 1982, Electromagnetic determination of soil water content using TDR: I. Applications to wetting fronts and steep gradients. Soil Science Society of America Journal, Vol. 46, pp. 672–678.

Tsui, F. and Matthews, S.L., 1997, Analytical modelling of the dielectric properties of concrete for subsurface radar applications. Construction and Building Materials, Vol. 11, No. 3, pp. 149–161.

Turner, G. and Siggins, A.F., 1994, Constant Q attenuation of subsurface radar pulses. Geophysics, Vol. 59, No. 8, pp. 1192–1200.

Ulaby, F.T., Bengal, T.H., Dobson, M.C., East, J.R., Garvin, J.B. and Evans, D., 1990, Microwave dielectric properties of dry rocks. IEEE Transactions on Geoscience and Remote Sensing, Vol. 28, No. 3, pp. 325–335.

Van Dam, R.L., Schlager, W., Dekkers, M.J. and Huisman, J.A., 2002, Iron oxides as a cause of GPR reflections. Geophysics, Vol. 67, pp. 536–545.

Van Damme, S., Franchois, A., De Zutter, D. and Taerwe, L., 2004, Non-destructive determination of the steel fibre content in concrete slabs with an open ended coaxial probe. IEEE Transactions on Geoscience and Remote Sensing, Vol. 42, No. 11, pp. 2511–2521.

Von Hippel, A.R., 1954, Dielectrics and Waves (second edition), Artech House, Boston, USA.

Von Hippel, A.R. (eds), 1995, Dielectric Materials and Applications, Artech House, Boston, USA.

Wang, X.M., Teo, Y.H., Chiu, W.K. and Foliente, G., 2006, Evaluation of moisture content in concrete with microwave. Key Engineering Materials, Vol. 312, pp. 311–318.

Weiler, K.W., Steenhuis, T.S., Boll, J. and Kung, K.J.S., 1998, Comparison of ground penetrating radar and time domain reflectometry as soil water sensors. Soil Science Society of America Journal, Vol. 62, pp. 1237–1239.

Wensink, W.A., 1993, Dielectric properties of wet soils in frequency range 1–3000 MHz. Geophysical Prospecting, Vol. 41, pp. 671–696.

West, L.J., Handley, K., Huang, Y. and Pokar, M., 2003, Radar frequency dielectric dispersion in sandstone: Implications for determination of moisture and clay content. Water Resources Research, Vol. 39, No. 2, Article no. 1026 doi:10.1029/2001WR000923.

Wraith, J.M. and Or, D., 1999, Temperature effects on soil bulk dielectric permittivity measured by time domain reflectometry: Experimental evidence and hypothesis development. Water Resources Research, Vol. 35, No. 2, pp. 361–369.

CHAPTER 3

Ground Penetrating Radar Systems and Design

Steven Koppenjan

Contents

3.1. Introduction and Background	73
3.2. Methodology – Types of Ground Penetrating Radar	74
3.2.1. Impulse	75
3.2.2. Swept frequency-modulated continuous wave	75
3.2.3. Stepped frequency-modulated continuous wave	76
3.2.4. Gated, stepped frequency-modulated continuous wave	76
3.3. Radio Frequency Specifications and Definitions	77
3.3.1. Dynamic range	77
3.3.2. Bandwidth	78
3.3.3. Range resolution	78
3.3.4. Lateral resolution	79
3.3.5. Unambiguous range	79
3.4. General Design Criteria for Ground Penetrating Radar	80
3.4.1. System performance	81
3.5. Impulse Ground Penetrating Radar	81
3.5.1. Theory of operation: Impulse radar	81
3.5.2. System design parameters: impulse radar	84
3.5.3. Implementation of an impulse ground penetrating radar	85
3.6. Continuous-Wave Ground Penetrating Radar	86
3.6.1. Theory of operation – stepped-frequency, continuous-wave radar	86
3.6.2. System design parameters: stepped-frequency radar	92
3.6.3. Implementation of a gated, stepped-frequency, ground penetrating radar	93
References	95

3.1. Introduction and Background

There are several main components of a ground penetrating radar (GPR): the radar electronics, antenna, data digitizer, computer, and display module (Figure 1.12). This chapter will primarily focus on the radar section of the GPR system and its design. Computer technology has advanced to the point where the computer, display, and data transfer do not limit the functionality or operation of the GPR system. The antenna is a very important element of a GPR and for its ultimate performance and thus requires a full dedicated chapter — Chapter 4. High-speed data sampling or analog-to-digital conversion (ADC) is one area that has advanced greatly over the past 20 years. The technological aspect of ADC and its role in GPR operation will be mentioned (also see Section 1.4). The theory of operation of the radar technique is discussed, and a basic analysis of its application toward GPR is presented from the source level to a signal-processing level. The fundamental signal processing of the radar return is introduced and more details can be found in Chapter 5. Specific design criteria are interpreted into specifications for the GPR. These include the system bandwidth and the system dynamic range, the two parameters that essentially characterize a radar.

There are several types of GPR; the difference is the manner in which the data are acquired, either the *time domain* or the *frequency domain*. Impulse radar operates in the time domain while continuous-wave (CW) radar operates in the frequency domain. Theoretically, impulse and CW radar with identical specifications and parameters will perform identically (Skolnik, 1980). These radar systems, and the basic designs of these two types, are presented along with the advantages and disadvantages (Poirier, 1993). Recent progress in GPR design will be listed with references.

The topics covered in this chapter are the following:

- Methodology – description of the types of GPR
- RF specifications and definitions
- General design criteria for GPR
- Theory of operation of impulse and CW radar
- GPR system design parameters
- Implementation of GPR systems

3.2. METHODOLOGY – TYPES OF GROUND PENETRATING RADAR

There are two common categories of GPR: impulse and CW. Most GPR systems are based on the impulse radar technique and are prevalent in the commercial market. Continuous-wave and stepped-frequency radars have been developed over the past decade (Noon et al., 1994), though most have involved research institutions, universities, and government-sponsored laboratories. Advanced techniques and variations such as ultra-wideband (UWB), synthetic-aperture radar (SAR), noise source (Narayanan et al., 2001; Xu et al., 2001), and arbitrary waveforms (Eide and Hjelmstad, 1999) have been applied to GPR, but are beyond the scope of this book. There are also many custom GPR systems designed for very specific applications such as borehole radar. Previous summaries on subsurface radar techniques can be found in Daniels et al. (1988) and updated in Daniels (1996). Recent progress (1999–2002) for various related subsurface remote sensing topics can be found in Noon and Narayanan (2002), as well as international GPR conference proceedings (Koppenjan and Lee, 2002; Slob and Yarovoy, 2004; Daniels et al., 2006).

3.2.1. Impulse

Radars that acquire data in the time domain are generally known as impulse. A time domain pulse is transmitted and the reflected energy is received as a function of time. The resulting waveform indicates the amplitude of energy scattered from subsurface objects versus time. Range information is based on the *time-of-flight* principle, and a simple display of the radar return (amplitude versus range) is an "A-scope" presentation.

The majority of GPR systems incorporating the impulse technique send a pulse to an antenna, which produces an electromagnetic (EM) wave. The characteristics of the antenna determine the center frequency of the EM wave and the associated bandwidth is determined by the pulse width. The antenna plays a major function in the impulse GPR system dynamics (see Chapter 4).

Ground penetrating radar incorporating the impulse technique was first manufactured for commercial purposes in the mid-1970s (Morey, 1974) and was demonstrated as a useful geophysical tool (Annan and Davis, 1976). Although the basic application of the impulse technique to GPR has remained the same, the display and recording of the radar return data has changed. Modern impulse GPR digitally samples the return waveform for display, data storage, and post-processing. Commercial-impulse GPR systems to date use a repetitive sampling method or *equivalent-time sampling* (ETS) (Hansen, 1942), where single successive samples are made after each transmitted pulse. An adjustable delay is used to sample along the received waveform. Prototype systems have been built with a flash conversion or *complete waveform sampling* where the entire received waveform, after one transmit pulse, is digitized (Wright et al., 1993). These offer high data rates but can be limited in dynamic range. An alternate implementation for acquisition involves a correlation-based receiver with the wideband signal transmitted being a digitally generated, pseudo-random binary sequence (Wills, 1992).

Some advantages of impulse radar are the simplicity of generating an impulse waveform and low-cost parts. The disadvantages include undesirable ringing, inefficient use of transmit power (low duty cycle), and the resolution limited by pulse width. Other difficulties involve sampling of wideband signals with slow-speed sequential digitizers.

3.2.2. Swept frequency-modulated continuous wave

Radars that acquire data in the frequency domain and transmit continuously (transmitter always on) are known as CW. If the carrier is frequency-modulated (FM), then it is referred to as FM-CW. The concept involves transmitting a frequency "sweep" over a fixed bandwidth, from a start frequency to a stop frequency. The reflected energy is received as a function of frequency and indicates the amplitude of energy scattered from subsurface objects. The received signal is mixed or heterodyned with a portion of the transmitted signal filtered and digitized or sampled during the sweep. The digitized waveform from the entire sweep is then transformed into the time domain. The result is known as a *synthesized pulse*.

Ground penetrating radar systems incorporating the synthesized pulse technique have been implemented with commercial network analyzers (Robinson et al., 1972; Kong, 1990) and a commercial synthesized source and spectrum analyzer (Oliver et al., 1982). The sweep rate was inherently slow due to the nature of the test equipment and issues with data storage. The implementation of these prototype-swept FM-CW systems was limited by the state of technology at the time of development. Test equipment was used until lower-cost components such as frequency sources could be used (Oliver and Cuthbert, 1988), faster digital samplers were developed, and fast Fourier transforms (FFT) could be performed on digital signal-processing (DSP) boards or PCs.

3.2.3. Stepped frequency-modulated continuous wave

A stepped-frequency radar is similar to a swept-frequency radar except that the transmitting frequency is stepped in linear increments over a fixed bandwidth, from a start frequency to a stop frequency. The received signal is mixed and sampled at each discrete frequency step. The digitized waveform is transformed into the time domain to create the synthesized pulse.

The ability to "step" faster led to the development of stepped FM-CW GPR systems. Some early examples of stepped-frequency GPR are Robinson et al. (1974) and Iizuka et al. (1984) and were implemented with high-quality, commercial network analyzers (Hamran and Aarholt, 1993; Kong and By, 1995). Technology advances in the radio frequency (RF) components, wideband frequency synthesizers, DSP boards, and portable computers led to the development and fielding of prototype step-frequency GPR systems (Koppenjan and Bashforth, 1993).

Several advantages of the stepped-frequency GPR are the controlled transmission frequencies, efficient use of power, and efficient sampling of wideband signals with low-speed ADCs. The nature of the system architecture allows the collection of coherent (real and imaginary) data, which allows complex processing and the implementation of SAR algorithms (Koppenjan et al., 2000). Disadvantages of stepped frequency include the complex electronics and the requirement of DSP, but it becomes practical with current technology. Also, time-varying gain cannot be applied to the return signal. A negative effect with the conversion from frequency to time is the introduction of sidelobes (from strong signals) that can mask out small signals from weak reflections.

The advantages of swept-frequency over stepped are simpler design and lower cost for implementation. However, swept frequency may have a lower performance in some cases due to frequency ambiguities of the sweep, i.e., if the timing of the sample with the instantaneous frequency cannot be maintained throughout the entire sweep.

3.2.4. Gated, stepped frequency-modulated continuous wave

In a stepped-frequency system, the masking of a weaker signal from a deep target by a strong signal is due to the receiver always being on. This can be caused by several factors including the following:

- Leakage from the transmitter to the receiver in the electronics
- EM waves on the ground surface
- Large reflections from shallow reflectors

A solution to this problem is to gate the transmitter and the receiver (Hamran et al., 1995).

3.2.4.1. What is gating?

Gating is the technique of timing the transmitter and receiver circuits in order to avoid the negative strong signal effects. At each frequency step, the transmitter is pulsed "on" and after a delay the receiver is gated "on." This reduces or "gates out" the strong signals from entering the receiver. The return signal at a particular frequency is the sum or integration of the received power over time (the receiver gate width). The description and system performance of a gated, stepped-frequency GPR can be found in Stickley et al. (2000).

3.3. Radio Frequency Specifications and Definitions

There are several important specifications of a GPR that characterize the performance: dynamic range, bandwidth, resolution, and unambiguous range.

3.3.1. Dynamic range

The radar receiver must be capable of handling large signals from surface reflections and short-range targets and also detect small signals near the noise floor. The ratio of the largest receivable signal to the minimal detectable signal is called the *dynamic range* and is defined as follows:

$$\text{Dynamic Range} = 20 \log\left(\frac{V_{\max}}{V_{\min}}\right) \tag{3.1}$$

It is usually expressed in decibels (dB) for a specified bandwidth in hertz. The largest receivable signal, $V_{\max}$ (in V), must not overload the radar front end and, assuming some gain has been applied to the received signal, is the maximum sample voltage of the ADC. Additionally, for a CW GPR, the largest signal is defined by the 1-dB compression point of the receive mixer where the gain deviates from linearity. This assumes no saturation after the mixer.

The minimal detectable signal, $V_{\min}$ (in V), must be above the receiver noise level and have a minimum signal-to-noise ratio (SNR) to be detected. In most radar detection applications, an SNR of 8 dB is required (Erst, 1985). In most GPR applications, $V_{\min}$ must also have a minimum signal-to-clutter ratio (SCR) in order to be detected and identified in a GPR profile.

The dynamic range of the system will affect the maximum range at which a target can be detected. Typically, radars will have a greater *system* dynamic range than *sampling* dynamic range. The dynamic range in decibels of an ADC is equal to

$20 \log(2^N)$, where N is the number of bits, or approximately N times 6 dB. Thus, a 16-bit ADC will have 96 dB of theoretical dynamic range, but other factors influence the actual realizable dynamic range (Smith, 2004). There are methods such as *stacking* (see Chapter 5) that can be used to improve the SNR and automatic gain control to overcome the problem of dynamic range variations (Erst, 1985).

3.3.2. Bandwidth

The *system bandwidth*, B, is defined as the inverse of the pulse width, τ_p, for impulse GPR. The bandwidth is generally centered about and for practical reasons is approximately equal to the impulse GPR *center frequency*, f_c. The center frequency is a very common GPR parameter and is the answer to the universal question: "what frequency did you use?" For more details on center frequency, see the discussions in Chapters 1 and 4. In a CW GPR, the bandwidth is the difference from the start frequency, f_{min}, to the stop frequency, f_{max}, and can also be calculated by multiplying the number of steps by the output frequency step size in a stepped-frequency GPR. The bandwidth ultimately determines the range resolution

$$B = \frac{1}{\tau_p} \text{ for impulse} \quad \text{or} \quad B = (f_{max} - f_{min}) \text{ for CW} \qquad (3.2)$$

3.3.3. Range resolution

The theory of operation sections that follow will assume a return from a single target at a fixed distance. In practical situations, this is not the case. Multiple interfaces within the ground produce multiple returns. The addition of targets increases the complexity. The result is a combination of signal returns at different times and of varying amplitudes. The ability of a radar to resolve between two closely spaced targets is called *range resolution*, R_{res}. Two targets (or pulses) separated in time can be distinguished if the envelopes of their respective transient returns are clearly separated. In general, the half-width (−6 dB in voltage) or half-power (−3 dB in power) point of the return signal is used as a reference point of clear separation. This definition is more theoretical and is presented in Section 1.3.4). A GPR return must account for frequency-selective dispersion of the system, antenna, ground, and target. By defining the range resolution as the half-power point of the normalized sinc function, it can be obtained from the theoretically achievable resolution (Eaves and Reedy, 1987). Then the range resolution is

$$R_{res} = \frac{1.39c}{2B\sqrt{\varepsilon_r}} \qquad (3.3)$$

where c = speed of light, B = bandwidth, and ε_r = dielectric constant or relative dielectric permittivity. The factor of 1.39 is associated with the deviation from the theoretical range resolution and is derived empirically. Range resolution is plotted versus ε_r in Figure 3.1 for several common bandwidths. The actual resolution can be expected to be equal or better but not worse using Equation (3.3) and the graphs.

Ground Penetrating Radar Systems and Design

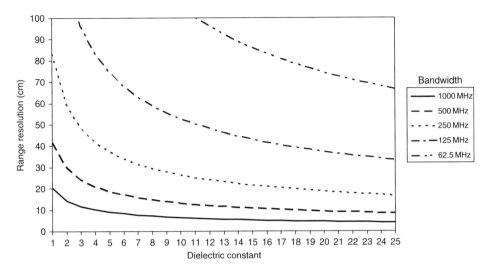

Figure 3.1 Range resolution versus dielectric constant for common GPR bandwidths.

3.3.4. Lateral resolution

Lateral resolution is the resolution in the cross-range direction (it is different from range resolution) and is described in Section 1.3.4.

3.3.5. Unambiguous range

The furthest distance that a target can be determined without *aliasing* occurring is called the *unambiguous range*, R_{max}. To avoid aliasing, the reflected energy should be received within the time period or range cells of its associated transmit pulse and before the next transmit pulse, or a range ambiguity will result. The rate at which transmit pulses are sent is the pulse repetition frequency (PRF), f_r, also expressed as the pulse repetition interval (PRI), T_r, which is the inverse of the PRF. T_r is commonly referred to as the *programmable time window*. For an impulse GPR, the PRI determines the maximum unambiguous range

$$R_{max} = \frac{cT_r}{2\sqrt{\varepsilon_r}} \quad (3.4)$$

where c = speed of light, T_r = PRI, and ε_r = dielectric constant.

For a stepped-frequency GPR, aliasing is an unavoidable result of sampling the data in the frequency domain. In effect, the reflections from deeper targets are folded back onto returns from shallow targets, making absolute distance indeterminate. This phenomenon is directly related to the *sampling theorem* and the *Nyquist rate*. Applying the results of a two-frequency CW radar (Skolnik, 1980; Eaves and

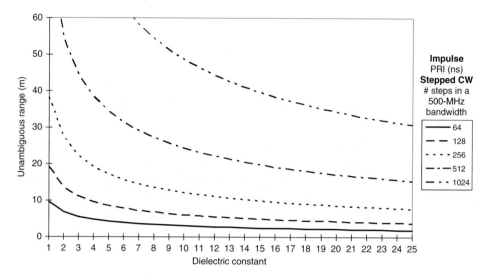

Figure 3.2 Unambiguous range versus dielectric constant for various PRI (impulse radar) and number of frequency steps in a 500 MHz bandwidth (CW radar).

Reedy, 1987) to a stepped-frequency CW radar, the unambiguous range can be calculated from the following equation:

$$R_{max} = \frac{Nc}{4B\sqrt{\varepsilon_r}} \qquad (3.5)$$

where N = number of frequency steps, c = speed of light, B = bandwidth, and ε_r = dielectric constant.

Unambiguous range is plotted versus ε_r in Figure 3.2 for various PRIs and, similarly, the number of frequency steps (fixed 500 MHz bandwidth). When a stepped-frequency GPR is also gated, the gate repetition interval (GRI), which is similar to the PRI, must be taken into account.

3.4. GENERAL DESIGN CRITERIA FOR GROUND PENETRATING RADAR

The GPR concept and design differs significantly from conventional radar primarily because of the short range of the targets and the lossy propagation media for the EM waves. The maximum range is considerably influenced by the path loss, $L_p(\lambda)$, which is a function of wavelength. L_p is generally the prevailing factor in the radar range equation for GPR

$$P_R = \frac{P_T G^2 \sigma (\lambda/\varepsilon_r)^2}{(4\pi)^3 R^4 L_p} \qquad (3.6)$$

where P_R = received power, P_T = transmitted power, G = antenna gain (transmit = receive), σ = radar cross section of the target, λ = wavelength, ε_r = dielectric constant, and R = range to target. P_R must be above the minimum detectable signal level of the system. There are many other forms of the range equation that can be found in radar handbooks, such as Skolnik (1980).

The properties of the ground, such as soil type and water content, affect the path loss (see Chapter 2), and the path loss is not always a linear function of depth. To overcome path losses and increase range, the operating frequency can be lowered, but this reduces bandwidth, which is directly proportional to resolution [Equation (3.3)]. High resolution is generally a desirable goal of a GPR system. In turn, if the bandwidth is reduced, the resolution will be reduced. This engineering trade-off between operating frequency and bandwidth (resolution) is the major challenge for the GPR designer. Additionally, antenna size increases as the frequency decreases (see Chapter 4). The PRF, sample rate, and timing must also be considered to reach the desired maximum range (see Section 1.5). Operators must evaluate the specific application, considering desired penetration depth, size of target, and required resolution to determine the optimal operating frequency and the resulting resolution.

3.4.1. System performance

The system performance, also referred to as the total dynamic range (TDR), is a parameter that relates the total loss that a GPR signal can have and still be detectable in the receiver. It includes the loss parameters in the radar range equation and also accounts for other system loss and gain, such as time-varying gain in the receiver (also see Section 1.4.5).

3.5. IMPULSE GROUND PENETRATING RADAR

3.5.1. Theory of operation: Impulse radar

This section describes the theory of operation of impulse radar (Harmuth, 1969; Annan, 2005) and the application of various sampling techniques (Xu et al., 1989; Annan, 2003). A diagram of an impulse radar is shown in Figure 3.3.

An impulse radar incorporates a timing unit, transmitter electronics (pulse generator), transducer element, and digitizing circuitry, i.e., ADC. The timing unit initiates a signal to the transmitter electronics, and a short DC pulse is fed to the transducer element (the antenna). The output signal is the convolution of the transfer function of the pulse, $p(t)$, with the transfer function of the transducer element, $a(t)$.

$$w(t) = 2\pi f(t) = p(t)a(t) \text{ or in the frequency domain, } \Omega(\omega) = P(\omega)A(\omega) \quad (3.7)$$

The pulse is band-pass filtered by the antenna's tuning characteristics and the shape of the pulse is determined by the transfer function of the antenna. Ideally, $w(t) = \delta(t)$, the dirac delta function or the perfect impulse. Realistically, the transmitted

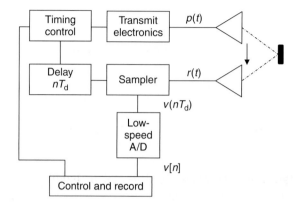

Figure 3.3 Simplified diagram of the impulse radar technique.

energy is centered at a frequency determined by the antenna. The bandwidth of the energy and associated transmitted pulse width is determined by a combination of the applied pulse width and antenna as represented in Equation (3.7). Typical transducer elements are bow-tie or resistively loaded dipole antennas, as these exhibit good broadband characteristics (Johnson and Jasik, 1984). The received energy is recorded or digitized as a function of time, and there are several methods to accomplish the sampling, as shown in Figure 3.4.

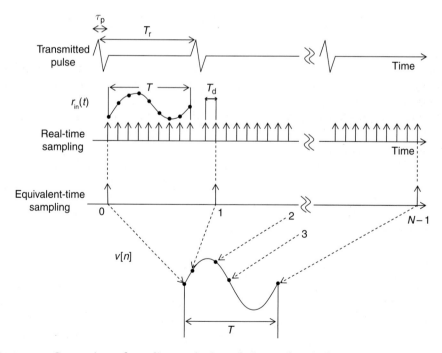

Figure 3.4 Comparison of sampling methods: real-time and equivalent-time.

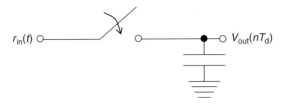

Figure 3.5 Typical sampling bridge with a fast switch and capacitor.

3.5.1.1. Equivalent-time sampling – sequential receiver

The most common method of sampling the received waveform is a sequential receiver, which utilizes ETS. This repetitive sampling method is where single successive samples are made after each transmitted pulse. An adjustable and precise delay, T_d, is used to sample along the received waveform (in time). T_d is commonly referred to as the *sampling interval*.

To sample the received signal, it is first *acquired* by the use of a sampling bridge, typically a sample and hold circuit, sketched in Figure 3.5. The sample and hold is a capacitor and fast switch, closed for a very short duration, Δ, and controlled by the timing unit. The capacitor is charged when the switch is closed, and the capacitor voltage is proportional to the sum or integral of the input voltage over this time period.

$$v_{out}(T_d) = \int_{T_d}^{T_d+\Delta} r_{in}(t)dt = Cr_{in}(T_d), \quad \text{where } C \text{ is a constant} \quad (3.8)$$

If $\Delta \ll$ the rate of change of $r_{in}(t)$, then $v_{out}(T_d)$ effectively "samples" $r_{in}(t)$ and "holds" it to be digitized. This DC voltage can be sampled with a low-speed ADC; thus $v_{out}(t)$ is now represented as $v[n]$, the sampled version at time T_d. The ETS method requires multiple transmitter pulses, N_p, to sample the entire waveform at different time positions. With the interpolation of N_p sampled points, a slow-varying replica of the high-frequency radar return signal is produced and can be expressed as a sequence:

$$v[n] = \sum_{i=0}^{N_p-1} \delta(n-i)v_{out}(nT_d) \quad (3.9)$$

where $\delta(n) =$ unit impulse.

3.5.1.2. Real-time sampling – complete waveform receiver

Another method of sampling the received waveform is a complete waveform receiver, which utilizes real-time sampling. This method is where all time positions or range cells are simultaneously sampled after a single transmit pulse (see Figure 3.4). To implement the real-time sampling, the block diagram in Figure 3.3 must be modified as follows: remove the sampler, change to high-speed ADC, and change the delay circuit feeding the sampling bridge to a high-speed clock

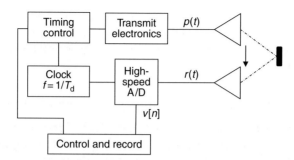

Figure 3.6 Simplified diagram of the impulse radar technique with real-time sampling.

with frequency $1/T_d$ as an input to the ADC. This block diagram is shown in Figure 3.6. The sampled received signal is as follows:

$$v[n] = \sum_{i=0}^{N_p-1} \delta(n-i) r_{in}(nT_d) \qquad (3.10)$$

Generally, high-speed ADCs have a lower number of sample bits than low-speed ADCs, and thus a lower dynamic range. Prototype systems incorporating real-time sampling have been built for several applications (Wright et al., 1994).

3.5.2. System design parameters: impulse radar

Several design parameters must be chosen with impulse radar. These involve pulse width and sampling interval versus resolution, and the PRI versus unambiguous range.

The desired range resolution will determine the required pulse width and affect the designs of the transmitter electronics and the antenna. The transmitter electronics will need to generate narrow pulses to achieve the bandwidth. The antenna design must be considered when a low center frequency, f_c, is chosen with a wide bandwidth, especially when B approaches f_c, as it may result in an impractical antenna realization or less than desired bandwidth. Additionally, the sampling interval should be 10 times the required resolution for sufficient waveform reconstruction.

For example, from Figure 3.1, with 500 MHz of bandwidth, a range resolution of 20.8 cm is obtained in $\varepsilon_r = 4$ (dry sand). To attain this bandwidth, a very common GPR frequency of $f_c = 500$ MHz (frequency range of 250–750 MHz) could be used and allows a realistic antenna. If $f_c = 250$ MHz were required, a bandwidth of 500 MHz would not be possible. The pulse width, τ_p, required for a 500 MHz bandwidth is 2 ns, but typically τ_p is smaller to overcome the pulse rise-time broadening effects, which reduce the bandwidth. The sampling interval should not be greater than 200 ps.

The unambiguous range versus ε_r is plotted in Figure 3.2 with the PRI varying from 64 to 1024 ns. The PRI and amount of *stacking* will affect how fast the GPR can be moved across the surface. The PRI can be increased, but R_{max} will be reduced. For a pavement evaluation application (see Chapter 13), R_{max} may be less than 1 m, but the data rate is high to accommodate vehicle velocities; thus the PRI must be fast. For geological applications, where range is of interest, the PRI can be slower with more stacking to reduce noise.

3.5.3. Implementation of an impulse ground penetrating radar

Applying the impulse technique to the GPR is a fairly direct process, with the main factor being the need to generate a high-voltage pulse. The impulse radar can be divided into three subassemblies: the timing source, the transmitter, and the sampler. Typically the transmitter electronics are integrated into the antenna housing.

3.5.3.1. Transmitter

In order to generate an EM wave with sufficient peak power, a high-voltage pulse is required. A block diagram of an impulse GPR system is shown in Figure 3.7 and includes a voltage source, a high-voltage supply, a pulse generator, and a waveform shaper.

The high-voltage supply and the pulse generator use a circuit where capacitors are charged in parallel from a low-voltage source such as a 12-V battery, then discharged in series to gain a voltage multiplication and produce a high-voltage pulse. This circuit typically uses an *avalanche transistor* as its switching source in order to produce a fast-rising edge of the pulse. A fast rise time, typically 10 times the pulse width, is needed to produce sufficient energy in the high-frequency components of the EM wave. Alternately, a DC-to-DC converter operating in a boost mode (i.e., 12 V input to 400 V output) can be used for the high-voltage source, or a combination of both can be used (Wright et al., 1990). There is generally a waveform shaper following the pulse generation, which is a matching network to the antenna. This matching network is used to minimize undesirable effects such as ringing and is usually incorporated into the antenna structure.

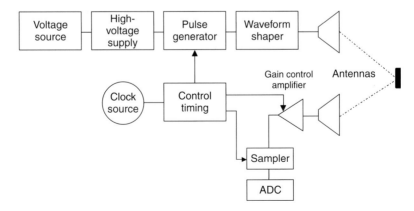

Figure 3.7 Impulse GPR system block diagram.

3.5.3.2. Timing source

The impulse GPR has an amplifier after the receive antenna, which provides automatic gain control, and can be used to implement time-varying gain functions (i.e., range gain). A clock source or a computer clock is needed to control the pulse generation, subsequently adjust the gain on the receiver amplifier, and time the sampler.

3.5.3.3. Sampler and signal processing

Low-speed sampling with the ETS method is most commonly used in impulse GPR, and 16-bit ADCs are readily available, which yield 96 dB of dynamic range. However, impulse GPR systems can have a much larger system dynamic range (>150 dB). The gain control compensation must be adjusted properly to place the return signal of interest in the 96-dB recording dynamic range. Additionally, time-adjusted gain or range gain should be set so that the recorded signal is properly compressed into the recording dynamic range and will not clip the limit of the ADC or be below the minimum recording level (see Figures 1.15–1.17). Detailed signal processing of impulse data is covered in Chapter 5.

3.6. CONTINUOUS-WAVE GROUND PENETRATING RADAR

3.6.1. Theory of operation – stepped-frequency, continuous-wave radar

This section describes the theory of operation of a stepped, FM-CW GPR (Koppenjan and Bashforth, 1993) and the application of a gating technique (Stickley et al., 2000).

3.6.1.1. Stepped-frequency technique – synthesized pulse

A stepped-frequency CW radar incorporates an RF source or a direct digital synthesis (DDS) source, and DSP. The source is stepped between a start frequency, f_0, and a stop frequency, f_{N-1}, in equal, linear increments. It is important to note that for a swept FM-CW radar, the source is swept from f_{min} to f_{max} and linearly sampled *on the fly*. In either case, the radar is continuously transmitting. A return signal is formed by mixing the received signal with a portion of the transmitted one. This return signal is digitized at each step and stored. After each complete *sweep* of N steps, a Fourier transform is performed to convert the data from the frequency domain to the time domain. This is the process of creating the synthesized pulse.

Range information is based on the *time-of-flight* principle, which is a phase path difference measurement. This concept is outlined with reference to Figure 3.8. Two paths are defined: 2–3–4–5–6 and 7–8. There is an associated phase, θ_{rf} and θ_{lo}, with each path, respectively. When the phase path lengths are equal, $\theta_{rf} = \theta_{lo}$,

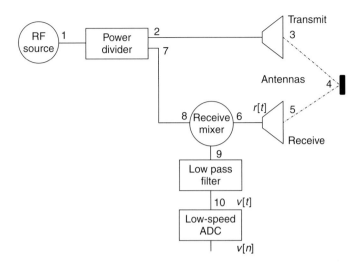

Figure 3.8 Simplified diagram of the stepped-frequency radar technique.

the length of 7–8 is equal to that of 2–3–5–6 (transmit is directly connected to receive). The phase paths of 7–8 and 2–3–4–5–6 are not equal when a target is present. A phase difference, ψ, occurs as a result of the time-of-flight difference and can be expressed as follows:

$$\psi = (2\pi f)\tau = (2\pi f)d/c \tag{3.11}$$

where f = frequency (Hz), τ = two-way time of flight (s), d = two-way distance to the target 3–4–5, and c = speed of light.

If a single frequency, f_0, from the RF source, 1, is power divided with one side transmitted and the other side connected to the receive mixer, the received signal at 6 is

$$r(0) = \cos\left(2\pi f_0(t+\tau_0) + \theta_{rf}\right) = \cos(2\pi f_0 t + \theta_{rf} + \psi_0) \tag{3.12}$$

and the output of the mixer, 9, will be

$$v(0) = A_0 \cos(2\pi f_0 t + \theta_{lo})\cos(2\pi f_0 t + \theta_{rf} + \psi_0) \tag{3.13}$$

where ψ_0 is the phase associated with the target path length, d (3–4–5) for f_0. A_0 is the amplitude of the return. If the phase lengths are preset so that $\theta_{rf} = \theta_{lo}$, then $v(0)$ reduces to

$$v(0) = A_0 \cos(2\pi f_0 t)\cos(2\pi f_0 t + \psi_0) \tag{3.14}$$

Using a trigonometric identity

$$v(0) = \frac{A_0}{2}\cos(\psi_0) + \frac{A_0}{2}\cos(2\pi(2f_0)t + \psi_0) \tag{3.15}$$

If $v(0)$ is low-pass filtered with a cutoff frequency of f_0, then the output of the filter, 10, will be

$$v(0) = \frac{A_0}{2}\cos(\psi_0) \tag{3.16}$$

Since the target's distance is fixed, the phase path, ψ_0, at a given frequency is also fixed. Thus $v(0)$ is a constant, which is represented by a DC voltage. This DC voltage can be sampled with a low-speed ADC; thus, $v(0)$ is now represented as $v[0]$, the sampled version at f_0. If the RF source is stepped by an amount Δf to a higher frequency, f_1, such that $f_1 = f_0 + \Delta f$, the phase path length, ψ_1, is longer. Then the sampled output of the low-pass filter, $v[1]$, is

$$v[1] = \frac{A_1}{2}\cos(\psi_1) \tag{3.17}$$

When the RF source is stepped in equal, linear increments of Δf from (f_0 to f_1 to f_2 to ... f_{N-1}), the output voltages ($v[0]$, $v[1]$, $v[2]$, ..., $v[N-1]$) resemble a sampled sine wave, as shown in Figure 3.9. This is due to the periodic nature of the phase. This sequence is the radar return signal and can be expressed as follows:

$$v[n] = \frac{1}{2}\sum_{i=0}^{N-1}\delta(n-i)A_i\cos(\psi_i) \tag{3.18}$$

where $\delta(n)$ is the dirac delta function or unit impulse. Rewriting this sequence, in a sampled sine wave format with frequency, ω, and amplitude, A, $v[n]$ becomes

$$v[n] = A\sum_{i=0}^{N-1}\delta(n-i)\cos(\psi_i n) \tag{3.19}$$

where $\psi_i = \omega_i \tau_0 = 2\pi(f_i + n\Delta f)\tau_0$.

At each step, n, an ADC is performed on the DC voltage, $v[n]$. The data from all N steps are then converted into the *time domain pulse response equivalent* with a discrete Fourier transform (DFT):

$$V(k) = \begin{cases} \frac{1}{N}\sum_{n=0}^{N-1}v[n]e^{-j(2\pi/N)kn}, & 0 \leq k \leq N-1 \\ 0, & \text{otherwise} \end{cases} \tag{3.20}$$

Taking the DFT of Equation (3.19) yields

$$V(k) = \begin{cases} \frac{A}{N}\frac{\sin\{(k-\omega)\Omega_o[(N+1)/2]\}}{\sin[(k-\omega)\Omega_o/2]}, & 0 \leq k \leq N-1 \\ 0, & \text{otherwise} \end{cases} \tag{3.21}$$

where $\Omega_o = 2\pi/N$.

$V(k)$ is plotted in Figure 3.10. Equation (3.21) is recognized as the discrete-time counterpart of the sinc function: $\text{sinc}(x) = \sin(x)/x$, which is the Fourier transform of a continuous-time rectangular pulse. Notice that the sinc function is shifted

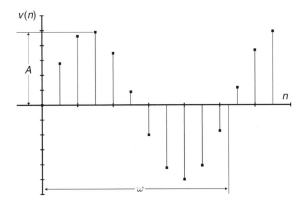

Figure 3.9 Sampled sine wave.

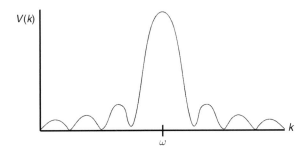

Figure 3.10 Sinc function.

(from 0) along the k axis by an amount equal to ω. This is a direct result of the sampled sine wave's frequency, ω, and in fact the range of a target is a function of ω. Closer targets produce smaller phase changes because the path, 3–4–5, is shorter. This results in low-frequency sine waves. More distant targets produce larger phase changes resulting in high-frequency sine waves. The amplitude of the sine wave is a function of the radar cross section of the target, the range, and the propagation loss of the ground.

3.6.1.2. Frequency modulation

In the previous scenario, the return signal, $v(t)$, would need to be amplified before an A/D conversion is performed because of the lossy propagation medium. When a radar return signal is mixed directly to baseband and amplified, a phenomenon known as $1/f$ noise or *flicker noise* exists ($1/f$ meaning one over frequency and spoken as "1 over f"). This noise is also amplified and is very prevalent at low frequencies. It is defined by the noise–temperature ratio of the receive mixer and varies inversely with frequency. Above approximately 500 kHz, the noise–temperature ratio approaches a constant value (Skolnik, 1980). To reduce the effects of $1/f$ noise, the transmitted RF signal is commonly modulated, thus generating an intermediate frequency (IF) in the receiver. Amplification takes place at this IF and not at baseband. The receive signal is then demodulated to obtain the baseband signal, A/D converted, and Fourier transformed. An alternative implementation, with a similar result, is to modulate the received signal, amplify at the IF, and demodulate to baseband.

3.6.1.3. Gating

To implement the gating concept, the diagram in Figure 3.8 can be modified to add gating switches to the transmit and receive paths, and a timing circuit block is shown in Figure 3.11. The typical timing sequence of the gating signals generated by the timing circuit is shown in Figure 3.12. A transmit pulse of pulse width W_t is generated and followed by a gating pulse to the receiver of pulse width W_r. The delay between the transmit pulse and the receive pulse (receiver gate delay, D_r) acts to gate out the undesired larger signal returns by keeping the receiver off for a brief time. A parameter of interest is the GRI, T_g. The GRI is equivalent to a PRI, which is the inverse of the PRF, more commonly referred to as the PRF in pulse radars.

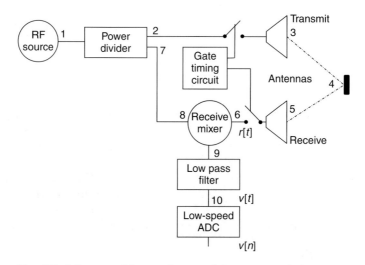

Figure 3.11 Simplified diagram of the gated, stepped-frequency technique.

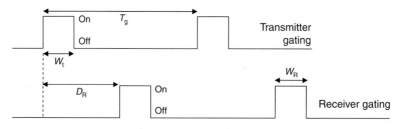

Figure 3.12 Typical timing sequence of the gate signals.

To determine the effect on the return signal, start with one fixed target at a fixed phase or fixed two-way time delay, $2\pi\tau_0$. The received signal, r, for single frequency, f_0, at 6 (ungated) is

$$r(0)_{\text{ungated}} = A_0 \cos\left(2\pi f_0(t - \tau_0)\right) \tag{3.22}$$

The receive signal at 6, after the gate timing signals are applied to obtain spatial filtering and assuming the transmit pulse and receiver pulse are equal, $W = W_t = W_r$, can be expressed using the following comb functions:

$$\text{III}(t) = \sum_{i=-\infty}^{\infty} \delta(t - i), \quad \text{for all integer } i \tag{3.23}$$

$$\Pi(t) = \begin{cases} 1, & -0.5 < t < 0.5 \\ 0, & \text{else} \end{cases} \tag{3.24}$$

Thus, the receive signal is the convolution of Equations (3.22)–(3.24)

$$r(0)_{\text{gated}} = \left[\text{III}\left(\frac{t - \tau_0 - D_{r/2}}{T_g}\right) * \Pi\left(\frac{t - \tau_0 - D_{r/2}}{W_c}\right)\right] * A_0 \cos\left(2\pi f_0(t - \tau_0)\right) \tag{3.25}$$

where $W_c = \begin{cases} W - |\tau_0 - D_r|, & W > |\tau_0 - D_r| \\ 0, & \text{otherwise} \end{cases}$

In the time domain, the reflection of the transmitted signal from the target at time delay, τ_0, must coincide with the time that the receiver is gated to produce a gated sine wave, $r(0)_{\text{gated}}$. If there is no coincidence with the delayed receiver gate, the result is no received signal, i.e., $r(0)_{\text{gated}} = 0$.

The frequency domain representation of Equation (3.25) is a series of line spectra centered at f_0 and spaced at $1/T_g$. When $r(0)_{\text{gated}}$ is mixed with the ungated frequency component, $\cos(2\pi f_0 t)$, and band-pass filtered to select only one of the frequency components in the spectrum, an approximate expression for $r(0)_{\text{gated}}$ can be made. The component of $r(0)_{\text{gated}}$ that has the same frequency content as the ungated local oscillator is

$$r(0)_{\text{gated}} \approx \left(\frac{W_c}{T_g}\right)_0 \cos\left(2\pi f_0(t - \tau_0)\right) \tag{3.26}$$

$$r(0)_{\text{gated}} \approx \left(\frac{W_c}{T_g}\right) r[0]_{\text{ungated}} \tag{3.27}$$

Thus, by replacing $r(0)_{\text{gated}}$ into Equation (3.14), the resulting output signal is

$$v(0)_{\text{gated}} = A_0 \left(\frac{W_c}{T_g}\right) \cos(2\pi f_0 t) \cos\left(2\pi f_0(t - \tau_0)\right) \tag{3.28}$$

which after filtering reduces to

$$v(0)_{\text{gated}} = \left(\frac{A_0 W_c}{2T_g}\right)\cos(2\pi f_0 \tau_0) = \frac{A_0 W_c}{2T_g}\cos\psi_0 \qquad (3.29)$$

This is equivalent to Equation (3.16) except for the decrease in amplitude. One effect of applying gating will be a reduction in the received power (and minimum detectable signal) compared to an ungated system, as the receiver is no longer a matched receiver (Hamran et al., 1995). The received power is reduced proportionally by the duty cycle of the gate, assuming equal duty cycles, $W_t/T_g = W_r/T_g$ (Stickley et al., 2000). A detailed system performance discussion on the mean transmitted power to minimum detected signal power can be found in Plumb et al. (1998).

If the similar procedure of stepping the frequency over N steps is done as for the ungated system, the sampled sine wave output is

$$v[n] = \frac{AW_c}{T_g}\sum_{i=0}^{N=1}\delta(n-i)\cos(\psi_i n), \quad \text{where } \psi_i = \omega_i \tau_0 = 2\pi(f_i + n\Delta f)\tau_0 \qquad (3.30)$$

And, after a DFT, the time domain pulse response equivalent is

$$V(k) = \begin{pmatrix} \dfrac{AW_c}{NT_g}\dfrac{\sin\{(k-\omega)\Omega_o[(N+1/2)]\}}{\sin[(k-\omega)\Omega_o/2]}, & 0 \leq k \leq N-1 \\ 0, & \text{otherwise} \end{pmatrix} \qquad (3.31)$$

where $\Omega_o = 2\pi/N$.

Equations (3.30) and (3.31) are similar to the ungated system, Equation (3.19) and (3.21), except for the decrease in amplitude.

3.6.2. System design parameters: stepped-frequency radar

Several design parameters must be chosen with a stepped-frequency radar. These involve bandwidth versus resolution (similar to pulse width/resolution with impulse systems) and the number of sample points, sweep time, and GRI versus unambiguous range.

The desired range resolution will determine the required bandwidth and affect the designs of the RF source and the antenna. The RF source will need to function over a frequency range to achieve the bandwidth. The antenna design must be considered when a low center frequency is chosen with a wide bandwidth, especially when B approaches f_c as it may result in an impractical antenna realization. Refer to the previous example using Figure 3.1.

The unambiguous range versus ε_r ($B = 500\,\text{MHz}$) is plotted in Figure 3.2, with the number of frequency steps, N, varying from 64, 128, 256, 512, and 1024. The number of points will also effect the overall sweep time, which is the time it takes to step from $f_{\min}$ to $f_{\max}$. The sweep time will affect how fast the GPR can be moved across the surface. The sweep time can be reduced with fewer steps, but R_{max} will also be reduced. For geologic applications where range is of interest, N must be large. For a pavement evaluation application, R_{max} may be less than 1 m, but the sweep time needs to be very fast to accommodate vehicle speeds; thus N can be small.

Additionally, for a gated, stepped-frequency GPR, the gate timing and the GRI need to be chosen to correspond with the unambiguous range determined by N. The GRI must be larger than the two-way travel time to a target at R_{max}. In practice, the T_g should be twice the minimum requirement:

$$T_g > \frac{2R_{max}\sqrt{\varepsilon_r}}{c} \qquad (3.32)$$

3.6.3. Implementation of a gated, stepped-frequency, ground penetrating radar

Applying the stepped-frequency technique to the GPR requires some RF design considerations and several additions to the basic block diagram. As previously mentioned, the RF signal is modulated to minimize the $1/f$ noise in the receiver, and a modulation frequency greater than 500 kHz should be used. The type of modulator scheme used in FM-CW GPR is actually *quadraphase modulation*. This is designed to maintain a coherent, complex waveform as a return signal. Coherent means both the phase and the magnitude information are maintained. Complex means that the phase and the magnitude are represented as real and imaginary numbers. Quadraphase modulation changes the phase of the transmitted signal between 0°/180° (inphase) and 90°/270° (quadrature). At each frequency step, an ADC is performed on two signals: the inphase and quadrature. This is known as I & Q data and it becomes the input to a complex FFT. The real and imaginary output of the FFT equation (3.21) is then converted to polar form. The magnitude, M, and phase, Φ, expressions are as follows:

$$M_i = \sqrt{I_i^2 + Q_i^2}, \quad \Phi_i = \tan^{-1}\left|\frac{Q_i}{I_i}\right| \qquad (3.33)$$

The sequence of magnitudes, $(M_0, M_2, \ldots, M_{N-1})$, from each FFT dataset, $[(I_0, Q_0), (I_1, Q_1), \ldots, (I_{N-1}, Q_{N-1})]$, will form the familiar time domain response.

The stepped-frequency radar can be divided into three subassemblies: the source, the radar front end (transmitter and receiver), and the sampler. Consideration to practical RF sources and components must be made in choosing the operating frequency range.

3.6.3.1. Frequency-synthesized source

Several important factors associated with the source can influence the performance of the radar: the purity of the source, the stability of the source, and the linearity of the *sweep*. A straightforward way to generate a wideband, multioctave source at low GPR frequencies (i.e., 200–700 MHz) is to mix two high-frequency RF sources.

The presence of inband intermodulation products will degrade the purity of the signal and generate undesirable false returns. In order to minimize this effect, a spurious-free source and high-level mixers with high third-order intercept points should be used. To maintain a stable output frequency, a voltage-controlled oscillator (VCO) with phase lock loop (PLL) circuitry can be used as the source. Stability over time is required for accurate signal processing. When stepping the output frequency, the step size, Δf, should be kept constant to obtain linear steps. Deviations of Δf will cause a distortion to the return sampled sine wave. These deviations will have the effect of broadening the frequency spectra or widening the sinc pulse. The result will be a loss in resolution. To obtain linear stepping, a crystal-controlled oscillator should be used as the reference frequency for the PLL. Typically the IF source for modulation is referenced to the crystal source. Digital interface and control circuitry are required for the programming of the PLL and gate timing circuit.

Alternately, a DDS source (contains a numerically controlled oscillator and digital-to-analog converter) can be used as the frequency synthesizer for typical GPR frequencies. Advances in DDS technology have allowed higher clock rates, and thus higher output frequencies. The maximum output is typically 40% of the clock rate, and the output will require proper filtering to remove digitally generated spurs and harmonics, as these will have similar negative effects as intermodulation products.

3.6.3.2. Transmitter and receiver

A block diagram of a gated, stepped-frequency GPR RF system is shown in Figure 3.13, and includes the RF source, the quadraphase modulator, the receive mixer, filters, the IF amplifier, and the demodulator. Additionally, it contains

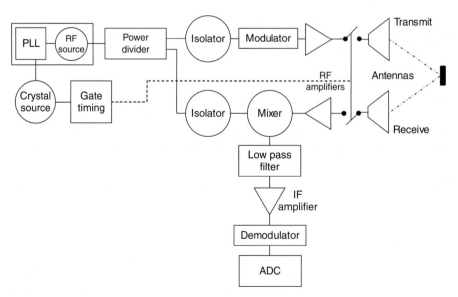

Figure 3.13 Gated, stepped, FM-CW GPR RF system block diagram.

transmit and receive amplifiers and isolators. The amplifiers attain the desired signal levels and help compensate for propagation losses. The receive amplifier should have a low noise figure (less than 5 dB), since it essentially determines the system noise figure. The isolators separate the modulated signals from the unmodulated. This minimizes RF signal leakage, which can generate undesirable false returns. Amplifiers with good reverse isolation can be used when isolators are impractical at the operating frequencies or bandwidth. Alternately, a quadraphase receiver/demodulator can be used on the receiver side (Stickley et al., 2000). In this case, either an offset frequency to the transmitted frequency is needed to generate the IF, which is then demodulated in the quadrature receiver, or in some applications, the RF is mixed directly to baseband (I&Q outputs) in the quadrature receiver.

3.6.3.3. Sampler and signal processing

Generally, low-speed sampling can be used to digitize the baseband data, and 16-bit ADCs are readily available. The timing of the actual sampling relative to each frequency step needs to be precisely controlled so as to allow sufficient time for the source oscillator to settle to the new frequency. The output of other circuits such as the IF amplifier and demodulator must also settle before the sample is made.

High-speed sampling can be performed on the quadrature (I&Q) signals instead of analog-mixing the IF signal to baseband. This is generally known as a *digital IF* and requires complex DSP to convert to baseband. The digital IF method allows optimization with digital filters and gain control.

Regardless of the sampling method, the conversion of the frequency data to the time domain is done with a complex, inverse FFT. As seen in Figure 3.10, sidelobes are generated because the sampled data is of limited bandwidth in the frequency domain. To minimize the sidelobes, the data is preprocessed with a *window* function. Typical types of window functions are Hanning, Hamming, and Kaiser-Bessel, each of which will have varying effects on increasing the sidelobe attenuation but will reduce resolution by increasing peak width (Skolnik, 1990).

REFERENCES

Annan, A.P., 2003, Introduction to GPR, Sensor & Software, Inc. — Tutorial notes.

Annan, A.P., 2005, Ground Penetrating Radar in Near-Surface Geophysics, in Investigations in Geophysics, No. 13, Society of Exploration Geophysicists, Dwain K. Butler (ed.), ISBN 1-56080-130-1, pp. 357–438.

Annan, A.P. and Davis, J.L., 1976, Impulse radar sounding in permafrost. Radio Science, Vol. 11, pp. 383–394.

Daniels, D.J., 1996, Surface-Penetrating Radar, The Institution of Electrical Engineers, London, United Kindom.

Daniels, D.J., Gunton, D.J., and Scott, H.F. (eds.), 1988, Introduction to subsurface radar. IEE Proceedings, F-Radar and Signal Processing, Vol. 135, No. 4, pp. 278–320.

Daniels, J.J., Chen, C., and Allred, B.J., 2006, Proceedings of the Eleventh International Conference on Ground Penetrating Radar (GPR 2006), Columbus, Ohio, available on CD-Rom only.

Eaves, J.L. and Reedy, E.K., 1987, Principles of Modern Radar, Van Nostrand Reinhold Company, New York, pp. 226–228, 409–420.

Eide, E.S. and Hjelmstad, J.F., 1999, A multi-antenna ultra wideband GPR system using arbitrary waveforms. Proceedings, IEEE International Geoscience and Remote Sensing Symposium, Hamburg, Germany.

Erst, S.J., 1985, Receiving Systems Design, The Receiver, Artech House, Inc., Dedham, Massachusetts.

Hamran, S.-E. and Aarholt, E., 1993, Glacier study using wavenumber domain synthetic aperture radar. Radio Science, Vol. 28, No. 4, pp. 559–570.

Hamran, S.-E., Gjessing, D.T., Hjelmstad, J., and Aarholt, E., 1995, Ground penetrating synthetic pulse radar: dynamic range and modes of operation. Journal of Applied Geophysics, Vol. 33, pp. 7–14.

Hansen, S., 1942, Electrical Wave Analysis, U.S. Patent 2,280,524.

Harmuth, H.F., 1969, Transmission of Information by Orthogonal Functions, Springer, New York.

Iizuka, K., Freundorfer, A.P., Wu, K.H., Mori, H., Ogura, H., and Nguyen, V-K., 1984, Step-frequency radar. Journal Applied Physics, Vol. 56, pp. 2572–2583.

Johnson, R.C. and Jasik, H., 1984, Antenna Engineering Handbook, McGraw-Hill, Inc., New York.

Kong, F.-N., 1990, Ground penetrating radar using a frequency sweeping signal. Proceedings, Third International Conference on GPR, Lakewood, Colorado.

Kong, F.-N. and By, T.L., 1995, Performance of a GPR system which uses step frequency signals. Journal of Applied Geophysics, Elsevier, Vol. 33, No. 1–3, pp. 15–26.

Koppenjan, S.K. and Bashforth, M.B., 1993, The Department of Energy's Ground Penetrating Radar, an FM-CW system, Underground and obscured object imaging and detection, Nancy K. Del Grande, Ivan Cindrich, and Peter B. Johnson (eds.). Proceedings SPIE, 1942, pp. 44–55.

Koppenjan, S.K. and Lee, H (eds.), 2002, Proceeding of the Ninth International Conference on Ground Penetrating Radar (GPR 2002). SPIE Proceedings, 4758, Santa Barbara, California.

Koppenjan, S.K., Allen, C.M., Gardner, D., Wong, H.R., Lee, H., and Lockwood, S.J., 2000, Multi-frequency synthetic-aperture imaging with a lightweight ground penetrating radar. Journal of Applied Geophysics, Special Issue on Ground Penetrating Radar (GPR 98), Christopher T. Allen and Richard G. Plumb (eds.), Elsevier, Vol. 43, Nos. 2–4, pp. 251–258.

Morey, R.M., 1974, Continuous subsurface profiling by impulse radar. Proceedings, Engineering Foundations Conference on Subsurface Exploration for Underground Excavations and Heavy Construction, Henniker, New Hampshire, pp. 213–232.

Narayanan, R.M., Xu, Y., Hoffmeyer, P.D., and Curtis, J.O., 2001, Design, performance, and applications of a coherent ultra-wideband random noise radar, J.D. Taylor (ed.), Ultra-Wideband Radar Technology, CRC Press, Boca Raton, Florida, pp. 181–203.

Noon, D.A. and Narayanan, R.M., 2002, Subsurface remote sensing, Chapter 24, W. Ross Stone (ed.), Review of Radio Science, 1999–2002, URSI, John Wiley & Sons, pp. 535–552.

Noon, D.A., Longstaff, I.D., and Yelf, R.J., 1994, Advances in the development of step frequency ground penetrating radar. Proceedings, Fifth International Conference on Ground Penetrating Radar (GPR 94), Kitchener, Ontario, Canada, pp. 117–132.

Oliver, A.D. and Cuthbert, L.G., 1988, FM-CW radar for hidden objects detection. IEE Proceedings-F, Vol. 135, No. 4, pp. 354–361.

Oliver, A.D., Cuthbert, L.G., Nicolaides, M., and Carr, A.G., 1982, Portable FM-CW radar for locating buried pipes. Proceedings, Radar 82 Conference, IEE Conference Publications, London, Vol. 216, pp. 413–418.

Plumb, R.G., Noon, D.A., Longstaff, I.D., and Stickley, G.F., 1998, A waveform-range performance diagram for ground-penetrating radar. Journal of Applied Geophysics, Vol. 40, Nos. 1–3, pp. 117–126.

Poirier, M.A., 1993, Comparison of band pulsed and stepped frequency pulsed radar concepts for GPR applications. Second Government Workshop on Advanced Ground Penetrating Radar, Technologies and Applications, Columbus, Ohio, USA, pp. 221–222.

Robinson, L.A., Weir, W.B., and Young, L., 1972, An RF time-domain reflectometer not in real time. IEEE Transactions on Microwave Theory and Techniques, Vol. 20, No. 12, pp. 855–857.

Robinson, L.A., Weir, W.B., and Young, L., 1974, Location and recognition of discontinuities in dielectric media using synthetic RF pulses. Proceedings, IEEE, Vol. 62, No. 1, pp. 36–44.

Skolnik, M.I., 1980, Introduction to Radar Systems, McGraw-Hill Book Company, New York, pp. 74, 95–97, 348–349.

Skolnik, M.I. (ed.), 1990, Radar Handbook, McGraw-Hill Book Company, New York, pp. 10, 27–36.

Slob, E.C., Yarovoy, A.G., and Rhebergen, J.B. (eds.) 2004, Proceedings of the Tenth International Conference on Ground Penetrating Radar (GPR 2004), Delft, The Netherlands, Delft University of Technology & TNO-FEL, Vol. 1–2.

Smith, D.T., 2004, Digital Signal Processing, The ARRL Handbook for Radio Communications, 81st edition, D.G. Reed (ed.), ISBN: 0-87259-196-4.

Stickley, G.F., Noon, D.A., Cherniakov, M., and Longstaff, I.D., 2000, Gated stepped-frequency ground penetrating radar. Journal of Applied Geophysics, Special Issue on Ground Penetrating Radar (GPR 98), Christopher T. Allen and Richard G. Plumb (eds.), Elsevier, Vol. 43, Nos. 2–4. pp. 259–269.

Wills, R.H., 1992, A digital phase coded ground probing radar, in J. Pilon (ed.), Ground Penetrating Radar, Geological Survey of Canada, pp. 231–235.

Wright, D.L., Hodge, S.M., Bradley, J.A., Grover, T.P., and Jacobel, R.W., 1990, Instruments and methods: a digital low-frequency, surface-profiling ice-radar system. Journal of Glaciology, Vol. 36, No. 122, pp. 112–121.

Wright, D.L., Olhoeft, G.R., Grover, T.P., and Bradley, J.A., 1993, High-speed digital radar systems and applications to subsurface exploration. Proceedings of the Fourth Tunnel Detection Symposium on Subsurface Exploration Technology, Golden, Colorado.

Wright, D.L., Bradley, J.A., and Grover, T.P., 1994, Data acquisition systems for ground penetrating radar with example applications from the air, the surface and boreholes. Fifth International Conference on Ground Penetrating Radar (GPR 94), Kitchener, Ontario, Canada, pp. 1075–1089.

Xu, Y., Narayanan, R.M., Xu, X. and Curtis, J.O., 1989, Sampling oscilloscope techniques: Tektronix – Technique Primer 47W-7209.

Xu, Y., Narayanan, R.M., Xu, X., and Curtis, J.O., 2001, Polarimetric processing of coherent random noise radar data for buried object detection. IEEE Transactions on Geoscience and Remote Sensing, Vol. 39, No. 3, pp. 467–478.

Xu, Y., Narayanan, R.M., Xu, X., and Curtis, J.O., 1989, Sampling oscilloscope techniques: Tektronix – Technique Primer 47W-7209, 1989.

CHAPTER 4

ANTENNAS

David J. Daniels

Contents

4.1. Introduction	99
4.2. Basic Antenna Parameters	102
4.2.1. Energy transfer from antennas	102
4.2.2. Gain	104
4.2.3. Directivity	105
4.2.4. Coupling energy into the ground	105
4.2.5. Antenna efficiency	106
4.2.6. Sidelobes and back lobes	106
4.2.7. Bandwidth	106
4.2.8. Polarisation – linear, elliptical, circular	107
4.2.9. Antenna phase centre	108
4.2.10. Antenna patterns	108
4.2.11. Time sidelobes and ring-down	109
4.2.12. Antenna footprint	110
4.3. Antennas For Ground Penetrating Radar	112
4.3.1. Introduction	112
4.3.2. Coupling into a dielectric	113
4.3.3. Time domain antennas	115
4.3.4. Frequency domain antennas	124
4.3.5. Array antennas	128
4.4. Summary	133
4.5. Definitions	133
References	135
Further Reading	136

4.1. INTRODUCTION

This chapter provides an introduction to the antennas used in ground penetrating radar (GPR). As most users of GPR are not primarily antenna specialists, it may be useful to define many of the key terms used in antenna engineering before describing the principal antennas used in GPR systems, and a list of commonly encountered terms is provided at the end of this chapter. The aim of this chapter is to provide an introduction to antennas and present some of the basic design types that have played a key role in GPR. More detailed information can be found in the bibliography and references at the end of the chapter.

The Institution of Engineering and Technology (IET) has given permission to include material contained in the chapter relating to antennas in the book *Surface Penetrating Radar* ISBN 0 85296 862 0 and much of this is very relevant and provides a more detailed reference for those interested in the design of antennas.

An antenna is a device for coupling energy from a source of radio frequency energy into a transmitting medium, which is normally air. For GPR, the radiation from the antenna is normally coupled into the ground, and this affects the radiation characteristics of the antenna to a considerable extent, if the latter is electrically close to the ground. An antenna can be used to transmit energy, receive energy, or both.

Antennas can be classified into two general classes: omni-directional and directional. Omni-directional antennas radiate energy in all directions simultaneously. This leads to the concept of an isotropic antenna, which is one that radiates energy uniformly in all directions and is a useful, if fictitious, concept to enable real antenna characteristics to be referenced as shown in Figures 4.1 and 4.2.

Directional antennas radiate energy in patterns of lobes or beams that extend outwards from the antenna in one direction for a given antenna position. The radiation pattern also contains minor lobes otherwise known as sidelobes or back lobes, but these lobes are weak and may have little effect on the main radiation pattern. In conventional radar systems, the main lobe may vary in angular width from 1° in some cases to 15° in others. For GPR antennas, the main lobe is much wider and can be in the region of 90°.

Directional antennas have two important characteristics: directivity and power gain. The directivity of an antenna refers to the degree of sharpness of its beam. If the beam is narrow in either the azimuth or the elevation plane, the antenna is said to have high directivity in that plane. Conversely, if the beam is wide in either plane, the antenna is said to have low directivity in that plane. Thus, if an antenna has a narrow azimuthal beam and a wide elevation beam, the horizontal directivity is high and the vertical directivity is low. When the directivity of an antenna is increased, that is, when the beam or the main lobe is narrowed, less power is required to cover the same range because the power is concentrated.

The power gain of an antenna is directly related to directivity and is the ratio of its radiated power to that of a reference (basic) dipole. In order to measure power gain, both the reference dipole antenna and the antenna under test (AUT) must have been excited or fed in the same manner and each must have radiated from the same position. A single point of measurement for the power–gain ratio must lie within the radiation field of each antenna. An antenna with high directivity has a high power gain and vice versa. The power gain of a single dipole with no reflector is unity, and in the azimuth plane, the radiation is omni-directional.

Ground penetrating radar presents the antenna designer with significant restrictions on the types of antennas that can be used. The propagation path consists, in general, of a lossy, inhomogeneous dielectric, which, in addition to being occasionally anisotropic, exhibits a frequency-dependent attenuation and hence acts as a low-pass filter. The upper frequency of operation of the system, and hence the antenna, is therefore limited by the properties of the material. The need to obtain a high value of range resolution requires the antenna to exhibit ultra-wide

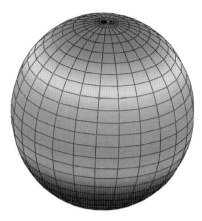

Figure 4.1 Isotropic surface radiation pattern.

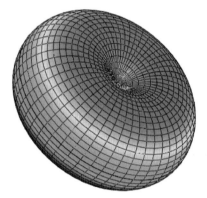

Figure 4.2 Radiation surface pattern of simple dipole.

bandwidth, and in the case of impulsive radar systems, linear phase response. The requirement for wide bandwidth and the limitations in upper frequency are mutually conflicting and hence a design compromise is adopted whereby antennas are designed to operate over some portion of the frequency range, 10 MHz–5 GHz, depending on the resolution and range specified. The requirement for portability of the operator means that it is normal to use electrically small antennas, which consequently result generally in a low gain and associated broad radiation patterns. The classes of antennas that can be used are therefore limited, and the following factors have to be considered in the selection of a suitable design: large fractional bandwidth, low time sidelobes and in the case of separate transmit and receive antennas, low cross-coupling levels. The interaction of the reactive field of the antenna with the dielectric material and its effect on antenna radiation pattern characteristics must also be considered.

4.2. BASIC ANTENNA PARAMETERS

4.2.1. Energy transfer from antennas

A GPR can be operated so that its antenna is very close to the ground surface and target such that the energy transfer is by quasi-stationary and induction fields, as well as radiated fields (the near field); alternatively, it can be operated such that the energy transfer is in the far-field region (radiated field). It is important to understand the implications of these modes on the overall detection capability of radars. Essentially GPRs operated in standoff mode are fully described by radiated field models, whereas radars operated in proximal mode may achieve better performance due to the increased contribution by the quasi-stationary and induction fields.

A small electric dipole radiates electric and magnetic fields as described in Equations (4.1)–(4.3), which are derived from Maxwell's equations

$$E_r = \frac{2Z_0 Idl\pi \cos\theta}{\lambda^2} \left[\left(\frac{\lambda}{2\pi r}\right)^3 \cos\psi + \left(\frac{\lambda}{2\pi r}\right)^2 \sin\psi \right] \quad (4.1)$$

$$E_\theta = \frac{Z_0 Idl\pi \sin\theta}{\lambda^2} \left[-\left(\frac{\lambda}{2\pi r}\right)^3 \cos\psi - \left(\frac{\lambda}{2\pi r}\right)^2 \sin\psi + \left(\frac{\lambda}{2\pi r}\right) \cos\psi \right] \quad (4.2)$$

$$H_\phi = \frac{Idl\pi \sin\theta}{\lambda^2} \left[\left(\frac{\lambda}{2\pi r}\right)^2 \sin\psi + \left(\frac{\lambda}{2\pi r}\right) \cos\psi \right] \quad (4.3)$$

where

dl is the length of the current element
$\psi = (2\pi\phi/\lambda) - \omega t$
ω is the radian frequency of the signal
t is the time ($=1/f$)
c is the speed of light ($1/\sqrt{\mu\varepsilon} = 3 \times 10^8$ m/s)
Z_0 is the free space impedance ($= 377 \,\Omega$)
I is the current in the element
θ is the zenith angle to radial distance r
λ is the wavelength of the signal
r is the distance from the element to the point of observation

The terms can be considered to fall into three basic components:

1. A term proportional to r^{-1}, which is called the radiation term. This term represents the flow of energy away from the conducting element of the antenna.
2. A term proportional to r^{-2}, which is called the induction term and represents energy stored in the field during one quarter of a cycle and then returned to the antenna in the next.
3. A term proportional to r^{-3}, which is called the quasi-stationary term or the electrostatic field term, and results from the accumulation of charges at the ends of the element.

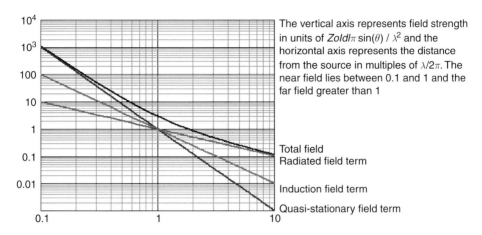

Figure 4.3 Signal level versus distance for components of a dipole field.

Figure 4.3 shows field strength as a function of distance from the source. At the distance $r = \lambda/2\pi$, all of these terms are equal, and this distance represents the boundary between the near fields and far fields where the contributions from the radiation, induction and the quasi-stationary terms are all of the same magnitude. The regions may also be described as the reactive near field, the radiating near field, and the far field.

In the reactive near field, energy is stored in the electric and magnetic fields very close to the source but not radiated from them. Instead, energy is exchanged between the signal source and the fields. If a target is capable of coupling energy from the source fields, a signal will be coupled or received in the antenna.

The approximate field boundaries are as follows:

Electrically small antenna, reactive near field	Electrically large antenna, reactive near field	Electrically large antenna, radiating near field
$r < \lambda/2\pi$	$r < 0.62\,(D^3/l)^{1/2}$	$r < 2D/\lambda$

Any consideration of the signal detected in a radar receiver should therefore fully account for the physical proximity of the antenna and the target. In the case of an electrically small antenna on the surface of a dielectric and radiating frequencies in the reactive near field centred at 0.5 GHz ($\lambda_m = 30$ cm for a dielectric whose $e_r = 4$), it can be seen that targets closer than 5 cm will have increased field contributions, whereas a radar radiating shorter wavelengths will lose out on the increased field contributions because it moves into the electrically larger zone Studies of accurate near-field measurements of GPR antennas are considered by Yarovoy et al. (2007) and Lenler-Eriksen (2005).

In more conventional radar systems, the fields of the antenna are classified as follows:

Near field, also called the reactive near-field region, is the region that is closest to the transmitting antenna and for which the reactive field dominates over the radiative fields.

Fresnel zone, also called the radiating near-field region, is that region between the reactive near-field and the far-field regions and is the region in which the radiation fields dominate and where the angular field distribution depends on the distance from the transmitting antenna.

Far Field or Fraunhofer zone is the region where the radiation pattern is independent of the distance from the transmitting antenna. In the far field, the power received per unit area from an isotropic antenna is calculated from the following equation:

$$P_r = \frac{P_t}{4\pi R^2} \tag{4.4}$$

This equation is also referred to as the inverse square law, since doubling the range gives a four-fold reduction in signal power. The region where the near field becomes the far field is a gradual transition; however, for practical applications, radio and optical engineers have defined the maximum radius to be r_{min} and corresponds to distance 1 on the x axis in Figure 4.3. This is also sometimes known as the *Rayleigh* distance.

Ground penetrating radar can thus be operated in several modes; most usually it is operated in a proximal mode whereby the antenna is electrically closer to the lossy dielectric so that the quasi-stationary or reactive fields are dominant or less often where the Fresnel or Fraunhofer regions dominate.

4.2.2. Gain

Gain is a widely used parameter directly measurable by substituting an antenna with known gain (generally a gain reference antenna) in for an Antenna Under Test (AUT). The output levels of the AUT and the gain reference can then be measured for the same incident field. The gain can then be determined by comparing those measured levels. Gain of an antenna is expressed in dB, 10 $\log_{10}$(numerical gain), which is generally referenced to an isotropic radiator and expressed as dBi. The gain expressed for an antenna is generally the maximum or peak gain.

The gain of an antenna is defined as follows:

$$G = eD = \frac{4\pi}{\lambda^2} A_{eff} \tag{4.5}$$

where

D is the directivity
A_{eff} is the antenna effective aperture

Note that the gain of an antenna also includes contribution due to resistive losses within the antenna, and this gives rise to the alternative concept of directivity.

4.2.3. Directivity

Directivity is similar to gain, but the resistive losses are not included. The directive gain of an antenna is given by

$$G_D = \frac{\text{maximum radiation intensity}}{\text{average radiation intensity}}$$

and an approximate expression for this is given below:

$$G_D = \frac{4\pi}{\Theta_A \Theta_E} \tag{4.6}$$

where

Θ_A is the azimuth 3-dB beamwidth
Θ_E is the elevation 3-dB beamwidth

The directivity of an antenna is generally combined with efficiency and expressed as gain as described above. The half-power beamwidth (HPBW), Θ_A, of an antenna is an expression, in degrees, of the width of the radiated beam between the half-power and 3 dB points (down from the peak of the beam). Many antennas will exhibit one HPBW in azimuth and a different HPBW in elevation written as $HPBW_A$ (Θ_A) and $HPBW_E$ (Θ_E). An antenna described as omni-directional will have equal coverage in all directions. A typical wide-band, omni-directional (in azimuth) antenna will have a $HPBW_A$ of 360° and $HPBW_E$ of 50°. A full characterisation of the radiation pattern of an antenna requires a 3D measurement whereas cuts in the principal planes provide only a partial understanding of the pattern. Radiation patterns in dielectrics may differ radically from those in free space and the measurement in the latter is not trivial. Probes for the measurement of radiated field and their associated means of transmitting energy to equipment should not perturb the field being measured. Fibre optically coupled probes have been used to minimise disturbance of the radiated field; however, such methods require extensive and sophisticated measurement instrumentation to be immersed in the dielectric. Simpler techniques using small loop antennas or small dipoles and orthogonally arranged feed cables have been used but great care must be taken in layout to avoid distorting the actual field by the measurement probes.

4.2.4. Coupling energy into the ground

The proximity of the antenna to the ground means that it is necessary to consider the coefficients of reflection and transmission as the wave passes through the dielectric to the target. Snell's laws describe the associated angles of incidence, reflection, transmission and refraction. For proximal operation, the efficiency of the coupling process is generally high, but this is not the case for standoff radar systems since, where lossy materials are involved, complex angles of refraction may occur. Buried targets pose a difficult detection problem for standoff radars, and their performance is strongly influenced by the ground conditions. With vertical polarisation at incidence angles less than the Brewster angle, transmission losses at the air/ground interface are relatively small, but at incidence angles larger than the Brewster angle, the losses increase more rapidly. The dependence of transmission loss on dielectric constant and angle of incidence suggests that this should be not more than 85°. Hence to maximise the operating range, the radar should be mounted as high off the ground as possible. Thus for a given height, the performance of the radar will be set by the relative dielectric constant of the ground. In addition to the problem of coupling energy into the ground, the effective cross section of all targets decreases when they are buried. Measurements and modelling suggest that under conditions of negligible attenuation losses, as are expected in very arid ground or for shallow burial depths, target-to-clutter ratios are expected to be degraded on burial by approximately 10 dB. Clutter is considered to be unwanted returns from sources other than the targets of interest. Under the same conditions, the cross section of non-metallic targets is reduced by a larger factor because of reduced dielectric contrast between it and the surrounding soil, so that non-metallic targets are more readily detected in wet sandy conditions than in dry conditions.

4.2.5. Antenna efficiency

Antenna efficiency relates to the fact that all antennas suffer from losses. A simple horn antenna, for example, will not be as efficient as a perfect aperture of the same size because of phase offset. The real efficiency of an antenna combines impedance match with other factors such as aperture and radiation efficiency to give the overall radiated signal for a given input. The best and most widely used expression of this efficiency is to combine overall efficiency with directivity (of the antenna) and express the efficiency times directivity as gain.

4.2.6. Sidelobes and back lobes

Sidelobe level: The energy radiated at lobes other that the main lobe on boresight is termed the sidelobe or sidelobes. The maximum sidelobe level is often dictated by the regulatory bodies for transmit antennas that could interfere with other systems if the sidelobe levels were excessive.

Front-to-back ratio: Often listed in dB, this specification is the difference between the peak gain of the antenna and the radiation at the back of the antenna (often 180° from the peak of the beam).

For GPR, a low level of sidelobes and back lobes is important to reduce the cross coupled energy, which could saturate the receiver and reduce reflections from structures behind the antenna.

4.2.7. Bandwidth

An antenna will radiate energy over a defined bandwidth. This is sometimes defined as that bandwidth where the input voltage–standing wave ratio (VSWR) is 2:1. Most antennas possess a fractional bandwidth, which is defined by the following expression:

$$B = 2\frac{(F_h - F_l)}{(F_h + F_l)} \qquad (4.7)$$

where

F_h is the highest frequency limit ($-10\,\text{dB}$ below maximum of the power spectral density envelope)

F_l is the lowest frequency limit ($-10\,\text{dB}$ below maximum of the power spectral density envelope)

An ultra-wide-band antenna has a bandwidth equal to or greater than 25% of the fractional bandwidth about the carrier frequency. Most GPR systems need to use antennas that have a minimum octave bandwidth and often up to decade bandwidth.

4.2.8. Polarisation – linear, elliptical, circular

The polarisation of an antenna is the orientation of the transmitted (or received) electric field (E field). The optimum polarisation for a system depends on the polarisation of other antennas in the system. An infinite number of polarisations exist, but the most common are linear, elliptical and circular. For a linear antenna, three possibilities are generally seen: vertical, horizontal and slant linear. It is important to match linear polarisations for transmit and receive antennas. A linear polarisation mismatch can result in up to a 20 dB loss (for cross-linear polarisation). Circular polarisation is generally given as right-hand circular polarisation (RHCP) or left-hand circular polarisation (LHCP).

Where the target is, for example, a planar surface, then linear polarisation is the obvious choice for the system designer. Where, however, the target is a buried pipe or cable, then the backscattered field exhibits a polarisation characteristic, which is independent of the state of polarisation of the incident field. For linear targets, it is possible to use orthogonally disposed transmit and receive antennas as a means of preferential detection. Essentially the received signal varies sinusoidally with angle between antenna pair and the target. As it is inconvenient to physically rotate the antenna, it is also possible to electronically switch (commutate) the transmit/receive signals to a set of multiple colocated antenna pairs. A further step along this overall strategy is to employ circular polarisation, which is essentially a means of automatically rotating the polarisation vector in space. However, circular polarisation inherently requires an extended time response of the radiated field and in consequence either hardware or software deconvolution of the received signal is needed.

It has been found that very large diameter pipes exhibit depolarising effects, not from the crown of the pipe but from the edges. The choice of polarisation-dependent schemes should thus be considered very carefully as it may not be possible to cover all possible sizes of targets with one antenna/polarisation scheme.

4.2.9. Antenna phase centre

The phase centre of an antenna is considered to be that point where the source of an expanding spherical shell of radiation is situated. The concept of a phase centre is somewhat artificial as all radiations have to originate from an oscillating current source, which occupies some physical space rather than a point. As noted earlier, the isotropic radiator is a useful fiction but a real source of uniform radiation in all directions is difficult to achieve.

The importance of the concept of antenna phase centre relates to the characteristics of the radiated field. For example, with an antenna such as a log periodic or spiral, the physical position of the source of radiation at a particular frequency will vary along either the length of the antenna or the position across the antenna. With horn antennas, the phase centre will depend on the aperture distribution and the taper of the horn and the resultant far field pattern will be affected by the variation of phase centre.

4.2.10. Antenna patterns

Antenna patterns show the gain of the antenna versus angle in a graphical manner. Generally two patterns, called principal plane cuts, that show the azimuth and elevation plot of antenna gain are provided. These patterns will show not only the gain and beamwidth for the main beam of the antenna but also sidelobes and the back lobe of the antenna. Such plots can be in polar form or linear form and an example of each is given in Figure 4.4 or in Cartesian form in Figure 4.5.

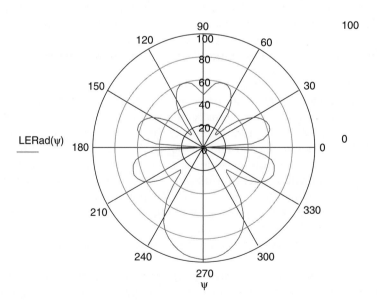

Figure 4.4 Typical far-field radiation pattern of an antenna showing main lobe, sidelobes and back lobes. Signal level in dB.

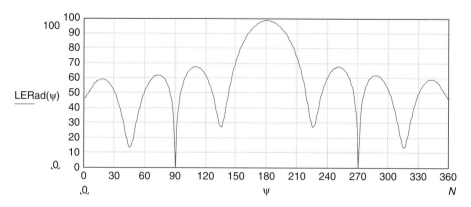

Figure 4.5 Typical linear plot of antenna radiation pattern (dB) versus degrees (shown shifted by 180° for convenience).

4.2.11. Time sidelobes and ring-down

Ground penetrating radar antennas often have to work at very short ranges of some tens of nanoseconds and therefore the rate of decay of energy stored within the antenna is a key parameter in defining the inherent self clutter of the complete radar system. These time sidelobes would obscure targets that are close in range to the target of interest; in other words, the resolution of the radar can become degraded if the impulse response of the antenna is significantly extended. Two examples of the effect of time sidelobes are shown. The linear amplitude of the time domain waveform is shown in Figures 4.6 and 4.7.

As can be seen, although the linear plot suggests that the waveform is marginally acceptable, the graph in dB shows that the sidelobes would limit this particular radar's sensitivity in that targets whose amplitude was less than the time sidelobe level would not be able to be detected unless some form of deconvolution was employed.

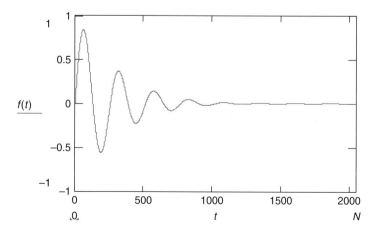

Figure 4.6 Linear amplitude of radiated impulse with time sidelobes.

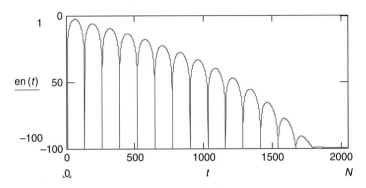

Figure 4.7 Amplitude (dB) of radiated impulse in Figure 4.6.

In contrast, the Ricker wavelet, which is the ideal function for a GPR to radiate, has a much better rate of decay and thus would enable a GPR to detect targets at much lower levels. Control of the range sidelobes in any type of GPR is a fundamental issue for the GPR system performance as shown in Figures 4.8 and 4.9.

4.2.12. Antenna footprint

The azimuth and elevation resolution of the GPR is defined by the characteristics of the antenna and the signal processing employed. In general, radar systems (apart from synthetic aperture radar (SAR)) require a high antenna gain to achieve an acceptable plan resolution. This necessitates a sufficiently large aperture at the lowest frequency to be transmitted. Therefore, to achieve small antenna dimensions and high gain, the use of a high carrier frequency is required, which may not penetrate the material to sufficient depth. When selecting equipment for a particular application, it is necessary to compromise between plan resolution, the size of antenna, the scope for signal processing and the ability to penetrate the material. Plan resolution improves as attenuation increases, provided that there is sufficient signal to discriminate under the prevailing clutter conditions. In low-attenuation media, the resolution obtained by the horizontal scanning technique is degraded, but only under these conditions do synthetic aperture techniques increase the plan resolution. Essentially the ground attenuation has the effect of placing a "window" across the SAR aperture, and the higher the attenuation, the more severe the window. Hence in high-attenuation soils, SAR techniques may not provide any useful improvement to GPR systems. SAR techniques have been applied to GPR but very often in dry soils with low attenuation.

The plan resolution of a subsurface radar system is important when localised targets are sought and when there is a need to distinguish between more than one at the same depth. Where the requirement is for location accuracy, which is primarily a topographic surveying function, the system requirement is less demanding.

The effect of the radiation footprint on the ground can be seen from Figure 4.10, which shows an energy projection on the ground surface, where the distance

Antennas

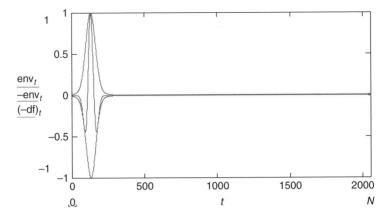

Figure 4.8 Ricker wavelet in the time domain showing envelope of the function.

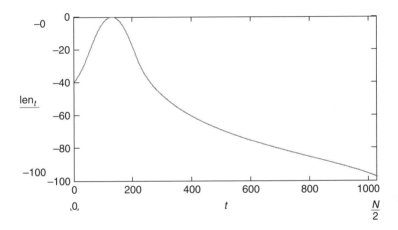

Figure 4.9 Amplitude (dB) of the envelope of the Ricker wavelet.

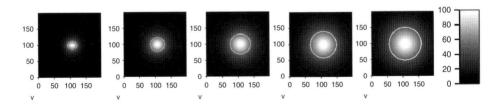

Figure 4.10 Radiation footprint on the ground from an isotropic source.

between the radiating source and the ground surface has been increased from 0.1 to 0.5 m (left to right). The ground area is 2 m × 2 m and it can be seen that the width of the 3-dB footprint increases considerably as the source is raised from the ground. The effect of this on the image resolution is also considerable as the

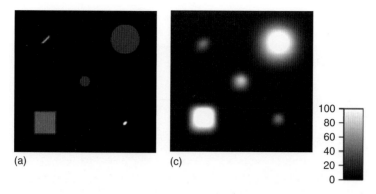

Figure 4.11 Effect of convolution of antenna footprint on ground penetrating radar (GPR) image.

convolution of the antenna pattern with the target causes a blurring of the target image as shown in Figure 4.11. The simulated targets represent buried targets of various shapes and sizes and are used to gain an appreciation of the effect of increasing antenna to ground spacing. The original targets are shown in A and the effect of antenna to ground spacing is shown in C.

The plan resolution is defined by the characteristics of the antenna and the signal processing employed. In general, to achieve an acceptable plan resolution, a high antenna gain is required. This necessitates an antenna with a significant aperture at the lowest frequency transmitted. Therefore, to achieve small antenna dimensions and high gain, the use of a high carrier frequency is required, which may not penetrate the material to sufficient depth. When choosing equipment for a particular application, it is necessary to compromise between plan resolution, the size of antenna, the scope for signal processing and the ability to penetrate the material.

In low-attenuation media, the use of advanced signal processing techniques is possible by means of measurements made using transmitter and receiver pairs at a number of antenna positions to generate a synthetic aperture or focus the image. Unlike conventional radars, which generally use a single antenna, most GPR systems use separate transmit and receive antennas in what has been termed a bistatic mode. However, as the antenna configuration is normally mobile, the term bistatic is not really relevant.

4.3. ANTENNAS FOR GROUND PENETRATING RADAR

4.3.1. Introduction

Where the radar system is a time domain system that applies an impulse to the antenna, there is a requirement for linear phase response and this means that only a limited number of types of antenna can be used unless the receiver uses a matched filter to deconvolve the effect of the frequency-dependent radiation characteristics of the antenna. Where the radar system is frequency-modulated or synthesised, the requirement for linear phase response from the antenna can be relaxed and log periodic, horn or spiral antennas can be used as their complex frequency response can be corrected if necessary by system calibration. The overall configuration is further complicated by the use of separate transmit and receive antennas, which causes a convolution of the separate radiation patterns to form a composite pattern.

The use of separate transmit and receive antennas is dictated by the difficulty associated with operation with a single antenna, which would require an ultra-fast transmit receive switch. As it is not yet possible to obtain commercially available, ultra-fast transmit–receive switches to operate in the subnanosecond region with sufficiently low levels of isolation between either transmit and receive ports or breakthrough from the control signals, most surface-penetrating radar systems use separate antennas for transmission and reception in order to protect the receiver from high level of transmitted signal.

Therefore the cross-coupling level between the transmit and the receive antenna is a critical parameter in the design of antennas for surface-penetrating radar, and satisfactory levels are usually achieved by empirical design methods. Typically, a parallel dipole arrangement achieves a mean isolation in the region of -50 dB, whereas a crossed dipole arrangement can reduce levels of cross-coupling to -60 dB to -70 dB. For the crossed dipole arrangement, such levels are highly dependent on the standard of mechanical construction and a high degree of orthogonality is necessary. The crossed dipole is sensitive to variations in antenna to surface spacing, and it is important to maintain the plane of the antenna parallel with the plane of material surface to avoid degrading the isolation.

4.3.2. Coupling into a dielectric

It is important to appreciate the effect of the material in close proximity to the antenna. In general, this material, which in most cases will be soil or rocks or indeed ice, can be regarded as a lossy dielectric and by its consequent loading effect can play a significant role in determining the low-frequency performance of the antenna and hence surface-penetrating radar. The behaviour of the antenna is intimately linked with the material, and in the case of borehole radars, the antenna actually radiates within a lossy dielectric, whereas in the case of the surface-penetrating radar working above the surface, the antenna will radiate from air into a very small section of air and then into a lossy half space formed by the material. The behaviour of antennas both within lossy dielectrics and over lossy dielectrics has been investigated by Junkin and Anderson (1988), Brewitt-Taylor et al. (1981), Burke et al. (1983) and Rutledge and Muha (1982) and is well reported. The propagation of electromagnetic pulses in a homogeneous conducting earth has been modelled by Wait (1960) and King and Nu (1993), and the dispersion of rectangular source pulses suggests that the time domain characteristics of the received pulse could be used as an indication of distance.

The interaction between the antenna and the lossy dielectric half space is also significant as this may cause modification of the antenna radiation characteristics both spatially and temporally and should also be taken into account in the system design.

Table 4.1 Power density patterns in air and dielectric

Plane	Power $S(\theta_a)$ Radiation pattern in air	Power $S(\theta_a)$ Radiation pattern in dielectric
H	$\alpha \left(\dfrac{\cos \theta_a}{\cos \theta_a + \eta \cos \theta_d} \right)^2$	$\alpha \eta \left(\dfrac{\eta \cos \theta_d}{\cos \theta_a + \eta \cos \theta_d} \right)^2$
E	$\alpha \left(\dfrac{\cos \theta_a \cos \theta_d}{\eta \cos \theta_a + \cos \theta_d} \right)^2$	$\alpha \eta \left(\dfrac{\eta \cos \theta_a \cos \theta_d}{\eta \cos \theta_a + \cos \theta_d} \right)^2$

In the case of an antenna placed on an interface, the two most important factors are the current distribution and the radiation pattern. At the interface, currents in the antenna propagate at a velocity, which is intermediate between that in free space and that in the dielectric. In general, the velocity is retarded by the factor $\sqrt{e_r}$.

The net result is that evanescent waves are excited in air, whereas in the dielectric, the energy is concentrated and preferentially induced by a factor of n^3:1. The respective calculated far-field power density patterns, in both air and dielectric, are given by Rutledge (1982) (as shown in Table 4.1) and these are plotted for relative dielectric constants of 2, 4, 6 and 8 in Figures 4.12 and 4.13. The lower plots are smallest for the relative dielectric constants of 2 and increase incrementally to 8 being the largest.

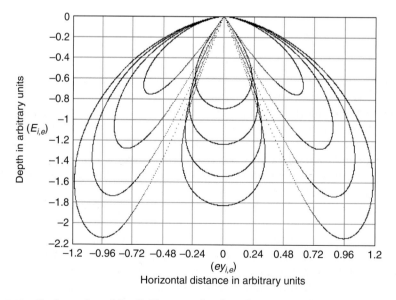

Figure 4.12 E-plane plot of far-field power density of a current element radiating into a dielectric.

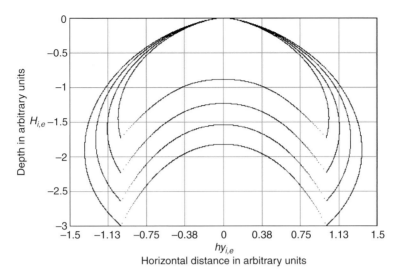

Figure 4.13 H-plane plot of far-field power density of a current element radiating into a dielectric.

The above expressions assume that the current source contacts the dielectric, whereas a more general condition is when the antenna is just above the dielectric. The sidelobes in the pattern are a direct result of reactive field coupling. A significant practical problem for many applications is the need to maintain sufficient spacing to avoid mechanical damage to the antenna. It can therefore be appreciated that the effect of changes in distance between the antenna and the half space causes significant variation in the resultant radiation patterns in the dielectric. Where the source interface space is increased, the antenna field patterns are modified by a reduction in the effect of the reactive field.

4.3.3. Time domain antennas

Examples of non-dispersive antennas are the TEM horn, the bicone, the bow-tie, the resistive, lumped element-loaded antenna or the resistive, continuously loaded antenna. A typical antenna used in an impulse radar system would be required to operate over a frequency range of a minimum of an octave and ideally at least a decade, for example, 100 MHz–1 GHz. The input voltage driving function to the terminals of the antenna in an impulse radar is typically a Gaussian pulse, and this requires the impulse response of the antenna to be extremely short. The main reason for requiring the impulse response to be short is that it is important that the antenna does not distort the input function and generate time sidelobes. These time sidelobes would obscure targets that are close in range to the target of interest; in other words, the resolution of the radar can become degraded if the impulse response of the antenna is significantly extended. All of the antennas used to date have a limited low-frequency performance unless compensated and hence act as high-pass filters; thus the current input to the antenna terminals is radiated as a differentiated version of the input function.

4.3.3.1. Dipole

Element antennas such as monopoles, dipoles, conical antennas and bow-tie antennas have been widely used for surface-penetrating radar applications. Generally they are characterised by linear polarisation, low directivity and relatively limited bandwidth, unless either end loading or distributed loading techniques are employed in which case bandwidth is increased at the expense of radiation efficiency. Various arrangements of the element antenna such as the parallel dipole and the crossed dipole, which is an arrangement that provides high isolation and detection of the cross-polar signal from linear reflectors, have been used.

It is useful to consider those characteristics of a simple, normally conductive dipole antenna that affect the radiation response to an impulse applied to the antenna feed terminals.

As shown in Figure 4.14, two current and charge impulses will travel along the antenna elements until they reach each end. At the end of the antenna, the charge impulse increases while the current collapses. The charge at the end of

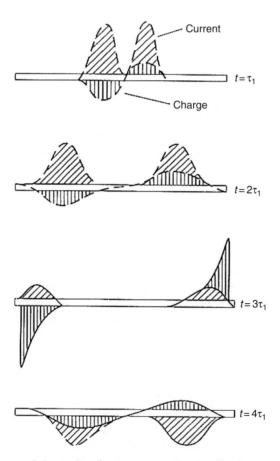

Figure 4.14 Current and charge distribution on a conducting dipole antenna due to an applied impulse.

the antenna gives rise to a reflected wave carried by a current travelling back to the antenna feed terminals. This process continues for a length of time defined by the ohmic losses within the antenna elements. As far as the radiation field is concerned, the relevant parameters, electric field strength, displacement current and energy flow, can be derived from consideration of Maxwell's equation.

The electric field component E_z is given by Kappen and Monich (1987) as

$$E_z = -\frac{1}{4\pi\varepsilon_0} \int_{l_1}^{l_2} \left\{ \frac{1}{c}\frac{dl}{dt} + \frac{\partial q}{\partial z'} \right\} \frac{1}{r} dz' \tag{4.8}$$

and must equal zero at the surface of the antenna. This condition can be satisfied only at certain points along the element and implies that for a lossless antenna, there is no radiation of energy from the impulse along the element. The radiated field is therefore caused by discontinuities, that is, the feed point, and end points are the prime sources of radiation. As would be expected, the time sequence of the radiated field can be visualised by the electric field lines as shown in Figure 4.15.

For this reason, the loaded dipole has become much used for time domain GPR systems.

4.3.3.2. Loaded antennas

As it is required to radiate only a single impulse, it is important to either eliminate the reflection discontinuities from the far end of the antenna by end loading or reduce the amplitude of the charge and current reaching the far end. The latter can be achieved either by resistively coating the antenna or by constructing the antenna from a material such as Nichrome, which has a defined loss per unit area. In this

Figure 4.15 Radiated field pattern from a conducting dipole element due to an applied impulse.

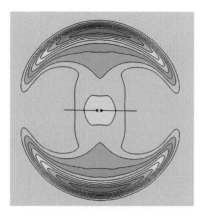

Figure 4.16 Radiated field pattern from a resistively loaded dipole element due to an applied impulse.

case, the antenna radiates in a completely different way as the applied charge becomes spread over the entire element length and hence the centres of radiation are distributed along the length of the antenna.

In essence, the electric field E_z must now satisfy the following condition:

$$-\frac{1}{4\pi\varepsilon_0}\int_{l_1}^{l_2}\left\{\frac{1}{c}\frac{dl}{dt}\frac{\partial q}{\partial z'}\right\}\frac{1}{r}dz' = R'I \quad (4.9)$$

which implies that some dispersion takes place. The electric field lines for the lossy element are now different from the lossless case and are shown in Figure 4.16. Further analysis of the radiation characteristics of a resistively coated dipole antenna is given by Randa et al. (1991) and Esselle and Stuchly (1990).

The parameters of the antenna such as input resistance and resistivity profile have all been extensively treated in a classic paper by Wu and King (1965). Lumped element resistors can be placed at a distance $\lambda/4$ from the end of the antenna; (Altshuler, 1961), and a travelling wave distribution of current can be produced by suitable values of resistance. The distribution of current varied almost exponentially with distance along the element. Instead of lumped element resistors, a continuously distributed constant internal impedance per unit length can be used. The parameters of a centre-fed cylindrical antenna can be characterised by a distribution of current equivalent to a travelling wave.

The cylindrical antenna with resistive loading has been shown by Wu and King (1965) to have the following properties:

The far-field pattern of the antenna comprised both real and imaginary components, i.e.

$$F_m = \sqrt{(F_r^2 + F_i^2)} \quad (4.10)$$

where for a quarter wave antenna

$$F_r = \frac{1+\cos^2\theta - 2\cos\theta\left(\frac{\pi}{2}\cos\theta\right)}{\left(\frac{\pi}{2}\right)\sin^3\theta} \quad (4.11)$$

$$F_t = \frac{-\frac{\pi}{2}\sin^2\theta + (1+\cos^2\theta)\cos\theta\left(\frac{\pi}{2}\cos\theta\right)}{\left(\frac{\pi}{2}\right)\sin^3\theta} \quad (4.12)$$

The efficiency of the antenna is given as follows:

$$\eta = \frac{P_r}{P_r + P_a} \quad (4.13)$$

where

P_r is the radiated power
P_a is the absorbed power

For a resistively loaded antenna of the Wu–King type, the efficiency is approximately 10% but rises to a maximum of 40% for antenna lengths of 40λ.

The resistivity taper profile for a cylindrical monopole has the following form given by Rao (1991):

$$R(z) = \frac{R_0}{1 - z/H} \quad (4.14)$$

where

R_0 is the resistivity at the drive point of the element
H is the element length
Z is the distance along the antenna

A graph of resistivity versus length for a 200-mm element is shown in Figure 4.17. The overall efficiency of this type of antenna can be improved by reducing the value of R_0 and an increase from 12 to 28% by reduction of R_0 to $0.3R_0$ was shown by Rao.

Further improvement in bandwidth can be gained by matching the antenna with a compensation network and a field probe has been developed, with a bandwidth of 20 MHz–10 GHz, by Esselle and Stuchly. Obviously the use of a compensation network further reduces efficiency, but with a high-impedance receiver probe, a frequency range of 10 MHz–5 GHz can be achieved.

Resistively loaded dipoles have been used as electric field probes for electromagnetic compatibility (EMC) measurement applications, and although the frequencies of operation are well in excess of that used for surface-penetrating radar applications, it is useful to consider the general approach adopted by Maloney and Smith (1991).

Antennas have been developed by Kanda, initially using 8-mm dipoles, to measure frequencies up to 18 GHz, and subsequently 4-mm dipoles were used to measure over the frequency range 1 MHz–40 GHz with an error of ±4 dB. The transfer function of this antenna is in the order of −50 dB, which illustrates the penalty that is paid for ultra-wideband width operation.

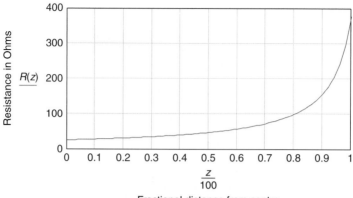

Figure 4.17 Resistivity profile of a 200-mm element.

A design that offers improved efficiency over the continuously loaded resistive antenna is based on a pair of segmented blade antennas arranged in a butterfly configuration and fed in phase. Each blade consists of a series of concentric conducting rings connected together by chip resistors. Radial cuts are used to reduce transverse currents. The efficiency of this class of antenna is higher than the continuously loaded dipole without serious degradation of the time domain response.

4.3.3.3. Biconical antennas

The biconical antenna is simple enough to evaluate and to generate an approximate solution to the wave equation and has been one of the most important in yielding useful results for the antenna impedance problem. Much of this is available in the literature but Schelkunoff gave particular attention. He concluded among other things that the fatter the antenna, the more broadband its properties. The bandwidth of the biconical antenna is a function of its length and angle and the design of the ends. Suitable parameters can provide a usable octave performance. A useful feature of the biconical antenna is that it can be developed into either a TEM horn or a bow-tie antenna as shown in Figures 4.18–4.20.

4.3.3.4. Bow-tie antennas

The triangular bow-tie antenna has been widely used in commercial surface-penetrating radar systems. A triangular bow-tie dipole of 35 cm length with a 60° flare angle can provide useful performance over an octave bandwidth of 0.5–1 GHz with a return loss of better than 10 dB as shown by Brown and Woodward (1952). Evidently, without some form of end loading, such an antenna would not be immediately suitable for use with impulse radar systems, and the triangular antenna normally uses end loading to reduce the ringing that would normally occur in an unloaded triangular-plate antenna. The technique can also be used with a folded dipole, and the use of terminating loads results in a transient response equivalent to one-and-a-half cycles (Young et al., 1977).

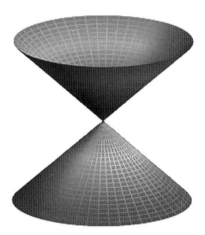

Figure 4.18 Basic biconical antenna.

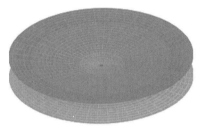

Figure 4.19 Biconical antenna with small angle – a sector of this can be used for TEM horn.

Figure 4.20 Biconical antenna realised as planar structure – leads to bow-tie antenna.

4.3.3.5. TEM horn antennas

The TEM horn antenna is important in GPR because of its time domain characteristics. In this section, we shall consider the use of antennas capable of supporting a forward-travelling TEM wave. In general, such antennas consist of a pair of conductors, either flat, cylindrical or conical in cross section forming a V structure in which radiation propagates along the axis of the V structure as shown in Figure 4.21 and is termed the tapered impedance travelling wave antenna (TWIT) and is a variant of the basic TEM structure. Although resistive termination is used, this type of antenna has directivity in the order of 10–15 dB; hence useful gain can still be obtained even with a terminating loss in the order of 3–5 dB. The travelling wave current in one of the cylindrical elements of a V antenna is given by Ilzuka (1967) and is also discussed by King.

$$I_t = I_a \, e^{-j\beta z} \tag{4.15}$$

Hence the azimuthal radiation field E is given by the following equation:

$$E = \frac{j\omega\mu \, e^{-j\beta r}}{4\pi} \int_0^l I_0 \, e^{-j\beta z(1-\cos\theta)} \sin\theta \, dz \tag{4.16}$$

where

R is the loading resistance
l is the length of the element
z is the distance from the radiating source
θ is the angle in H plane

This simplifies to

$$E_T(\theta) = \frac{1 - \exp(-j\beta(l_0 - l_1)(1 - \cos\theta))}{(1 - \cos\theta)} \sin\theta \tag{4.17}$$

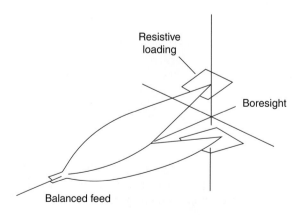

Figure 4.21 Basic tapered impedance travelling wave antenna (TWIT) horn antenna for time domain signals.

However, a standing wave caused by the resistive termination at the end of the antenna also exists and the contribution from this is given by the following equation:

$$E_S(\theta) = j\frac{\exp(-j\beta(l_0 - l_1)(\cos\theta))}{\sin\theta}\left[\cos\left(\frac{\pi}{2}\cos\theta\right) + j\left(\sin\left(\frac{\pi}{2}\cos\theta\right) - \cos\theta\right)\right] \quad (4.18)$$

The resultant field from one element can be derived from the sum of the contribution

$$E'(\theta) = E_T(\theta) + E_S(\theta) \quad (4.19)$$

and hence the field from both elements is

$$E = E'_U(\theta) + E_L(\theta) \quad (4.20)$$

where

U denotes the upper element
L denotes the lower element

The antenna will in fact radiate an impulse, which is extended in time as a consequence of the geometry of the antenna. The pulse distortion on boresight is given by Theodorou et al. (1981).

$$t = \frac{L}{\nu}(1 - \cos\alpha) \quad (4.21)$$

where

α is the half angle between the elements
L is the element length
ν is the phase velocity of waves along the antenna

Evidently a small flare angle and a short element length help in reducing pulse extension.

The electric field on boresight is related to the time derivative of input current and is given by

$$E = \frac{-\mu_0 L \sin\alpha}{2\pi r}\frac{\partial I_1(t - r/u)}{\partial t} \quad (4.22)$$

The impedance of the antenna should vary in such a way that the derivative of impedance at the feed and end parts is a minimum and along the antenna is low.

Typically the characteristic impedance is given as a function of distance x as

$$Z_0(x) = Z_x \exp(-K_1 \cos K_2 x) \quad (4.23)$$

Hence

$$\frac{dZ_0(x)}{dx} \to 0 \text{ for } K_2 x = 0 \vee \pi \tag{4.24}$$

Usually the feed impedance is in the order of 50 Ω and the end impedance is desired to be equal to that of free space (377 Ω). However, there is usually a difference between the transmission line wave impedance characteristic and that of wave in free space, and a design to meet the given criteria in terms of return loss must take this effect into account.

Graphs of both typical antenna impedance and rate of change of impedance as a function of length are shown in Figures 4.22 and 4.23. Using this characteristic, a typical antenna–antenna time domain response is shown in Figure 4.24.

Improved directivity can be obtained using a V-conical antenna as shown by Shen et al. (1988). A pair of triangular metal plates bent around a cone forms this. The antenna is characterised by two angles, the flare half angle and the azimuthal angle Ø. Further developments of the TEM horn design from the original design first described by Wohlers (1970) are found in papers by Daniels (1977), Evans and Kong (1993), Reader et al. (1985) and Foster and Tun (1993).

4.3.4. Frequency domain antennas

This class of antennas has a geometry entirely defined by angles and exhibits a performance over a range of frequencies set by the overall dimensions of the structure. Typical examples are the biconical dipole, equiangular spiral and conical

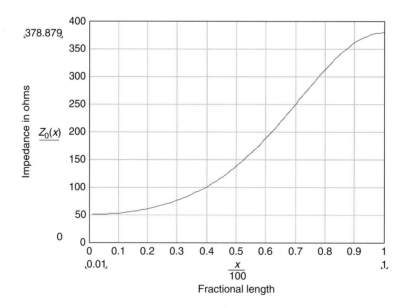

Figure 4.22 Characteristic impedance of a travelling wave antenna.

Antennas

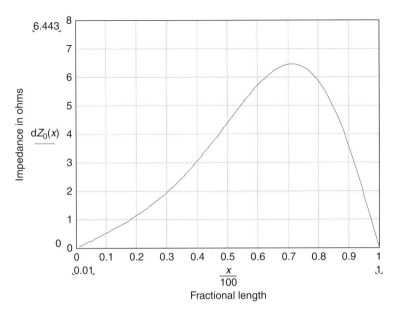

Figure 4.23 Rate of change of impedance of a travelling wave antenna.

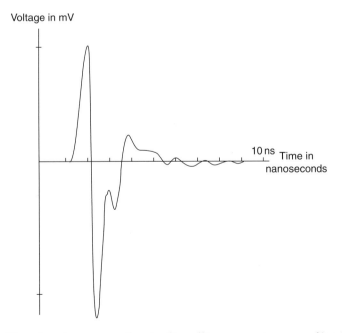

Figure 4.24 Time domain response of a pair of travelling wave antennas used in a face-to-face configuration.

spiral. Log periodic structures can also provide broadband performance but are not completely defined in terms of angles (Rumsey, 1966).

Various developments of the spiral antenna or conical spiral antenna have been carried out by Miller and Landt (1977), Pastol et al. (1990), Dyson (1959), Morgan (1985), Kooy (1984), Deschamps (1959), Bower and Wolfe (1960) and Goldstone (1983).

Examples of dispersive antennas that have been used in surface-penetrating radar are the exponential spiral, the Archimedean spiral, the logarithmic planar antenna, the Vivaldi antenna and the exponential horn.

4.3.4.1. Vivaldi

In theory, the bandwidth of Vivaldi antenna is infinite. In practice, the main bandwidth limitations are the aperture size, which defines the low-frequency limit, and the slotline-to-microstrip transition, which defines the high-frequency limit. The Vivaldi antenna (Gibson, 1979) also falls into the class of a periodic, continuously scaled antenna structure and within the limiting size of the structure has unlimited instantaneous frequency bandwidth. It provides end-fire radiation and linear polarisation and can be designed to provide a constant gain–frequency performance.

The Vivaldi antenna consists of a diverging, slot-form guiding conductor pair as shown in Figure 4.25. The curve of one of the guiding structures follows the following equation:

$$z = A\, e^{kx} \qquad (4.25)$$

Radiation is produced by a non-resonant travelling wave mechanism by waves travelling down a curved path along the antenna. Where the conductor separation is small, the travelling wave energy is closely coupled to the conductor but becomes less so as the conductor separation increases. The Vivaldi antenna provides gain when the phase velocity of the travelling wave on the conductors is equal or greater than that in the surrounding medium. Typical radiation patterns for an elemental Vivaldi are shown in Figure 4.26.

The lower cutoff frequency is defined by the dimensions of the conductor separation, being a half wavelength, and the gain is proportional to the overall length. The impulse response of the antenna is extended due to the non-stationary phase centre but can of course be corrected by the use of a matched filter. Note that the sidelobe and back lobe radiation is significant unless suitable absorbers and screening are used.

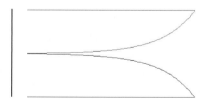

Figure 4.25 Vivaldi antenna.

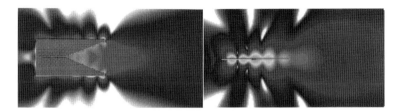

Figure 4.26 Calculated radiation patterns of elemental Vivaldi antenna (*Courtesy: ERA technology*).

Figure 4.27 Equiangular planar spiral antenna.

4.3.4.2. Equiangular antennas

The impulse response of this class of antennas is extended and generally results in a 'chirp' waveform if the input is an impulse. The main reason for this is that the high frequencies are radiated in time before the low frequencies as a result of the time taken for the currents to travel through the antenna structure and reach a zone in which radiation can take place as shown in Figure 4.27.

The geometry of the equiangular spiral is defined by the following equation:

$$\rho = \kappa \, e^{a\phi} \tag{4.26}$$

4.3.4.3. Horn antennas

Horn antennas have found most use with frequency modulated continuous wave (FMCW) surface-penetrating radars where the generally higher frequency of operation and relaxation of the requirement for linear phase response permits the consideration of this class of antenna. Exponentially flared TEM horns with dielectric loading have been developed to operate over decade bandwidths (Kerr, 1973). The design of horn antennas is well covered in the literature but of particular interest is the short axial-length, double-ridged horn (Daniels, 1991) as shown in Figure 4.28.

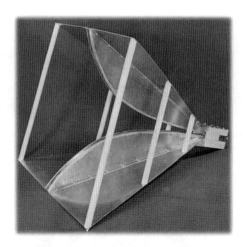

Figure 4.28 Short axial–length, double-ridged horn antenna.

This design can provide useful gain over a decade bandwidth using a logarithmic characteristic curve for the ridges. Typically the short axial-length horn provides a VSWR of better than 2:1 and a gain of 10 dB over a frequency band of 0.2–2 GHz for an axial length of 0.76 m. The concept of the ridged horn design can be adapted to form a quad-ridged horn operating from 0.3 to 1.9 GHz. A return loss of better than 10 dB and a cross-coupling level of better than 35 dB can be obtained. The quad-ridged horn can be used to extract information on the polarisation state of the reflected signal.

An FMCW radar has been developed using an offset paraboloid fed by a ridged horn (Sun and Rusch, 1991). The arrangement was designed to focus the radiation into the ground at a slant angle to reduce the level of reflection from the ground. Care needs to be taken in such arrangements to minimise the effect of back and sidelobes from the feed antenna, which can easily generate reflection from the ground surface. Although horn antennas have been mostly used with FMCW systems, it is possible to radiate pulses.

4.3.5. Array antennas

Arrays of antennas are an obvious method of increasing the rate of survey of areas of the ground and have been designed and built for road survey and mine detection. Some examples of antenna configurations are given below. There are two approaches to array design. The first is simply to take a number of single-channel radars, slave them with a master logger and arrange for the data to be appropriately logged. The other approach is to design the system as an integral array design and exploit the increased capability offered by combining multiple looks and SAR processing. Consideration as to whether the antenna/system should be downward look or forward look is important. A number of antenna array designs are discussed in Chapter 14. With all array systems, it is important that the surface clutter is properly removed. Close coupling of the antenna to the ground surface is one method. An alternative relies on coherent subtraction but this often means that the ground topography must be relatively smooth. Where removal of the surface clutter is not easy then antennas operating off-normal incidence can be used. However, this in turn brings other problems, and antenna near-field effects must be accounted for and the effect of grazing angle can be a limitation as it limits the potential for full 3D imaging because of the refractive index of the ground compressing the beam within the soil.

All array systems are geared to generate an image of the buried targets, and for that, accurate positioning of the array elements is crucial. This can be achieved with differential GPS (DGPS) systems coupled with inertial navigation systems (INS).

An example of the first approach to array design is that taken by the CART Imaging System[*] ('CART' stands for 'computer-assisted radar tomography') from Witten Technologies, Inc. See www.wittentech.com/products_CART.html

This uses a fixed array of nine transmitters and eight receivers. Each radar element in the array is a standard ultra-wideband GPR from Mala Geoscience, which broadcasts an impulse with a frequency spectrum from about 50 to 400 MHz. The array is controlled by special electronics that fires the transmitter elements and controls the receivers in sequence to create 16 standard bistatic GPR channels covering a 2 m swath on the ground. In this standard 'bistatic' mode of operation, each transmitter fires twice in sequence, with each firing being recorded by an adjacent receiver. A multi-static mode, in which each transmitter fires once in sequence and is recorded by all the receivers, is also possible. The array can be towed by a vehicle or pushed in front of a modified commercial lawnmower at speeds up to about 1 km/h (30 cm/s).

Alternative approaches have been adopted by companies in the United States (Planning Systems, GeoCenters, BASystems (ex GDE), Mirage, ARL, Jaycor, SRI, Coleman), United Kingdom (ERA Technology, Thales and PipeHawk), France (Thales, Satimo), Germany (Rheinmetall) and Israel (Elta), who have developed array systems as an integral design rather than combining existing single channel radars. Much of the interest in high-speed array radar systems is for mine detection. Work is being carried out on various national (US, UK, Canada, Germany, France, Netherlands etc.) as well as international CEU programmes particularly for mine detection. Arrays are typically between 1 and 4-m in width and can operate at speeds up to 10 km/h.

The key issues for the design of multi-element GPR systems lie in the channel-to-channel performance and tracking over the desired operational environmental range. A 32-channel GPR system, for example, must maintain calibration of both start time and time linearity for all channels within demanding limits. In addition, the relative gain and, if used, time-varying gain profile must also match to within-close tolerances. Where the antenna array is spaced off the ground, there may also be the need to compensate for variations in surface topography. A further aspect to be considered is the antenna element spacing. This needs to be adequate to provide proper resolution of the wanted target, and it can be shown that the probability of detection with respect to small targets is closely related to the density of the elements of the antenna array. An example section showing 16 elements of a 4-m-wide swathe radar system with a total of 32 antenna elements is shown in

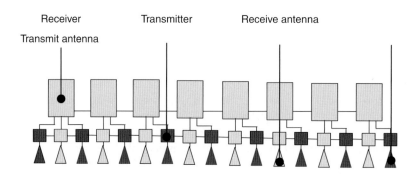

Figure 4.29 Block diagram of 16-channel radar system (*Courtesy: ERA technology*).

Figure 4.29. This was developed as part of the UK Minder CAP programme on behalf of the UK MoD as shown in Figure 4.30.

The architecture of the system is based around 16 receivers (8 only shown) each of which sequentially sample the signal incident on receive antenna elements. The transmitters are synchronised by adjacent receivers and a central master clock. The system is designed to be modular in that it can be increased either in width or alternatively in density and up to 64 channels can be configured. The transmitter–receiver modules can be used either singly or up to $N/2$, where N is the number of antenna elements.

However, the ability to carry out beam forming by means of inverse synthetic aperture processing could be potentially valuable in many applications. Rutledge and

Figure 4.30 Photograph of downward-look, 32-channel array system used on MINDER system (*Courtesy: ERA technology*).

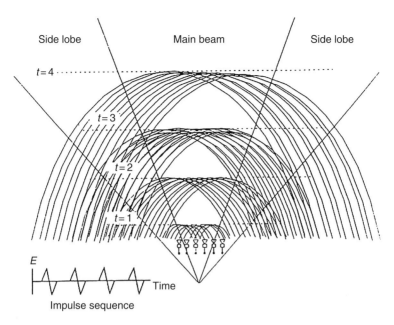

Figure 4.31 Timed-array antenna array.

Muha consider the general situation of imaging antenna arrays while Anderson et al. consider the specific problem of wideband beam patterns from sparse arrays.

If an array of emitters is driven by a sequence of impulses without any differential time delay, the radiated time sequence is as shown in Figure 4.31.

Note the gradual disappearance of the sidelobes as radiated wavefront propagates away from the array. If the sequences of impulses are controlled in time by means of a differential time delay between each element, the beam position can be steered as shown in Figure 4.32.

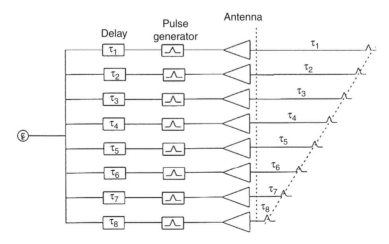

Figure 4.32 Beam steering by differential time delay.

The possibility of beam steering by means of time control exists although the inter-element time delay is limited to a maximum equivalent to the distance between each element. An alternative means of beam forming is by means of an array as shown in Figure 4.33. Here the objective was to create directivity in the azimuth plane and was achieved by means of a beam-forming network using wideband hybrid elements.

The use of timed transmitter arrays and inverse synthetic aperture processing of the signals from a receiver array offers the possibility of achieving considerably improved directivity over all the previous types of antennas discussed in this chapter. A 10 × 10 element array will have peak sidelobe amplitude of −26 dB of the main lobe (Anderson et al.) and beam steering of up to 50° is feasible.

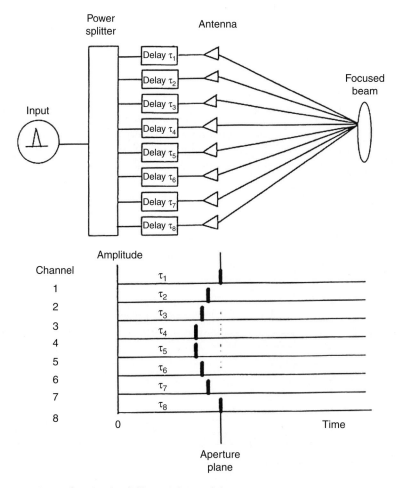

Figure 4.33 Beam forming by differential time delay.

4.4. Summary

The antennas used in surface-penetrating radar systems are, for reasons of portability, usually electrically small and consequently exhibit low gain. This has a profound effect on the performance of the overall system and is probably the only example of a radar system where antenna gain is, in general, so low. However, the bandwidth of the antennas is very much greater than that normally used in conventional radar systems, and surface-penetrating radars generally demonstrate very high-range resolution.

The choice of antenna is generally straightforward. The resistively loaded dipole, bow-tie and TEM travelling wave antenna have been primarily used for the impulse-based radar. Where matched filtering can be incorporated in impulse radars, then either horn or frequency-independent antennas can also be considered.

All the classes of antenna discussed can be used in synthesised, FMCW or noise-modulated radars. Attention must also be given to the means by which the antenna is fed from the transmitter. Generally an antenna is a balanced structure but where cables are used to connect the antenna to the transmitter or the receiver, some means is needed for transforming from the unbalanced configuration of the feed cable to the balanced structure of antenna. On this frequency range, baluns are generally commercially available or alternatively purpose-designed units can be constructed.

It may be found that multiple reflections between the transmitter and the antenna can be troublesome and these can be avoided by making the feed cable long enough to place the reflection outside the time window of interest. This, however, may be undesirable as the cable will act as a low-pass filter unless compensated. An alternative is to mount the transmitter and the receiver immediately adjacent to the antenna and this, if correctly designed, can remove the need for either a balun or a feed cable.

The anticipated main developments in the field of antennas appear to be related to array antennas. The current interest in the development of free-space, ultra-wideband radar systems may result in the transfer of useful developments.

4.5. Definitions

The following definitions are taken from *IEEE Standard Definitions of Terms for Antennas, IEEE STD 145-1983* and are relevant to some of the materials in this chapter.

Adaptive (smart) antenna: An antenna system having circuit elements associated with its radiating elements such that one or more of the antenna properties are controlled by the received signal.

Antenna polarisation: In a specified direction from an antenna and at a point in its far field, it is the polarisation of the (locally) plane wave, which is used to represent the radiated wave at that point.

Antenna: That part of a transmitting or a receiving system, which is designed to radiate or to receive electromagnetic waves.

Coaxial antenna: An antenna composed of an extension to the inner conductor of a coaxial line and a radiating sleeve, which in effect is formed by folding back the outer conductor of the coaxial line.

Collinear array antenna: A linear array of radiating elements, usually dipoles, with their axes lying in a straight line.

Copolarisation: That polarisation which the antenna is intended to radiate.

Cross-polarisation: In a specified plane containing the reference polarisation ellipse, the polarisation orthogonal to a specified reference polarisation.

Directional antenna: An antenna having the property of radiating or receiving electromagnetic waves more effectively in some directions than others.

Effective radiated power (ERP): In a given direction, the relative gain of a transmitting antenna with respect to the maximum directivity of a half-wave dipole multiplied by the net power accepted by the antenna from the connected transmitter.

E-plane: For a linearly polarised antenna, the plane containing the electric field vector and the direction of maximum radiation.

Far-field region: That region of the field of an antenna where the angular field distribution is essentially independent of the distance from a specified point in the antenna region.

Frequency bandwidth: The range of frequencies within which the performance of the antenna, with respect to some characteristics, conforms to a specified standard.

Front-to-back ratio: The ratio of the maximum directivity of an antenna to its directivity in a specified rearward direction.

Half-power beamwidth: In a radiation pattern cut containing the direction of the maximum of a lobe, the angle between the two directions in which the radiation intensity is one-half the maximum value.

Half-wave dipole: A wire antenna consisting of two straight collinear conductors of equal length, separated by a small feeding gap, with each conductor approximately a quarter wave length long.

H-plane: For a linearly polarised antenna, the plane containing the magnetic field vector and the direction of maximum radiation.

Input impedance: The impedance presented by an antenna at its terminals.

Isolation: A measure of power transfer from one antenna to another.

Isotropic radiator: A hypothetical, lossless antenna having equal radiation intensity in all directions.

Log-periodic antenna: Any one of a class of antennas having a structural geometry such that its impedance and radiation characteristics repeat periodically as the logarithm of frequency.

Major/main lobe: The radiation lobe containing the direction of maximum radiation.

Microstrip antenna: An antenna that consists of a thin metallic conductor bonded to a thin-grounded dielectric substrate.

Omni-directional antenna: An antenna having an essentially non-directional pattern in a given plane of the antenna and a directional pattern in any orthogonal plane.

Radiation efficiency: The ratio of the total power radiated by an antenna to the net power accepted by the antenna from the connected transmitter.

Sidelobe suppression: Any process, action or adjustment to reduce the level of the sidelobes or to reduce the degradation of the intended antenna system performance resulting from the presence of sidelobes.

REFERENCES

Altshuler, E.E., 1961, The travelling-wave linear antenna. IRE Trans., Vol. Ap-9, No. 4, 324–329.

Bower, R. and Wolfe, J.J., 1960, The apiral antenna. IRE National Convention Record, Part J, pp. 84–95.

Brewitt-Taylor, C.R., Gunton, D.J. and Rees, H.D., 1981, Planar antennas on a dielectric surface. Electron. Lett., Vol. 17, No. 20, pp. 729–731.

Brown, G.H. and Woodward, O.M., June 1952, Experimentally determined radiation characteristics of conical and triangular antennas. RCA Rev., Vol. 13, pp. 425–452.

Burke, G.J., Johnson, I.A., and Miller, E.K., 1983, Modelling of simple antennas near to and penetrating an interface. Proc. IEEE., Vol. 71, No. 1, pp. 174–175.

Daniels, D.J., 1977, The use of radar in geophysical prospecting, IEE International Conference "Radar 77", IEE, Institution of Electrical Engineers, pp. 540–546.

Daniels, D.J., 1991, Radar for non-destructive testing of materials IEE Colloquium "Measurements, Modelling and Imaging for Non-Destructive Testing", London, 27 March 1991.

Deschamps, G., 1959, Impedance properties of complementary multiterminal planar structures. IRE Trans., AP-7, 5371–5378.

Dyson, J.D., 1959, The equiangular spiral antenna. IRE Trans., Ap-7, 181–187.

Esselle, K.P. and Stuchly, S.S., 1990, A new broadband antenna for transient electromagnetic field measurements. Proceedings of IEEE Symposium on Antennas and propagation – Merging Technologies for 90s, Dallas, TX, USA, pp. 1584–1587.

Evans, S. and Kong, F.N., 1993, TEM horn antenna: Input reflection characteristics in transmission. IEE Proc. H, Vol. 130, No. 6, pp. 403–409.

Foster, P.R. and Tun, S.M., 1993, Design and test of two tem horns for ultrawideband use. IEE Colloquium on Antennas and Propagation Problems of Ultrawideband Radar IEE Colloquium Digest, IEE.

Gibson, P.J., 1979, The Vivaldi aerial. Proceedings of the Seventh European Microwave Conference, Microwave Exhibitions & Publishers Ltd, pp. 101–105.

Goldstone, L.L., 1983, Termination of a spiral antenna. IBM Tech Disclosure Bulletin. Disclosure Bull. Vol., 25, No. 11a, pp. 5714–5715.

Ilzuka, K., 1967, The travelling-wave V-antenna and related antennas. IEEE Trans AP-15, No. 2, pp. 23–43.

Junkin, G., and Anderson, A.P., 1988, Limitations in microwave holographic imaging over a lossy half space. IEE Proc, F, Vol. 135, No. 4, pp. 321–329.

Kappen, F.W. and Monich, C. (1987): Single pulse radiation from a resistive coated dipole antenna. Proceedings of Fifth International Conference on Antennas and Propagation, ICAP "87, York, UK, IEE Conf, Publ. 274, Vol. 1, pp. 90–93.

Kerr, J.L., 1973, Short axial length broadband horns. IEEE Trans, Ap-21, 710–714.

King, R.W.P. and Nu, T.T., 1993, The propagation of a radar pulse in sea water. J. Appl. Phys., Vol. 73, No. 4, pp. 1581–1589.

Kooy, C., 1984, Impulse response of a planar sheath equi-angular spiral antenna. Arch. Elektron. Uebertrag.tech, Vol. 38, No. 2, pp. 89–92.

Lenler-Eriksen, H.-R., Planar near-field measurements of GPR antennas and applications to imaging, PhD Thesis, Technical University of Denmark, August 2005.

Maloney, J.G. and Smith, C.S., 1991, The role of resistance in broadband, pulse-distortionless antennas. Proceedings of IEEE Antennas & Propagation Society Symposium, 24–28 June 1991, Vol. 2, pp. 707–710.

Miller, E.K. and Landt, J.A., 1977, Short-pulse characteristics of the conical spiral antenna. IEEE Trans., Ap-25, No. 3, 621–626.

Morgan, T.E., 1985, Spiral antennas for ESM. IEE Proc. F, Vol. 132, No. 4, pp. 245–251.

Pastol, Y., Arsavalingam, G., and Halbout, J.-M., 1990, Transient radiation properties of an integrated equiangular spiral antenna. Proceedings of IEEE Symposium on Antennas & Propagation – Merging Technologies for 90s, Dallas, TX, USA, pp. 1934–1937.

Randa, J., Kanda, M., and Orr, K.D., 1991, Resistively-tapered-dipole electric-held probes up to 40 GHz. Proceedings of IEEE International Symposium on Electromagnetic Compatibility, Cherry Hill, Ni, USA, pp. 265–266.

Rao, B.R., 1991, 1 Optimised Tapered Resistivity Profiles for wideband HF monopole antenna. Proceedings of the IEEE Antennas & Propagation Society Symposium, Vol. 2, pp. 711–713.

Reader, H.C., Evans, S., and Young, W.K., 1985, Illumination of a rectangular slot radiator over a 3 octave bandwidth. Proceedings of the Fourth International Conference on Antennas and Propagation, Cap "85, Coventry, UK IEE Conf Publ. 248, pp. 223–226.

Rutledge, D.B. and Muha, M.S., 1982, Imaging antenna arrays. IEEE Trans., Ap-30, No. 4), pp. 535–540.

Shen, H.M., King, R.W.P. and Wu, T.T., 1988, V-conical antenna. IEEE Trans AP-36 No. 11, 1519–1525.

Sun, E.-Y. and Rusch, W.V.T., 1991, Transient analysis of large reflector antennas under pulse-type excitation. Proceedings of IEEE Antennas & Propagation Society Symposium, pp. 674–677.

Theodorou, E.A., Gorman, M.R., Rig, R.R. and Kong, F.N., 1981, Broadband pulse optimised antenna. IEE Proc H, Vol. 128, No. 3, pp. 124–130.

Wait, J.R., 1960, Propagation of electromagnetic pulses in a homogeneous conducting earth. Appllied Science Research B, Vol. 8, pp. 213–253.

Wohlers, R.J., 1970, The CWIA, an Extremely Wide bandwidth low dispersion antenna. Abstracts of 20th Symposium, US Air Force R&D Programme.

Wu, T.T. and King, R.W.P., 1965, The cylindrical antenna with non-reflecting resistive loading. IEEE Trans., AP-13, No. 5, 369–373.

Yarovoy, A.., Qiu, W., Yang, B. and Aubry, P.J., 2007, Reconstruction of the field radiated by GPR antenna into ground. The Second European Conference on Antennas and Propagation ELICAP 2007, 11–16 November 2007, The IECC, Edinburgh, UK, pp. 1–6.

Young, M. et al., 1977, Underground Pipe Detector. US Patent 4 062 010.

FURTHER READING

The following references may be useful to those interested in the design of antennas for GPR.

Balanis, C.A., 1997, Antenna Theory and Design, John Wiley & Sons, New York, 2nd edition.

Baum, C.E. and Farr, E.G., 1993, Impulse radiating antennas, in H. L. Bertoni et al. (eds.), Ultra Wideband/Short-Pulse Electromagnetics, Plenum Press, New York.

Baum, C.E., 1998, Intermediate field of an impulse-radiating antenna. Ultra-Wideband Short-Pulse Electromagnetics, Vol. 4, 14–19 June 1998, pp. 77–89.

Baum, C.E., 1992, Aperture efficiencies for IRAs. Antennas and Propagation Society International Symposium. AP-S. Digest. Held in Conjunction with: URSI Radio Science Meeting and Nuclear EMP Meeting, Vol. 3. IEEE, 18–25 July 1992, pp. 1228–1231.

Buchenauer, C.J., Tyo, J.S. and Schoenberg, J.S.H., 1998, Aperture efficiencies of impulse radiating antennas. Ultra-Wideband Short-Pulse Electromagnetics, Vol. 4, 14–19 June 1998, pp. 91–108.

Burke, G.J., Johnson, W.A., and Miller, E.K., 1983, Modelling of simple antennas near to and Penetrating an Interface. Proc. IEEE, Vol. 71, No. I, pp. 174–175.

Chen, C.-C., Rama Rao, K. and Lee, R., 2001, A tapered-permittivity rod antenna for ground penetrating radar applications. Journal of Applied Geophysics, Vol. 47/3–4, pp. 309–316, September 2001.

Chen, C.-C., 2003, Broadband dielectric probe prototype development for near-field measurements, ElectroScience Laboratory Technical Report 743119, The Ohio State University, January 2003.

Chen, C.-C., Rama Rao K. and Lee R., 2003, A new ultra-wide bandwidth dielectric rod antenna for ground penetrating radar applications. IEEE Trans. on Antennas and Propagation, March 2003.

Chen, C.-C., Higgins, M.B., O"Neill, K., and Detsch, R.,2000, UWB full-polarimetric horn-fed bow-tie GPR antenna for buried unexploded ordnance (UXO) discrimination Geoscience and remote sensing symposium. Proceedings IGARSS 2000. IEEE 2000 International Conf, Vol. 4, pp. 1430–1432.

Cloude S., 1995, An Introduction to Electromagnetic Wave Propagation and Antennas, UCL Press, ISBN 1-85728-240-X HB (241 PB).

Collin, R.E. and Zucker F.J, 1969, Antenna Theory, McGraw-Hill, New York.

Collin, R.E., 1985, Antennas and Radio Wave Propagation, McGraw-Hill, New York.

Daniels D.J., 2004, Ground Penetrating Radar, 2nd edition, IEE Radar Sonar, Navigation and Avionics, ISBN: 0 86341 360 9.

De Jongh, R.V., Yarovoy, A., Ligthart, L. P., 1998, Experimental Set-up for Measurement of GPR Antenna Radiation Patterns, 28th European Microwave Conference, Amsterdam (experiment).

Diez, P.A., Chen, C.-C. and Burnside, W.D., 2001, Broadband, near-field dielectric probe. Proceeding of 2001 Antenna Measurement Techniques Association (AMTA) Symposium, Boulder, Colorado.

Elliott, R.S., 1981, Antenna Theory and Design, Prentice Hall Inc., Englewood Cliffs, NJ.

Farr, E.G.,1992, Analysis of the impulse radiating antenna. Antennas and Propagation Society International Symposium, AP-S. 1992 Digest. Held in Conjunction with: URSI Radio Science Meeting and Nuclear EMP Meeting. IEEE, Vol. 3, 18–25 July 1992, pp. 1232–1235

Farr, E.G. and Baum, C.E., 1993, The Radiation Pattern of Reflector Impulse Radiating Antennas: Early-Time Response, Sensor and Simulation Note 358, June 1993.

Farr, E.G., Bowen, L.H., 2002, Recent progress in impulse radiating antennas. Ultra Wideband Systems and Technologies. Digest of Papers. 2002 IEEE Conference on, 21–23 May 2002, pp. 337–340.

Farr, E.G., Baum, C.E. and Buchenauer, C.J., 1995, Impulse Radiating Antennas, Part II, in L. Carin et al. (eds.), Ultra Wideband/Short-Pulse Electromagnetics 2, pp. 159–178, Plenum Press, New York.

Farr, E.G. and Frost, C.A., 1996, Development of a Reflector IRA and a Solid Dielectric Lens IRA, Part I: Design, Predictions and Construction, Sensor and Simulation Note 396, April 1996.

Farr, E.G., Baum, C.E., Prather, W.D. and Bowen, L.H., 1998, Multifunction impulse radiating antennas: Theory and experiment. Ultra-Wideband Short-Pulse Electromagnetics, Vol. 4, 14–19 June 1998, pp. 131–144.

Fujimoto, K., Henderson, A., Hirasawa, K. and James, J.R., 1987, Small Antennas, Research Studies Press Ltd, Letchworth, England.

Harmuth, H.F., 1983, Antennas for non-sinusoidal waves. Part III Arrays. IEEE Trans EMC, Vol. 25, No. 3, pp. 346–357.

Harmuth, H.F. and Ding-Rong, S., 1983a, Antennas for non-sinusoidal waves. Part II Sensors. IEEE Trans EMC, Vol. 25, No. 2, pp. 107–115.

Harmuth, E-I.F. and Ding-Rong, S., 1983b, Large-current, short-length radiator for nonsinusoidal waves. Proceedings of IEEE International Symposium on Electromagnetic Compatibility, Arlington, TX, USA, pp. 453–456.

Hoffmann, R.K., 1987, Handbook of Microwave Integrated Circuits, Artech House, pp. 152–154.

Jasik, H., 1961, Antenna Engineering Handbook, McGraw-Hill, New York.

Johnson, R.C. and Jasik, H., 1984, Antenna Engineering Handbook, McGraw-Hill, New York.

Johnson, R.C., 1993, Antenna Engineering Handbook, McGraw-Hill, New York.

Kraus, J.D., 1988, Antennas, McGraw-Hill, New York.

Kraus, J.D. and Marhefka, R.J., 2002, Antennas for all Applications, McGraw-Hill, New York.

King, R.W.P. and Smith, G.S., 1981, Antennas in Matter, MIT Press, USA.

King, R.W.P., 1971, Table of Antenna Characteristics, Plenum Press, New York.

Lai, A.K.Y. and Sinopoli, L., 1992, A novel antenna for ultra-wide-band applications. IEEE Trans., Vol. Ap40, No. 7, pp. 755–760.

Lestari, A.A., Yarovoy, A.G. and Ligtharl, L.P., 2002, Improvement of bow-tie antennas for pulse radiation Antennas and Propagation Society International Symposium, 2002. IEEE, Vol. 4, pp. 566–569.

Lestari, A.A., Yarovoy, A.G. and Ligthart, L.P., 2004a, Ground influence on the input impedance of transient dipole and bow-tie antennas. IEEE Transactions Oil Antennas and Propagation, Vol. 52, No. 8, pp. 1970–1975, August 2004.

Lestari, A.A., Yarovoy, A.G. and Ligthart, L.P., 2004b, Numerical and experimental analysis of circular-end wire bow-tie antennas over a lossy ground. IEEE Transactions on Antennas and Propagation, Vol. 52, No. 1, pp. 26–35, 77, January 2004.

Lestari, A.A., Yarovoy, A.G. and Ligthart, L.P., 2004c, RC-loaded bow-tie antenna for improved pulse radiation. IEEE Transactions on Antennas and Propagation, Vol. 52, No. 10, pp. 2555–2563, October 2004.

Lo, Y.T. and Lee, S.W., 1988, Antenna Handbook: Theory, Applications and Design. Van Nostrand Reinhold, New York.

Lowe, W., 1976, Electromagnetic Horn Antennas, IEEE Press, New York.

Maloney, J.G. and Smith, G.S., 1993, Optimisation of a conical antenna for pulse radiation: An efficient design using resistive loading. IEEE Trans. Antennas Propagation, Vol. 41, pp. 940–947, July 1993.

Marcuvitz, N., 1951, Waveguide handbook. IEE Electronic- Wave Series, pp. 180–183.

Martel, C., Philippakis, M. and Daniels, D.J., 2000, Time domain design of a TEM horn antenna for ground penetrating radar, AP2000 Millennium Conference on Antennas & Propagation Davos, Switzerland, 9–14 April.

Meincke, P. and Hansen, T.B., 2004, Plane-wave characterization of antennas close to a planar interface. IEEE Transactions on Geoscience and Remote Sensing, Vol. 42, No. 6, June 2004.

Milligan, T.A., 1985, Modern Antenna Design, McGraw-Hill, New York.

Morrow, I.L., Persijn, J. and van Genderen, P., 2002. Rolled edge ultra-wideband dipole antenna for GPR application antennas and propagation society international symposium, 2002. IEEE, Vol. 3, pp. 484–487.

Mueller, G.E. and Tyrrell, W.A., 1947, Polyrod antennas. Bell Systems. Tech. J., Vol. XXVI, pp. 837–851.

Nishioka, Y., Maeshima, O., Uno, T., and Adachi, S., 1999, FDTD analysis of resistor-loaded bow-tie antennas covered with ferrite-coated conducting cavity for subsurface radar Antennas and Propagation, IEEE Transactions on, June 1999, Vol. 47, No. 6, pp. 970–977.

Rao, B.R., 1991. 1 Optimised Tapered Resistivity Profiles for wideband HF monopole -antenna. Proceedings of the IEEE Antennas & Propagation Society Symposium, Vol. 2, pp. 711–713.

Scott, H.E. and Gunton, D.J., 1987. Radar detection of buried pipes and cables. Institution of gas engineers. 53rd Annual Meeting, London, UK, Communication 1345.

Phelan, M., Su, H. and LoVetri, J., 2002. Near field analysis of a wideband log-spiral antenna for 1-2 GHz GPR Electrical and Computer Engineering. IEEE CCECE 2002. Canadian Conference on, Vol. 1, pp. 336–341.

Simon,R., Whinnery, J., Van Duzer, T., 1965, Fields and Waves in Communications and Electronics, LCC 65-19477.

Rudge, A.W., Milne, K., Olver, A.D. and Knight, P. (eds), 1982, The Handbook of Antenna Design, Peter Peregrinus, Stevenage.

Rumsey, V.H., 1966, Frequency Independent Antennas. Electrical Science Series, Academic Press, New York.

Schelkunoff, S.A., 1952, Advanced Antenna Theory, John Wiley & Sons, New York.

Schelkunoff, S. and Friis, H., 1952, Antennas Theory and Practice, John Wiley &Sons, New York, LCC 52-5083.

Silver, S. (ed.), 1984, Microwave Antenna Theory and Design, MIT Radiation Lab. Series, Vol. 12, McGraw-Hill, New York, 1949, reprinted b) Peter Peregrinus, Stevenage, 1984.

Tyo, J.S., 2001. Optimisation of the TEM feed structure for four-arm reflector impulse radiating antennas. Antennas and propagation. IEEE Transactions on April 2001, Vol. 49, No. 4, pp. 607–614.

Tyo, J.S., Buchenauer, C.J. and Schoenberg, J.S.H., 1999, Beamforming in time-domain arrays. Antennas and propagation society. IEEE International Symposium, Vol. 3, 11–16 July 1999, pp. 2014–2017.

Valle, S., Zanzi, L., Sgheiz, M., Lenzi, G. and Friborg, J., 2001, Ground penetrating radar antennas: Theoretical and experimental directivity functions Geoscience and Remote Sensing, IEEE Transactions on April 2001, Vol. 39, No. 4, pp. 749–759.

Venkatarayalu, N.V., Chen, C.C, Teixeira, F.L.and Lee, R., 2002, Impedance characterization of dielectric horn antennas using FDTD antennas and propagation society international symposium. IEEE, Vol. 4, pp. 482–485.

Wait, J.R., 1959, Electromagnetic Radiation from Cylindrical Structures, Pergammon Press, New York.

Wait, J.R., 1960, Propagation of electromagnetic pulses in a homogeneous conducting earth. Applied Science Research B, Vol. 8, p. 213–253.

Wait, J., 1986, Introduction to Antennas and Propagation, Peter Peregrinus, ISBN 0-86341-054.

Walter, CAL, 1965, Travelling Wave Antennas, McGraw-Hill, New York.

Walton, K.L. and Sunberg, V.C., 1964, Broadband ridged horn design. Microwave. J., Vol. 7, No. 3 96-101.

Yarovoy, A.G., Schukin, A.D., Kaploun, I.V. and Ligthart, L.P., 2002. The dielectric wedge antenna Antennas and propagation. IEEE Transactions on October 2002, Vol. 50, No. 10, pp. 1460–1472.

CHAPTER 5

Ground Penetrating Radar Data Processing, Modelling and Analysis

Nigel J. Cassidy

Contents

5.1. Introduction	141
5.2. Background and Practical Principles of Ground Penetrating Radar Data Processing	143
5.3. Ground Penetrating Radar Data Processing: Developing Good Practice	145
5.4. Basic Ground Penetrating Radar Data Processing Steps	148
5.4.1. Data/trace editing and 'rubber-band' interpolation	148
5.4.2. Dewow filtering	150
5.4.3. Time-zero correction	150
5.4.4. Filtering	152
5.4.5. Deconvolution	158
5.4.6. Velocity analysis and depth conversion	158
5.4.7. Elevation or topographic corrections	159
5.4.8. Gain functions	161
5.4.9. Migration	164
5.4.10. Advanced imaging and analysis tools	166
5.4.11. Attribute analysis	167
5.4.12. Numerical modelling	168
5.5. Processing, Imaging and Visualisation: Concluding Remarks	171
Acknowledgements	172
References	172

5.1. Introduction

Of all the topics associated with ground penetrating radar (GPR), data processing and signal analysis are, arguably, the ones that cause the most controversy amongst GPR users. How far a user should go beyond the basic processing steps of dewow, time-zero correction, band-pass filtering, gain control and topographic correction is a matter of personal opinion, experience and, ultimately, the nature of the individual dataset. Even the list above will be too simplistic for some users who will advocate that more sophisticated steps, such as spatial or frequency–wavenumber (FK) filters or migration, are vital for their specific application area. As such, one group of practitioners will always consider advanced processing steps compulsory for good GPR practice, whilst others will question the need for such 'complicated' and costly processing. Even something as mundane as finding the 'best' gain function can result in wide differences in opinion. Personally, I prefer to use an automatic gain correction (AGC) for initial on-site data interpretation, spreading and exponential correction (SEC) gains to evaluate the nature of the signal post-collection and a spatially varying, user-defined gain function in the final post-processed, published sections. This is in contrast to my colleagues who say that I am wasting valuable interpretation time by 'playing' with the data and that the final result is often no different, visually, from the basic AGC-gained section viewed on-site. In some cases, they are right. However, it is the way I prefer to operate and other users, who will have their own processing idiosyncrasies, will work differently from me but produce the same results. What really matters is that the final interpretation is valid, and although processing is important, ultimately, the key to good data interpretation is good data collection in the first place.

There are many different GPR processing and analysis techniques out there, but if the collected data are of poor quality to start with, no amount of processing will rescue it – hence the adage *rubbish in, rubbish out*. In general, it is always best to go with the simplest processing options first and stop when there is nothing else to gain from the process. A good practical mantra for most users to adopt is *if it cannot be seen in the raw data – is it really there?* As such, processing steps should be used to improve the raw-data quality, therefore, making interpretation easier. In practice, this means increasing the signal-to-noise ratio of coherent responses and presenting the data in a format that reflects the subsurface conditions accurately. Many users will interpret directly from the screen (or printout), particularly on-site. The dynamic range of the information produced on a screen is about 10–20 dB, whilst the dynamic range of a GPR system is at least 60 dB. This means that only a small component of the available information is present on-screen and that potentially, additional information can be extracted from the dataset by advanced signal processing. As such, the goal of advanced signal processing methods is to extract information from the signal that can help characterise the physical/natural properties of the subsurface rather than just help the user to "see something in a radargram". This does not mean that a user can apply sophisticated processing steps blindly until they get a 'picture' or an answer that they like (or want to believe) as these methods can introduce user-dependent biases into the process. At best, this will 'distort' the final interpretation, and at worst, will introduce features that are a product of the ill-posed nature of that particular processing step. However, in specific application areas (UXO, pavement evaluation and utilities for instance), targets are better 'defined' and the prudent use of application-related, sophisticated processing steps can enhance data interpretation considerably (see Daniels, 2004 for some relevant examples). Unfortunately, it is very easy to overprocess GPR data, and the novice user, who armed with the latest software and good-quality data lovingly collected in the field, can find the whole subject very daunting. All users want to process their sections correctly whilst producing a 'good picture' that is easy to interpret. Fortunately, the two are not mutually exclusive, and with good practice, it is possible to obtain well-processed data that accurately reflects the true quality of the collected data. Therefore, the aim of this chapter is to provide a practical overview of the common GPR processing steps, from basic dewow to more sophisticated convolution and migration, without being too mathematical or theoretical. However, it is not intended to be an in-depth 'GPR or signal processing manual' but instead will explain the basic concepts and practical aspects/limitations of each technique.

Further reading and appropriate references are provided but, because the subject area is so vast (it would easily make a book in its own right), omissions and selectivity in the citations are inevitable. For readers with an engineering bent, Chapter 7 of Daniels (2004) provides detailed information on GPR data processing, including some of the more sophisticated techniques, whilst the practical notes of Annan (1999, 2002) and the articles by Young et al. (1995) and Olhoeft (2000) are invaluable for anyone. The 'bible' of seismic data processing by Yilmaz (2001) is a must for the geophysically minded as many of the GPR processing steps have their background in seismic processing. For those readers who wish to delve into the murky world of digital signal processing methods, Ifeachor and Jervis (2002) and/or Lyons (2004) are a good place to start, although, in general, it is unnecessary to be an expert in signal processing to understand the intricacies of GPR data processing. Purely due to length constraints, only the processing methods of the most common, commercially based GPR systems and applications have been considered (i.e., time domain, bistatic/monostatic impulse GPRs with simple antenna). Many of these processing steps are applicable to other modulation types and antenna configurations [e.g., multi-antenna array systems, frequency modulation continuous wave (FMCW) systems] but they have not been dealt with specifically in the text. This is an unfortunate, but unavoidable, concession and, as such, means that an immensely interesting area of GPR data processing has been left uncovered. Daniels (1996, 2004) will provide more detailed information on these specialist techniques along with related papers from journals such as the *IEEE Transactions on: Geoscience and Remote Sensing; Antenna and Propagation; Microwave Theory and Techniques; Radar Sonar and Navigation;* plus *Geophysics, Near-surface Geophysics* and the *Journal of Applied Geophysics*.

Ultimately, most users process GPR data with dedicated proprietary software, either system-specific (e.g., RADAN^{TM} – GSSI, 2008) or independent packages that can import a range of different data types (e.g., IXGPR^{TM} – Interpex, 2008; Radar Unix – Grandjean and Durand H., 1999; Radexplorer – Deco Geophysical, 2008; ReflexW – Sandmeier Software, 2008). These programs usually come with very comprehensive help and tutorial files, and these are always the first place to look for help on the specifics of data processing with that particular program.

5.2. Background and Practical Principles of Ground Penetrating Radar Data Processing

In order to appreciate the nuances of modern GPR data processing, it is worth spending some time discussing the background, development and practical principles of the subject, plus the relationship between seismic data processing and GPR methods. In the early days of GPR, systems were analogue and the GPR traces were shown on a screen and printed directly on paper. As such, data processing was more concerned with the quality of the recorded trace and the removal of noise/artefacts than obtaining better interpretations. Data processing was the realm of the electronics engineer with the processing 'steps' built into the system in the form of analogue electronics. With the development of lower-cost analogue-to-digital converters and personal computing in the early 1980s, GPR data processing passed into the digital age and modern signal processing approaches became relevant. At the same time, significant advances were being made in the digital processing of seismic data, and many of the techniques were adapted for use with GPR. From a processing perspective, the two techniques are very similar as the recorded data is a simply a spatially distributed collection of time-domain, voltage signals containing discrete sets of pulses (or 'wiggles'). Consequently, the basic processing steps are the same and significant insights into GPR processing can be gained from the seismic processing developments of the period (Yilmaz, 2001). However, GPR is *not seismics* and there are a number of key differences between the two, which are important for the validity of the more advanced processing methods. One of the most central is the nature and form of the transmitted wavelet, with the GPR pulse being more complex than its seismic counterpart. Attenuation and dispersion effects are more extreme with GPR, and therefore, the frequency component (and phase relationship) of the signals can change markedly with recorded time and depth. Wavelets that exhibit this behaviour are called 'non-stationary' or dispersive and their processing is more complex. Seismic signals do not suffer from the same degree of alteration during propagation and many of the more advanced seismic processing methods, particularly inversion, are less successful with GPR. Another significant difference between seismics and GPR is the assumption about the nature of the subsurface conditions. In comparative terms, GPR signals undergo a greater degree of scattering and 'interference' during propagation as the natural scale of heterogeneity in the subsurface is closer to the incident wavelength. In seismics, the signals undergo less scattering and the data usually contain significantly more coherent reflections and much less clutter than GPR. Velocity and travel time variations are also less severe in seismics, and in general, the materials exhibit less frequency dependence in their elastic properties. Above all else, the spatial variation in the strength and polarisation of the propagating energy is different between seismics and GPR. In basic terms, the S & P components of a propagating seismic wave front can be considered as non-polarising (although they do have a specific orientation), relatively uniform in amplitude across the whole wave front (non-directional) and spherical in their spatial form. A GPR wave front is likely to be highly directional (amplitude varies significantly across the wave), exhibits varying and complex electrical and magnetic field vector polarisations as the wave passes from near-field to far-field conditions and becomes non-spherical as it encounters pathways with significant differences in velocity. To add to this, antenna type (dipole, bow-ties, etc.), frequency, configuration, degree of antenna ground coupling and specific material at the surface all have an influence on the nature of GPR waves. The resultant complexity of all these factors means that many of the more advanced seismic-based processing steps can perform poorly with GPR data (migration is a good example of this) and that the degree of data processing required for a section is very much site-dependent.

5.3. Ground Penetrating Radar Data Processing: Developing Good Practice

As the previous discussion illustrates, GPR data processing (and its subsequent interpretation) is not easy, and to many new users it can seem a bit of a 'dark art'. There have been many times I have processed/interpreted a section to death only to revisit it some months later as I have seen something new in the data or have developed a new processing approach. Not everyone has this luxury, and therefore, it is important to follow good practice guidelines from the beginning. This way, at least the processing will be consistent, efficient and realistic.

1. *Keep it simple* – In general, the amount of processing is dependent on the raw-data quality, the specific needs of the user and the time/cost limitations of the project. Therefore, if quick, approximate interpretations are satisfactory (as for target location), then little processing is necessary. Ninety percent of all data collected needs only basic processing. If more advanced processing is being considered, then ask yourself why. Are you trying to achieve something that is not possible with the quality of the data you have (i.e., is it too noisy?)? Is the time effort worth it? If so, what can be achieved realistically and what benefit will it provide? It may mean that the data is good enough to warrant only simple quick processing and no matter what you do, the interpretation will be the same. Always remember, it often takes much longer to process and interpret the data than it does to collect it! That said, it does not mean that more advanced processing methods are unsuitable, it is just that there may not be enough time to apply them properly.
2. *Keep it real* – Avoid the temptation to overprocess in an indiscriminate fashion. The more sophisticated the processing technique, the more likely it is to introduce potential artefacts or bias into the data. Remember the basic adage from before "if it cannot be seen in the raw data, is it really there?" Have you tested this? How can you be sure what you see in the processed data is real? These are quality control issues that must be considered before attempting methods that are more sophisticated. Novice GPR users are often concerned that their processed sections do not "look good", and, therefore, it must be their fault for not processing the data properly. We are all to blame for this as nobody ever publishes bad data (believe me, I have plenty). The data you have is the data you have – simple as that. If it looks good, great, if not, well that's the way it is. At the end of the day, it is the final interpretation that counts, not how pretty the section looks.
3. *Understand what you are doing* – Make sure that you understand what the processing steps are doing to the data. This is the only way to check what is real and what is not. This applies to automated processing steps as well as the

user-defined processes. Data processing must enhance GPR data interpretation not control it.
4. *Be systematic and consistent* – Make sure that you are consistent with your processing steps and follow a defined processing flow route through the processing steps (e.g., dewow, then time-zero correction, then filters). If possible, use the same parameters on equivalent datasets and record the details of each processing step in a data processing sequence log, or similar processing diary. This is vital when attempting to batch process a large number of similar GPR sections. Processing parameters (e.g., the gain window) should be selected on some physical or practical criteria, such as pulse length, rather than an arbitrary guess at the best value.

This last point is very important, as good-quality, realistic data processing relies on a systematic approach to the processing sequence and the accurate recording of all processing steps and parameters. In practice, the specific nature of each step and its position within the processing flow sequence is often dictated by the processing software being used. Many modern, system-related programs are tailored for defined applications (e.g., utility detection) and/or set antenna-system configurations. This is not a bad thing, per se, but it can mean that the user loses some of the flexibility to modify their processing approaches to match unique circumstances.

Figure 5.1 illustrates the typical processing flow sequence for a typical set of 2D bistatic, common-offset, reflection mode GPR data with each of the steps in their most relevant order. Not everyone will process in this fashion, but in general, it is representative of good GPR practice.

The processing steps highlighted in bold can be considered as essential for decent interpretations (e.g., post-collection filtering and depth conversion). However, it is still possible to undertake basic processing on-site (often automated) to obtain a working GPR data section that can be used for initial, real-time interpretations. The more sophisticated and complex steps (such as deconvolution and migration) are optional and their use must be balanced with the project's needs and costs. Analysis tools, such as image/pattern recognition, modelling and attribute analysis, have been included in the flow sequence, but in practice, they can be considered as interpretational aids rather than specific processing features. As such, they are very powerful and can be used to extract useful information from the data (such as spectral content and phase relationships), but they are 'bonus' features and it is usually unnecessary to include them in the processing strategy in order to obtain good interpretations.

A brief description of each processing step has been provided in the figure and more specific details will follow later. However, it is worth noting that many software packages include a much wider range of processing features such as arithmetic functions and spectral analysis. These can be considered as extra features, in the same way as the modelling and attribute/image analysis tools, and tend to provide more advanced interpretational information rather than specific processing benefits.

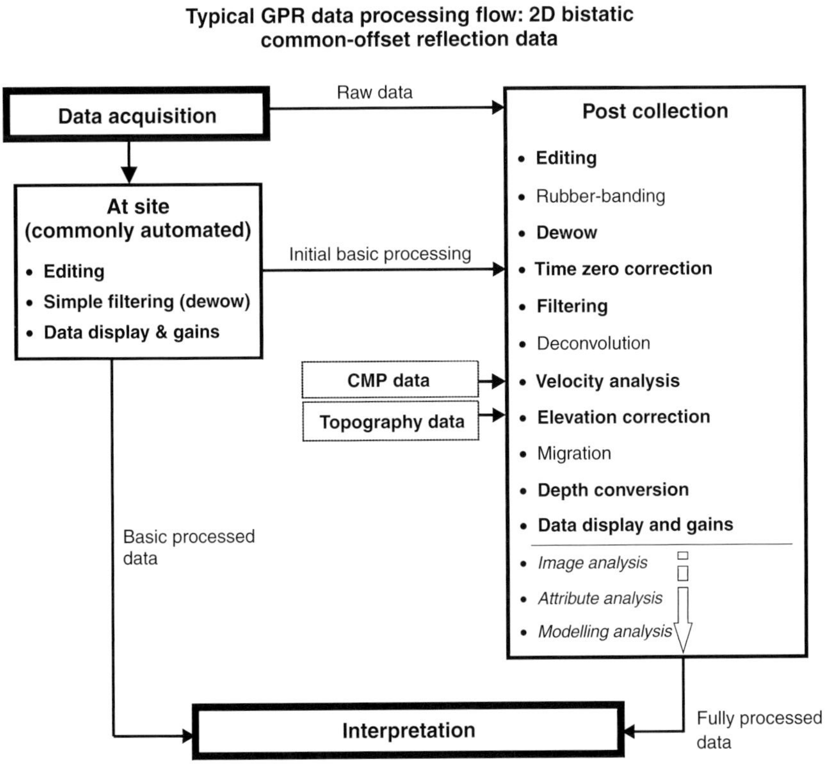

Typical GPR data processing flow: 2D bistatic common-offset reflection data

Data acquisition → Raw data → **Post collection**

At site (commonly automated):
- Editing
- Simple filtering (dewow)
- Data display & gains

Initial basic processing →

Post collection:
- Editing
- Rubber-banding
- Dewow
- Time zero correction
- Filtering
- Deconvolution
- Velocity analysis
- Elevation correction
- Migration
- Depth conversion
- Data display and gains
- *Image analysis*
- *Attribute analysis*
- *Modelling analysis*

CMP data, Topography data → (feed into Velocity analysis / Elevation correction)

Basic processed data → **Interpretation** ← Fully processed data

GPR data processing and analysis steps: Basic descriptions

Editing - Removal and correction of bad/poor data and sorting of data files.
Rubber-banding - Correction of data to ensure spatially uniform increments.
Dewow - Correction of low-frequency and DC bias in data.
Time-zero correction - Correction of start time to match with surface position.
Filtering - 1D & 2D filtering to improve signal to noise ratio and visual quality.
Deconvolution - Contraction of signal wavelets to "spikes" to enhance reflection events.
Velocity analysis - Determining GPR wave velocities.
Elevation correction - Correcting for the effects of topography.
Migration - Correction for the effects of survey geometry and spatial distribution of energy.
Depth conversion - Conversion of two-way travel times into depths.
Display gains - Selection of appropriate gains for data display and interpretation.

Image analysis - Using pattern or feature recognition tools.
Attribute analysis - Attributing signal parameters or functions to identifiable features.
Modelling analysis - Simulation of GPR responses.

Figure 5.1 Typical processing flow sequence for 2D bistatic, common-offset, reflection mode ground penetrating radar (GPR) data. Processing steps in bold can be considered as essential for interpretation; the rest are optional depending on time, data quality and cost implications.

5.4. Basic Ground Penetrating Radar Data Processing Steps

These are usually applied directly to the raw data (often automatically) without the need for additional subsurface information and are generally applicable to most collection modes. Typically, they take the form of trace editing, filtering or data correction and introduce minimal operator bias into the data. In practice, most, if not all, of these steps are required before a basic interpretation can be made. If the sections are to be viewed in a three-dimensional (3D) context (e.g., with time slicing), then an element of manual control is vital in order to retain both data quality and consistency between each section.

5.4.1. Data/trace editing and 'rubber-band' interpolation

Data editing is the first, and often most time-consuming, task in any processing sequence as the files usually need sorting, rearranging and, if being used in a 3D sense, locating correctly. Effective maintenance of the data from the start is vital for good-quality interpretation, particularly with large volumes of data. The inevitable errors that occur in the field mean that sections need reversing, merging, splicing or other manipulations. The incorrect recording of survey parameters (e.g., spatial increment, section interval and start position) can result in data file headers being incorrect, which, in turn, will affect how the data is imported into the processing program. Therefore, it is good practice to view the header files of each section before starting processing in order to check whether the survey parameters are consistent and correct. In most cases, header files can be edited post-collection. Incoherent, noisy or missed traces require editing out and/or filtering to improve the visual nature of the section. This is commonly caused by overenthusiastic triggering, external noise sources, equipment failure/problems or, if an odometer is being used, by traversing too quickly. In most cases, only the occasional trace is corrupted and a simple interpolation between traces is sufficient. However, in difficult terrain or with poorly trained personnel, missed/corrupted traces can become common and repeated interpolation is necessary. In extreme cases, this will adversely affect the structure of the data and the subsurface features can easily become distorted. The same is true for 'rubber-band' interpolation. This is a common processing step usually applied to data collected in continuous trigger mode where the GPR unit is 'fired' at regular time intervals. As there is no direct spatial measurement, the operator attempts to maintain a constant towing speed in order to keep sample points equidistant. Having equidistant, regular data is important for some of the more sophisticated processing steps (e.g., migration), but it is almost impossible to achieve in practice. Variations in towing speed are inevitable and often evident by the 'stretching' of the GPR image, particularly at the end of a section (e.g., Olhoeft, 2000). To correct this, a series of marker points at known distances are usually recorded and the whole section interpolated to a higher spatial density. The data are then resampled to produce a section with equally spaced traces between the markers. This can work well if regular, close-spaced markers are present, but if not, distortions will be produced that can affect subsequent processing and imaging.

One particularly useful editing tool is the 'desaturation' or 'declipping' feature available in some programs. Under certain circumstances, the initial ground wave signal can become 'clipped' as the GPR receiver system saturates with strong ground coupling, and therefore, the recorded trace does not represent the true peak amplitude of the returning signal (Figure 5.2). If trace normalising is being used to account for differences in antenna–ground coupling (i.e., each trace is normalised to the peak amplitude – usually the ground wave wavelet), then the saturated traces will have their later arrivals artificially enhanced in comparison to the non-saturated traces. The desaturation function attempts to correct for this effect by reconstructing the form of the ground wave pulse by spline interpolation (see Figure 5.2). As long as the trace in not too noisy and the form of the saturated ground wave wavelet is 'clean', this process can work well and the subsequent normalised traces will be realistic and representative. However, if the saturation is strong enough, the electronics of the GPR receiver will not recover in time and the majority of the trace will suffer from significant ringing or other noise effects. In extreme cases, the saturation can produce a time-delayed, high-frequency 'pulse', which can be seen in the later part of the recorded trace as the receiving electronics attempt to obtain signal stability.

At this point, it is also worth making a general comment about interpolation methods, in either time or space, as their overuse can introduce errors into the data, no matter which technique is being used (i.e., linear, polynomial, cubic spline;

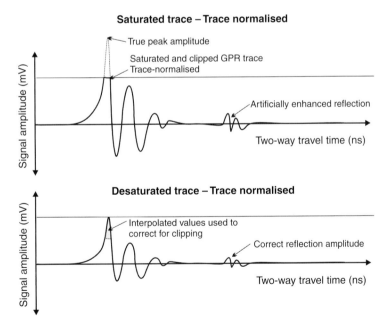

Figure 5.2 Example of how the desaturation function can be used to correct the clipped form of saturated traces.

Fruhwirth et al., 1994). This is particularly so when attempting to rubber-band interpolate over a large number of irregularly spaced traces. To retain spectral integrity, the data should be interpolated in the frequency–wavenumber (FK) domain with infinite Nyquist frequencies and then inversely transformed back into the time domain with a finite spectral length to give a resultant section of the same frequency spectrum as the original data. Unfortunately, most interpolation algorithms do not do this and, therefore, it is possible for spectral errors to be introduced into the interpolated data. Consequently, care must be taken when interpolating across large numbers of traces to ensure that the interpolation process produces reliable, realistic results.

5.4.2. Dewow filtering

This is the removal of the initial DC signal component, or DC bias, and subsequent decay of 'wow' or low-frequency signal trend present in the data (Dougherty et al., 1994). 'Wow' is caused by the swamping or saturation of the recorded signal by early arrivals (i.e., ground/air wave – Annan, 1993) and/or inductive coupling effects and requires the subtraction of the DC bias from the signal and the application of an optimised, low-cut or median filter for effective correction (Gerlitz et al., 1993; Fisher et al., 1994). Dewowing is a vital step as it reduces the data to a mean zero level and, therefore, allows positive–negative colour filling to be used in the recorded traces (Figure 5.3). If applied incorrectly, the data will contain a decaying, low-frequency component that distorts the spectrum of the whole trace. This can affect subsequent spectral processing methods as well as the visual nature of the section (Gerlitz et al., 1993). Fortunately, most modern GPR systems now apply dewow to each trace automatically with the filter parameters set to the optimal conditions. If a manual dewow correction is required, it is good practice to attempt a simple DC subtraction first, then either a median filter with a short filter window (typically the same length as the GPR pulse wavelet) and/or a low-cut filter with a cut-off frequency that is below the bandwidth of the recorded data (e.g., a 10 MHz cut-off for a 500-MHz antenna).

5.4.3. Time-zero correction

Thermal drift, electronic instability, cable length differences and variations in antenna airgap can cause 'jumps' in the air/ground wavelet first arrival time (usually referred to as the time-zero point; Olhoeft, 2000; Nobes, 1999; Young et al., 1995). This has an effect on the position of the ground interface in the section, the time sequence of later events and the degree of parity across adjacent traces/sections (Figure 5.4). Therefore, traces require adjusting to a common time-zero position before processing methods can be applied. This is usually achieved using some particular criteria (e.g., the air wave first break point or first negative peak of the trace) and is often done automatically by the processing software. Under normal circumstances, this works well, but noisy traces can be problematic as there is often no definitive start or peak to the initial part of the air–ground wavelet. As such, care

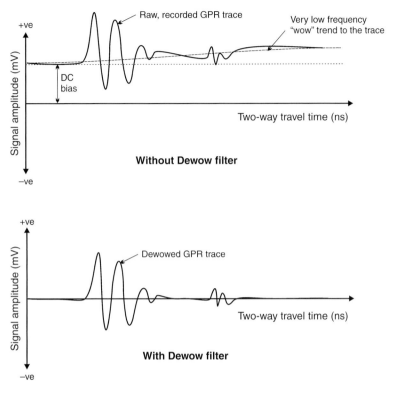

Figure 5.3 Dewow filter correction on a raw ground penetrating radar (GPR) trace.

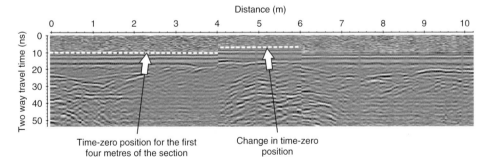

Figure 5.4 Example of time-zero variations in a 450-MHz ground penetrating radar (GPR) section.

must be taken to ensure that the time-zero correction produces results that are consistent across the traces. This is particularly so when attempting to correlate reflections across individual sections.

These three processing steps are the minimum that is required to view the data in an intelligible form. Although processing has been applied, the data are still considered as very 'basic' and further methods must now be used to improve the section interpretation.

5.4.4. Filtering

Filters are generally applied to the data to remove cultural (i.e., human-induced) or system noise and improve the visual quality of the data (e.g., the removal of high-frequency 'speckle' from radio transmissions – Olhoeft, 2000 or the striping effect from antenna ringing – Lehmann et al., 1996). However, they are also useful in extracting particular aspects of the data and, therefore, aiding interpretation (e.g., emphasising flat-lying reflectors, diffractions; Annan, 1999; Gerlitz et al., 1993). There are many different filter types, from simple band-pass filters to sophisticated domain and transform filters (e.g., Lehmann et al., 1996; Pipan et al., 1999; Young and Sun, 1999). Simple filters are often very effective at removing high/low-frequency noise, whilst sophisticated methods are more appropriate for specific problems (e.g., excessive ringing or noise spikes; Malagodi et al., 1996; Annan, 1993). However, complex filters may not always be necessary (Basile et al., 2000) and, as the filtering process introduces an element of subjective bias, they must be applied judicially. This is particularly true in complex heterogeneous environments where coherent reflection returns are less common and scattering/clutter is more prevalent. In this case, filtering may actually remove important information from the section, making interpretation more difficult (Conyers and Goodman, 1997). In practice, some form of filtering is nearly always required, and the simple methods are often the most effective (Annan, 1993). Filters can be applied before or after gains but pre-gain filters do operate on the data in its truest form. If filters are being applied post-gain, then the effect of the gain on the amplitude and spectral content of the data must be fully understood first.

In general, filters can be classified into two basic types: temporal (down the individual traces in time) or spatial (across a number of traces in distance). These are often combined to produce advanced 2D filters that operate on the data in both time and space simultaneously. In general, basic filters alter the data by removing, suppressing or enhancing signals of given frequency(ies) or across a specific number of traces in space or samples (time). There is a wide range of different GPR filters, each with specific advantages and disadvantages and not all are successful or appropriate for use with every GPR dataset. The most basic are the 1D temporal filters that operate along each trace individually, either universally across all traces in a section or over a selected range traces. Typical 1D temporal filters include the following:

Simple mean – takes the mean of the data across a specified time window and smoothes the data. Good for removing excessive higher-frequency noise from the data such as radio frequency interference from communication devices.

Simple median – takes the median of the data across a specified time window. Good for power spike removal and are often referred to as despiking or clean-up filters.

Low- or *high-pass (frequency domain filters)* – lets through either the low-frequency components of the data (low-pass; good for noise) or the high-frequency components (high-pass; good for removing signal drift and low frequencies).

Band-pass (frequency domain filters) – combination of both high- and low-pass filters and lets through a specific range of frequency components that are defined by a 'pass region'. Band-pass filters are very common and there are a range of types, each with different shaped filter operators that define the shape and form of the pass region (Figure 5.5).

In practice, these temporal filters are good only for removing noise at frequencies either higher or lower than the main GPR signal bandwidth and, as such,

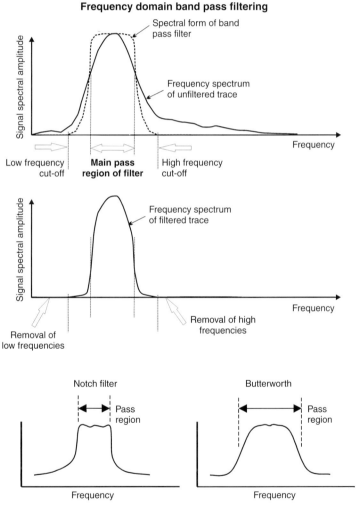

Figure 5.5 Principle of a simple band-pass filter in the frequency domain and the form of two common filter functions: a notch filter and a Butterworth filter.

are ultimately just 'clean-up' filters that make the GPR section visually better. If a too narrow pass region is selected, then the filter will remove components of the actual recorded signal and the resultant GPR section will look worse. A good 'rule of thumb' is to set the pass region symmetrically around the peak signal frequency with a bandwidth that is equal to 1.5 times its value (e.g., for a peak frequency of 400 MHz, the pass region should be at least 100–700 MHz). It should be noted that the peak frequency of the recorded data is often lower than the stated antenna frequency due to the frequency "downshifting" effect of antenna–ground coupling. This should be taken into account when setting the filter parameters.

Spatial filters operate in the same manner as temporal filters but, instead, operate across the traces in distance instead of in time. They tend to take the form of simple mean or averaging filters that span a defined number of traces and are usually used to emphasis or suppress specific features in the section. The most common are the following:

Simple running average – takes the mean of a number of traces. Smoothes the data horizontally and emphasises flat-lying reflectors whilst suppressing dipping reflectors and/or diffractions. Operates best across a relatively large number of traces in the filter window and is good for emphasising stratigraphic horizons (e.g., geological bedding).

Average subtraction – takes the mean of a number of traces in a window and subtracts it from each individual trace in sequence. Suppresses flat-lying reflectors and emphasises diffractions. Operates best with a fairly small trace window and is good for emphasising dipping reflectors (e.g., fractures and dipping bedding).

Background removal – takes the mean of all traces in a section and subtracts it from each trace. Removes background noise and is good for antenna ringing. Is a very useful filter for removing 'ringing' in data but can remove continuous flat-lying reflectors. The judicial use of background removal filters is a key step in the processing and interpretation of GPR data in relatively lossy materials (e.g., wet soils). In these environments, strong antenna–ground coupling and shallow near-surface layers can cause significant reverberation in the signal that can mask later signals.

Spatial high-, low- and band-pass filters (wavenumber domain) – These filters are equivalent to their temporal counterparts in that they convert the data from the distance domain to the wavenumber domain. In essence, the wavenumber represents the spatial size, in metres, of the features in the data. High wavenumbers relate to small, spatially restricted responses (clutter, diffractions, etc.) whilst low wavenumbers are associated with large continuous, spatially coherent responses such as flat-lying reflectors.

Spatial filters operate well with good-quality, low-clutter data and, in general, tend to be good at removing the strong air/ground wave response and ringing from the datasets. Simple running averages and low-pass spatial filters are good for geological and sedimentological GPR sections where the features of interest (bedding reflectors, etc.) are often low-angled and spatially extensive. In more heterogeneous environments, such as archaeological and engineering GPR applications, sections can contain a range of features including flat-lying and dipping reflectors, diffractions, clutter and regions of high and low attenuation. In these cases, spatial filtering can be less helpful as the scale of the subsurface features is often the same as the 'noise' that the filter is trying to remove. However, spatial filtering should not be completely dismissed as there is always some element of interpretational reward that can be gained from the process.

A realistic example of both temporal and spatial filtering is shown in Figure 5.6 where a Butterworth temporal band-pass filter and a spatial background removal filter have been applied to a 450-MHz GPR section collected over a reinforced concrete roadway that contains a speculated subsurface void/collapse feature between 3 and 10 m. The section is realistic in that it contains both high-frequency noise in the form of late-time 'speckle' and signal ringing effects from the presence of the shallow concrete basal layer near the surface. The effect of the temporal band-pass filter can be seen in the 'deeper' parts of the section with the late-time speckle being removed. The background removal spatial filter does a reasonable job of reducing the signal ringing and removing the air–ground wave. However, the filtering is not perfect and elements of the ringing remain. To improve this, more advanced specialist processing methods have developed (e.g., Kim et al., 2007) but, in general, these approaches are highly site-specific and are yet to be incorporated into commercial processing software programs. It should be noted that the main features of the section (the rebar diffractions and semi-coherent reflection/diffractions from the collapse structure) are evident in the unfiltered data and that the filtering process has only slightly improved the visibility and definition of the collapse feature's responses. This highlights the good practice principles stated earlier in that any processing should help emphasise features that are evident in the unfiltered data.

At this point, it is worth discussing some general aspects of filters and the effect that their phase characteristics have on the processed data. In the previous example, a Butterworth temporal band-pass filter was used across a relatively wide frequency range. Butterworth filters have the advantage of having a flat amplitude response over the pass band, meaning that the filtered data retain the correct relative amplitude information. They belong to a class of filters referred to as infinite impulse response (IIR) filters, which tend to have flat, or low-ripple, pass-band amplitude characteristics (Ifeachor and Jervis, 2002). Unfortunately, their phase characteristics are only approximately linear within the pass band, which can cause dispersion or ringing of the processed signals in the time domain. This is less problematic for lower-frequency surveys (<200 MHz) but can be an issue with high-frequency GPR data (i.e., signal bandwidths greater than 500 MHz). In these instances, finite impulse response (FIR) filters are more suitable as they exhibit phase linearity within the filter's frequency range (i.e., no signal distortion). However, not all proprietary data processing programs include FIR filters and the user is often stuck with IIR filters such as the Chebyshev Butterworth or Bessel filters. From a practical aspect, it is ultimately a case of being aware of potential distorting effect of the filtering process and identifying it in the resultant data.

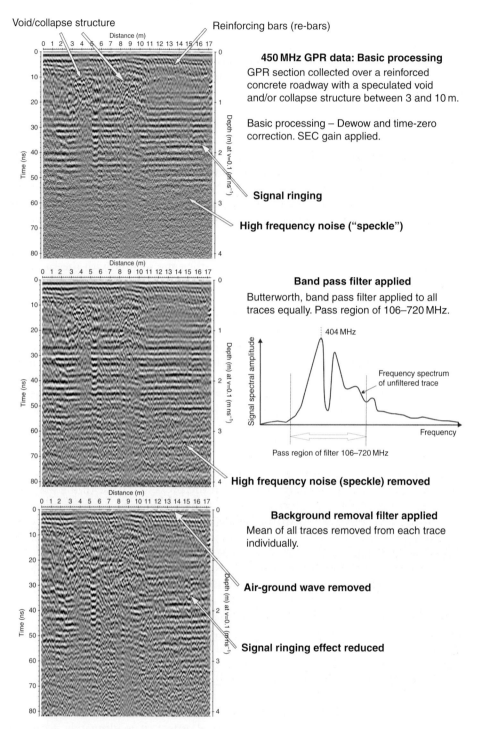

Figure 5.6 Realistic example of the use of temporal band-pass and background removal filtering on a 450-MHz ground penetrating radar (GPR) section collected over a reinforced concrete roadway with a speculated subsurface void/collapse feature between 3 and 10 m.

Two-dimensional filters – Two-dimensional (2D) filters operate on the data in both time and space (*t*–*x*) and, in general, produce similar results to their singular temporal and spatial counterparts. The common 2D filters are the following:

t–*x* average *(2D filter)* – takes the mean of a region in time and space. Good for general noise removal.

t–*x* median *(2D filter)* – takes the median of a region in time and space. Good for general noise and spike removal.

Frequency–wavenumber (FK) filters – A more advanced 2D filter that acts as a combined time–space band-pass filter. The data is transformed into the frequency–wavenumber domain and a combined frequency–distance high-, low- or band-pass filter is applied to the data in a similar manner to the equivalent 1D filters. The key advantage of this filter is that it can be used to suppress events with particular dip directions as the transformed wavenumbers are positive for one dip direction and negative for another. As such, it is a specialist filter and not always better than the basic 1D filters used in sequence. However, it can be a very powerful interpretational tool for specific applications. An example of its use is shown in Figure 5.7, where a 225-MHz GPR section, collected over a buried river channel, has had a FK filter applied to it with the positive wavenumber (left-hand dipping) features removed from the section. The effect of the filter is immediately evident with the strong, right-hand dipping part of the bedrock interface reflection being emphasised and the left-hand section (between 60 and 75 m) being removed. The filter has also emphasised some weak, right-hand dipping reflectors just below the bedrock interface (which are consistent with the bedding orientation of the sandstones) and the right-hand dipping form of the gravel infill in latter parts of the section. Unfortunately, the filter has also

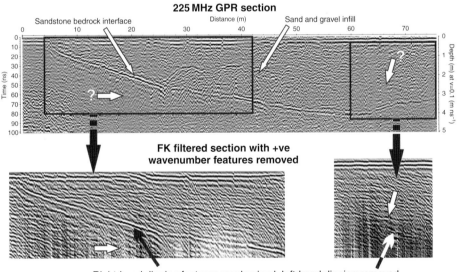

Figure 5.7 Example of the use of frequency–wavenumber (F–K) 2D filtering on a 225-MHz GPR section collected over a buried river channel where the erosional, sandstone bedrock channel has been progressively in-filled with coarse sands and gravels.

introduced a whole series of right-hand dipping features that are not evident in the unfiltered data. To the inexperienced eye, it would be tempting to overinterpret the section and start adding in geological interfaces that do not exist. This highlights the need for care when using more sophisticated processing tools.

5.4.5. Deconvolution

In basic terms, deconvolution is a temporal, "inverse filtering" process that improves the resolution of sections by compressing the recorded GPR wavelet into a narrow, distinct form (Yilmaz, 2001). In other words, it is used to remove the effect of the source wavelet from the recorded data (Neves et al., 1996) and leave the impulse response of subsurface layers only. It is extremely effective in the seismic exploration industry where the primary assumptions are that the propagating wavelet is a simple pulse that is minimum phase (i.e., its energy is concentrated at the beginning of the wavelet), stationary, non-dispersive and emanates from plane-wave source. The deconvolution algorithm also assumes that the subsurface is horizontally layered, has uniform intra-layer velocities and that reflections emanate from interfaces with coherent, regular signals (i.e., do not scatter energy). For GPR, these are very restricting assumptions as the subsurface is more complex and the propagating GPR wavelet is generally mixed phase, non-stationary and vectoral with non-planar, spatially complex fields. This has led to a debate on the usefulness of deconvolution techniques (Annan, 1993; Conyers and Goodman, 1997) and the opinion that other predictive filtering methods are better suited for GPR (Daniels, 1996). However, deconvolution can be reasonably successful if the form of the propagating source wavelet can be determined accurately. Simple, regularised, deconvolution methods have been developed by Savelyev et al. (2007) based on the free-space reflection signal from a large metal plate. By adding additional components to the deconvolution model, Turner (1994) used a constant Q attenuation model to account for the non-stationary nature of the wavelet, whilst Malagodi et al. (1996) and Neves et al. (1996) developed source wavelet models from an analysis of the direct air wave. More recently, predictive and deterministic deconvolution has been used in conjunction with other advanced processing methods to successfully improve the interpretation of GPR data from land mines (Roth et al., 2005), fuel tanks (Porsani and Sauck, 2007) and geological successions (Xia et al., 2003, 2004; Chen and Chow, 2007). Despite these successes, deconvolution methods are, in general, limited in their practical application for most users and tend to be successful only in well-defined and or less-complex subsurface environments.

5.4.6. Velocity analysis and depth conversion

So far, the processing steps have operated in the time domain only and the data has not been related directly to depth. In order to convert the sections to a depth scale, which is required for realistic interpretations and the application of elevation corrections, an accurate estimate of the average subsurface velocity must be obtained through ground truthing, common midpoint surveys (CMPs) and/or

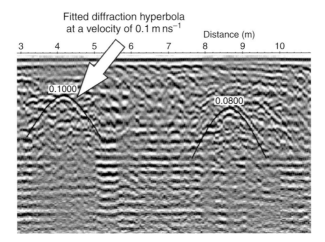

Figure 5.8 An example of hyperbolic velocity matching where the hyperbolic functions have been fitted to diffraction hyperbola from a 450-MHz ground penetrating radar (GPR) section.

hyperbolic velocity analysis. The practical aspects of collecting and analysing CMPs have been covered earlier in this book and most data processing programs have some form of CMP analysis tool in their armoury. These usually take the form of a semblance analysis procedure (Yilmaz, 2001) where the subsurface velocity profile is determined either automatically or manually from the optimal stacking velocity. Hyperbolic matching can be performed on any section that contains diffraction or reflection hyperbola and is achieved by matching the ideal form of a velocity-specified hyperbolic function to the form of the observed data. An example of this is shown in Figure 5.8, where two hyperbolic functions have been fitted to diffraction hyperbola from a 450-MHz GPR section with a 0.1 and 0.08 m/ns velocity, respectively. More advanced programs allow the user to vary the size, radius and velocity of the hyperbolic function to match a hyperbola anywhere in the section but, in general, these options tend to be overkill as basic fitting parameters are all that is normally required. In practice, both CMP analysis and hyperbolic matching techniques tend to produce approximate velocity values with errors and variance of $\pm 10\%$ or worse. Therefore, it is not worth wasting time trying to get a highly accurate velocity–depth profile when a constant, average velocity for all depths will produce the same interpretational results. Most processing programs allow the user to manually input a uniform velocity value or create a velocity profile verses two-way travel time from the hyperbolic matching data. Alternatively, velocity profile information can usually be imported from the CMP analysis via a suitably formatted text file. Sections are then converted to a depth scale and 'stretched' visually to match the variation in velocity with depth.

5.4.7. Elevation or topographic corrections

Unless the dataset is collected over a level, flat surface, some form of topographic correction is required to 'position' the data in its correct spatial context. This is vital for 3D and pseudo-3D surveys where an accurate correlation between sections and/or time slicing is to be attempted. Topographic corrections (often referred to as elevation static corrections) are normally performed with a simple constant velocity correction that acts in a vertical sense (Yilmaz, 2001). This corrects the two-way travel time of the traces to a flat datum level some distance above the air/ground interface. This is adequate for relatively smooth varying topography but if topographic changes are in the same order as the subsurface features, then more advanced topographic correction routines are required (e.g., Lehman and Green, 2000). Strictly speaking, topographic variations exhibiting dips of greater than approximately 10^0 should be corrected using more advanced methods, otherwise travel times will become inaccurate and the image will suffer from 'blurring'. However, most commercial processing programs include only standard topographic correction algorithms, and therefore, the visual qualities of extreme topography GPR sections will be degraded. In addition, topographic corrections assume normal-incidence survey modes and although this is reasonable for deeper targets, it is not applicable for shallow features due to the antenna separation. Normal move-out corrections (NMO) can be applied (Gerlitz et al., 1993) but these generally assume direct ray paths, flat local topography, planar reflectors, homogeneous layers and spherical wave fronts (Yilmaz, 2001). In heterogeneous subsurface environments, this may cause additional errors to be introduced into the data.

Effective topographic correction relies on accurate surveying, particularly with 3D datasets. A reasonable rule of thumb for gentle to rugged terrain is a surveying accuracy of approximately 10% of the dominant wavelength and a sampling interval of two/three times the same value (Lehman and Green, 2000). Table 5.1 illustrates typical surveying accuracies for a damp, sandy soil (i.e., $\varepsilon_{ave} = 16$) at the common GPR frequencies of 100–900 MHz. In practice, this level of accuracy is rare and the surface topography is commonly interpolated between sparse surveying points. This is acceptable for smooth terrains, but as the topographic gradient increases, inadequate interpolation will lead to marked correction errors. This is particularly important for complex target geometries and it is vital that accurate surveying is performed to maintain data integrity.

Data processing programs approach elevation correction differently, but most require the user to specify the subsurface velocity profile independently and the data is shifted to positive (or negative) times from a given datum reference, usually either the lowest or the highest point in the whole survey line. Sensible topographic corrections are important as they place the GPR reflections/diffractions in their

Table 5.1 Sampling interval and spatial accuracy requirements for the topographic surveying of 100–900-MHz ground penetrating radar (GPR) surveys across gentle to rugged terrain (based on a typical damp, sandy soil of $\varepsilon_{ave} = 16$)

Frequency (MHz)	Sampling interval (m)	Spatial accuracy (cm)
100	~2	~5
225	~1	~3
450	~0.5	~1.5
900	~0.25	~0.8

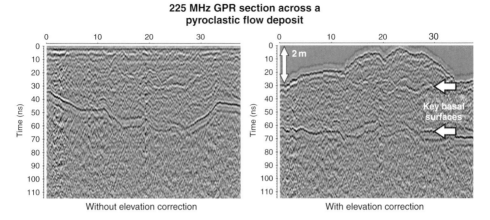

Figure 5.9 Example of elevation (or topographic) corrections on a 225-MHz ground penetrating radar (GPR) section collected across a pyroclastic flow lobe. Note the key flat-lying basal reflectors identified with the arrows in the topographically corrected section.

correct stratigraphical context, thereby ensuring that the interpretations are realistic. A good example of this is shown in Figure 5.9, where a 225-MHz GPR section, collected across a pyroclastic flow lobe, is shown in its corrected and uncorrected form. From the figure, it is clear that a realistic interpretation is very difficult to achieve without elevation correction, particularly with respect to the key flat-lying basal reflectors. The topography varies only by a maximum of 2 m but the slopes are quite steep. This illustrates that despite the fear that some visual 'burring' may occur, a corrected section is still easier to interpret than an uncorrected section.

5.4.8. Gain functions

The data are now in a 'processed' form suitable for the direct application of gains for interpretation and data analysis. Gains improve the visual form of the GPR sections and most techniques alter the data structure in some way (i.e., relative amplitudes and/or phase relationships are changed). Therefore, it is important that the effects of gain functions are understood before they are applied and that the data are treated with care when interpreting (Annan, 1999). Temporal gains are required to enhance the appearance of later arrivals due to the effect of signal attenuation and geometrical spreading losses. There are different types, e.g., constant gain, exponential gain, SEC, AGC, each having different characteristics. All gain functions tend to operate in a similar fashion by applying some multiplying factor to successive regions of the trace in time (referred to as the time window). Gain functions can be easily changed, typically, the window length (in ns), the gain function (linear, exponential, user-defined, etc.) and the maximum gain allowed. However, it must be remembered that applying a gain effectively alters the data, so some operator bias is inevitable. In general, both noise and coherent signals are usually amplified together in an indiscriminate way.

The most common gain functions are the following:

SEC or energy decay – automatically corrects the signal amplitude for loss of energy due to the geometrical spreading effect of the propagating wavefront (approximately a $1/r^2$ relationship). In more advanced modes, material attenuation losses can be included as an additional factor in the function (usually in dB/m) but these require accurate attenuation information to be effective. This can be difficult to obtain even with CMP or WARR surveys and an appropriate a priori value or a 'best guess' is normally used. Spreading and exponential correction gains do retain the relative amplitude information both in time and space, which is important for detailed interpretations.

User-defined, constant, linear or exponential gains – these speak for themselves and are systematically applied gain functions that have a specific mathematical or multiplication operator defined by the user or the system automatically. In general, they retain some relative amplitude information but how much depends on the type and mode of the function applied.

AGC – An automatic gain function applied to each trace, which is based on the difference between the mean amplitude of the signal in a particular time window and the maximum amplitude of the trace as a whole. This function is very convenient for displaying deeper, weak events but amplifies noise as well as coherent signals. It is also important to set the right window length. A very small window time (typically less than 3% of the total sample length or 25% of the propagating wavelet length) results in both the noise and the signal being amplified equally and the section becoming 'messy'. If the window is too long (typically greater than 10% of the total sample length or two times the wavelet length), then the trailing edge of the high-amplitude pulses tends to dominate the gain calculation (Horstmeyer et al., 1996). As a result, 'shadow zones' of reduced amplitude are generated at the rear of the reflections and the section tends to look washed out (see example in Figure 5.10). To avoid these effects, a prudent selection of sample window is required or the maximum gain of the AGC function must be scaled to match the decaying amplitude of the data (Horstmeyer et al., 1996).

An example of different gain functions is provided in Figure 5.10, where some 'typical-quality', archaeological 450-MHz GPR data have had SEC, AGC (both small and large windows) and user-defined gain functions applied equally to all traces in the section. The data were collected at a ruined, twelfth-century monastic burial site comprising damp, sandy–clayey soils, reworked ground (containing stone and masonry rubble), an upper horizon of landscaped topsoil and, potentially, a number of undisturbed monastic graves. As such, the subsurface is highly heterogeneous and the data are likely to be noisy and full of clutter. The region of interest is between 0.25 and 2 m and the basal interface of the upper-landscaped, soil layer is at a depth of between 0.1 and 0.2 m. The effect of the SEC gain is evident, with the features in the region of interest being identifiable but weak. However, the SEC gain does retain relative amplitude information and, because the ungained (or raw) section is trace-normalised, the reflection amplitudes realistically represent the true strength of the returned signal. This relative amplitude information is vital for determining variations in the degree of signal attenuation across the section and/or the amount of relative reflectivity (and, therefore, material property contrasts

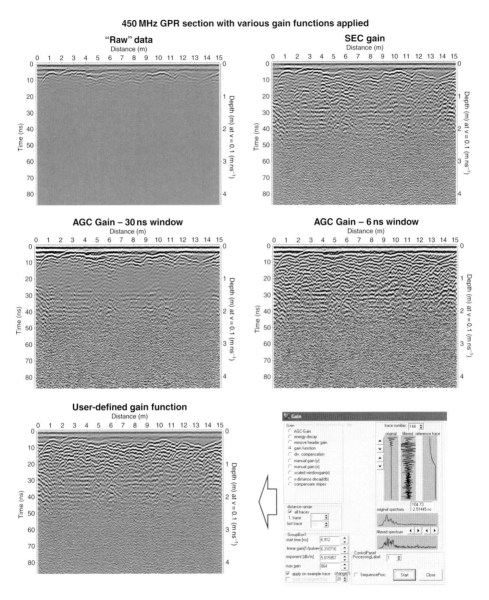

Figure 5.10 Examples of different gain functions (AGC, SEC and user-defined) applied to a 450-MHz ground penetrating radar (GPR) section collected over an archaeological burial site. The 'raw' data has been dewowed, time-zero corrected and each trace is shown as individually trace-normalised.

along reflectors, etc.). Unfortunately, the AGC loses this amplitude information, and with a long time window (30 ns and approximately six times longer than the propagating GPR wavelet), the shadowing effect becomes evident. Features immediately below the topsoil interface have been "washed out" and barely visible in the section. With a short time window (6 ns and approximately one wavelet length), all signals are amplified equally, and although the region of interest is now visible, it is difficult to identify the relationship between adjacent features. Clearly, a compromise value nearer to 10 ns for the time window would be about right for this section. The final example in the figure is a user-defined gain function that has been designed to show the region of interest in its clearest visual form. The filter function is based on a combination of linear and exponential gain components whose parameters can be adjusted manually to suit a given set of traces, section or other parameters. The specific parameters from the example are shown in the figure along with the form of the gain function with time. The ability to set different gain functions is invaluable for both interpretation and final section publication, but getting the parameters right is time-consuming and fiddly. Ultimately, it is a choice between interpretational needs and time/cost, but it is always worth spending a little time playing around with different gains to help improve the visual 'look' of the sections and, therefore, the interpretational process.

5.4.9. Migration

The final weapon in the processing armoury of the GPR user is migration. Migration is generally used for improving section resolution and developing more spatially realistic images of the subsurface and is, arguably, the most controversial of the GPR processing techniques. Like deconvolution, migration techniques were originally developed for the seismic industry where they are considered as vital for even basic interpretations. Unfortunately, migration tends to be less successful with GPR, and although it can be used in relatively uniform environments (e.g., deep geological and glacial, pavements), it is not so good with complex, heterogeneous sites. That said, migration is not completely useless and classical techniques have been applied successfully to a range of different applications. Examples include reverse time migration (Fisher et al., 1992a, 1999b; Sun and Young, 1995; Meats, 1996), F–K migration (Fisher et al., 1994; Pettinelli et al., 1994; Pipan et al., 1996; Yu et al., 1996; Hayakawa and Kawanaka, 1998) and Kirchhoff migration (in Moran et al., 1998). More appropriately, specific GPR-based methods have been developed to overcome some of the limitations in the seismic routines. Examples include matched filter migration (Leuschen and Plumb, 2000); Kirchhoff migration modified for radiation patterns and interface reflection polarisation (Moran et al., 1998; Van Gestel and Stoffa, 2000); eccentricity migration for pipe hyperbola collapsing (Christian and Klaus-Peter, 1994), 3D-based vector and topographic migration (Lehman and Green, 2000; Heincke et al., 2006; Streich et al., 2007); frequency domain migration for lossy soils (Di and Wang, 2004; Sena et al., 2006; Oden et al., 2007) and variations of the F–K-based migration technique for landmine detection (Song et al., 2006). These new methods are yet to be incorporated into mainstream GPR processing packages, although most do contain some form of relatively sophisticated (if classical) migration algorithm. The most common are diffraction stack migration, F–K migration (or Stolt migration), Kirchhoff migration and wave equation or finite-difference modelling migration. These can be performed on 2D sections or across 3D volumes of data. The specific details of how each method works (and their underlying theory) are sadly beyond the scope of this chapter, but Yilmaz (2001) should provide enough information for most readers. Although each of these methods operates differently, they all attempt to 'reconstruct' the GPR section in a spatially accurate form using a model of the subsurface velocity. Ideally, diffraction hyperbolae will be collapsed to a point source and dipping reflections repositioned into their correct location (Figure 5.11).

Unfortunately, all migration methods have limiting assumptions that are often violated by the nature of more complex, heterogeneous environments. These can generalised as follows:

- The velocity structure of the subsurface must be known (or accurately estimated) and the stratigraphy is constructed of laterally invariant, constant velocity layers.
- The source is spatially uniform and propagates spherically.
- The far-field conditions of a radial, uniformly propagating scalar field are assumed.
- Data are collected in normal incidence or monostatic mode – i.e., there is no antenna separation.
- No dispersion or attenuation is allowed – i.e., the materials are lossless and have frequency-independent properties.

Of all of these, the constant, lateral velocity assumption is probably the most important. In simple homogeneous environments, it is reasonably valid and the features will be correctly migrated. However, in complex environments, lateral velocities can change two-fold over centimetre scales and the assumption of

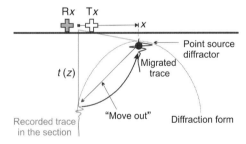

Methodological principles of migration

Diffractors – Migration tries to corrects the position and amplitude of the recorded signal on the "legs" of the diffraction hyperbola by reconstructing the wavelet at the target point source.

The algorithm attempts to either match an ideal hyperbolic function to the diffractor's form, or calculate/model the degree of recorded signal "move out" of the trace from its true position. The calculation needs to know the horizontal distance (x) and depth (z) from time (t) using an accurate velocity-depth relationship. The Migration results in the diffraction hyperbola's legs being collapsed into a single point.

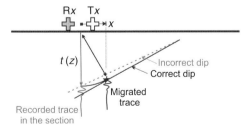

Reflections – Migration tries to corrects the position and, therefore, dip of the recorded reflection by moving the reflected signal back to its original position.

This, as with the diffractor, is due to the fact that the recorded signal is plotted beneath the mid-point of the antenna pair but, in fact, the reflection signal emanates from a position which is "ahead" of the antenna. Again, the horizontal distance (x) and depth (z) needs to be known. The Migration results in the dip of sloping reflectors being corrected to a stepper angle.

Figure 5.11 Basic methodological principle of migration: diffracted/scattered energy is collapsed back to its point-source location whilst the true dip angle of the sloping reflectors is restored.

constant velocity will be violated. Although migration schemes can cope with some degree of variation, this amount of inhomogeneity is too much for most techniques (Yilmaz, 2001) and the migration process will be erroneous. Even if a constant, lateral velocity assumption is accepted, the success of the migration process is highly dependent on the accuracy of the velocity analysis. In an ideal world, highly accurate CMP data would be collected along each section line at regular, overlapping intervals. Unfortunately, surveying time/equipment cost constraints mean that only a limited number of CMP surveys are ever conducted, and at best, only sparse velocity information is available. Other methods can be used, such as hyperbola matching, but these determine only the average velocity to isolated horizons or spatially restricted zones. Alternatively, iterative migration can be used to find the optimal velocity conditions by finding a velocity that gives 'the best-looking picture'. This can work well as long as there are known or easily established subsurface features (Annan, 1993; Hayakawa et al., 1996; Jaya et al., 1999), but if this is not the case, then reaching an optimal solution is difficult and a strong subjective bias is introduced into the data. This illustrates how difficult it is to get the subsurface velocity profile right, and no matter how comprehensive the velocity information is, some degree of error will always be present. Nevertheless, migration can be successful with good velocity information, and therefore, the practical question must be "what is an acceptable degree of velocity error?" Although each migration scheme is different, in general, velocity errors in excess of 5–10% will cause blurring, defocusing of the target features and the misalignment of reflectors (Yilmaz, 2001). If this error exceeds 20%, then the migrated sections will be more difficult to interpret than the unmigrated ones (refer to the seismic examples in Yilmaz, 2001); hence, the additional time/effort involved in applying migration is not justified. When it is and the data is of good quality, then the results can be worth the effort (see example in Figure 5.12).

5.4.10. Advanced imaging and analysis tools

In practical terms, migration can be considered as the final step in the processing sequence and the resultant sections can now be thought of as 'fully processed' and ready for final interpretation. That said, there are a number of additional analysis tools (such as pattern recognition, attribute analysis and forward modelling) that are commonly found in the more comprehensive GPR processing packages. Although not really data processing in the strictest sense, these analysis methods are often used in conjunction with advanced processing techniques and are good at helping the user extract additional information from the data (e.g., relative attenuation, velocity, reflection amplitude and phase information). A number of these methods are highly specialised, such as image and pattern recognition analysis, and tend to be associated with specialist software programs or specific application areas (landmine detection, for instance). Nevertheless, it is worth mentioning two of the more common GPR analysis tools: attribute analysis and numerical forward modelling. Both of these methods have gained in popularity over the past few years (particularly numerical modelling), thanks, in part, to the rapid development of cost-effective, high-performance, PC-based computational resources. Used judiciously,

High-frequency GPR data collected over a concrete roadway

Unmigrated

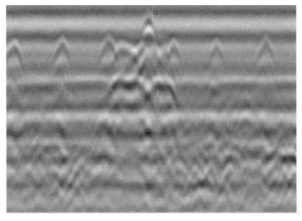

Migrated

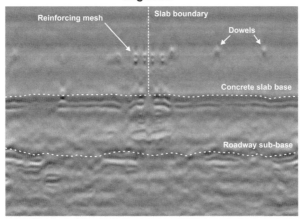

Figure 5.12 Example of successful migration on a high-frequency (1.5 GHz) ground penetrating radar (GPR) section collected over a concrete roadway slab. Note how the migration routine has collapsed the diffractions back to their point sources (reinforcing bars and dowels) and sharpened up the form and shape of the slab and basal reflectors. (Data reproduced in part from RGSL Geophysical Application Sheet 9, Reynolds Geosciences, 2008).

both techniques can be very powerful but as with all advanced data processing/ analysis tools, they must be applied with care and understanding.

5.4.11. Attribute analysis

Attribute analysis is common in the seismic industry where sections are displayed with certain functions applied to the colour rendering of their reflection signals. Information about the relative reflectivity, amplitude, frequency and phase relationships have all been used in aid interpretation and obtain additional information about the subsurface directly from the seismic data (Yilmaz, 2001). Until fairly recently, attribute analysis methods have been less common in the GPR community because of the inherent vectoral nature of the GPR signals and the more complex conditions of the subsurface. However, attribute analysis can be useful (Annan, 1993) and a number of papers have shown that the method can be successfully applied to relatively homogeneous environments (Falak, 1998; Goodman et al., 1998; Lui and Oristaglio, 1998; Dérobert and Abraham, 2000). Unfortunately, other studies have shown that in complex environments, the attribute analysis findings are erroneous or, at best, inconclusive (Burton et al., 2004). A simpler form of attribute analysis involves the selective analysis of amplitude information (Daniels et al., 1997) and has been used successfully with utility-based 3D GPR data and for the determination of subsurface contaminant/water contents in hydrological studies (Schmalz et al., 2002; Cassidy, 2007a).

Most comprehensive GPR processing programs contain some attribute analysis tools, with the majority of methods attempting to analyse the data contained within a specified, but moving, time–space window $(t - x)$ that is incrementally marched across the whole GPR section. At any given position, the data within the window (usually a set number of traces and a specific time length, e.g., 10 traces and 10 ns) have some mathematical attribute function applied to it before it is moved across to the next position in the section. On the completion of the cycle, a new attribute-based GPR section is generated that represents the true spatial distribution of that particular attribute parameter (e.g., mean signal amplitude, peak amplitude, phase relationship, etc.). Some functions operate on only one trace at a time (e.g., instantaneous amplitude, instantaneous phase and instantaneous frequency – Annan, 1999), whilst others use a larger spatial window (e.g., mean amplitude and peak amplitude). Under favourable conditions, such as relatively uniform subsurface environments and flat-lying undulose reflectors, attribute analysis methods can help identify thin layers and trace the continuity of reflection arrivals (instantaneous phase), separate reflections that arrive at similar times (instantaneous frequency) and evaluate the reflectivity and signal strength of the data (instantaneous, mean and peak amplitudes). Unfortunately, attribute analysis methods have not been subjected to the same degree of user-based practical scrutiny as the other GPR processing methods and the jury is still out over their effectiveness in the real world. Nevertheless, the increasing number of published successes illustrates that attribute/amplitude analysis should not be completely dismissed and if the sections show lateral variations in reflection amplitude (commonly referred to as 'bright' or 'dim' spots) and/or areas of discrete signal attenuation, then it is worth trying the technique.

5.4.12. Numerical modelling

Of all the current research areas in GPR, numerical modelling is arguably one of the most popular, with increasing numbers of publications containing some form of numerical modelling in their content. Unfortunately, there is no scope or space in this chapter to cover the topic in full, but, instead, a brief overview of the subject that discusses the background and practical applicability of the technique from a user perspective (and without the maths) will be provided. For a more in-depth review of the subject, refer to Cassidy (2007b), Giannopoulos (2005) and Daniels (2004 – Chapter 3).

There are many different numerical modelling methods available to GPR practitioner with schemes ranging from basic ray-tracing and one-dimensional (1D) transmission–reflection techniques (Goodman, 1994; Olhoeft, 2001) through to more sophisticated finite-difference (Bergmann et al., 1998; Teixeria et al., 1998; Teixeira and Chew, 2000; Cassidy, 2001; Giannopoulos, 2005), finite-volume, Z-transform and discrete-element techniques (e.g., Bourgeois and Smith, 1996; Roberts and Daniels, 1997; Yee and Chen, 1997; Nishioka et al., 1999) and their hybrids (Weedon and Rappaport, 1997; Carcione et al., 1999; Huang et al., 1999). Although varied in their individual approaches, they all attempt to simulate the propagation of the GPR wave from the surface downwards with the emphasis on the interaction of the electromagnetic wave with the subsurface materials. Basic ray-tracing methods and simple transmission–reflection techniques rely on the modeller describing objects as geometrical 'objects' of contrasting relative permittivity (ε) and conductivity (σ) with sharp interfacial boundaries. The modelled GPR response is then computed directly from the reflection–transmission properties of each interface and the form of a basic source wavelet (commonly a Ricker wavelet with a specific antenna-related central frequency). More sophisticated commercial versions use ray tracing, simple finite-difference modelling (REFLEX-WIN^{TM} – Sandmeier Software, 2008) or frequency–wavelength inversions (EKKO_MODEL^{TM} – Sensors and Software, 2005) to produce a 2D, synthetic 'radargram' from a geometrically based subsurface model of contrasting materials. Although relatively basic, these user-friendly schemes can produce comprehensive and sophisticated results that are perfectly adequate for the basic interpretation of GPR responses from simple structures and targets. Being quick, computationally undemanding and adaptable, these more schemes are very well-suited to practical GPR applications where identification and/or verification are/is key to the interpretation. If more sophisticated interpretations are required, then the finite-difference, time domain (or FDTD) technique has evolved into one the most popular of the advanced modelling tools, particularly in more complex environments (Taflove and Hagness, 2005). These techniques are able to model the subsurface properties more accurately than ray-tracing methods (particularly in lossy, dispersive environments) and, more importantly, realistically represent the true 3D geometry and structure of both subsurface targets and the GPR antennae. This allows the user to simulate GPR propagation through quite complex environments and, therefore, extract subtle interpretational information from the real data (such as the timing and presence of multiples, the dispersive characteristics of the subsurface media, target material properties).

To model the propagating GPR wave, the subsurface is usually subdivided into a 3D, orthogonal grid of individual 'field cells' ranging in size between 1/8 and 1/15 of the dominant signal wavelength. Within each individual cell, the electric field (E) and the magnetic field (H) are described by discrete field components (Ex, Ey, Ez, Hx, Hy, Hz) that are uniformly staggered in space and time to form the 'Yee Cell' (Yee, 1966). With the use of a finite-difference approximation to the differential form of Maxwell's electromagnetic field equations, it is possible to calculate the electric field at any point in space (and time) from a knowledge of its neighbouring magnetic fields and vice versa. If an incremental, time-varying electric 'source' is used to simulate the radiating GPR antenna signal, then for a specific time increment, the total 3D EM field is calculated by computationally stepping through each of the alternating *E* and *H* field calculations for every individual cell in sequence. This procedure is then repeated for the next time increment, etc., resulting in the time-dependent propagation of the EM wave throughout the whole of the modelled volume. In order to replicate the subsurface physically, individual permittivity, magnetic permeability and conductivity properties are assigned to each cell in turn, and geometrically complex structures are constructed by grouping together cells of equal properties. With appropriate antenna designs and source wavelets, almost any GPR situation can be modelled, however complex, as long as there are enough computational resources to throw at the model. The flexibility of the method has led to examples from across the whole range of GPR applications from forensics (Hammon et al., 2000) through to landmine detection (Montoya and Smith, 1999; Oğuz and Gürel, 2003), borehole GPR (Holliger and Bergmann, 2002), hydrocarbon contamination (Cassidy, 2008) and NDT applications (Giannopoulos, 2005). A representative example of the method's use is shown in Figure 5.13, where a 950-MHz GPR bistatic reflection survey over two plastic anti-tank mines, buried in a heterogeneous sandy soil, is modelled in three dimensions (the figure shows a 2D section through the model). The model

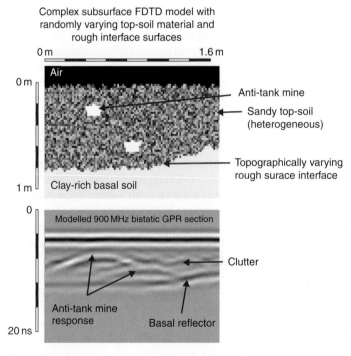

Figure 5.13 Finite-difference, time domain (FDTD) modelling example of a 900-MHz ground penetrating radar (GPR) bistatic reflection survey collected over two plastic anti-tank mines buried in a heterogeneous sandy soil. The model includes shielded antenna, dispersive, lossy materials and a complex subsurface environment with randomly varying topsoil materials and rough interface surfaces.

includes shielded antennae, dispersive, lossy materials and a complex subsurface environment with randomly varying topsoil materials and rough interface surfaces. Despite the presence of significant clutter from the heterogeneous nature of the subsurface conditions, the response of the two landmines is evident in the simulated section with each one having slightly more coherent returns than the background material. What the model has allowed us to do is assess the likely amplitude, phase, location, extent and form of the anti-tank mine responses in an environment that, although simplified, is realistic. This is an invaluable interpretational information and with the ability to change the material parameters and subsurface geometry to fit any situation, we have the potential to simulate potential anti-tank mine responses from anywhere in the world. Even with this relatively simple example, the interpretational benefits of the modelling technique are clear. However, the downside is that this level of modelling sophistication is not available in commercially available GPR processing packages and that a fair bit of computational power is required to run the simulations. That said, accurate 3D, FDTD models at realistic scales (e.g., centimetre grid resolutions and cubic metre scales for 900-MHz antennae) can easily be run in a few hours on a modern dual-core PC. Ultimately, such complex and time-consuming techniques are unlikely to be appropriate for the majority of GPR users, but if modelling could have some potential in your own application area (and you have a modern PC), then the freely available GPR-based, FDTD software program GPRMax is for you. It can be obtained over the internet from http://www.gprmax.org (GPRMax – Giannopoulos, 2008).

5.5. Processing, Imaging and Visualisation: Concluding Remarks

All GPR data require some form of processing before interpretation can be attempted, but to be appropriate and successful, the techniques must be used judicially, particularly the more advanced ones. Often, processing steps are applied blindly without regard for their consequences and with little improvement in the data. Therefore, each method should be evaluated before its use and, if there is little to gain from its application, then it should be deemed unnecessary. This is particularly true in complex environments where the use of more advanced methods may actually be detrimental to the interpretation (Conyers and Goodman, 1997).

Ultimately, processing, analysis and modelling are interrelated as they all attempt to improve data quality and/or interpretation. In practice, processing methods are necessary for interpretations whilst forward modelling and analysis methods can be considered "extras" that are required only if additional information is needed for advanced interpretation, material analysis or target classification. Realistic interpretations are generally guaranteed if the data have been collected and processed by trained, competent personnel in the first place, but it is worth remembering that the 'processing of GPR data tends to improve the appearance of data, but rarely does processing substantially change the interpretation' (Daniels et al., 1997). In reality, the ethos advocated at the start of the chapter, "processing is an aid to data interpretation and if a feature is absent in the raw data, it is likely to be absent in the processed data", is a good one to adopt under general circumstances. Of course, there is always more that can be done with the data, and improved signal processing methods are being continuously developed to help the user extract subtle, yet important, interpretational information from the received signal. However, there is a long way to go before signal processing and data interpretation become 'automated' and, as such, users still need to develop the necessary skills and experience to know what works best in their own application areas.

 ACKNOWLEDGEMENTS

I would like to thank the two anonymous reviewers for their constructive and helpful comments.

REFERENCES

Annan, A.P., 1993, Practical processing of GPR data. Proceedings of the Second Government Workshop on Ground-Penetrating Radar, Columbus, Ohio.

Annan, A.P., 1999, Practical processing of GPR data, Sensors and Software Inc. Mississauga, ON, Canada.

Annan, A.P., 2002, Ground-penetrating radar workshop notes, Sensors and Software Inc. Mississauga, ON, Canada.

Basile, V., Carrozzo, M.T., Negri, S., Nuzzo, L., Quarta, T. and Villani, A.V., 2000, A ground-penetrating radar survey for archaeological investigations in an urban area (Lecce, Italy). Journal of Applied Geophysics, Vol. 44, pp. 15–32.

Bergmann, T., Robertsson, J.O.A. and Holliger, K., 1998, Finite-difference modelling of electromagnetic wave propagation in dispersive and attenuating media. Geophysics, Vol. 63, No. 3, pp. 856–867.

Bourgeois, J.M. and Smith, G.S., 1996, A fully three-dimensional simulation of a ground-penetrating radar: FDTD theory compared with experiment. IEEE Transactions on Geoscience and Remote Sensing, Vol. 34, No. 1, pp. 36–44.

Burton, B.L., Olhoeft, G.R. and Powers, M.H., 2004, Frequency spectral analysis of GPR data over a crude oil spill. Proceedings of the Tenth International Conference on Ground Penetrating Radar, Delft, The Netherlands, pp. 267–270.

Carcione, J.M., Lenzi, G. and Valle, S., 1999, GPR modelling by the Fourier method: Improvement of the algorithm. Geophysical Prospecting, Vol. 47, pp. 1015–1029.

Cassidy, N., 2001, The Application of Mathematical Modelling in the Interpretation of Ground Penetrating Radar Data. Ph.D. Thesis, Keele University.

Cassidy, N.J., 2007a, Evaluating LNAPL contamination using GPR signal attenuation analysis and dielectric property measurements: Practical implications for hydrological studies. Journal of Contaminant Hydrology, Vol. 94, No. 1–2, pp. 49–75.

Cassidy, N.J., 2007b, A review of practical numerical modelling methods for the advanced interpretation of ground-penetrating radar in near-surface environments. Near Surface Geophysics, Vol. 5, No.1, pp. 5–22.

Cassidy, N.J., 2008, Characterizing GPR signal attenuation and scattering in a mature LNAPL spill: A FDTD modeling and dielectric analysis study. Vadoze Zone Journal, Vol. 7, No. 1, pp.140–159.

Chen Y.L. and Chow J.J., 2007, Ground penetrating radar signal processing improves mapping accuracy of underground voids and seawater table: An application in deteriorating coastal structure, Nanfangao Port, Taiwan, Environmental Geology, Vol. 53, No. 2, pp. 445–455.

Christian, S. and Klaus-Peter, N., 1994, Eccentricity-migration: A method to improve the imaging of pipes in radar reflection data. Proceedings of the 5th International Conference on Ground Penetrating Radar (GPR'94), Canada, pp. 723–733.

Conyers, L.B. and Goodman, D., 1997, Ground-Penetrating Radar: An Introduction for Archaeologists, AltaMira Press, Sage Publications, California, USA.

Daniels, D.J., 1996, Surface Penetrating Radar, Radar, Sonar, Navigation and Avionics Series 6, Institute of Electrical Engineers, London, UK.

Daniels D.J., 2004, Ground Penetrating Radar, 2nd edition, Radar, Sonar, Navigation and Avionics Series 15, Institute of Electrical Engineers, London, UK.

Daniels, J.J., Grumman, D.L. and Vendl, M.A., 1997, Coincident Antenna Three-Dimensional GPR. Journal of Environmental and Engineering Geophysics, Vol. 2, No.1, pp. 1–9.

Deco Geophysical, 2008, Radexplorer^{TM}, http://www.radexpro.ru, Deco Geophysical, Moscow, Russia.

Dérobert, X. and Abraham, O., 2000, GPR and seismic imaging in a gypsum quarry. Journal of Applied Geophysics, Vol. 45, pp.157–169.

Di, Q.Y. and Wang, M.Y., 2004, Migration of ground-penetrating radar data method with a finite-element and dispersion, Geophysics, Vol. 69, No. 2, pp. 472–477.

Dougherty, M.E., Micheals, P., Pelton, J.R. and Liberty, L.M., 1994, Enhancement of ground penetrating radar data through signal processing. Proceedings of the Symposium on the Application of Geophysics to Engineering and Environmental Problems (SAGEEP '94), Boston, USA, pp. 1021–1028.

EKKO_MODEL^{TM}, 2005, GPR Modelling Software, Sensors and Software^{TM}, Canada.

Falak, R.E., 1998, GPR investigations at two prehistoric chert mining quarry sites in the northern Madison range, Southwestern Montana. Proceedings of the 7th International Conference on Ground-Penetrating Radar (GPR'98), USA.

Fisher, E., McMechan, G.A. and Annan, A.P., 1992a, Acquisition and processing of wide-aperture ground penetrating radar data. Geophysics, Vol. 57, No. 3, pp. 495–504.

Fisher, E., McMechan, G.A., Annan, A.P. and Cosway, S.W., 1992b, Examples of reverse-time migration of single-channel, ground-penetrating radar profiles. Geophysics, Vol. 57, No. 4, pp. 577–586.

Fisher, C.S., Stewart, R.R. and Jol, H.M. 1994, Processing ground penetrating radar data. Proceedings of the 5th International Conference on Ground Penetrating Radar (GPR'94), Canada, pp. 661–675.

Fruhwirth, R.K., Müller, R.E. and Schmöller, R., 1994, Resampling in the frequency domain, a method for interpolation of time series. Proceedings of the 5th International Conference on Ground Penetrating Radar, Canada, pp. 677–687.

Gerlitz, K., Knoll, M.D., Cross, G.M., Luzitano, R.D. and Knight, R., 1993, Processing ground penetrating radar data to improve resolution of near-surface targets. Proceedings of the Symposium on the Application of Geophysics to Engineering and Environmental Problems (SAGEEP '93), San Diego, USA. pp. 561–575.

Giannopoulos, A. 2005, Modelling ground penetrating radar by GprMax. Construction and Building Materials, Vol. 19, pp. 755–762.

Giannopoulos, A., 2008, GprMax, Electromagnetic simulator for ground penetrating radar, http://www.gprmax.org, University of Edinburgh.

Goodman, D.J., 1994, Ground-penetrating radar simulation in engineering and archaeology. Geophysics, Vol. 59, No. 1, pp. 224–232.

Goodman, D., Nishimura, Y., Hongo, H. and Okita, M., 1998, GPR amplitude rendering in archaeology. Proceedings of the 7th International Conference on Ground Penetrating Radar (GPR'98), USA.

Grandjean, G. and Durand, H., 1999, Radar unix: A complete package for GPR data processing. Computers and Geosciences, Vol. 25, No. 2, pp. 141–149.

GSSI, 2008. RADAN^{TM}, http://www.geophysical.com/software.htm, Geophysical Survey Systems, Inc. Salem, NH, USA.

Hammon, W.S., McMechan, G.A. and Zeng, X., 2000, Forensic GPR: Finite-difference simulations of responses from buried human remains. Journal of Applied Geophysics, Vol. 45, pp. 171–186.

Hayakawa, H. and Kawanaka, A., 1998, Radar imaging of underground pipes by automated estimation of velocity distribution versus depth. Journal of Applied Geophysics, Vol. 40, pp. 37–48.

Hayakawa, H., Niigawa, T. and Kawanaka, A., 1996, Radar imaging of underground objects in inhomogeneous soil using X-T-V data matrix. Proceedings of the 6th International Conference on Ground Penetrating Radar (GPR'96), Japan, pp. 579–584.

Heincke, B., Green, A.G., van der Kruk, J. and Willenberg, H., 2006, Semblance-based topographic migration (SBTM): A method for identifying fracture zones in 3D georadar data. Near Surface Geophysics, Vol. 4, No. 2, pp. 79–88.

Holliger, K. and Bergmann, T., 2002, Numerical modelling of borehole georadar data. Geophysics, Vol. 67, No. 4, pp. 1249–1257.

Horstmeyer, H., Gurtner, M., Büker, F. and Green, A., 1996, Processing 2-D and 3-D georadar data: Some special requirements. Proceedings of the Second Meeting of the Environmental and Engineering Geophysical Society, European section, Nantes, France.

Huang, Z., Demarest, K.R. and Plumb, G., 1999, An FDTD/MOM hybrid technique for modeling complex antennas in the presence of heterogeneous grounds. IEEE Transactions on Geoscience and Remote Sensing, Vol. 37, No. 6, pp. 2692–2699.

Ifeachor, E.C. and Jervis, B.W., 2002, Digital Signal Processing: A Practical Approach, 2nd edition, Pearson Education, Harlow, UK.

Interpex, 2008, IXGPR^{TM}, http://www.interpex.com/ixeterra/gprixeterra.htm, Interpex Ltd, Golden, Colorado, 80402, USA.

Jaya, M.S., Botello, M.A., Hubral, P. and Liebhart, G., 1999, Remigration of ground-penetrating radar data. Journal of Applied Geophysics, Vol. 41, pp.19–30.

Kim, J.H., Cho, S.J. and Yi, M.J., 2007. Removal of ringing noise in GPR data by signal processing. Geosciences Journal, Vol. 11, No. 1, pp. 75–81.

Lehmann, F. and Green, A.G., 2000, Topographic migration of georadar data: Implications for acquisition and processing. Geophysics, Vol. 65, No. 3, pp. 836–848.

Lehmann, F., Horstmeyer, H., Green, A., Sexon, J. and Coulybaly, M., 1996, Georadar data from the northern Sahara Desert: Problems and processing strategies. Proceedings of the 6th International Conference on Ground Penetrating Radar (GPR'96), Japan, pp. 51–56.

Leuschen, C. and Plumb, R., 2000, A matched-filter approach to wave migration. Journal of Applied Geophysics, Vol. 43, pp. 271–280.

Lui, L. and Oristaglio, M., 1998, GPR signal analysis: Instantaneous parameter estimation using the wavelet transform. Proceedings of the 7th International Conference on Ground Penetrating Radar (GPR'98), USA, pp. 219–223.

Lyons, R.G., 2004, Understanding Digital Signal Processing, 2nd edition, Prentice Hall, NJ, USA.

Malagodi, S., Orlando, L. and Piro, S., 1996, Approaches to increase resolution of radar signal. Proceedings of the 6th International Conference on Ground Penetrating Radar (GPR'96), Japan, pp. 83–88.

Meats, C. 1996, An appraisal of the problems involved in three-dimensional ground penetrating radar imaging of archaeological features. Archaeometry, Vol. 38, No. 2, pp. 359–379.

Montoya, T.P. and Smith, G.S., 1999, Land mine detection using a ground-penetrating radar based on resistively loaded vee dipoles. IEEE Transactions on Antennas and Propagation, Vol. 47, No. 12, pp.1795–1806.

Moran, M., Arcone, S.A., Delaney, A.J. and Greenfield, R., 1998, 3-D migration/array processing using GPR data. Proceedings of the 7th International Conference on Ground Penetrating Radar (GPR'98), USA, pp. 225–231.

Neves, F.A., Miller, J.A. and Roulston, M.S., 1996, Source signature deconvolution of ground penetrating radar data. Proceedings of the 6th International Conference on Ground Penetrating Radar (GPR'96), Japan, pp. 573–578.

Nishioka, Y., Maeshima, O., Uno, T. and Adachi, S., 1999, FDTD analysis of resistor-loaded antenna covered with ferrite-coated conducting cavity for subsurface radar. IEEE Transactions on Antennas and Propagation, Vol. 47, No. 6, pp. 970–977.

Nobes, D.C., 1999, Geophysical surveys of burial sites: A case study of the Oaro urupa. Geophysics, Vol. 64, No. 2, 357–367.

Oden, C.P., Powers, M.H., Wright, D.L. and Olhoeft, G.R., 2007, Improving GPR image resolution in lossy ground using dispersive migration. IEEE Transactions on Geoscience and Remote Sensing, Vol. 45, No. 8, pp. 2492–2500.

Oğuz, U. and Gürel, L., 2003, Frequency responses of ground-penetrating radars operating over highly lossy grounds. IEEE Transactions on Geoscience and Remote Sensing, Vol. 40, No. 6, pp. 1385–1394.

Olhoeft, G.R., 2000, Maximizing the information return from ground penetrating radar. Journal of Applied Geophysics, Vol. 43, pp. 175–187.

Olhoeft, G.R., 2001, GRORADAR™, Acquisition, Processing, Modeling and Display of Dispersive Ground Penetrating Radar Data, version 2001.01.

Pettinelli, J.D., Redman, J.D., Endres, A.L., Annan, A.P. and Johnston, G.B., 1994, GPR response quantification: Initial processing and model testing. Proceedings of the 5th International Conference on Ground Penetrating Radar (GPR'94), Canada.

Pipan, M., Baradello, L., Forte, E., Prizzon, A. and Finetti, I., 1999, 2-D and 3-D processing and interpretation of multi-fold ground penetrating data: A case history from an archaeological site. Journal of Applied Geophysics, Vol. 41, pp. 271–292.

Pipan, M., Finetti, I. and Ferigo, F., 1996, Multi-fold GPR techniques with applications to high-resolution studies: Two case histories. European Journal of Environmental and Engineering Geophysics, Vol. 1, pp. 83–103.

Porsani, J.L. and Sauck, W.A., 2007, Ground-penetrating radar profiles over multiple steel tanks: Artifact removal through effective data processing. Geophysics, Vol. 72, No. 6, pp. J77–J83.

Reynolds Geosciences, 2008, Geophysical surveys for highway maintenance and reconstruction, RGSL Geophysical Application Sheet no. 9, Reynolds Geosciences, http://www.geologyuk.com/geophysics/Geophysical_Applications_Sheets.htm, RGSL, Flintshire, Wales, UK.

Roberts, R.L. and Daniels, J.J., 1997, Modelling near-field GPR in three dimensions using the FDTD method. Geophysics, Vol. 62, No. 4, pp. 1114–1126.

Roth, F., van Genderen, P. and Verhaegen, M., 2005, Convolutional models for buried target characterization with ground penetrating radar. IEEE Transactions on Antennas and Propagation, Vol. 53, No. 11, pp.3799–3810.

Sandmeier Software, 2008, ReflexW™, http://www.sandmeiergeo.de/Reflex/gpr.htm, Karlsruhe Germany.

Savelyev, T.G., Kobayashi, T., Feng, X. and Sato, M., 2007, Robust target discrimination with UWB GPR, in Sabath, F., Mokole, E.L., Schenk, U. and Nitsch, D. (eds), Ultra-Wideband, Short-Pulse Electromagnetics 7, Springer-Verlag, New York, USA, pp.732–739.

Schmalz, B., Lennartz, B. and Wachsmuth, D., 2002, Analyses of soil water content variations and GPR attribute distributions. Journal of Hydrology, Vol. 267, Nos. 3–4, pp. 217–226.

Sena, A.R., Stoffa, P.L. and Sen, M.K., 2006, Split-step Fourier migration of GPR data in lossy media. Geophysics, Vol. 71, No. 4, pp. K77–K91.

Song, J.Y., Liu, Q.H., Torrione, P. and Collins, L., 2006, Two-dimensional and three-dimensional NUFFT migration method for landmine detection using ground-penetrating radar. IEEE Transactions on Geoscience and Remote Sensing, Vol. 44, No. 6, pp.1462–1469.

Streich, R., van der Kruk, J. and Green, A.G., 2007, Vector-migration of standard copolarized 3D GPR data. Geophysics, Vol. 72, No. 5, pp. J65–J75.

Sun, J. and Young, R.A., 1995, Recognizing surface scattering in ground-penetrating radar data. Geophysics, Vol. 60, No. 5, pp. 1378–1385.

Taflove, A. and Hagness, S., 2005, Computational Electrodynamics: The Finite-Difference Time-Domain Method, 3rd edition, Artech House Antennas and Propagation Library, Artech House, London, UK.

Teixeira, F.L. and Chew, W.C., 2000, Finite-difference computation of transient electromagnetic waves for cylindrical geometries in complex media. IEEE Transactions on Geoscience and Remote Sensing, Vol. 38, No. 4, pp. 1530–1543.

Teixeria, F.L., Chew, W.C., Straka, M., Oristaglio, M.L. and Wang, T., 1998, Finite-difference time-domain simulation of ground penetrating radar on dispersive, inhomogeneous and conductive soils. IEEE Transactions on Geoscience and Remote Sensing, Vol. 36, No. 6, pp. 1928–1937.

Turner, G., 1994, Subsurface radar propagation deconvolution. Geophysics, Vol. 59, No. 2, pp. 215–223.

Van Gestel, J. and Stoffa, P.L., 2000, Migration using multi-configuration data. Proceedings of the 8th International Conference on Ground Penetrating Radar (GPR 2000), Australia, pp. 448–452.

Weedon, W.H. and Rappaport, C.M., 1997, A general method for FDTD modeling of wave propagation in arbitrary frequency-dispersive media. IEEE Transactions on Antennas and Propagation, Vol. 45, No. 3, pp. 401–409.

Xia, J.H., Franseen, E.K., Miller, R.D., Weis, T.V. and Byrnes, A.P., 2003, Improving ground-penetrating radar data in sedimentary rocks using deterministic deconvolution. Journal of Applied Geophysics, Vol. 54, Nos. 1–2, pp. 15–33.

Xia, J.H., Franseen, E.K., Miller, R.D. and Weis T.V., 2004, Application of deterministic deconvolution of ground-penetrating radar data in a study of carbonate strata. Journal of Applied Geophysics, Vol. 56, No. 3, pp. 213–229.

Yee K.S. and Chen J.S., 1997. The finite-difference time-domain (FDTD) and the finite-volume time-domain (FVTD) methods in solving Maxwell's equations, IEEE Transactions on Antennas and Propagation, Vol. 45, No. 3, 354–363.

Yilmaz, Ö., 2001, Seismic Data Analysis: Processing, Inversion, and Interpretation of Seismic Data, Investigations in Geophysics No. 10 (vol 1 & 2), Society of Exploration Geophysicists, Tulsa, USA.

Young, R.A. and Sun, J., 1999, Revealing stratigraphy in ground-penetrating radar data using domain filtering. Geophysics, Vol. 64, No. 2, pp. 435–442.

Young, R.A., Deng, Z. and Sun, J., 1995, Interactive processing of GPR data, The Leading Edge, April 1995, pp. 275–280.

Yu, H., Ying, X. and Shi, Y., 1996, The use of FK techniques in GPR processing. Proceedings of the 6th International Conference on Ground Penetrating Radar (GPR'96), Japan, pp. 595–600.

PART II

ENVIRONMENTAL APPLICATIONS

CHAPTER 6

SOILS, PEATLANDS, AND BIOMONITORING

James A. Doolittle *and* John R. Butnor

Contents

6.1. Introduction	179
6.2. Soils	180
6.2.1. Soil properties that affect the performance of ground penetrating radar	180
6.2.2. Soil suitability maps for ground penetrating radar	181
6.2.3. Ground penetrating data and soil surveys	185
6.2.4. Uses of ground penetrating radar in organic soils and peatlands	190
6.3. Biomonitoring	193
References	197

6.1. INTRODUCTION

Soils are three-dimensional (3D) natural bodies consisting of unconsolidated mineral and organic materials that form a continuous blanket over most of the earth's land surface. At all scales of measurements, soils are exceedingly complex and variable in biological, chemical, physical, mineralogical, and electromagnetic properties. These properties influence the propagation velocity, attenuation, and penetration depth of electromagnetic energy, and the effectiveness of ground penetrating radar (GPR). Knowledge of soils and soil properties is therefore useful, and often essential, both in the design and operation of GPR surveys. In this chapter, soil properties that influence the use of GPR are discussed. Ground penetrating radar soil suitability maps are introduced. These maps can aid GPR users who are unfamiliar with soils in assessing the likely penetration depth and relative effectiveness of GPR within project areas. This chapter cites studies that have used GPR to investigate soils. Also discussed are the uses of GPR to measure root biomass, distribution and architecture, and detect internal defects in trees.

6.2. SOILS

6.2.1. Soil properties that affect the performance of ground penetrating radar

The resolution and penetration depth of GPR are determined by antenna frequency and the electrical properties of earthen materials (Olhoeft, 1998; Daniels, 2004). Because of high rates of signal attenuation, penetration depths are greatly reduced in soils that have high electrical conductivity. The electrical conductivity of soils increases with increasing water, soluble salt, and/or clay contents (McNeill, 1980). These soil properties determine electrical charge transport and storage (Olhoeft, 1998). In soils, the most significant conduction-based energy losses are due to ionic charge transport in the soil solution and electrochemical processes associated with cations on clay minerals (Neal, 2004). These losses can seriously impact the performance of GPR (Campbell, 1990; Olhoeft, 1998).

Electrical conductivity is directly related to the amount, distribution, chemistry, and phase (liquid, solid, or gas) of the soil water (McNeill, 1980). Electrical conductivity, dielectric permittivity, and energy dissipation increase with increasing soil water content (Campbell, 1990; Daniels, 2004). Water is a polar molecule. When an alternating electrical field is applied to the soil, water molecules experience a force that acts to align their permanent dipole moments parallel to the direction of the applied electrical field (Daniels, 2004). The small displacement of bound water molecules results in the loss of some energy as heat (Neal, 2004). Polarization processes result in the storage of some electrical field energy and dielectric relaxation losses. At frequencies above 500 MHz, the absorption of energy by water is the principal loss mechanism in soils (Daniels, 2004). Even under very dry conditions, capillary-retained water is sufficient to influence electrical conductivity and energy loss.

Electrical conductivity and energy loss are also affected by the amount of salts in the soil solution (Curtis, 2001). All soil solutions contain some salts, which increase the conductivity of the electrolyte. In general, soluble salts are leached to a greater degree from soils in humid than in semiarid and arid regions. In semiarid and arid regions, soluble salts of potassium and sodium, and less soluble carbonates of calcium and magnesium are more likely to accumulate in the upper part of soils. These salts increase the electrical conductivity of the soil solution and consequent attenuation of electromagnetic energy (Doolittle and Collins, 1995). Because of their high electrical conductivity, saline (electrical conductivity $> 4\,dS/m$) and sodic (sodium absorption ratio ≥ 13) soils are considered unsuited to most GPR applications. In these soils, effective GPR penetration is usually restricted to the surface layers and depth of less than 25 cm.

Calcareous and gypsiferous soils mostly occur in base-rich, alkaline environments in semiarid and arid regions. These soils are characterized by layers with secondary accumulations of calcium carbonate and calcium sulfate, respectively. High concentrations of calcium carbonate and/or calcium sulfate imply less-intense leaching, prevalence of other soluble salts, greater quantities of inherited minerals from parent rock, and accumulations of specific mineral products of weathering (Jackson, 1959).

Typically, soils with higher calcium carbonate contents have higher dielectric permittivity (Lebron et al., 2004). Grant and Schultz (1994) observed a reduction in the depth of GPR penetration in soils that have high concentrations of calcium carbonate.

The electrical conductivity of soils is governed by the amount of clay particles (particles <0.002 mm in diameter) and the types of clay minerals present (McNeill, 1980). Clay particles have greater surface areas and can hold more water than the silt (particles 0.002–0.05 mm in diameter) and sand (particles 0.05–2.0 mm in diameter) fractions at moderate and high water tensions. Because of isomorphic substitution, clays minerals have a net negative charge. To maintain electrical neutrality, exchangeable cations occupy the surfaces of clay particles and contribute to energy losses (Saarenketo, 1998). These cations concentrate in the *diffuse double layer* that surrounds clay minerals and provide an alternative pathway for electrical conduction. Surface conduction is directly related to the amount of clay particles in the soil and the concentration and mobility of the adsorbed cations on the clay particles (Shainberg et al., 1980). In general, the contribution of clay particles and surface conduction to electrical conductivity and energy loss is more evident in soils that have low rather than high salt concentrations (Klein and Santamarina, 2003).

Because of their high adsorptive capacity for water and exchangeable cations, clays increase the dissipation of electromagnetic energy. As a consequence, the penetration depth of GPR is inversely related to clay content. Olhoeft (1986), using a 100-MHz antenna, observed a penetration depth of about 30 m in some clay-free sands. However, with the addition of only 5% clay (by weight), the penetration depth was reduced by a factor of 20 (Olhoeft, 1986). Doolittle and Collins (1998) noted that depending on antenna frequency and the specific conductance of the soil solution, penetration depths range from 5 to 30 m in dry, sandy (>70% sand and <15% clay) soils, but average only 50 cm in wet, clayey (>35% clay) soils.

Soils contain various proportions of different clay minerals (e.g., members of kaolin, mica, chlorite, vermiculite and smectite groups). The size, surface area, cation-exchange capacity (CEC), and water-holding capacity of clay minerals vary greatly. Variations in electrical conductivity are attributed to differences in CEC associated with different clay minerals (Saarenketo, 1998). Electrical conductivity and energy loss increase with increasing CEC (Saarenketo, 1998). Soils with clay fractions dominated by high CEC clays (e.g., smectitic and vermiculitic soil mineralogy classes) are more attenuating to GPR than soils with an equivalent percentage of low CEC clays (e.g., kaolinitic, gibbsitic, and halloysitic soil mineralogy classes). Soils classified as belonging to the kaolinitic, gibbsitic, and halloysitic mineralogy classes characteristically have low CEC and low base saturation. As a general rule, for soils with comparable clay and moisture contents, greater depths of penetration can be achieved in highly weathered soils of tropical and subtropical regions than in soils of temperate regions.

6.2.2. Soil suitability maps for ground penetrating radar

Increasingly, GPR is being used in agronomic, archaeological, engineering, environmental, crime scene, and soil investigations. A common concern of GPR users is whether or not the radar will be able to achieve the desired depth of penetration. Ground penetrating radar is highly suited to most applications in dry sands, where penetration depths can exceed 50 m with low-frequency antennas (Smith and Jol, 1995). However, a thin, conductive soil horizon or layer will cause high rates of signal attenuation, severely restricting penetration depths and limiting the suitability of GPR for a large number of applications. In saline and sodic soils, where penetration depths are typically less than 25 cm (Daniels, 2004), GPR is an inappropriate tool. In wet clays, where penetration depths are typically less than 1 m (Doolittle et al., 2002), GPR has a very low potential for many applications.

Knowledge of soils and soil properties is important for the effective use of GPR. Most radar users have limited knowledge of soils and are unable to foretell the relative suitability of soils for GPR within project areas. Soil survey reports and databases provide information on soil properties that affect GPR and are available for most areas of the United States. Hubbard et al. (1990) developed a GPR suitability map for the state of Georgia based on information contained in published soil survey reports. Collins (1992) used the US soil taxonomic classification system to create GPR suitability maps based on properties within the upper 2 m of soils. Doolittle et al. (2002, 2003, 2007) developed and later revised a thematic map, the *Ground Penetrating Radar Soil Suitability Map of the Conterminous United States* (GSSM-USA) (Figure 6.1), which shows the relative suitability of soils for GPR applications. The *GSSM-USA* is based on field observations made throughout the United States and soil attribute data contained in the USDA-Natural Resources Conservation Service (NRCS) State Soil Geographic (STATSGO) database. The STATSGO database was developed by the USDA-NRCS for broad land use planning encompassing state, multi-state, and regional areas (National Soil Survey Center, 1994). The STATSGO database consists of digital map data, attribute data, and Federal Geographic Data Committee compliant metadata. The database is linked to soil interpretation records that contain data on the physical and chemical properties of about 18,000 different soils.

The lack of adequate data on soil moisture and the high spatial and temporal variations in the degree of soil wetness precluded the use of moisture content in the preparation of this map. As a consequence, properties selected to prepare the *GSSM-USA* principally reflect variations in the clay and soluble salt contents of soils. Attribute data used to determine the suitability indices of soils include taxonomic criteria, clay content and mineralogy, electrical conductivity, sodium absorption ratio, and calcium carbonate and calcium sulfate contents. Each soil attribute was rated and assigned an index value ranging from 1 to 6. Lower attribute index values are associated with lower rates of signal attenuation, greater penetration depths, and soil properties that are characteristically more suited to GPR. For each soil attribute, the most limiting (maximum) index value within depths of 1.0 or 1.25 m was selected. These limiting soil attribute indices were summed for each soil. For each soil map unit, the relative proportions of soils with the same index values were summed. The dominant index value (value with the most extensive representative area in each map unit) is selected as the GPR suitability index for each soil map unit. The dominant suitability index for each soil map unit is joined to the map unit identifiers in the digital map for classification and visualization.

Soils, Peatlands, and Biomonitoring 183

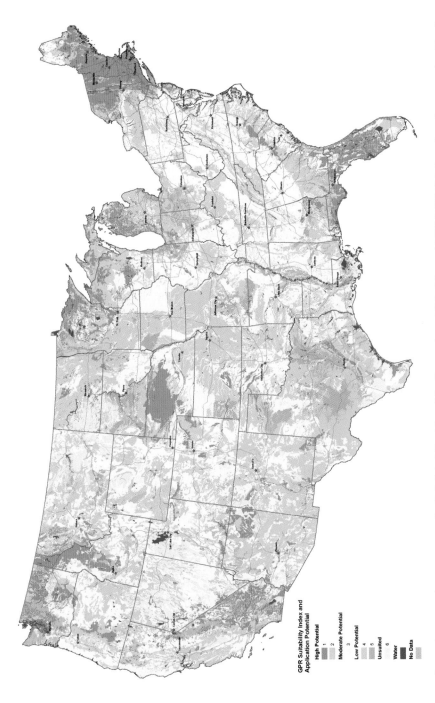

Figure 6.1 The *Ground Penetrating Radar Soil Suitability Map of the Conterminous United States* (GSSM-USA) is based on data contained in the state soil geographic (STATSGO) database.

Soil attribute index values and relative soil suitability indices are based on observed responses from antennas with center frequencies between 100 and 200 MHz. For mineral soils, the inferred suitability indices are based on unsaturated conditions and the absence of contrasting materials within depths of 1 m. Penetration depths and the relative suitability of mineral soils will be less under saturated conditions.

The *GSSM-USA* provides an indication of the relative suitability of soils to GPR within broadly defined soil and physiographic areas of the conterminous United States. Within any broadly defined area, the actual performance of GPR will depend on the local soil properties, the type of application, and the characteristics of the subsurface target. Because of the small compilation scale (1:250,000) of the *GSSM-USA*, the minimum polygon size is about 625 ha. As a consequence of this small map scale, field soil data have been generalized and much spatial information omitted.

Ground penetrating radar users would benefit from larger-scale, less-generalized maps, which show in greater detail the spatial distribution of soil properties that influence the penetration depth of GPR. Larger-scale GPR soil suitability maps have been prepared on a state basis using the Soil Survey Geographic (SSURGO) database (Doolittle et al., 2006). The SSURGO database contains the most detailed level of soil geographic data developed by the USDA-NRCS (1995). Base maps are USGS 7.5-min topographic quadrangles and 1:12,000 or 1:24,000 orthophotoquads. Soil maps in the SSURGO database duplicate the original soil survey maps, which were prepared at scales ranging from 1:12,000 to 1:63,360 (minimum delineation size ranging from about 0.6 to 16.2 ha, respectively) (Soil Survey Staff, 1993). The same soil properties attribute index values, and processing programs used to prepare the *GSSM-USA* are used with the SSURGO database to produce these larger-scale state maps.

An example of a state GPR soil suitability map, the *Ground Penetrating Radar Soil Suitability Map of Wisconsin* (*GSSM-WI*), is shown in Figure 6.2. The *GSSM-WI* was prepared at a display scale of 1:700,000. Compared with the *GSSM-USA* (see Figure 6.1), information contained on the *GSSM-WI* (see Figure 6.2) is less generalized, soil patterns are more intricate, and soil polygons are shown in greater detail. Broad spatial patterns, which correspond to major soil and physiographic units within Wisconsin, are evident on both thematic maps (see Figures 6.1 and 6.2). However, the *GSSM-WI* provides a more detailed overview of the spatial distribution of soil properties that influence the depth of penetration and effectiveness of GPR. As soil delineations are not homogenous and contain dissimilar inclusions, on-site investigations are needed to confirm the suitability of each soil polygon for different GPR applications. The spatial information contained on GPR soil suitability maps can aid investigators who are unfamiliar with soils in assessing the likely penetration depth and relative effectiveness of GPR within project areas. In addition, these maps can help radar users evaluate the relative appropriateness of using GPR, select the most suitable antennas and survey procedures, and assess the need and level of data processing. Ground penetrating radar soil suitability maps are available for most states and can be accessed at http://soils.usda.gov/survey/geography/maps/GPR/index.html. These maps are periodically updated as additional areas are surveyed and soil information is collected and certified.

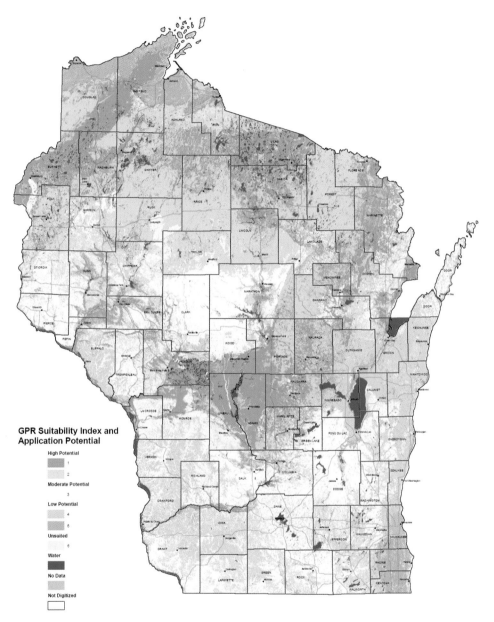

Figure 6.2 The *Ground Penetrating Radar Soil Suitability Map of Wisconsin* (GSSM-WI) is based on data contained in the Soil Survey Geographic (SSURGO) database.

6.2.3. Ground penetrating data and soil surveys

Soil surveys are the "systematic examination, description, classification, and mapping of soils" (Soil Science Society of America, 2001). The nature, composition, and boundaries of soil polygons that appear on soil maps were inferred by soil scientists from a limited number of point observations made with augers, probes, and shovels. Soil mapping is a slow and labor-intensive process. As a consequence, observations are generally sparse and a very large portion of the soil continuum below the surface is not observed. Constrained by limited exposures and burdened by partial or detached information, inferences on the nature and properties of soils must be extended across the more expansive areas between observation points. Because of these limitations, alternative methods are being explored to complement traditional soil survey techniques, provide more comprehensive coverage, and improve the assessment of soil properties. To be effective, these methods must be relatively fast, accurate, and inexpensive. Different geophysical tools are being used to characterize soil properties and variability at different scales and level of resolutions. Ground penetrating radar has been used to help characterize the soil continuum and support soil survey investigations.

Since the late 1970s, GPR has been used as a quality control tool for soil surveys in the United States. In 1979, the use of GPR for soil surveys was successfully demonstrated in Florida (Benson and Glaccum, 1979; Johnson et al., 1979). Because of the ubiquity of sandy soils with favorable characteristics and contrasting soil horizons, GPR has been extensively used to update soil surveys in Florida (Schellentrager et al., 1988).

In the United States, mineral soils are typically observed, described, and classified to a depth of 2 m or to bedrock (if within depths of 2 m) (Soil Survey Staff, 1999). Ground penetrating radar is principally used by soil scientists as a quality control tool to verify the taxonomic composition of soil map units, document the presence and depth to diagnostic soil horizons and features, and assess spatial and temporal variations in soil properties.

For most GPR soil investigations, a transect line or a small grid is established across a representative soil area. Typically, reference points are located at uniform intervals along transect or grid lines. The interval between reference points varies with the purpose of the survey and the anticipated variability of soil features under investigation but typically ranges from 0.5 to 15 m. A suitable radar antenna is towed or dragged along these lines. After reviewing the radar record in the field, soils are observed and described at selected reference points to verify GPR depth measurements and interpretations. Based on these observations, diagnostic subsurface horizons, contrasting layers, and/or soil features are identified and traced laterally across the radar record. The presence and depth to diagnostic subsurface horizons or soil features is used to determine the taxonomic classification and name of the soil at each reference point.

The most commonly used antennas for soil investigations have center frequencies between 100 and 500 MHz. Higher-frequency (400–500 MHz) antennas often provide more satisfactory results in relatively dry, electrically resistive soils. In highly attenuating soils, where the depth of penetration is very limited, these higher-frequency antennas often provide comparable depths and greater resolution than lower-frequency antennas. Antennas with frequencies of 900 MHz–1.5 GHz have been used for some shallow investigations in sandy soils. For organic soils, where greater depths of penetration are often needed, lower-frequency (70–200 MHz) antennas are commonly used.

Ground penetrating radar has been effectively used to provide data on the presence, depth, lateral extent, and variability of diagnostic subsurface horizons that are used to classify soils (Collins et al., 1986; Doolittle, 1987; Schellentrager et al., 1988; Puckett et al., 1990). Provided soil conditions are suitable, GPR is used to determine the depth to contrasting master (B, C, and R) subsurface horizons. Other soil horizons and layers (e.g., buried genetic horizons, dense root-restricting layers, frozen soil layers, illuvial accumulations of organic matter, and cemented or indurated horizons) have also been identified with GPR. Ground penetrating radar does not image subtle changes in soil properties (e.g., color, mottles, structure, porosity, and slight changes in texture), transitional horizons (e.g., AB, AC, BC), or vertical divisions in master horizons.

Radar interpretations provide fairly accurate measurements of the depth and thickness of some soil horizons. Johnson et al. (1979), working in sandy soils with well-expressed horizons, observed that radar-interpreted depths were within $\pm 2.5–5.0$ cm of the measured depths. Asmussen et al. (1986) observed an average difference of 19.2 cm between the interpreted and measured depths to argillic (Bt) horizons, which ranged in depth from approximately 20 to 450 cm. Rebertus et al. (1989) observed that the difference between the interpreted and measured depths to a discontinuity, which ranged in depth from 0 to about 230 cm, was less than 15 cm in 94% of the observations. Collins et al. (1989) observed an average difference of 6 cm between the interpreted and measured depths to bedrock, which ranged in depth from about 80 to 240 cm.

Typically, strong radar reflections (high-amplitude reflections) are produced by soil interfaces that have abrupt boundaries and separate contrasting soil materials. These interfaces often correspond to boundaries that separate soil horizons. Contrast between soil horizons is often associated with differences in moisture contents, physical (texture and bulk density), and/or chemical (organic carbon, calcium carbonate, and sesquioxides) properties. Ground penetrating radar has been used to estimate the depth to argillic (Asmussen et al., 1986; Collins and Doolittle, 1987; Doolittle, 1987; Truman et al., 1988; Doolittle and Asmussen, 1992), spodic (Collins and Doolittle, 1987; Doolittle, 1987; Burgoa et al., 1991), and placic (Lapen et al., 1996) horizons. These horizons generally have well-defined upper boundaries that display abrupt increases in bulk density and illuviated silicate clays (argillic horizon), humus and free sesquioxides (spodic horizon), or cemented Fe, Mn, or Fe–humus complexes (placic horizon). Ground penetrating radar has also been used to determine the thickness of albic horizons and chart the depth, lateral extent, and continuity of duripans, petrocalcic, and petroferric horizons (Doolittle et al., 2005), fragipans (Olson and Doolittle, 1985; Lyons et al., 1988; Doolittle et al., 2000), ortstein (Mokma et al., 1990a), and traffic pans (Raper et al., 1990). Duripans, petrocalcic, and petroferric horizons are indurated (primarily cemented with secondary SiO_2, $CaCO_3$, and, Fe_2O_3, respectively). Fragipans and traffic pans have higher bulk densities and are less permeable than overlying or underlying horizons. Ortstein is a cemented spodic horizon. Ground penetrating radar has been used to infer distinct changes in soil color associated with abrupt and contrasting changes in organic carbon contents (Collins and Doolittle, 1987). Ground penetrating radar has also been used to infer the concentration of lamellae (Farrish

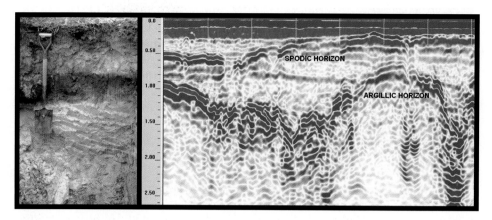

Figure 6.3 The spodic and argillic horizons of Pomona soil are well expressed in this picture and radar record from north-central Florida. (Picture of soil profile is courtesy of Dr. Mary Collins, University of Florida.)

et al., 1990; Mokma et al., 1990b; Tomer et al., 1996) and plinthite (Doolittle et al., 2005) in soils. In areas of permafrost, GPR has been used to estimate the thickness of active layers (Doolittle et al., 1990b).

Figure 6.3 shows a soil profile and radar record from an area of Pomona soil (sandy, siliceous, hyperthermic Ultic Alaquods) in north-central Florida. The Pomona soil formed in sandy overlying loamy (10 to 27 percent clay) marine sediments on the Lower Coastal Plain. The shovel in the picture of the soil profile (left) is about 90 cm in length. The depth scale on the radar record (right) is in meters. The white vertical lines at the top of the radar record represent equally spaced (3 m) reference points. The upper boundaries of the spodic and argillic horizons are abrupt and separate contrasting soil materials and therefore produce high-amplitude reflections. The spodic horizon is the dark subsurface horizon in the upper part of the soil profile (midway along the shovel handle). Spodic horizons are illuvial layers of active amorphous materials composed of organic matter and aluminum, sometimes with iron (Soil Survey Staff, 1999). Because of differences in their bulk density and water retention capacity, spodic horizons are detectable with GPR. On the radar record, the spodic horizon provides a continuous reflection that varies in depth from about 20 to 60 cm.

On the soil profile (see Figure 6.3), the argillic horizon appears as a grayish colored, subsurface horizon with an irregular upper boundary near the base of the shovel blade. Argillic horizons are illuvial layers that contain significant accumulations of silicate clay (Soil Survey Staff, 1999). Because of abrupt and substantial increases in clay content and bulk density, the upper boundary of argillic horizons is usually detectable with GPR. On the radar record (see Figure 6.3), the upper boundary of the argillic horizon is highly irregular and varies in depth from about 60 to 160 cm. Generally, argillic horizons provide smooth, continuous reflectors that occur at more uniform depths. The irregularly upper boundary of the argillic horizon is attributed to underlying dissolution features that are associated with karst.

Figure 6.4 contains a soil profile and radar record from an area of Enfield soil (coarse–silty over sandy or sandy-skeletal, mixed, active, mesic Typic Dystrudepts)

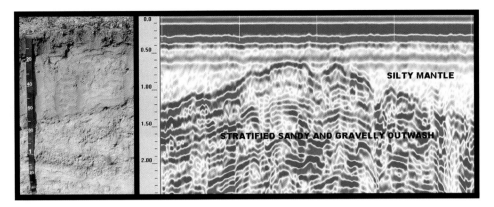

Figure 6.4 A discontinuity separating a loamy eolian mantle from sandy glacial outwash is evident in this picture and radar record from southern Rhode Island. (Picture of soil profile is courtesy of Jim Turenne USDA-NRCS.)

in southern Rhode Island. The depth scales are in centimeters on the soil profile (left) and meters on the radar record (right). The white vertical lines at the top of the radar record represent equally spaced (3 m) reference points. In both the soil profile and the radar record, an abrupt and contrasting *discontinuity* separates the loamy eolian mantle from the underlying sandy outwash. Discontinuities represent contrasting soil materials. Soil materials on both sides of this discontinuity differ substantially in particle size distribution, bulk density, pore size distribution, and mineralogy. On the radar record shown in Figure 6.4, the discontinuity affords an easily identified, high-amplitude reflector that ranges in depth from about 70 to 140 cm. Linear reflectors in the materials underlying the discontinuity helped to confirm that the substratum consists of glacial outwash rather than till. Tills represent unsorted and unstratified materials deposited by glacial ice. Typically, on radar records, tills display chaotic graphic signatures characterized by an abundance of point reflectors from cobbles and boulders and the absence of linear reflectors, which would suggest layering and the flow of water. Other than parallel bands of reverberated signals, the eolian mantle is relatively free of reflectors.

In many upland areas, it is difficult to excavate and examine soil profiles and determine the depths to bedrock. Rock fragments and irregular or weathered bedrock surfaces limit the effectiveness of conventional probing techniques. Ground penetrating radar has been used extensively to chart the depths to bedrock (Collins et al., 1989; Davis and Annan, 1989), changes in rock type (Davis and Annan, 1989), characterize internal bedding, cleavage and fracture planes (Holloway and Mugford, 1990; Stevens et al., 1995; Toshioka et al., 1995; Lane et al., 2000; Grasmueck et al., 2004; Nascimento da Silva et al., 2004; Porsani et al., 2005), and cavities, sinkholes, and fractures in limestone (Barr, 1993; Pipan et al., 2000; Al-fares et al., 2002).

In many upland soils, GPR is more reliable and effective than traditional soil-surveying tools for determining the depth to bedrock and the composition of soil map units based on soil depth criteria (Collins et al., 1989; Schellentrager and

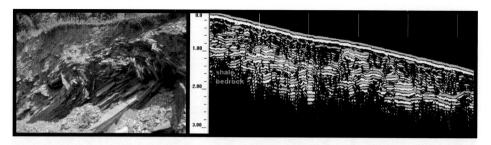

Figure 6.5 The irregular topography of the soil/bedrock interface can be traced laterally on this picture and radar record from an area of Berks and Weikert soils in central Pennsylvania.

Doolittle, 1991). The soil/bedrock interface often provides an abrupt and well-expressed, easily identifiable reflector on radar records. Often, this interface provides smooth, continuous, and high-amplitude reflections. However, the soil/bedrock interface is not always easy to identify on radar records. Coarse fragments in the overlying soil, irregular bedrock surfaces, fracturing, and the presence of saprolite make the identification of the soil/bedrock interface more ambiguous on some radar records.

Figure 6.5 shows a soil profile and a radar record from an area of Weikert and Berks soils (loamy-skeletal, mixed, active, mesic Lithic and Typic Dystrudepts, respectively) in central Pennsylvania. The depth scales is about 3 m. The white vertical lines at the top of the radar record represent equally spaced (3 m) reference points. Weikert and Berks soils are shallow (0–50 cm) and moderately deep (50–100 cm) to shale bedrock, respectively. On the picture of the soil exposure, the shale bedrock appears highly fractured with noticeably inclined, twisted, and convoluted bedding and fracture planes. On the radar record, a green-colored line has been used to identify the interpreted soil/bedrock interface. This interface is highly irregular and segmented. Because of the lack of a single, well-expressed, continuous, high-amplitude reflection, the picking of the soil/bedrock interface is unclear on this radar record, and the accuracy of interpreted soil depth measurements is lessened.

Ground penetrating radar has also been used by soil scientists and geomorphologists to improve soil–landscape models and soil map unit design on glacial-scoured uplands (Doolittle et al., 1988), wetland catena (Lapen et al., 1996), and coastal plain sediments (Rebertus et al., 1989; Puckett et al., 1990). Recent advancements in processing technologies have facilitated the manipulation of large datasets and the creation of 3D radar images. These displays can provide unique perspectives into the subsurface but have been infrequently used in soil–landscape investigations.

6.2.4. Uses of ground penetrating radar in organic soils and peatlands

Peatlands occupy an estimated area of $3.46 \times 10^6 \, km^2$ and comprise more than 50% of the global wetlands (Bridgham et al., 2001). Within the United States, peatlands cover an estimated area of $231,781 \, km^2$ (Bridgham et al., 2001). Globally, peatlands represent a significant soil carbon reserve and methane reservoir. Once avoided or overlooked, today many peatlands are managed to meet increasing agricultural, mining, and urban needs (Johnson and Worley, 1985). A prerequisite for the effective use and management of these peatlands is knowledge of the thickness, distribution, and volume of peat. Ground penetrating radar has been used to inventory and map peatlands. Compared to traditional surveying methods, GPR is faster and requires significantly less time and effort to obtain similar information on the thickness, volume, and geometry of peatlands (Jol and Smith, 1995).

Ground penetrating radar can provide information on the depth and geometry of organic deposits at a level of detail and accuracy that is comparable to information obtained with manual methods (Ulriksen, 1980). In a comparative study with traditional methods, Ulriksen (1982) found GPR to be a more efficient tool for estimating the thickness and characterizing the subsurface topography of organic deposits. Ground penetrating radar has been used to estimate the thickness and volume of organic deposits (Ulriksen, 1982; Shih and Doolittle, 1984; Tolonen et al., 1984; Collins et al., 1986; Worsfold et al., 1986; Welsby, 1988; Doolittle et al., 1990a; Pelletier et al., 1991; Hanninen, 1992; Turenne et al., 2006), to distinguish layers having differences in degree of humification and volumetric water content (Ulriksen, 1982; Tolonen et al., 1984; Worsfold et al., 1986; Chernetsov et al., 1988; Theimer et al., 1994; Lapen et al., 1996), and to classify organic soils (Collins et al., 1986). Lowe (1985) used GPR to assess the amount of logs and stumps buried in peatlands. Holden et al. (2002) used GPR to locate subsurface piping in organic deposits. Ground penetrating radar has also been used to provide information for the placement of roads, pipelines, and dikes on peatlands (Ulriksen, 1982; Saarenketo et al., 1992; Jol and Smith, 1995). Moorman et al. (2003) discussed GPR surveys of peatlands located in areas of permafrost. Ground penetrating radar has also been used in peatlands to characterize subsurface deposits and look for communalities in substrate formations and sequences, which may be used for their hydrologic classification.

Although profiling depths as great as 8–10 m have been reported in some peatlands (Ulriksen, 1980; Worsfold et al., 1986), GPR does not provide similar results on all organic soils. In organic soils, the penetration depth and resolution of subsurface features is limited by the specific conductivity and the concentration of solutes in the pore water (Theimer et al., 1994). In general, penetration depths are greater in ombrogenous bogs than in minerogenous fens (Malterer and Doolittle, 1984). Ombrogenous bogs receive inputs only from precipitation and therefore have lower pH and basic cation (Ca, Mg, Na, and K) contents. Minerogenous fens receive significant inputs from groundwater and/or overland runoff, which contain varying amounts of soluble salts. As a consequence, the groundwater in minerogenous fens often has higher ionic conductivity and pH than the groundwater in ombrogenous bogs (Bridgham et al., 2001). Ground penetrating radar is more effective in acidic, low-nutrient peatlands than in alkaline, high-nutrient peatlands. However, because of variations in the specific conductivity of the groundwater, wide ranges in minerotrophy exist (Bridgham et al., 2001).

Organic soils that are classified as *sulfidic* or *halic* are unsuited to GPR. Typically, these organic soils form in coastal marshes that are inundated by brackish waters and are either enriched with acid sulfates (*sulfidic*) or salt (*halic*) (Soil Survey Staff, 1999).

The high salinities and ionic solute levels in these fens rapidly absorb the radar's electromagnetic energy and restrict observation depths to less than 0.5 m.

Organic deposits often display considerable anisotropy in moisture content and bulk density. Differences in moisture contents have allowed some to distinguish organic layers that are different in degree of humification, bulk density, and dielectric permittivity (Tolonen et al., 1982; Chernetsov et al., 1988; Hanninen, 1992; Nobes and Warner, 1992; Theimer et al., 1994). Some peatlands consist of organic layers that are interstratified with mineral soil layers. These mineral layers may have high clay contents that rapidly attenuate the radar's energy and limit penetration depths.

Lower-frequency (<200 MHz) antennas are typically used to profile peatlands. Survey procedures vary with site conditions and survey objectives. In higher latitudes, peatlands are often surveyed during winter months when the upper organic soil layers are frozen and the surface is snow covered. Under these conditions, the use of snowmobiles or tracked vehicles facilitates GPR surveys. In lower latitudes, grass and reed-covered peatlands have been successfully surveyed in all seasons with airboats. Pelletier et al. (1991) described the use of helicopters to survey extensive peatlands in remote areas of Ontario.

Figure 6.6 shows a soil profile and a radar record from an area of Freetown soil (dysic, mesic Typic Haplosaprists) in southeastern Massachusetts. In Figure 6.6, the depth scales are in meters: 0–2 m on the soil profile and 0–7.2 m on the radar record. The white vertical lines at the top of the radar record represent equally spaced (10 m) reference points. Abrupt and strongly contrasting differences in water content makes the organic/mineral interface distinguishable on radar records. In Figure 6.6, this interface forms a conspicuous reflector that varies in depth from about 1.0 to 5.1 m. Weak planar reflectors are evident and suggest layering within the organic materials. The layering within the organic materials represents differences in degree of decomposition and associated water contents. On the soil profile shown in Figure 6.6, layers of lighter-colored, less-decomposed organic soil materials (fibric materials) alternate with darker-colored layers of more decomposed organic soil materials (sapric materials). No variations in signal attenuation,

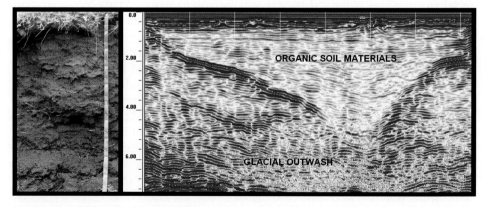

Figure 6.6 The organic/mineral soil material provides a high-amplitude reflector that can be traced laterally across a peatland formed in a kettle in southeastern Massachusetts.

penetration depths, or the effectiveness of GPR have been associated with differences in the degree of organic matter decomposition (e.g., fibric, hemic, and sapric organic materials).

6.3. BIOMONITORING

Ground penetrating radar can be used to detect and monitor below-ground biological structures, provided there is sufficient electromagnetic contrast with the surrounding soil matrix. Forest researchers are interested in measuring root biomass, distribution, and architecture to evaluate forest productivity and health. Tree root systems are commonly evaluated via labor-intensive, destructive, time-consuming excavations. Ground penetrating radar has been used to resolve roots and buried organic debris, assess root size, map root distribution, and estimate root biomass (Butnor et al., 2001). Being noninvasive and nondestructive, GPR allows repeated measurements that facilitate the study of root system development. Root biomass studies provide insight into the effectiveness of varying water and fertilizer treatments and are an indicator of tree health. Although live tree roots are the most common targets for biomonitoring studies, GPR can be used to detect internal tree defects (Miller and Doolittle, 1990; Schad et al., 1996).

Hruska et al. (1999) first used GPR to nondestructively map the distribution of coarse (>3 cm diameter) root systems. In this study, a 450-MHz antenna and an image analysis system were used to produce 3D graphics showing the distribution of roots of several oak trees (*Quercus petraea*) within a 6 × 6 m plot. Woody roots often present very complex reflective surfaces, which require some degree of verification. This may be accomplished with root excavations (Stokes et al., 2002) or soil core samples (Butnor et al., 2003) to confirm that root distribution maps are accurate for a particular site. When data collected with a 450-MHz antenna were compared to excavations, large roots were accurately profiled, while smaller structures (<2 cm) were not detectable (Stokes et al., 2002). Surface-based GPR systems can provide useful information on lateral roots; however, the distribution of large roots extending vertically or near-vertically in the soil is not possible (Stokes et al., 2002).

Roots, as small as 0.5 cm in diameter, have been detected at depths of less than 30 cm with a 1.5-GHz antenna in well-drained, sandy soils (Butnor et al., 2001). However, without detailed, methodical scanning of small grids, it is not possible to separate roots by size class or depth under field conditions (Wielopolski et al., 2000; Butnor et al., 2001). Under optimal conditions in a sand test bed, enhanced migration filtering methods have allowed accurate determination of root diameter (Barton and Montagu, 2004). This work represents an important advance in postcollection processing of root data, but the ideal conditions (widely spaced, nonoverlapping roots, scanned at 90°) are quite different from the orientation and geometry of root reflective surfaces found in a forest. More work is needed to parameterize this type of analysis for real-world conditions. Since forests and tree plantations are often found on soils that are marginal for agriculture, there are many surface and textural conditions, which can confound interpretation. Root detection is ineffective in soils with high clay or water contents, having large number of coarse fragments, or in most unimproved, forested terrains where presence of herbaceous vegetation, fallen trees limbs, and irregular soil surfaces impede the travel of the antenna (Butnor et al., 2001).

The estimation of root mass and root distribution in forests has been successful on sites amenable to radar investigations. Butnor et al. (2003) correlated GPR-based estimates of root biomass within the upper 30 cm of soil profiles with harvested root samples. With advanced image processing, high-amplitude areas and reflector tally were directly proportional to the actual root biomass. A highly significant ($r = 0.86$, $p < 0.0001$) relationship was observed between actual biomass in cores and GPR estimates in a loblolly pine (*Pinus taeda* L.) plantation. Transect-based root biomass surveys combined with small destructive samples (soil cores) are the most widely adopted application of biomonitoring with GPR. The USDA Forest Service, Southern Research Station has partnered with universities to include GPR root biomass surveys in forest productivity studies in North Carolina, South Carolina, Georgia, Florida, and Ontario. Other practical applications of this methodology include monitoring residual root materials that harbor root disease fungi (*Armillaria* spp.) following the clearing of an old peach orchards (Cox et al., 2005) and evaluating the mass of coarse roots, burls, and lignotubers in a scrub-oak ecosystem that had been exposed to elevated carbon dioxide at the Kennedy Space Center (Stover et al., 2007).

Postcollection processing is necessary to reduce clutter on radar records containing root data. Tree roots typically appear as hyperbolic reflections on radar records (Figure 6.7a), unless the root follows the same path as the antenna. Background removal filters are required to eliminate parallel echoes from plane reflectors such as the ground surface or soil horizons (Oppenheim and Schafer, 1975). Background removal is helpful to distinguish roots near the soil surface from the surface reflection generated at the soil–air interface (see Figure 6.7b). Reflected GPR data may not be representative of the actual size and shape of the buried anomaly. Migration techniques are essential for developing a 3D representation of roots. Kirchoff migration is a filter technique (see Figure 6.7c) that uses the geometry of a hyperbolic reflection to guide decomposition to a representative size (Oppenheim and Schafer, 1975; Barton and Montagu, 2004). However, Kirchoff migration may be confused by the variable orientations of roots. An alternative approach is the Hilbert transformation (see Figure 6.7d), which uses the magnitude of the return signal to decompose multiple hyperbolic reflections into a more compact and representative form (Oppenheim and Schafer, 1975; Berkhout, 1981; Daniels, 2004). Both techniques can be very valuable for assessing tree roots (Butnor et al., 2003; Stover et al., 2007). The Hilbert transform is useful when the orientations of the roots are unknown, but may be affected by moisture content in poorly drained sites.

There has been considerable interest in mapping tree root systems to understand root architecture and soil volume utilization (Hruska et al., 1999; Cermak et al., 2000; Stokes et al., 2002). Compared with simple transects for biomass analysis, 3D datasets are tedious to collect and process for interpretation. As long as the grid line spacing is kept small (2–5 cm between scans), larger roots that are continuous across several two-dimensional (2D) radar records are distinguishable. Reconstructing

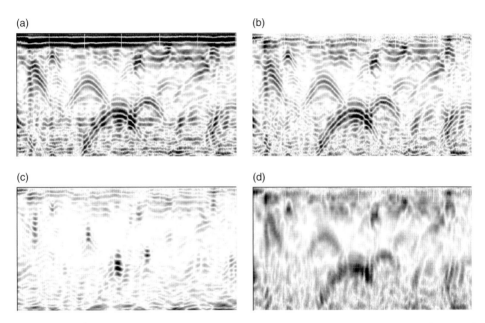

Figure 6.7 Radar profiles collected with a 1500-MHz antenna in the North Carolina Sand Hills ($Z = 0$–0.6 m, $X = 3$ m). In this well-drained, sandy soil, there is sufficient contrast to resolve tree roots. Interpretation may be enhanced by digital signal processing: (a) raw data, (b) background removal, (c) background removal and migration, and (d) background removal and Hilbert transform.

the location of roots is straightforward, but successfully modeling size, shape and root volume is not. Examples of mapping loblolly pine (*P. taeda* L.) roots are shown in Figure 6.8, where a series of Z slices (X and Y coordinates projected at specific Z depth) illustrate the location of several tree roots located between two rows of trees. For most forest survey projects, root biomass transects yield-sufficient information. Three-dimensional root mapping is useful when detailed root location information is required for a small area, provided there is sufficient time to collect and process the data.

Surface-based GPR can provide excellent records of lateral roots. However, some forest trees have a significant allocation to large, vertical tap roots (i.e. loblolly pine, *P. taeda* L., longleaf pine, *Pinus palustris* Mill.), which cannot be accurately assessed by surface measures (Butnor et al., 2003). A collaborative project between the USDA Forest Service, Southern Research Station (Research Triangle Park, NC), Radarteam AB (Boden, Sweden), and the SLU, Vindeln Experimental Forest System (Vindeln, Sweden) was undertaken in 2003 to assess the potential of high-frequency borehole radar to detect vertical near-surface reflectors (0–2 m). Cross-hole tomography provided excellent information on the depth of electromagnetic anomalies but was less useful for imaging near-surface features. Borehole to surface data provided the best information on the near surface, where the bulk of roots are found (0–0.3 m). Cross-hole and borehole to surface data may be combined to further define vertical root systems.

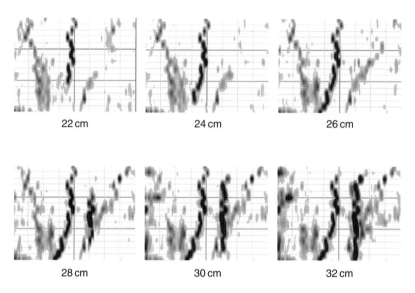

Figure 6.8 A series of parallel radar scans (2 cm interval) were combined using RADAN 4.0 to map loblolly pine roots. Each Z slice presented above is reconstructed from the raw data and centered (±1 cm) at the specified depth. The x-axis is 3 m and the y-axis is 2 m.

Ground penetrating radar has been used to detect internal defects in forest and urban trees (Miller and Doolittle, 1990; Detection Sciences, Inc., 1994; Nicolotti et al., 2003). Internal decay, which results in changes in moisture content or wood density, can provide a detectable target for electromagnetic techniques (Nicolotti et al., 2003). Miller and Doolittle (1990) were able to detect hollow areas, decayed wood, and brown rot in several species of forest trees. Using a 500-MHz antenna in bistatic mode, healthy trees were generally void of internal reflections, with the exception of weak parallel bands attributed to variations in moisture and wood density near the heartwood/sapwood interface. Miller and Doolittle (1990) found that areas of hollowness and decay were correlated with cluttered reflections and discontinuities on radar records. Four trees were destructively ground-truthed and found to have a high degree of accuracy with the GPR assessment. High-frequency radar (1.5 GHz) has been employed to identify areas of decay in a plane tree (*Platanus hybrida* Brot.) in an urban setting (Nicolotti et al., 2003). By advancing the antenna around the circumference of the tree, researchers were able to acquire data in single reflection mode from a bistatic antenna. The linear, 2D data were transformed into polar coordinates for ready comparison to tree sections. There was good agreement between radar assessment of decay and destructive sampling via physical means; areas of decay exhibited increased dielectric properties. The greatest difficulty with using GPR to evaluate defects is the difficulty in coupling the antenna to the curved bark surface of the tree and interpretation of complex data (Schad et al., 1996; Nicolotti et al., 2003). Differences between tree species, stem diameters, moisture gradients related to heartwood development, and environmental conditions may make interpretation between trees complicated. This area of research is rapidly advancing, and applications of GPR designed specifically for trunk evaluations are now commercially available.

REFERENCES

Al-fares, W., Bakalowicz, M., Guerin, R. and Dukhan, M., 2002, Analysis of the karst aquifer structure of the Lamalou area (Herault, France) with ground-penetrating radar. Journal of Applied Geophysics, Vol. 51, pp. 97–106.

Asmussen, L.E., Perkins, H.F. and Allison, H.D., 1986, Subsurface descriptions by ground-penetrating radar for watershed delineation. Georgia Agricultural Experiment Stations, University of Georgia, Athens, Georgia, USA, Research Bulletin 340, 15 p.

Barr, G.L., 1993, Application of ground-penetrating radar methods in determining hydrogeologic conditions in a karst area, West-Central Florida. U.S. Geological Survey Water Resources Investigation Report 92-4141, 26 p.

Barton, C.V.M. and Montagu, K.D., 2004, Detection of tree roots and determination of root diameters by ground-penetrating radar under optimal conditions. Tree Physiology, Vol. 24, No. 12, pp. 1323–1331.

Benson, R. and Glaccum, R., 1979, Test report. The application of ground-penetrating radar to soil surveying for National Aeronautical and Space Administration (NASA). Technos Inc. Miami, Florida, USA, 18 p.

Berkhout, A.J., 1981, Wave field extrapolation techniques in seismic migration, a tutorial. Geophysics, Vol. 46, pp. 1638–1656.

Bridgham, S.C., Ping, C-L., Richardson, J.L. and Updegraff, K., 2001, Soils of northern peatlands: Histosols and Gelisols, in Richardson, J.L. and Vepraskas, M.J. (eds), Wetland Soils: Genesis, Hydrology, Landscapes, and Classification, Lewis Publishers, Boca Raton, Florida, USA, pp. 343–370.

Burgoa, B., Mansell, R.S., Sawka, G.J., Nkedi-Kizza, P., Capece, J. and Campbell, K., 1991, Spatial variability of depth to Bh horizon in Florida Haplaquods using ground-penetrating radar. Soil and Crop Science Society of Florida Proceedings, Vol. 50, pp. 125–130.

Butnor, J.R., Doolittle, J.A., Kress, L., Cohen, S. and Johnsen, K.H., 2001, Use of ground-penetrating radar to study tree roots in the southeastern United States. Tree Physiology, Vol. 21, pp. 1269–1278.

Butnor, J.R., Doolittle, J.A., Johnsen, K.H., Samuelson, L., Stokes, T. and Kress, L., 2003, Utility of ground-penetrating radar as a root biomass survey tool in forest systems. Soil Science Society of America Journal, Vol. 67, pp. 1607–1615.

Campbell, J.E., 1990, Dielectric properties and influence of conductivity in soils at one to fifty megahertz. Soil Science Society of America Journal, Vol. 54, pp. 332–341.

Cermak, J., Hruska, J., Martinkova, M. and Prax, A., 2000, Urban tree root systems and their survival near houses analyzed using ground-penetrating radar and sap flow techniques. Plant and Soil, Vol. 219, pp. 103–116.

Chernetsov, E.A., Beletsky, N.A. and Baev, M.Y., 1988, Radar profiling of peat and gyttja deposits. Proceedings, 8th International Peat Congress, Leningrad, USSR, International Peat Society, Jyska, Finland, pp. 15–21.

Collins, M.E., 1992, Soil taxonomy: A useful guide for the application of ground-penetrating radar. Proceedings, Fourth International Conference on Ground-Penetrating Radar, June 8–13, 1992, Rovaniemi, Finland, Geological Survey of Finland, Special Paper 16, pp. 125–132.

Collins, M.E. and Doolittle, J.A., 1987, Using ground-penetrating radar to study soil microvariability. Soil Science Society of America Journal, Vol. 51, pp. 491–493.

Collins, M.E., Schellentrager, G.W., Doolittle, J.A. and Shih, S.F., 1986, Using ground-penetrating radar to study changes in soil map unit composition in selected Histosols. Soil Science Society of America Journal, Vol. 50, pp. 408–412.

Collins, M.E., Doolittle, J.A. and Rourke, R.V., 1989, Mapping depth to bedrock on a glaciated landscape with ground-penetrating radar. Science Society of America Journal, Vol. 53, pp. 1806–1812.

Cox, K.D., Scherm, H. and Serman, N., 2005, Ground-penetrating radar to detect and quantify residual root fragments following peach orchard clearing. HortTechnology, Vol. 15, No. 3, pp. 600–607.

Curtis, J.O., 2001, Moisture effects on the dielectric properties of soils: Institute of Electrical and Electronics Engineers. Transactions on Geoscience and Remote Sensing, Vol. 39, No. 1, pp. 125–128.

Daniels, D.J., 2004, Ground-penetrating Radar, 2nd edition, The Institute of Electrical Engineers, London, UK, 726 p.

Davis, J.L. and Annan, A.P., 1989, Ground-penetrating radar for high resolution mapping of soil and rock stratigraphy. Geophysical Prospecting, Vol. 37, pp. 531–551.

Detection Sciences, Inc., 1994, Inspection of wood with impulse radar, Final Report, Research Joint Venture Agreement FPL-94-2326: U.S. Department of Agriculture, Forest Service, Forest Products Laboratory, Madison, Wisconsin, USA.

Doolittle, J.A., 1987, Using ground-penetrating radar to increase the quality and efficiency of soil surveys. Soil Survey Techniques, Soil Science Society of America, Madison, Wisconsin, USA, Special Publication No 20, pp. 11–32.

Doolittle, J.A., Rebertus, R.A., Jordan, G.B., Swenson, E.I. and Taylor, W.H., 1988, Improving soil-landscape models by systematic sampling with ground-penetrating radar. Soil Survey Horizons, Vol. 29, No. 2, pp. 46–54.

Doolittle, J.A. and Asmussen, L.E., 1992, Ten years of applications of ground-penetrating radar by the United States Department of Agriculture. Proceedings, Fourth International Conference on Ground-Penetrating Radar, June 8–13, 1992, Rovaniemi, Finland, Geological Survey of Finland, Special Paper 16, pp. 139–147.

Doolittle, J.A. and Collins, M.E., 1995, Use of soil information to determine application of ground-penetrating radar. Journal of Applied Geophysics, Vol. 33, pp. 101–108.

Doolittle, J., Fletcher, P. and Turenne, J., 1990a, Estimating the thickness of organic materials in cranberry bogs. Soil Survey Horizons, Vol. 31, No. 3, pp. 73–78.

Doolittle, J.A., Hardisky, M.A. and Gross, M.F., 1990b, A ground-penetrating radar study of active layer thicknesses in areas of moist sedge and wet sedge tundra near Bethel, Alaska, U.S.A. Arctic and Alpine Research, Vol. 22, No. 2, pp. 175–182.

Doolittle, J.A. and Collins, M.E., 1998, A comparison of EM induction and GPR methods in areas of karst. Geoderma, Vol. 85, pp. 83–102.

Doolittle, J., Hoffmann, G., McDaniel, P., Peterson, N., Gardner, B. and Rowan, E., 2000, Ground-penetrating radar interpretations of a fragipan in Northern Idaho. Soil Survey Horizons, Vol. 41, No. 3, pp. 73–82.

Doolittle, J.A., Minzenmayer, F.E., Waltman, S.W. and Benham, E.C., 2002, Ground-penetrating radar soil suitability map of the conterminous United States. Proceedings, Ninth International Conference on Ground-Penetrating Radar, 30 April–2 May, 2002, Santa Barbara, California, USA, Proceedings of SPIE, Vol. 4158, pp. 7–12.

Doolittle, J.A., Minzenmayer, F.E., Waltman, S.W. and Benham, E.C., 2003, Ground-penetrating radar soil suitability maps. Environmental and Engineering and Geophysics Journal, Vol. 8, No. 2, pp. 49–56.

Doolittle, J., Daigle, J., Kelly, J. and Tuttle, W., 2005, Using GPR to characterize plinthite and ironstone layers in Ultisols. Soil Survey Horizons, Vol. 46, No. 4, pp. 179–184.

Doolittle, J.A., Minzenmayer, F.E., Waltman, S.W., Benham, E.C., Tuttle, J.W. and Peaslee, S., 2006, State ground-penetrating radar soil suitability maps. Proceedings, Eleventh International Conference on Ground-Penetrating Radar, June 19–22, 2006, Columbus, Ohio, USA, Paper HYD; 13_mdx., pp. 1–8.

Doolittle, J.A., Minzenmayer, F.E., Waltman, S.W., Benham, E.C., Tuttle, J.W. and Peaslee, S., 2007, Ground-penetrating radar soil suitability map of the conterminous United States. Geoderma, Vol. 141, pp. 416–421.

Farrish, K.W., Doolittle, J.A. and Gamble, E.E., 1990, Loamy substrata and forest productivity of sandy glacial drift soils in Michigan. Canadian Journal of Soil Science, Vol. 70, pp. 181–187.

Grant, J.A. and Schultz, P.H., 1994, Erosion of ejecta at Meteor Crater: Constraints from ground-penetrating radar. Proceedings, Fifth International Conference on Ground-Penetrating Radar, June 12–14, 1994, Kitchner, Ontario, Canada, Waterloo Centre for Groundwater Research and the Canadian Geotechnical Society, pp. 789–803.

Grasmueck, M., Weger, R. and Horstmeyer, H., 2004, Three-dimensional ground-penetrating radar imaging of sedimentary structures, fractures, and archaeological features at submeter resolution. Geology, Vol. 32, No. 11, pp. 933–936.

Hanninen, P., 1992, Application of ground-penetrating radar techniques to peatland investigations. Proceedings, Fourth International Conference on Ground-Penetrating Radar, June 8–13, 1992, Rovaniemi, Finland. Geological Survey of Finland, Special Paper 16, pp. 217–221.

Holden, J., Burt, T.P. and Vilas, M., 2002, Application of ground-penetrating radar to the identification of subsurface piping in blanket peat. Earth Surface Processes and Landforms, Vol. 27, pp. 235–249.

Holloway, A.L. and Mugford, J.C., 1990, Fracture characterization in granite using ground probing radar. CIM Bulletin, Vol. 83, No. 940, pp. 61–70.

Hruska, J., Cermak, J. and Sustek, S., 1999, Mapping tree root systems with ground penetrating radar. Tree Physiology, Vol. 19, pp. 125–130.

Hubbard, R.K., Asmussen, L.E. and Perkins, H.F., 1990, Use of ground-penetrating radar on upland coastal plain soils. Journal of Soil and Water Conservation, Vol. 45, No. 3, pp. 399–405.

Jackson, M.L., 1959, Frequency distribution of clay minerals in major great soil groups as related to the factors of soil formation. Clays and Clay Minerals, Vol. 6, pp. 133–143.

Johnson, C.W. and Worley, I.A., 1985, Bogs of the northeast. University Press of New England, Hanover and London, New Hampshire, USA, 289 p.

Johnson, R.W., Glaccum, R. and Wojtasinski, R., 1979, Application of ground penetrating radar to soil survey. Soil and Crop Science Society of Florida Proceedings, Vol. 39, pp. 68–72.

Jol, H.M. and Smith, D.G., 1995, Ground penetrating radar surveys of peatlands for oilfield pipelines in Canada. Journal of Applied Geophysics, Vol. 34, pp. 109–123.

Klein, K.A. and Santamarina, J.C., 2003, Electrical conductivity of soils: Underlying phenomena. Journal of Environmental and Engineering Geophysics, Vol. 8, No. 4, pp. 263–273.

Lane, J.W., Buursink, M.L., Haeni, F.P. and Versteeg, R.J., 2000, Evaluation of ground-penetrating radar to detect free-phase hydrocarbons in fractured rocks – results of numerical modeling and physical experiments. Ground Water, Vol. 38, No. 6, pp. 929–938.

Lapen, D.R., Moorman, B.J. and Price, J.S., 1996, Using ground penetrating radar to delineate subsurface features along a wetland catena. Soil Science Society of America Journal, Vol. 60, pp. 923–931.

Lebron, I., Robinson, D.A., Goldberg, S. and Lesch, S.M., 2004, The dielectric permittivity of calcite and arid zone soils with carbonate minerals. Soil Science Society of America Journal, Vol. 68, pp. 1549–1559.

Lowe, D.J., 1985, Application of impulse radar to continuous profiling of tephra-bearing lake sediments and peats: An initial evaluation. New Zealand Journal of Geology and Geophysics, Vol. 28, pp. 667–674.

Lyons, J.C., Mitchell, C.A. and Zobeck, T.M., 1988, Impulse radar for identification of features in soils. Journal of Aerospace Engineering, Vol. 1, pp. 18–27.

Malterer, T.J. and Doolittle, J.A., 1984, Evaluation of ground-penetrating radar for characterizing Minnesota Histosols, Annual Meeting of Soil Science Society of America, Anaheim, California, USA, Agronomy Abstracts, p. 235.

McNeill, J.D., 1980, Electrical conductivity of soils and rock. Technical Note TN-5, Geonics Limited, Mississauga, Ontario, Canada, 21 p.

Miller, W.F. and Doolittle, J.A., 1990, The application of ground-penetrating radar to detection of internal defect in standing trees. Proceedings, 7th International Nondestructive Testing of Wood Symposium, September 27–29, 1989, Madison, Wisconsin, USA, Washington State University, pp. 263–274.

Mokma, D.L., Schaetzl, R.J., Doolittle, J.A. and Johnson, E.P., 1990a, Ground-penetrating radar study of ortstein continuity in some Michigan Haplaquods. Soil Science Society of America Journal, Vol. 54, pp. 936–938.

Mokma, D.L., Schaetzl, R.J., Johnson, E.P. and Doolittle, J.A., 1990b, Assessing Bt horizon character in sandy soils using ground-penetrating radar: Implications for soil surveys. Soil Survey Horizons, Vol. 30, No. 2, pp. 1–8.

Moorman, B.J., Robinson, S.D. and Burgess, M.M., 2003, Imaging periglacial conditions with ground-penetrating radar. Permafrost and Periglacial Processes, Vol. 14, pp. 319–329.

Nascimento da Silva, C.C., de Medeiros, W.E., Jarmin de Sa, E.F. and Neto, P.X., 2004, Resistivity and ground-penetrating radar images of fractures in a crystalline aquifer: A case study in Caicara farm – NE Brazil. Journal of Applied Geophysics, Vol. 56, pp. 295–307.

National Soil Survey Center, 1994, State Soil Geographic Database; Data use information: U.S. Department of Agriculture-Natural Resources Conservation Service, Miscellaneous Publication No. 1492, National Soil Survey Center, Lincoln, Nebraska, USA, 110 p.

Neal, A., 2004, Ground-penetrating radar and its use in sedimentology: principles, problems and progress. Earth-Science Reviews, Vol. 66, pp. 261–330.

Nicolotti, G., Socco, L.V., Martinis, R., Godio, A. and Sambuelli, L., 2003, Application and comparison of three tomographic techniques for detection of decay in trees, Journal of Arboriculture, Vol. 29, No. 2, pp. 66–78.

Nobes, D.C. and Warner, B.G., 1992, Application of ground penetrating radar to a study of peat stratigraphy: preliminary results, in Pilon, J. (ed.), Ground Penetrating Radar, Geological Survey of Canada, Paper 90-4, pp. 133–138.

Olhoeft, G.R., 1986, Electrical properties from 10-3 to 10+9 Hz – physics and chemistry. Physics and Chemistry of Porous Media II, American Institute of Physics Conference Proceedings, Ridgefield, Connecticut, USA, pp. 281–298.

Olhoeft, G.R., 1998, Electrical, magnetic, and geometric properties that determine ground-penetrating radar performance. Proceedings, Seventh International Conference on Ground-Penetrating Radar, May 27–30, 1998, University of Kansas, Lawrence, Kansas, USA, pp. 177–182.

Olson, C.G. and Doolittle, J.A., 1985, Geophysical techniques for reconnaissance investigations of soils and surficial deposits in mountainous terrain. Soil Science. Society of America Journal, Vol. 49, pp. 1490–1498.

Oppenheim, A.V. and Schafer, R.W., 1975, Digital Signal Processing: Prentice Hall, Inc., Englewood Cliffs, NJ, USA, 784 p.

Pelletier, R.E., Davis, J.L. and Rossiter, J.R., 1991, Peat analysis in the Hudson Bay lowlands using ground-penetrating radar. Proceedings, International Geoscience and Remote Sensing Symposium, June 1991, Helsinki, Finland, pp. 2141–2144.

Pipan, M., Baradello, L., Forte, E. and Prizzon, A., 2000, GPR study of bedding planes, fractures and cavities in limestone. Proceedings, Eight International Conference on Ground-Penetrating Radar, May 23–26, 2000, Gold Coast, Queensland, Australia. Proceedings of SPIE – The International Society of Optical Engineering, Bellingham, Washington, pp. 682–687.

Porsani, J.L., Elis, V.R. and Hiodo, F.Y., 2005, Geophysical investigations for the characterization of fractured rock aquifer in Itu, SE Brazil. Journal of Applied Geophysics, Vol. 57, pp. 119–128.

Puckett, W.E., Collins, M.E. and Schellentrager, G.W., 1990, Design of soil map units on a karst area in West Central Florida. Soil Science Society of America Journal, Vol. 54, pp. 1068–1073.

Raper, R.L., Asmussen, L.E. and Powell, J.B., 1990, Sensing hard pan depth with ground-penetrating radar. Transaction of the American Society of Agricultural Engineers, Vol. 33, pp. 41–46.

Rebertus, R.A., Doolittle, J.A. and Hall, R.L., 1989, Landform and stratigraphic influences on variability of loess thickness in northern Delaware. Soil Science Society of America Journal, Vol. 53, pp. 843–847.

Saarenketo, T., 1998, Electrical properties of water in clay and silty soils. Journal of Applied Geophysics, Vol. 40, pp. 73–98.

Saarenketo, T., Hietala, K. and Salmi, T., 1992, GPR applications in geotechnical investigations of peat for road survey purposes. Fourth International Conference on Ground-Penetrating Radar, June 7–12, 1992, Rovaniemi, Finland, Geological Survey of Finland, Special Paper 16, pp. 293–305.

Schad, K.C., Schmoldt D.L. and Ross, R.J., 1996, Nondestructive methods for detecting defects in softwood logs. U.S. Department of Agriculture, Forest Service, Forest Products Laboratory, Madison, Wisconsin, USA, Research Paper FPL-RP-546, 13 p.

Schellentrager, G.W. and Doolittle, J.A., 1991, Systematic sampling using ground-penetrating radar to study regional variation of a soil map unit, in Mausbach, M.J. and Wilding, L.P. (eds), Spatial Variabilities of Soils and Landforms, Soil Science Society of America, Madison, Wisconsin, USA, Special Publication No. 28, pp. 199–214.

Schellentrager, G.W., Doolittle, J.A., Calhoun, T.E. and Wettstein, C.A., 1988, Using ground-penetrating radar to update soil survey information. Soil Science Society of America Journal, Vol. 52, pp. 746–752.

Shainberg, I., Rhoades, J.D. and Prather, R.J., 1980, Effect of exchangeable sodium percentage, cation exchange capacity, and soil solution concentration on soil electrical conductivity. Soil Science Society of America Journal, Vol. 44, pp. 469–473.

Shih, S.F. and Doolittle, J.A., 1984, Using radar to investigate organic soil thickness in the Florida Everglades. Soil Science Society of America Journal, Vol. 48, pp. 651–656.

Smith, D.G. and Jol, H.M., 1995, Ground penetrating radar: Antenna frequencies and maximum probable depths of penetration in quaternary sediments. Journal of Applied Geophysics, Vol. 33, pp. 93–100.

Soil Science Society of America, 2001, Glossary of Soil Science Terms: Soil Science Society of America, Madison, Wisconsin, USA.

Soil Survey Staff, 1993, Soil Survey Manual: US Department of Agriculture-Soil Conservation Service, Handbook No. 18, US Government Printing Office, Washington, DC, USA.

Soil Survey Staff, 1999, Soil Taxonomy, A Basic System of Soil Classification for Making and Interpreting Soil Surveys, Agriculture Handbook No. 436, 2nd edition: USDA-Natural Resources Conservation Service, U.S. Government Printing Office, Washington, DC, USA.

Stevens, K.M., Lodha, G.S., Holloway, A.L. and Soonawala, N.M., 1995, The application of ground penetrating radar for mapping fractures in plutonic rocks within the Whiteshell Research Area, Pinawa, Manitoba, Canada. Journal of Applied Geophysics, Vol. 33, pp. 125–141.

Stokes, A., Fourcaud, T., Hruska, J., Cermak, J., Nadyezdhina, N., Nadyezhdin, V. and Praus, L., 2002, An evaluation of different methods to investigate root system architecture of urban trees in situ: 1. Ground-penetrating radar. Journal of Arboriculture, Vol. 28, No. 1, pp. 2–9.

Stover, D.B., Day, F.P., Butnor, J.R. and Drake, B.G., 2007, Effect of elevated CO_2 on coarse-root biomass in Florida scrub detected by ground-penetrating radar. Ecology, Vol. 88, No. 5, pp. 1328–1334.

Theimer, B.D., Nobes, D.C. and Warner, B.G., 1994, A study of the geoelectric properties of peatlands and their influence on ground-penetrating radar surveying. Geophysical Prospecting, Vol. 42, pp. 179–209.

Tolonen, K., Tiuri, M., Toikka, M. and Saarilahti, M., 1982, Radiowave probe in assessing the yield of peat and energy in peat deposits in Finland. Suo, Vols. 4–5, pp. 105–112.

Tolonen, K., Rummukainen, A., Toikka, M. and Marttilla, I., 1984, Comparison between conventional and peat geological and improved electronic methods in examining economically important peatland properties. Proceedings, 7th International Peat Congress, June 18–23, 1984, Dublin, Ireland, pp. 1–10.

Tomer, M.D., Boll, J., Kung, K.-J.S., Steenhius, T. and Anderson, J.L., 1996, Detecting illuvial lamellae in fine sand using ground-penetrating radar. Soil Science, Vol. 161, pp. 121–129.

Toshioka, T., Tsuchida, T. and Sasahara, K., 1995, Application of GPR to detecting and mapping cracks in rock slopes. Journal of Applied Geophysics, Vol. 33, pp. 119–124.

Truman, C.C., Perkins, H.F., Asmussen, L.E. and Allison, H.D., 1988, Some applications of ground-penetrating radar in the southern coast plain region of Georgia. The University of Georgia, Athens, Georgia, USA, Georgia Agricultural Experiment Stations Research Bulletin 362.

Turenne, J.D., Doolittle, J.A. and Tunstead, R., 2006, Ground-penetrating radar and computer graphic techniques are used to map and inventory Histosols in southeastern Massachusetts. Soil Survey Horizons, Vol. 47, No. 1, pp. 13–17.

Ulriksen, P., 1980, Investigation of peat thickness with radar. Proceeding of 6th International Peat Congress, Duluth, Minnesota, USA, pp. 126–129.

Ulriksen, C.P.F., 1982, Application of impulse radar to civil engineering [Ph.D. thesis]: Department of Engineering Geology, Lund University of Technology, Lund, Sweden, 175 p.

USDA-NRCS, 1995, Soil Survey Geographic (SSURGO) Database-Data Use Information, Misc. Publication No. 1527: National Soil Survey Center, Lincoln, Nebraska, USA.

Welsby, J., 1988, The utilization of geo-radar in monitoring cutover peatlands. Proceedings, 8th International Peat Congress, Leningrad, USSR, International Peat Society, Jyska, Finland, pp. 99–107.

Wielopolski, L., Hendrey, G. and McGuigan, M., 2000, Imaging tree root systems in situ. Proceedings, Eight International Conference on Ground-Penetrating Radar, May 23–26, 2000, Gold Coast, Queensland, Australia. Proceedings of SPIE – The International Society of Optical Engineering, Bellingham, Washington, pp. 642–646.

Worsfold, R.D., Parashar, S.K. and Perrott, T., 1986, Depth profiling of peat deposits with impulse radar. Canadian Geotechnical Journal, Vol. 23, No. 2, pp. 142–145.

CHAPTER 7

The Contribution of Ground Penetrating Radar to Water Resource Research

Lee Slater *and* Xavier Comas

Contents

7.1. Introduction	203
7.2. Petrophysics	206
7.3. Hydrostratigraphic Characterization	209
7.4. Distribution/Zonation of Flow and Transport Parameters	214
7.5. Moisture Content Estimation	217
7.6. Monitoring Dynamic Hydrological Processes	224
7.6.1. Recharge/moisture content in the vadose zone	225
7.6.2. Water table detection/monitoring	228
7.6.3. Solute transport in fractures	229
7.6.4. Studies of the hyporheic corridor	231
7.6.5. Studies of the rhizosphere	232
7.6.6. Carbon gas emissions from soils	232
7.7. Conclusions	237
References	238

7.1. Introduction

Water is our most valuable resource; yet the thirst of the growing planetary population for water, combined with our ever increasing industrial consumption of water, is driving a global water shortage that is pressurizing countries to exploit their groundwater resources in an unsustainable manner. In parallel with this water availability crisis, anthropogenic activities at the Earth's surface are threatening the quality of this precious resource, it being susceptible to contamination by agricultural, industrial, and municipal activities. Major problems include inadequately treated domestic sewage, poor controls on the discharges of industrial waste waters, loss and destruction of catchment areas, ill-considered placement of industrial plants, deforestation, and poor agricultural practices (UNCE, 1993). In addition to such direct impacts of human activities, our water resources are also indirectly threatened by human industrial activities that are largely considered the primary cause of the current rapid warming of our planet. In changing our climate, global warming is altering the hydrological cycle and thus patterns of precipitation, reducing our ability to predict patterns and amounts of rainfall. Many arid parts of the globe that face acute water supply shortages are getting not just hotter but also drier.

The 1992 Earth Summit at Rio was of paramount importance in raising awareness of the impending water crisis on an international platform. It stressed the need for a holistic management of freshwater as a finite and vulnerable resource, with water planning integrated into a framework of national economic and social policy (UNCE, 1993). The concerns of the Rio Summit were largely reinforced by the Ministerial Declaration of the 2nd World Water Forum, which advocated the common goal to provide water security in the 21st century based on integrated water resource management plans that "ensure that freshwater, coastal and related ecosystems are protected and improved; that sustainable development and political stability are promoted, that every person has access to enough safe water at an affordable cost to lead a healthy and productive life and that the vulnerable are protected from the risks of water-related hazards" (WWC, 2000). The declaration from the recent 4th World Water Forum reasserted the critical importance of water, "in particular freshwater, for all aspects of sustainable development, including poverty and hunger eradication, water-related disaster reduction, health, agricultural and rural development, hydropower, food security, gender equality as well as the achievement of environmental sustainability and protection" (WWC, 2006).

More than any other industry, the oil exploration industry has historically served as the financial engine driving the development of geophysical technologies that have been so critical to the search for new oil reservoirs. Yet, as oil fields continue to dry out and society becomes ever more reliant on alternative energy sources, water scarcity is likely to exert a dominant control on economic and political stability (WWC, 2000). Perhaps the greatest geophysical challenge on the horizon then is the evaluation of our water (rather than oil) resources. The need for geophysics in water resource evaluation is reflected in the fact that the depletion and degradation of our surface water resources is driving societies to increasingly rely on their groundwater reserves to provide water security. Extractable groundwater is a precious resource that comprises only about 0.3% of the total water resources on Earth. Furthermore, the extent and severity of contamination of unsaturated zones and aquifers has long been underestimated owing to the relative inaccessibility of aquifers and the lack of reliable information on aquifer systems (UNCE, 1993).

The focus of this chapter then is on applications of ground penetrating radar (GPR) in water resource research. We stress that it is widely accepted that GPR is a valuable geophysical methodology in water resource-related studies. This is primarily a consequence of the importance of moisture content (θ) in determining the measured relative dielectric permittivity (ε_r) of porous media (see Chapter 2, this volume for review). Moisture content impacts water resources at multiple scales; at the regional/continental scale, it can regulate atmospheric moisture and thus influence climate; at the catchment scale, it determines the relative contributions of surface water flow and recharge, impacts flood hydrographs and erosion, whereas at the field scale, it is important to agricultural productivity (Huisman et al., 2003).

Ground penetrating radar studies in hydrogeology have increased dramatically in recent years and the applications have diversified to include both novel applications and innovative measurement methodologies. A number of recent review papers have already addressed the subject of hydrogeophysical applications of GPR. Huisman et al. (2003) focused on GPR estimation and monitoring of θ in the vadose zone, recognizing four types of radar survey procedures that have been employed in this work. Moisture content estimation is an important part of our review and we particularly examine work conducted in the last 4 years since the review of Huisman et al. (2003). However, here we aim to provide a broader review of the application of GPR in water resource studies, of which vadose zone θ estimation and monitoring is one important component (Figure 7.1). We primarily construct our review around the following four water resource-related applications of GPR that we recognize to have emerged in the last 15 years (see Figure 7.1):

1. The characterization of aquifer geometries, hydrogeological units, and hydraulic length scales
2. The estimation of aquifer-saturated properties (hydraulic conductivity and porosity) from static and time-lapse measurements
3. Moisture content estimation
4. Monitoring dynamic hydrological processes

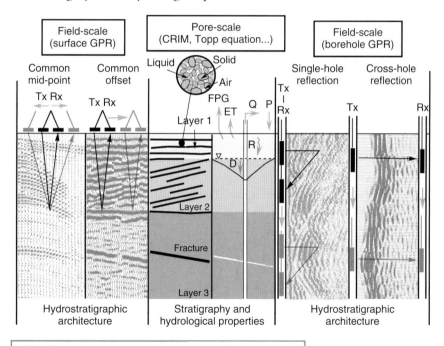

Figure 7.1 Overview of water resource-related applications of ground penetrating radar (GPR). Characterization of hydrostratigraphic architecture at the field scale using surface (common midpoint and common offset) and borehole (single- and cross-hole reflection) GPR surveys; and characterization of hydrological properties and dynamics at the pore scale combining GPR measurements and petrophysical models (e.g. complex refractive index model, CRIM and/or Topp equation).

Some of these applications have been the subject of recent reviews. For example, Hyndman and Tronicke (2005) review how GPR (and electrical resistivity) has been utilized in the structural characterization of aquifers, in the estimation of flow and transport parameters and in the monitoring of water table fluctuations. In the same volume, Daniels et al. (2005) describe how GPR (and electrical resistivity) has contributed to the characterization of the vadose zone, covering the detection of the water table and capillary fringe, the estimation of soil properties, and θ measurement and monitoring. In a more focused review, Davis and Annan (2002) considered the technical assumptions regarding the application of GPR to θ estimation. However, we are unaware of any recent reviews that consider the full range of GPR applications in water resource research.

In this review, we assume the reader has a good working knowledge of the theory/principles of the GPR technique and various measurement modes in which GPR has been deployed. For the nonexpert reader, we recommend reading the recent review of Annan (2006) and Part I of this volume. Annan (2005) provides a more concise treatment of this subject, with examples focused on water resource issues. The strong theoretical and empirical petrophysical relationships between ε_r and θ are the foundations of all water resource-related applications of GPR. We therefore preface our review of GPR applications in water resources with a short petrophysics section summarizing these relations. After reviewing established water resource-related applications of GPR, we briefly consider emerging areas of hydrology where GPR research is having an impact. Here we examine novel applications of GPR such as examining groundwater–surface water interactions, studies of the rhizosphere and investigation of carbon gas dynamics. Table 7.1 summarizes four primary applications of GPR that we identify, along with the most frequently applied data acquisition/processing strategies, and also gives the most common alternative geophysical methods that have been used to address the same problems. We conclude by discussing the future possibilities that may arise as improvements are made in GPR instrumentation, petrophysical relations and processing techniques.

7.2. Petrophysics

This section considers the problem of extracting subsurface properties from GPR measurements using petrophysical relationships. We need to know how the GPR data we collect are related to critical hydraulic parameters in water resource studies. The application of GPR in water resources is supported by strong petrophysical relations that link measured ε_r to θ and porosity (ϕ). A detailed treatment of these petrophysical relations is given in Lesmes and Friedman (2005), Knight and Endres (2006) and Chapter 2 of this volume. Here we summarize the primary petrophysical relations that are most often exploited in water resource-related applications of GPR.

Table 7.1 Depth applicability, lateral coverage, resolution, and proposed alternatives for ground penetrating radar (GPR) methods as related to water resource applications described in this chapter

Method	Depth applicability (m)	Lateral coverage (m)	Resolution (cm)	Alternatives
Hydrostratigraphic characterization				
Common-offset data acquisition (*multi-channel acquisition and 3D migration*)	0–30	100s–1000s	Sub 10 cm	Shallow seismic; electrical imaging
Distribution/Zonation of flow and transport parameters				
GPR tomography (*full-waveform inversion*)	~100	~100s	~10	Electrical resistivity tomography; seismic tomography
Moisture content estimation				
Common-midpoint velocity analysis	~20	~10s of m	Sub 10 cm	None
Common-offset, direct-wave analysis	<1.0	100s–1000s	Sub 10 cm	Synthetic aperture radar (SAR); resistivity profiling
Surface reflection	<0.5	100s of m	cm	None
GPR tomography	~100	~100s	~10	Electrical resistivity tomography
Monitoring dynamic hydrological processes				
Time-lapse, common-offset data acquisition	~20	100s–1000s	Sub 10 cm	Resistivity profiling
Time-lapse GPR tomography	~100	~100s	~10	Electrical resistivity tomography

By far the most commonly applied relationship for estimating θ of soils and rocks is the empirical Topp equation (Topp et al., 1980):

$$\theta = -5.3 \times 10^{-2} + 2.92 \times 10^{-2}\varepsilon_r - 5.5 \times 10^{-4}\varepsilon_r^2 + 4.3 \times 10^6 \varepsilon_r^3 \qquad (7.1)$$

The calibration parameters in this equation have been found to satisfy measurements on mineral soils, although organic-rich soils (e.g. peat) tend to deviate from this relationship. Furthermore, these relationships were derived based on time domain reflectometry (TDR), which is typically a 500–1000 MHz measurement. Clay minerals may generate a significant dispersion in the permittivity at the lower GPR frequencies. For example, West et al. (2003) found significant dispersions impacting ε_r–θ relationships for sandstone samples containing up to 5% θ. Field-scale estimates of ε_r (down to ~2 m) also exhibit frequency dispersion in clay-rich agricultural soils (Roth et al., 2004). Blonquist et al. (2006) investigated porous media characterized by dual porosity associated with inter-aggregate and intra-aggregate pores and showed that the ε_r–θ relation is characterized by an abrupt change of slope when all intra-aggregate pores are saturated. Recent studies have demonstrated how the ε_r–θ relation depends on moisture history and displays hysteresis between wetting and drying cycles (Lai et al., 2006). Such studies suggest that it will often be necessary to generate site-specific calibrations of θ for use in the interpretation of low-frequency GPR measurements.

Theoretical approaches are also frequently used to relate ε_r to θ. The simplest theoretical approaches utilize dielectric mixing models, whereby the volume fractions and ε_r of the components making up the soil are used to derive a relationship (see for review Lesmes and Friedman, 2005, and Chapter 2 of this volume). The complex refractive index model (CRIM) is a commonly applied mixing formula for estimating the measured ε_r from the dielectric and volumetric properties of the soil constituents (w: water; s: solid; and a: air)

$$\varepsilon_r = [\theta \varepsilon_w^\alpha + (1 - \phi)\varepsilon_s^\alpha + (\phi - \theta)\varepsilon_a^\alpha]^{1/\alpha} \qquad (7.2)$$

where α is a fitting parameter depending on the orientation of the electric field relative to the geometry of the medium. A simpler, two-component expression follows for a two-phase medium such as a fully saturated soil.

An important issue regarding petrophysics is the scale dependency of the ε_r–θ relationship that determines how local-scale petrophysical relationships (e.g. based on TDR or neutron probe data) relate to the field-scale relationships needed to interpret GPR data (Figure 7.2). Mixing models such as CRIM do not account for the geometric distribution of subsurface water, although this distribution is well-known to impact ε_r (Moysey and Knight, 2004). This dependence on geometry means that the petrophysical relationship linking ε_r to θ is scale-dependent where subsurface architecture is complex (see Figure 7.2). Field-scale relationships that capture this geologic heterogeneity and account for the effect of geologic complexity on the ε_r–θ relationship have been predicted using geostatistical characterizations of subsurface structure (Moysey and Knight, 2004). Cassiani and Binley

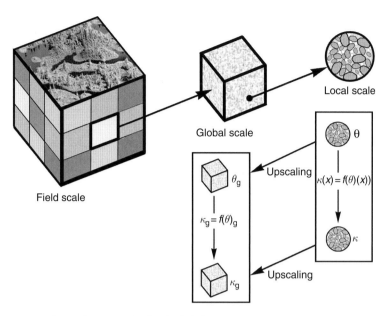

Figure 7.2 Correlation between the field, global and local scale. At the local scale, every point in space is described by a rock physics equation that relates moisture content (θ) and dielectric constant of a composite medium (κ). At the global scale, a different rock physics equation relates the effective block parameters θg and κg. The upscaling from local to global properties is problem-specific (e.g. θg is the volumetric average of θ). Figure modified from Moysey and Knight (2004).

(2005) showed that accounting for the scale of GPR measurements was critical to fitting the geophysical data using one-dimensional solute transport modeling.

7.3. HYDROSTRATIGRAPHIC CHARACTERIZATION

This section considers the challenge of imaging the sedimentary architecture of aquifers. Knowledge of this architecture is fundamental for the development of conceptual models of groundwater flow and solute transport. Information on the particle size distribution and spatial continuity of sedimentary packages is required to evaluate the likely effect of architecture on conditioning flow and transport. Ground penetrating radar is an excellent technique for mapping near-surface stratigraphy as θ changes are typically associated with boundaries between sedimentary layers at basin and field scales (Figure 7.3). Variations in grain size characteristics between different layers result in θ variation due to the increase in θ retention capacity with decreasing grain size. Ground penetrating radar stratigraphy is an advanced science, and a thorough review of the topic can be found in Neal (2004) and the chapters making up Part III of this volume. Here we consider first how GPR has been used to define the hydrostratigraphy of near-surface aquifers. A promising extension of this work is the quantitative analysis of images

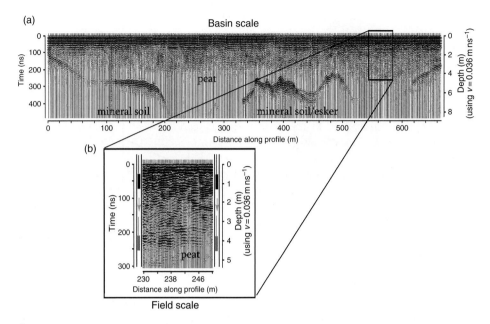

Figure 7.3 Constant-offset ground penetrating radar (GPR) image across a northern peatland basin (Kanonokolus Bog in Maine). Ground penetrating radar profiling clearly resolves the boundary between sedimentary layers (peat versus mineral soil/esker) at most places at both (a) the basin scale and (b) the field scale.

of hydrostratigraphy, whereby identification of the correlation structure of GPR images has been investigated as a means to obtain hydraulic length scales required for stochastic groundwater models. This work is also reviewed here. We end this section with a focus on fractured rock aquifers. The hydraulic properties of fractured rock aquifers are highly heterogeneous, and modeling of flow and transport in fractured rock remains a fundamental challenge in hydrogeology. Yet, we are increasingly turning to these aquifers both for water supply and as vaults for high-level industrial waste. The high resolution (relative to other geophysical techniques) of GPR makes it well-suited to fracture detection.

A classic application of geophysics in hydrogeology is the mapping of stratification and/or structural boundaries in aquifers. Characterization of large-scale heterogeneity is a necessary step of groundwater model calibration if accurate estimates of flow and transport are to be made (Moysey et al., 2003). Where sufficient depth penetration is achieved, GPR is a very effective hydrostratigraphic characterization geophysical technique due to its high resolution, relatively rapid data acquisition rates and, most importantly, the strong dependence of θ on stratigraphy. However, GPR depth penetration is typically limited to a few tens of meters (primarily being controlled by the electrical conductivity) such that other evolving geophysical methods (e.g. resistivity imaging, land-based seismic using towed arrays, and airborne shallow electromagnetic) will usually be more appropriate for investigating hydrostratigraphy of deeper aquifers.

There are numerous examples of how GPR can be used to image subsurface zonation and hydrostratigraphy (Beres and Haeni, 1991; Beres et al., 1995; Lesmes et al., 2002; Carreon-Freyre et al., 2003; Green et al., 2003; Skelly et al., 2003), it being possible to use the data to develop conceptual hydrogeological models (Kostic et al., 2005; Ezzy et al., 2006) in a wide range of depositional environments. The method is particularly effective in coarse, electrically resistive sedimentary aquifers, e.g. alluvial sands and gravels, where surface GPR can yield striking images of aquifer stratigraphy within the upper ~ 10 m (e.g. Beres et al., 1999), although stratigraphic information can also be obtained in finer-grained deposits (e.g. lacustrine deposits; Carreon-Freyre et al., 2003) when conductive clay minerals do not excessively attenuate the GPR signature. In many cases, GPR images of subsurface stratigraphy can greatly assist in the development of conceptual models of flow and transport in the subsurface. For example, groundwater transport into the Columbia River from the Department of Energy's Hanford Complex, a uranium-contaminated site, is believed to be accelerated within paleochannels that have been mapped with GPR (Kunk et al., 1993). Bowling et al. (2005) used GPR at the macrodispersion experiment (MADE) site to evaluate the stratigraphy of this fluvial aquifer system, defining a geological model of the site based on delineation of four stratigraphic units (meandering fluvial system, braided fluvial system, fine-grained sands, clay-rich interval, and a palaeochannel). By comparison with a unique, dense set of hydraulic conductivity measurements, they were able to show that groundwater flow is primarily in the braided system.

Ground penetrating radar images of stratigraphy have also yielded valuable insights into the hydrological controls on the formation of pool systems in peatlands (Comas et al., 2005a). Figure 7.4 shows a constant-offset GPR image of stratigraphy and structural boundaries beneath a portion of Caribou Bog, a well-studied, northern-raised bog in Maine. The GPR image illuminates an esker deposit that follows the line of a sequence of pools that exist within this bog. The formation of such pool systems in peatlands has been the subject of a considerable volume of ecological and hydrological research. The GPR images suggest that suborganic soil lithology exerts a control on the formation of these pools. One conceptual model for explaining these pools patterns, developed as a result of these GPR studies, is that vertical flow of relatively oxic waters into the highly anaerobic peat soils thereby locally accelerates rates of decomposition of organic soils.

Ground penetrating radar stratigraphy has advanced to incorporate the sequence stratigraphy concepts developed to assist in the interpretation of seismic reflection datasets acquired in oil exploration. Ground penetrating radar stratigraphy is described in detail by van Overmeeren (1998) and, more recently, in the review provided by Neal (2004). Also borrowing from developments made in oil exploration, outcrop analogue studies have been performed to provide physical models for assessing the hydrostratigraphic content of GPR data (Szerbiak et al., 2001; Tronicke et al., 2002, Van Dam et al., 2002). The study by Tronicke et al. is interesting as they excavated the study site after collecting GPR data. The GPR-based structural interpretation was found to be generally consistent with the major lithologic units mapped in the excavation. New approaches to the assessment of aquifer architecture from stratigraphic GPR data continue to be developed.

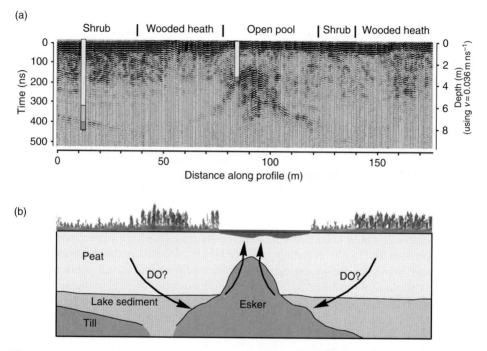

Figure 7.4 (a) Constant-offset ground penetrating radar (GPR) image of stratigraphy and structural boundaries beneath a portion of a northern peatland (Caribou Bog, Maine) showing the correspondence between an open pool and an esker deposit; (b) conceptual model to explain pool formation as a result of vertical flow of relatively oxic waters through the esker deposit and subsequent enhancement of peat soil decomposition above the esker crest.

For example, Moysey et al. (2003) describe a novel approach that uses neural networks to predict stochastic, facies-based models required in groundwater modeling from the subsurface architecture captured in GPR data.

An interesting extension of GPR-based hydrostratigraphy is the estimation of the spatial correlation structure of lithologic and hydrostratigraphic units that impart a major control on the distribution of hydraulic properties (Rea and Knight, 1998; Tercier et al., 2000; Oldenborger et al., 2003). Spatial correlation parameters are used to produce realizations of aquifer parameter distributions required in stochastic models of flow and transport. Inferring correlation structures in the horizontal direction is difficult from hydrogeological measurements as sampling points are typically sparsely distributed. Rea and Knight (1998) estimated correlation lengths and the direction of maximum correlation in a gravel pit. The GPR-estimated correlation lengths and direction of maximum correlation was found to compare well with lithologic structure mapped in the vertical face of an exposure of the gravel pit. Oldenborger et al. (2003) found that the correlation structure of stacked velocities from a common midpoint (CMP) survey was consistent with that obtained from hydraulic conductivity (K) measurements on core samples. Corbeanu et al. (2002) used similar concepts to identify GPR attributes used to refine a fluid permeability model of sandstone.

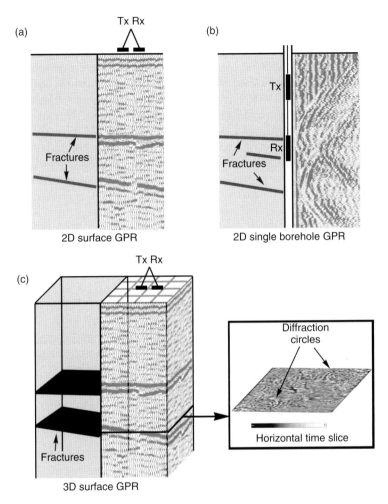

Figure 7.5 Schematic representation of fractured bedrock characterization using (a) 2D surface ground penetrating radar (GPR); (b) 2D single-borehole GPR; and (c) high-resolution 3D surface GPR. Inset in (c) (modified from Grasmueck et al., 2005) shows a horizontal time slice displaying diffraction circles associated with the presence of fractures.

Ground penetrating radar is also an effective method for examining architecture of fractured bedrock as depicted in Figure 7.5 (Stevens et al., 1995; Grandjean and Gourry, 1996; Grasmueck, 1996; Grasmueck et al., 2005; Jeannin et al., 2006; Porsani et al., 2006; Theune et al., 2006; Van den Bril et al., 2007). The accurate characterization of fractured-rock aquifer heterogeneity remains one of the most challenging and important problems in groundwater hydrology (Day-Lewis et al., 2003). Ground penetrating radar imaging of fractured rock may reveal the orientation and continuity of fractures that provide the primary pathways for fluid flow and solute transport. Fractures in bedrock aquifers represent sharp interfaces in ε_r that can generate high-amplitude reflections. The fracture is a water-filled feature, usually embedded in a host rock of low ϕ and therefore low θ. The high velocity of EM waves in bedrock means that the resolution of GPR in fractured rock environments is higher than that of other geophysical techniques. Depending on the nature of the rock mass and the sophistication of the survey method, GPR may reveal strong, laterally continuous reflections from large fractures and/or fracture zones (e.g. Stevens et al., 1995; Porsani et al., 2005, 2006) or multiple diffraction patterns from complex fracture distributions that can be correctly quantified only in high-resolution 3D GPR surveys (Grasmueck et al., 2005). Ground penetrating radar surveys in crystalline rock environments have been conducted to identify densely fractured zones suitable for the siting of water supply wells (da Silva et al., 2004). Most GPR studies of fractured rock have used surface measurements, although borehole GPR measurements, using directional antennae, have been used to map the orientation and continuity of fractures away from the borehole wall (see Figure 7.5b), as done in the study of a proposed nuclear waste site in Sweden, where fractures were clearly identified in GPR data and a methodology for determining the azimuth and orientation of these features described (Wänstedt et al., 2000). The preferential strike direction of fractures, typically exerting a dominant control on hydraulic anisotropy in fractured media, has been investigated using GPR measurements made as a function of azimuth (Seol et al., 2001) and with directional borehole GPR antennas (Seol et al., 2004). More recently, new methods for the detection of vertical thin reflectors, often undetected in conventional (single polarization) GPR studies, have been developed based on phase analysis of multi-polarization methods (Tsoflias et al., 2004).

7.4. Distribution/Zonation of Flow and Transport Parameters

This section considers the problem of predicting the spatial distribution of the physical properties controlling flow and transport in the subsurface. Hydrogeologists require accurate knowledge of the spatial distribution of hydraulic conductivity in order to reliably calibrate groundwater flow and transport models. The emerging science of hydrogeophysics is based on the premise that physical properties of the subsurface that control flow and transport are related, via reliable petrophysical models, to geophysical properties sensed with geophysical instrumentation. Geophysicists have strived for decades to estimate permeability (k), the fundamental property of the subsurface that governs fluid flow, from geophysical data. Despite these efforts, it is clear that there is no simple relationship between geophysical properties and k, and the proxy measures of k that have been derived from geophysical experimentation are uncertain. However, a plethora of recent hydrogeophysical research shows that the unique spatial richness of tomographic geophysical data may significantly improve the spatial delineation and zonation of hydraulic properties relative to estimates based on sparsely distributed direct measurements. Linde et al. (2006a) provide a thorough review of hydraulic parameter estimation using tomographic geophysical data. The subject is also treated in the review of Hyndman and Tronicke (2005). Here we briefly review the use of GPR tomography within this framework to quantify distributions, or zonation, of hydraulic properties. Such work relies on the premise that petrophysical relationships link ε_r to physical properties of porous media, as described in Section 7.2. For example, a common assumption is that some relation between ε_r and k can be expected due to both being dependent on ϕ.

One approach to imaging the zonation of aquifer properties is to invert the images of the spatial variation in geophysical properties and subsequently convert these images to distributions of a hydrological parameter based on an assumed petrophysical relationship, possibly established with direct hydrological measurements at the same site (Figure 7.6). This conversion may be obtained via direct mapping, geostatistical or Bayesian techniques (Chen et al., 2001; Linde et al.,

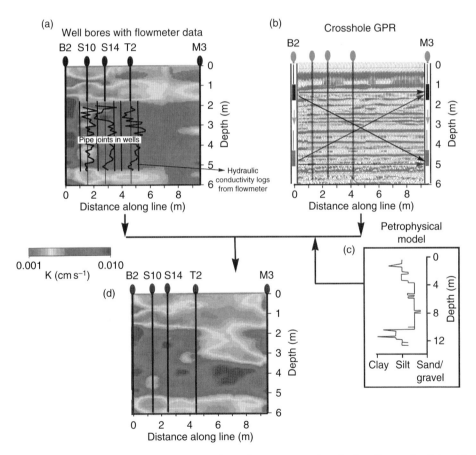

Figure 7.6 Comparison of the imaging of aquifer properties (e.g. hydraulic conductivity, K) using (a) flowmeter data; (b) constant-offset and cross-hole GPR data; (c) cone penetrometer log; and (d) combination of flowmeter, ground penetrating radar (GPR), and penetrometer data. The image illustrates how the spatial distribution of K differs when geophysical datasets are included in the estimation. All data modified from Hubbard et al. (2001).

2006a). Chen et al. (2001) describe the use of a Bayesian framework to improve estimates of K distribution relative to those obtained from hydrological methods alone, based on a correlation between ε_r derived from GPR data and ϕ. They showed that despite a relatively small variation in log K and ε across the site, the inclusion of geophysical data improved the estimation of K distribution, particularly where data coverage of the direct hydrological measurements was low. Hubbard et al. (2001) adopt this technique for the hydrogeophysical characterization of the US Department of Energy (DOE) South Oyster Bacterial Transport Site. Here a Bayesian framework, allowing for complex petrophysical relationships between the geophysical parameters and K, was used to condition K estimates from direct flowmeter hydraulic measurements combined with both tomographic GPR and seismic data. Figure 7.6 illustrates how estimates of the spatial distribution of K obtained using geostatistical mapping techniques differ when the geophysical datasets are included in the estimation. The incorporation of the geophysical datasets appears to allow the spatial variability in estimated K to be mapped in regions remote from the flowmeter wells. However, others have questioned the value of geophysical tomograms for inferring geostatistics as tomographic resolution and hence image structure depends on geophysical data acquisition, regularization, data error, and the physics of the underlying measurements (Day-Lewis and Lane, 2004). A thorough treatment of the role of geophysical images in defining geostatistics is beyond the scope of this review but is found in the recent review by Linde et al. (2006a).

Although documented examples of such direct mapping approaches exist, errors in data acquisition/inversion, combined with errors associated with the adopted petrophysical relationships, can limit the value of the geophysical realizations of hydraulic properties so acquired (Kowalsky et al., 2006; Linde et al., 2006a). Hydrogeophysical research is thus turning to joint interpretation/inversion strategies whereby multiple geophysical datasets, and/or hydrological datasets, are processed simultaneously to produce more realistic estimates of hydraulic parameters that satisfy all the available datasets. The inclusion of hydrological data obviously has the potential to better constrain K estimates from geophysical inversion. It is intuitive to expect that inversion of geophysical data for aquifer properties will benefit from hydrogeological constraints. Early work in this field focused on the use of seismic geophysical datasets, combined with petrophysical relationships between seismic data and k, to constrain the inversion of k distributions or "zonations" using hydrological datasets (e.g. Rubin et al., 1992; Copty et al., 1993). In the "split-inversion method", seismic geophysical data were used to extract the lithologic zonation matching both geophysical data and tracer concentration data (Hyndman et al., 1994; Hyndman and Gorelick, 1996). These concepts have more recently been applied to GPR datasets. In a synthetic study, Linde et al. (2006b) illustrated that tomographic GPR could be used to obtain a lithologic zonation (assuming a relationship between ε and K), which could then be applied to improve the K models obtained from inversion of direct-measured, tracer breakthrough data. One limitation of such approaches has been the questionable validity of assumed petrophysical relationships at the field scale as discussed earlier.

A promising joint inversion strategy, excluding the requirement of the existence of a specific site petrophysical relationship, is the "structural approach" that assumes that changes in different physical properties occur in the same direction for a given position (Gallardo and Meju, 2004). This approach is appealing as it recognizes that different geophysical properties of the subsurface depend on the same minerals and fluids without specifying an assumed petrophysical relationship. Linde et al. (2006c) used the structural approach on GPR and dc resistivity data to define the 3D zonation of a sandstone aquifer and estimated effective petrophysical properties of each zone. Characterizing the K distribution of the subsurface using such an approach would require that the K structure is controlled by the identified zonation. This situation was assumed in a very recent paper (Paasche et al., 2006) whereby fuzzy c-means cluster analysis was used to define a subsurface zonation that again assumes the existence of petrophysical relationships within the zones without defining them. Paasche et al. (2006) applied the method to cross-hole GPR and cross-hole seismic, constrained by limited gamma log and slug test data, to generate plausible K realizations.

The studies cited above all used static distributions of geophysical properties (i.e. ε) inferred from tomography to improve realizations of the spatial distribution of hydraulic properties. An attractive alternative approach is to jointly utilize time-lapse geophysical and hydrological data in order to calibrate flow and transport models for spatial variation/zonation in hydraulic properties. The changes in a geophysical dataset as a result of solute transport through an aquifer can be expected to depend only on the active volume influenced by fluid flow. For example, although ε_r distributions obtained from GPR tomography may reflect the distribution of total porosity (ϕ), changes in ε_r as a result of θ change will presumably reflect only the active porosity (ϕ_{eff}) through which fluid flows. In a synthetic study, Kowalsky et al. (2004) utilized an assumed petrophysical (ε_r–ϕ–K) relation to generate multiple parameter distributions that reproduced point measurements of K, contained pre-specified patterns of spatial correlation and were consistent with synthetic neutron probe and GPR travel time datasets. Kowalsky et al. (2006) illustrated how this approach could be used to generate realizations of log k at the DOE Hanford site through joint inversion of available neutron probe and GPR datasets. Others have used time-lapse tomographic GPR data to image the movement of the center of mass of a tracer, subsequently fitting these predicted mass distributions to flow and transport models and allowing estimation of the field-scale hydraulic conductivity (Binley et al., 2002a; Deiana et al., 2007).

7.5. Moisture Content Estimation

We now turn our attention to the problem of estimating moisture content variability in the subsurface. Information on soil moisture is critical to quantify recharge and predict crop yields. As evident from Section 7.2 of this review, the electrical properties of porous media in the GPR frequency range are almost uniquely dependent on θ (Topp et al., 1980; Davis and Annan, 2002). Moisture content estimation using GPR has therefore been extensively investigated at both the field and the laboratory scales for a range of applications spanning high-resolution imaging of θ in pavement (e.g. Al-Qadi et al., 2004; Grote et al., 2005) and concrete (Laurens et al., 2005) to estimation of decimeter to tens of meters scale average θ in the vadose zone (e.g. Binley et al., 2001; Alumbaugh et al., 2002; Cassiani et al., 2004). Thus ground penetrating radar has some distinct advantages over more traditional hydrological methods of θ estimation in that field-scale support volume of the measurement may be more appropriate, relative to point-based measurements, for input into hydrological models of unsaturated processes (Binley et al., 2001; Huisman et al., 2001). However, we recognize that the application of GPR to θ mapping is likely to be limited to the field scale, as the application of airborne/satellite active or passive microwave-based methods is well established for providing large-scale catchment information (e.g. Cognard et al., 1995; Quesney et al., 2000). Given instrumentation/data acquisition constraints, it is hard to imagine that GPR could ever compete with these methods at the catchment scale.

Huisman et al. (2003) present a review that is dedicated to θ estimation from GPR. Here we do not aim to cover the subject in the same detail, but our summary includes significant advances in this field that have been made since publication of this review. In covering this subject, Huisman et al. (2003) recognized that θ estimation has been performed using five GPR surveying techniques: (1) single-offset reflection methods (Figure 7.7a); (2) ground wave measurements (see Figure 7.7d); (3) multiple-offset reflection methods (see Figure 7.7b); (4) borehole transmission measurements (see Figure 7.7c); and (5) surface reflection methods (see Figure 7.7e). Here we group the relevant literature under the same categories, although the basis of each measurement configuration is not covered here as this information can be found in Chapter 1. The basic steps in GPR-based θ estimation are common to the first four of the five surveying techniques identified above; the travel time of an electromagnetic wave (either a reflection or a direct wave) is recorded, the ε_r at the scale of the GPR measurement is estimated and a petrophysical relationship is used to convert ε_r to θ (Figure 7.8). The surface reflection method is different in that it is based on amplitude analysis of the reflection at the air–earth interface. In this section, we focus on the estimation of static distributions of θ and cover the monitoring of θ changes with GPR in Section 7.6.

Single-offset reflection measurement of θ has mostly focused on modeling the reflection hyperbolae originating at scattering objects (often buried at known depths) in the subsurface (Grote et al., 2002; Loeffler and Bano, 2004) or from zero-offset, two-way travel time to horizontal reflectors at known depths (Lunt et al., 2005). With a single, zero-offset measurement, it is not possible to predict the reflector depths from the GPR data, and this must instead be available *a priori* to convert measured travel times into θ estimates. Grote et al. (2002) embedded scattering targets at known depths in test pits, such that the reflection hyperbolae recorded when moving the single-offset GPR instrument could be fit to a single average velocity representing the material between the target and known scatterer depth. They showed that θ estimates determined from application of GPR data to a site-specific petrophysical relationship were not statistically different from estimates based on collocated, gravimetric θ measurements. Good correspondence between

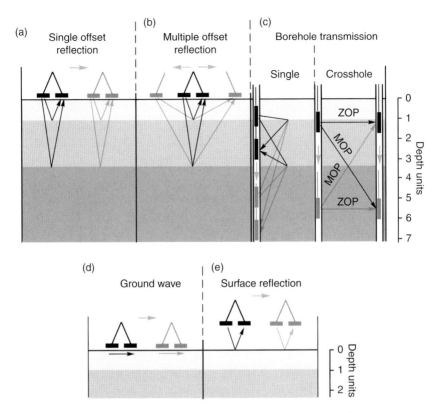

Figure 7.7 Ground penetrating radar (GPR) surveying techniques for moisture content estimation: (a) single-offset reflection; (b) multiple-offset reflection; (c) single- and crosshole borehole transmission; (d) ground wave; and (e) surface reflection. Each technique is characterized by a different array geometry, resolution and depth of penetration.

GPR-estimated θ and water volume added/removed from a tank was reported for various known reflectors in an experimental tank (Loeffler and Bano, 2004). Grote et al (2005) applied similar concepts to examine the subasphalt θ in pavements using horizontal reflectors at known depths in the pavement structure. Xia et al. (2004) used conductive rods drilled into the sides of a quarry face to develop velocity models from GPR measurements made at the quarry surface. Lunt et al. (2005) used the reflection from a well-constrained (in depth) clay layer to map moisture content variations over an 80 × 180-m area at a winery (Figure 7.9). By assuming that the clay layer was horizontal, they were able to translate spatial variability in the two-way travel time for the reflection from the layer into a map of θ variation (see Figure 7.9). Laboratory measurements using such an approach have been also been performed to determine θ in soil blocks where the block size is accurately known (Comas and Slater, 2007).

The measurement of the direct, ground wave travel time can be used to estimate θ of near-surface soils (Huisman et al., 2001). Using multiple-offset measurements, it is possible to extract θ from the linear relationship between travel time and offset

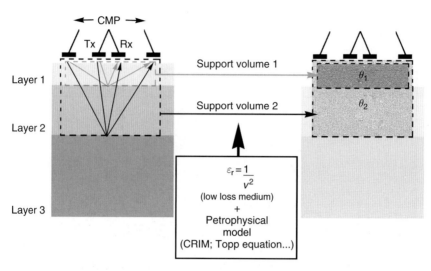

Figure 7.8 Schematic representation of the procedure for moisture content (θ) estimation using a multiple-offset reflection survey (common midpoint, CMP). Travel time of an electromagnetic wave to a certain stratigraphic interface allows for θ estimation for a particular support volume by using a petrophysical relationship (e.g. CRIM, Topp equation) to convert ε_r to θ.

for the direct wave. However, it is also possible to perform rapid, single-offset, reconnaissance-style mapping of θ after determining the approximate arrival time of the ground wave from a multiple-offset measurement. Recent experiments show that this is a promising method for mapping the spatial distribution of near-surface moisture (Huisman et al., 2002; Grote et al., 2003). However, uncertainty in the interpretation of the measurements exists primarily as the support volume of the method is poorly constrained (Huisman et al., 2003). Recent studies aimed at evaluating the effective depth averaged in the ground wave measurement suggest that the technique is sensitive to θ in the top 20 cm of soil (Grote et al., 2003), although this may depend on instrumentation settings and ground conditions. Galagedara et al. (2005) conducted numerical studies that showed that the sampling depth increased linearly with the wavelength. Huisman et al. (2003) suggest that disadvantages of the direct-wave method include (1) difficulties in distinguishing the direct wave from refracted/reflected events; (2) difficulties in selecting an appropriate antenna separation for rapid reconnaissance mapping; and (3) excessive attenuation of the ground wave. Weihermuller et al. (2007) found that the technique failed (due to poor data quality) when the soil contained a high silt and clay content. Recent promising applications of direct-wave measurements include evaluation of moisture content in pavement (Laurens et al., 2005).

Multiple-offset GPR reflection datasets (CMP and Wide Angle Reflection and Refraction (WARR), see Chapter 1) recorded over horizontal interfaces can be fit to a series of reflection hyperbolae from which average velocity between reflector and surface and depth can be estimated (Figure 7.10). The precision of the velocity estimate from normal moveout (NMO) GPR data has been estimated as ± 0.001 m ns^{-1} at the

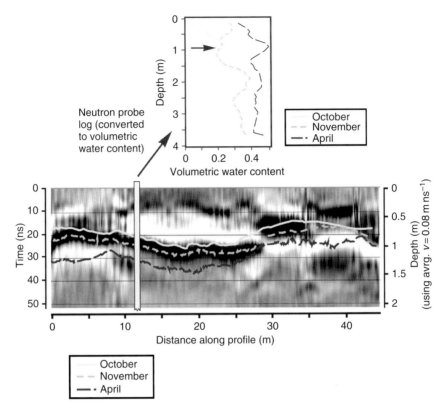

Figure 7.9 Constant-offset ground penetrating radar (GPR) reflection profile during November 2002 across vineyard soils in California. Solid lines across the profile show differences in travel time picks to a prominent reflection from a clay layer during three different times (October, November and April). Inset shows differences in volumetric water content estimated from neutron probe logs during the same time periods. All data modified from Lunt et al. (2005).

95% confidence level (Jacob and Hermance, 2004). NMO velocity estimation can be done manually or through processing algorithms, such as semblance analysis (see Figure 7.10d), that predict the best estimate of average velocity above a given reflector (Yilmaz, 1987). When a series of reflections are recorded from a sequence of reflectors, it may be possible (depending on the velocity contrasts) to transform these average θ values into a model of interval moisture contents (see Figure 7.10c), e.g. using the Dix formulation (Dix, 1955). A single CMP or WARR therefore can provide a 1D vertical velocity model for the earth, whereas multiple CMP datasets along a line can provide 2D velocity models. Numerous examples of this application of GPR for θ estimation exist (Tillard and Dubois, 1995; Greaves et al., 1996; van Overmeeren et al., 1997; Nakashima et al., 2001). However, the method has been cited as of limited value in spatial mapping of θ as it is cumbersome to make multi-offset GPR measurements as, unlike seismic, the multi-channel capabilities of GPR instruments are minimal (Davis and Annan, 2002; Huisman et al., 2003).

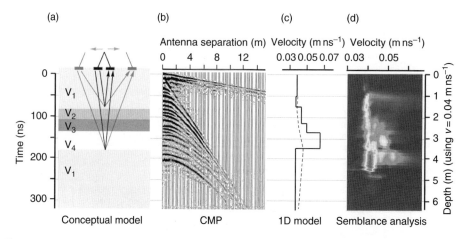

Figure 7.10 Multiple-offset ground penetrating radar (GPR) reflection survey across a northern peatland (Red Lake, Minnesota) showing (a) conceptual model with interpreted stratigraphy (based on EM wave velocities); (b) common midpoint GPR survey; (c) 1D vertical velocity model; and (d) semblance analysis.

Borehole transmission methods are primarily based on the estimation of θ from the measurement of the travel time of direct-waves traveling between a transmitting antenna in one borehole and a receiving antenna placed in a second, nearby borehole. The two most common forms of the method are (1) the zero-offset profiling (ZOP) technique, where the 1D distribution of θ, averaged over the scale of the borehole separation, is determined and (2) GPR tomography, where a large number of direct-wave travel times between transmitter and receiver at multiple locations along the borehole lengths are inverted for the estimation of the 2D (or 3D) θ distribution between boreholes. Vertical radar profiles (VRP), whereby the direct-wave travel time between surface transmitter and receiver in a borehole is measured, have also been used to quantify vertical θ profiles (Cassiani et al., 2004; Clement and Knoll, 2006). Although the method is in essence quite simple (particularly the ZOP technique), care is required in the correct identification of the fist arriving ray path. In layered soils with a high degree of θ variation, the first arrival may not be a direct wave when the antennas are in the low-velocity zone as it is possible (depending on the θ contrast and the thickness of low velocity layers) for waves refracted at the boundary of an over/underlying layer to arrive earlier as depicted in Figure 7.11a (Rucker and Ferré, 2003, 2004). Irving and Knight (2005) described problems arising from the assumption that all first arrivals travel directly between the centers of borehole antennas (see Figure 7.11b). Examples of θ estimation in the vadose zone appear in the recent literature (Binley et al., 2001; Alumbaugh et al., 2002).

Measurement of the surface wave reflection using high-frequency, off-ground GPR antennae can be used to estimate θ by comparing the reflection coefficient with that of an ideal reflector (Davis and Annan, 2002; Serbin and Or, 2004, 2005; Ghose and Slob, 2006). This method of θ estimation is thus distinctly different from those described previously in this section, as the latter are all based on the measurement of a

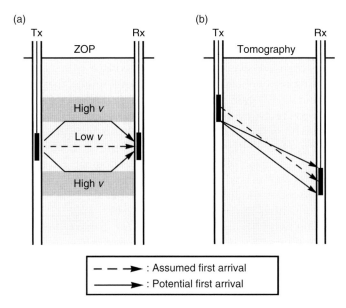

Figure 7.11 Schematic representation of nondirect-wave first arrivals when (a) strong contrasts between high- and low-velocity areas are present and (b) energy does not travel directly between the centers of antennas during borehole transmission surveys (e.g. Irving and Knight, 2005).

transmission time through the medium. Similar to the ground wave technique, measurement of the surface reflection can be interpreted in terms of θ for up to the top 20 cm of soil (Huisman et al., 2003), although others have suggested a sensitivity to only top ~ 1 cm of soil (Serbin and Or, 2004). However, θ estimates from surface wave reflections appear dependent on the surface roughness (e.g. resulting from vegetation) and the soil θ profile with depth (Huisman et al., 2003). Both surface roughness and varying θ profiles can contribute to scattering, which reduces the reflection coefficient, leading to an underestimation of θ values. Furthermore, this technique has a higher sensitivity to θ variation at lower θ values. Despite these limitations, recent applications of the technique demonstrate the potential for remote mapping of near-surface soil moisture at a high spatial resolution, offering clear advantages over TDR probe measurements (Lambot et al., 2004a, 2006a, 2006b). Efforts to invert a vertical soil moisture profile from such surface wave measurements have met with limited success due to the high frequency of the measurement and primary sensitivity to the very near-surface soil layer (Lambot et al, 2006c). Furthermore, Weihermuller et al. (2007) reported failure of the method to map the spatial distribution of θ at a site where the uppermost surface layer was very dry.

A critical consideration that is generic to all GPR methods for θ estimation is the accuracy of the θ estimates. This depends on both the accuracy of the velocity measurement and the validity of the petrophysical relationship applied to convert ε_r to θ. The accuracy of the velocity measurement has been assessed for CMP measurements. Jacob and Hermance (2004) used statistical tests of significance to show that the likely accuracy of a CMP velocity estimate is ± 0.001 m ns^{-1} at the 95% confidence interval. However, the accuracy of θ estimates from GPR will likely primarily depend on the validity of the petrophysical relationship applied. This accuracy has been assessed by comparison of GPR θ estimates with gravimetric measurements, or estimates obtained from other, point source, methods such as TDR or neutron probe data. Caution must be applied in correlating GPR estimates with TDR and/or neutron probe estimates as the support volumes of the techniques are very different, and these point source techniques also rely on petrophysical relationships in the calculation of θ. Overall, the reported results on the accuracy of θ estimates from GPR are encouraging. Single-offset reflection measurements have been validated as accurate to within 0.01 m^3 m^{-3} when compared with gravimetric measurements (Grote et al., 2002) and lysimeter measurements (Stoffregen et al., 2002). Grote et al. (2003) also reported a root-mean-square error of 0.01 m^3 m^{-3} for θ derived from ground wave measurements when compared to gravimetric measurements. Calibration of multi-offset, ground wave-based θ estimates with gravimetric measurements yielded root-mean-square errors of <0.024 m^3 m^{-3}, with lower errors at higher frequencies (Huisman et al., 2001). Based on site-specific relationships established with neutron probe data, Alumbaugh et al. (2002) reported a root-mean-square error for borehole transmission tomographic θ estimates of 0.02–0.03 m^3 m^{-3} when compared with neutron probe measurements from proximal boreholes, whereas Lunt et al. (2005) gave an RMS of 0.018 m^3 m^{-3} for common-offset measurements calibrated against neutron probe measurements at a winery. It is therefore clear that GPR is a powerful technique for estimating θ with the advantage of being deployable in multiple configurations that sense a range of spatial scales appropriate for a wide range of hydrological and agricultural applications.

The estimation of soil moisture is critical to many fields of research and thus new applications of GPR sensing of moisture continue to arise. Gish et al. (2005), in a study motivated by the need to understand corn grain yield patterns, described a method based on GPR combined with analysis of digital terrain maps to identify zones in the subsurface where water converged into discrete flow pathways. Serbin and Or (2005) described surface reflection GPR measurements to assess crop canopy properties and developed relationships between reflection coefficient and canopy biomass. A primary motivation for this work is that such surface data could be used to calibrate and validate much more spatially rich radar datasets obtained from the air and/or remote-sensing platforms. Leucci et al. (2006) report on GPR estimation of moisture content to assist in the characterization of archaeological features in danger of accelerated deterioration as a result of rising moisture levels.

7.6. Monitoring Dynamic Hydrological Processes

We next consider the problem of monitoring temporal hydrological processes occurring in the subsurface. There is a clear need for geophysical monitoring methods that can noninvasively observe the movement of fluids and solutes in response to natural and artificial (human-induced) loading. Up to this point, the majority of this review has focused on the use of static GPR datasets to infer hydrogeological structures and/or hydraulic properties. Imaging changes in physical properties impacting geophysical data present new challenges and opportunities. In fact, the concept of time-lapse geophysical imaging has been widely adopted as a noninvasive methodology for monitoring mechanical properties, fluid transport, and biogeochemical processes and is the subject of a recent review (Snieder et al., 2007). Noninvasive geophysical monitoring is an important, growing subdiscipline of hydrogeophysics that is finding application in the study of moisture dynamics in the vadose zone, solute transport in the saturated and unsaturated zone, saltwater intrusion into coastal aquifers, and contaminant transport. Certainly, GPR is one of the most promising hydrogeophysical tomographic monitoring technologies due to its unique sensitivity to θ. A well-recognized advantage of time-lapse GPR monitoring is that time-independent factors (e.g. ϕ variation) impacting the spatial distribution of velocity captured in a static image are removed in the difference image. Here we review the water resource-related applications of time-lapse GPR monitoring. We start with established monitoring applications that include studies of recharge in the vadose zone, water table dynamics, and solute transport in fractures. We then turn our attention to emerging hydrogeophysical monitoring efforts that include groundwater–surface water interactions, rhizosphere studies and monitoring of carbon gas dynamics in peatlands. The reader is referred to Chapter 8 of this book for a discussion of GPR monitoring of contaminants.

7.6.1. Recharge/moisture content in the vadose zone

The high sensitivity of GPR to θ makes it uniquely suited to the monitoring of vadose-zone moisture dynamics. Applications of GPR in monitoring studies of vadose-zone moisture dynamics are extensively reported in the hydrogeophysical literature and have inspired a great deal of recent research in the GPR method (see for reviews, Huisman et al, 2003; Daniels et al., 2005). Common-offset GPR monitoring of θ of near-surface soils, based on observed velocity changes computed from the travel time to targets at known depths, has been reported for monitoring of infiltration through permeable aggregates underlying pavement (Grote et al., 2002) and within lysimeters-containing sandy soils (Stoffregen et al., 2002). Near-surface soil water monitoring using the ground wave was reported by Grote et al. (2003) for a vineyard over a 1-year period. Truss et al. (2007) describe high-resolution 2D and 3D GPR studies (using laser positioning) of the vadose zone of limestone rock (the Miami limestone), which showed how rapid drainage occurs in buried, sand-filled dissolution sinks. Hourly, high-density data acquisition revealed both rapid vertical transport through a 5-m vadose zone and lateral migration of water into the rock driven along stratigraphic boundaries. Figure 7.12 is part of a time series of amplitude changes for a 2D GPR profile crossing two dissolution sinks taken at 3-min intervals. Transport via the dissolution sinks results in high-amplitude diffraction patterns interpreted as the signature of the wetting front. The diffractions were considered indicative of subwavelength-scale heterogeneity most likely caused by subwavelength irregularities such as fingering flow. Using 3D data acquisition and

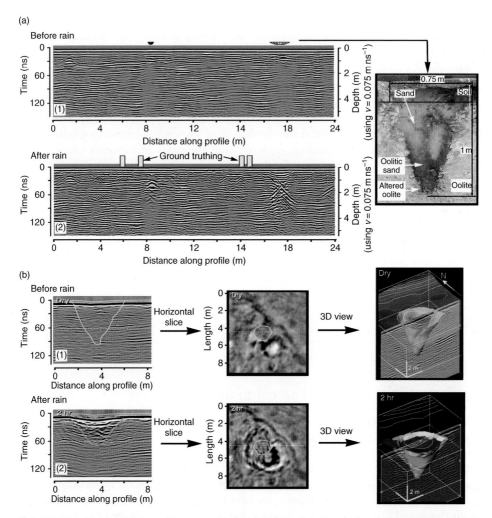

Figure 7.12 (a) Time-lapse, constant-offset ground penetrating radar (GPR) profiles across two sinkholes within oolitic limestone in Florida before and after a rain event. Ground penetrating radar reflection attributes show drainage patterns of the vadose zone. Inset shows the filling of highly permeable, well-sorted sand within the dissolution sinkhole. Modified from Truss et al. (2007). (b) Full-resolution 3D GPR time-lapse imaging of a surface infiltration experiment in a sinkhole; the figure displays a vertical cross section from the center of the sinkhole, followed by a horizontal slice extracted from 3D data, and a 3D view of interpreted sink boundary (yellow), wetting front (blue), and surface water puddles (red) before and after a rain event. All data modified from Truss et al. (2007).

processing, Truss et al. (2007) also showed how GPR could be used to define the geometry of such sinkholes and capture the wetting front resulting from an artificial infiltration experiment. Figure 7.12b shows a vertical cross section (top row), horizontal slice (at 2.4 m depth), and 3D views of interpreted sink boundary (yellow), wetting fronts (blue), and extent of surface water after 3 h of infiltration.

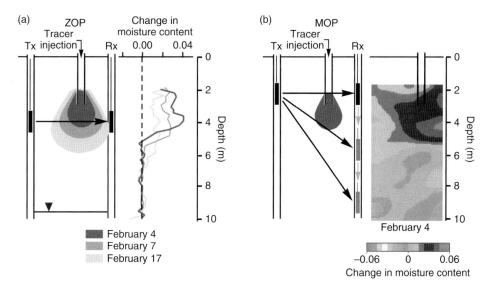

Figure 7.13 Schematic representation of a tracer injection experiment and change in moisture content from time-lapse ground penetrating radar (GPR) measurements in a sandstone aquifer in the UK from (a) zero-offset profile (ZOP) measurements and (b) multiple-offset profile (MOP) measurements. All data modified from Binley et al. (2001).

The possibility of monitoring soil moisture dynamics at depths within the vadose zone of near-surface aquifers has motivated considerable work on the use of borehole transmission methods to capture moisture dynamics at a larger scale than that typically captured with invasive methods, e.g. neutron probe data (Figure 7.13). Binley et al. (2002b) converted time-lapse ZOP measurements to 1D θ profiles that revealed seasonal moisture dynamics within the unsaturated zone of an important sandstone aquifer in the UK. They were able to capture relatively small (0.5–2%) changes in θ that were well-correlated with rainfall events. Cassaini et al. (2004) report on the use of vertical radar profiling (VRP), wherein the transmitter is at the surface and the receiver is lowered down a borehole, to investigate moisture content variations within the vadose zone to tens of meters below the surface. Cassiani et al. (2004) used GPR data to calibrate a hydrological model for vadose-zone moisture transport and thereby estimated the hydraulic conductivity of the subsurface. Cassiani and Binley (2005) modeled vadose-zone moisture dynamics using an 18-month time sequence of ZOP data from a UK sandstone aquifer. They highlighted some limitations of the ZOP-estimated θ profiles, most significantly that the flow parameters could not be uniquely defined as the system was overparameterized relative to the information inherent in the ZOP data.

A potentially powerful method for monitoring moisture dynamics is to perform GPR tomography over time and invert images of estimated velocity change that can then be converted to images of estimated θ change using the appropriate petrophysical relations. These difference images can, under the right conditions, reveal subtle variations in θ that would not be perceived by visual comparison of the static distributions of θ at two specific times. Early examples of the use of time-lapse GPR imaging to infer spatial patterns of θ change are reported in Hubbard et al. (1997a, 1997b) and Eppstein and Dougherty (1998). Parkin et al. (2000) used the same approach to monitor θ changes in a 2D horizontal plane using a pair of horizontal boreholes located below a shallow wastewater trench. Binley et al. (2001) describe the use of time-lapse GPR imaging to monitor the migration of moisture under both controlled tracer injection and longer-term periods of natural loading (see Figure 7.13). The images demonstrated how subtle contrasts in lithology within the sandstone aquifer impacted changes in θ as a function of time. Despite such successes, it is important here to again sound a cautionary note on the interpretation of GPR tomograms. Errors in data acquisition/inversion, combined with errors associated with the adopted petrophysical relationships, can seriously limit the value of the geophysical realizations of θ change acquired (Kowalsky et al., 2006; Linde et al., 2006a).

7.6.2. Water table detection/monitoring

The capillary fringe/water table in an unconfined aquifer represents an interface across which a relatively sharp change in θ occurs. It is therefore intuitive to expect that the level of the water table can be detected with GPR (under the right conditions) due to the relatively high reflection coefficient associated with this interface. However, Daniels et al. (2005) caution that the detection and, particularly, monitoring of the water table is nontrivial. They describe results of a tank experiment clearly illustrating that the primary GPR reflection was not from the water table per se but from the top of the capillary fringe. This is to be expected, as it is the top of the capillary fringe where θ changes dramatically (the capillary fringe is held 100% saturated by capillary tension and there is thus no θ change in soil at the water table measured in a well). More problematic was the observation that water level changes induced in the tank resulted in residual soil moisture redistribution within the soil. This resulted in a progressive change in the ε_r of the unsaturated portion of the tank that complicated tracking of the true water table (capillary fringe) location (i.e. accurate conversion from observed two-way travel time to depth) from the observed GPR reflection. Bano (2006) also sound a note of caution regarding the ability of GPR to detect the water table. They conducted measurements in an experimental tank that showed how a transition zone, characterized by a continual decrease in velocity with depth, could result in the absence of a recordable reflection at the water table. Despite these complications, numerous studies report on the detection of the water table using GPR data (Porsani et al., 2004; Roth et al., 2004; Clement and Knoll, 2006; Clement et al., 2006; Doolittle et al., 2006).

One interesting hydrogeophysical application of GPR is in monitoring of pumping tests (Figure 7.14). The accuracy of pumping test analyses depends on the availability of wells to monitor drawdown as a function of time and adequately define the cone of depression caused by pumping. The high cost of drilling typically limits the number of monitoring wells used in a pumping test. Ground penetrating

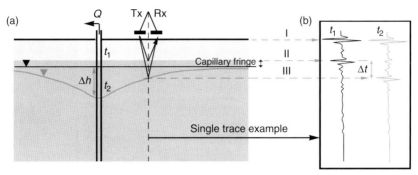

I: Direct wave (ground); II: Capillary fringe (at t_1); III: Drawdown (at t_2)

Figure 7.14 (a) Schematic representation of a pumping test before (t_1) and after pumping (t_2). Pumping induces a cone of depression reflected in a difference in depth to the water table (Δh) and (b) change in reflection patterns due to increased travel time (Δt) of the reflection from the capillary fringe after pumping.

radar-based monitoring of drawdown could provide an inexpensive alternative to a monitoring well installation and might be particularly effective in evaluating whether the shape of the cone of depression is consistent with the simplifying assumptions (e.g. radial flow to a well) used in common pumping test analyses. Endres et al. (2000) tested this concept on a pumping test conducted in a well-sorted, medium–fine-grained sand aquifer. Consistent with the tank experiment described in Daniels et al. (2005), they detected a reflection at the top of the capillary fringe. They monitored the "drawdown" of this reflection during a pumping test and compared it to drawdown directly measured in a monitoring well. They found that the GPR-estimated drawdown was smaller, and delayed, compared to the drawdown in the monitoring well. They attributed this observation to the moisture redistribution within the sands induced by the water table drawdown. Bevan et al. (2003) utilized the sensitivity of GPR to the top of the capillary fringe in order to understand capillary fringe dynamics, specifically an apparent extension of the capillary fringe that occurs during drawdown induced by pumping (Figure 7.15). Tsoflias et al. (2001) used GPR to monitor a pumping test in a carbonate-fractured rock aquifer, focusing on the response of a permeable, subhorizontal fracture plane. However, rather than examining variations in travel times, they analyzed the dependence of the amplitude and shape of the waveform reflected from this fracture plane on fracture saturation to improve the understanding of fractured formation fluid flow properties at the field scale.

7.6.3. Solute transport in fractures

Solute transport within fractures and fracture networks is a subject of active hydrogeological research. As previously noted, the accurate characterization of fractured-rock aquifer heterogeneity remains one of the most challenging and important problems in groundwater hydrology (Day-Lewis et al., 2003). Recent studies have focused on how amplitude analysis can be used to study (1) the spatial

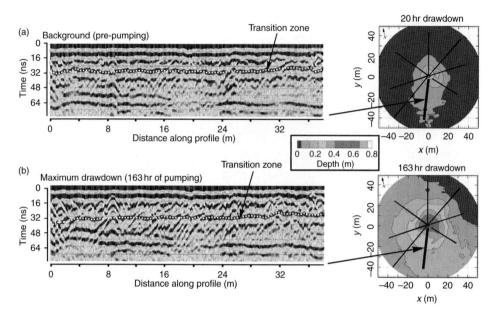

Figure 7.15 Multiple-offset ground penetrating radar (GPR) reflection survey during a pumping test across a sandstone aquifer in Canada (a) before and (b) after pumping. A white dotted line in both profiles indicates the reflection from the transition zone (e.g. capillary fringe). Insets exemplify plan views of GPR-derived transition zone drawdown at the beginning (20 h) and the end (163 h) of the pumping. All data modified from Bevan et al. (2003).

distribution of fractures imaged in bedrock (Day-Lewis et al., 2003) and (2) thickness/fluid properties of near-surface fractures (Tsoflias et al., 2004). Day-Lewis et al. (2003) used measurements of the change in amplitude resulting from the forced injection of a saline tracer into bedrock to infer preferential flow within transmissive fracture zones (Figure 7.16). They developed a novel, constrained inversion strategy that imaged changes in amplitude local to fracture zones. They also demonstrated how time-lapse attenuation data can be interpreted in terms of tracer breakthrough behavior based on the temporal response for a chosen voxel within the imaged volume (see Figure 7.16c). This concept suggests the possibility of treating such time-lapse GPR images of solute transport as a high-density set of solute breakthrough curves from which spatial variability in advective–dispersive transport behavior could be inferred.

Recent work has examined the information on fractures that can be retrieved from careful analysis of multi-polarization GPR measurements. In a synthetic study, Tsoflias et al. (2004) showed that at large oblique angles of incidence to a fracture, when the electric field component is oriented perpendicular to the plane of the fracture (H-pol), the transmitted through-the-fracture electric field leads in phase compared to when the incident electric field is oriented parallel to the plane of the fracture (E-pol). This observation allowed for the determination of vertical fracture location and azimuth from multi-polarization GPR measurements based on the presence of a phase lead in H-pol data relative to E-pol data. Tsoflias and

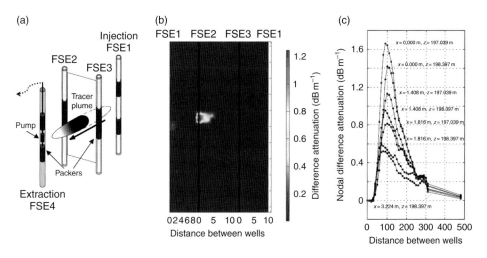

Figure 7.16 (a) Cross section of an experimental setup for a tracer experiment using crosshole GPR in fractured rock; (b) difference attenuation tomogram from constrained inversion after 2 h of tracer injection; and (c) nodal difference attenuation history in the FSE2–FSE3 image plane. All data modified from Day-Lewis et al. (2003).

Hoch (2006) used a novel, physical-scale model set up to investigate how high-incident–angle multi-polarization measurements of phase and amplitude were a function of the width and fluid properties (air versus water, salinity) of thin layers. In a field experiment, Talley et al. (2005) showed how amplitude variations associated with a reflection from a prominent subhorizontal bedrock fracture could be used to interpret the movement of a saline tracer within the fracture. The study demonstrated how GPR might be employed in a tracer-test monitoring format to improve the understanding of transport within discrete fractures.

7.6.4. Studies of the hyporheic corridor

There is growing interest in hydrological studies of the hyporheic corridor along river channels as it is recognized that groundwater–surface water exchanges regulate nutrient levels and stream temperature and thereby regulate the river ecosystem. Furthermore, where contaminated aquifers discharge into rivers, spatial and temporal variations in this mixing zone can result in a complex distribution of contaminants that is hard to quantify (Conant et al., 2004). For example, groundwater of the uranium-contaminated aquifer underlying the US DOE Hanford facility is hydrologically coupled to the Columbia River, which is a unique ecological habitat. The hyporheic interface along the Columbia River corridor is recognized as an important location for temporally and spatially complex hydrological processes impacting uranium transport. However, understanding of (1) the rate and extent of the processes, their interrelatedness across spatial and temporal scales as well as the influence of sediment characteristics on these processes, (2) gradients in physio-chemical characteristics between the river and the aquifer, and (3) temporal changes in discharge enhanced by river stage variations remains poor. Ground penetrating

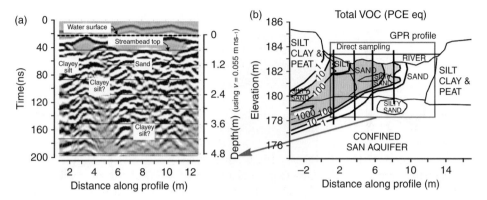

Figure 7.17 (a) Multiple-offset GPR reflection survey and (b) interpreted geologic cross section (confirmed with direct sampling) across a river bed in Canada where a groundwater tetrachloroethene plume is known to discharge into the river. All data modified from Conant et al. (2004).

radar has recently been used to assist in the generation of stratigraphic and hydrological models of the hyporheic corridor within rivers and streams (Cardenas and Zlotnik, 2003; Conant et al., 2004). Figure 7.17 shows a GPR survey across a river bed (Pine River, Toronto) where a groundwater tetrachloroethene plume is known to discharge into the river (Conant et al., 2004). Along-channel variability in the lithofacies identified in the GPR in part explains the complex spatiotemporal behavior of the plume discharging into the riverbed. The current interest in the hyporheic corridor processes will likely motivate new studies in this area.

7.6.5. Studies of the rhizosphere

Recent research shows that GPR may have a significant role to play in studies of the root zone and root zone processes. High-frequency 3D GPR has been used to image the coarse roots (20 cm or greater) of large trees, as well as the volume of total root zone, down to a depth of 2 m (Nadezhdina and Cermak, 2003). Soil moisture monitoring, coupled with sap flow measurements, is critical to studying root function and hydraulic redistribution of flow in the soil (Nadezhdina and Cermak, 2003). In another rhizosphere-related study, Hanafy and al Hagrey (2006) describe the development of a ray-bending inversion that was applied to the study of moisture changes within the root zone of a poplar tree. Al Hagrey (2007) further describes the use of GPR to discriminate woody root zones from soft root zones based on difference in attenuation, as well as the mapping of individual large roots from GPR diffraction hyperbolas.

7.6.6. Carbon gas emissions from soils

It follows that due to the high sensitivity of GPR to θ and changes in θ, GPR should also be equally sensitive to free-phase gas content (ϕ_{FPG}) and changes in free-phase gas content (as $\phi_{FPG} = \phi - \theta$). One topical and relevant application of GPR for monitoring changes in ϕ_{FPG} is in studies of carbon gas emissions from wetlands. Wetlands are, under current climatic conditions, typically net sinks of carbon dioxide (CO_2) but net sources of methane (CH_4). Methane is the third most important greenhouse gas after water vapor and CO_2. Its concentration has risen by 150% since the pre-industrial era (IPCC, 2007), and currently 20% of the enhanced greenhouse effect is attributed to CH_4 (IPCC, 2007). Although the atmospheric concentration of CH_4 is relatively small compared to CO_2, it is a relatively strong greenhouse gas. The global warming potential (GWP) of CH_4 is 25 whereas, by definition, the GWP of CO_2 is 1 (IPCC, 2007).

Wetlands represent the largest natural source of atmospheric methane (CH_4), accounting for approximately 24% of total emissions from all sources (Whalen, 2005). Scientists have studied methane production and methane emissions from northern peatlands over the past 20 years. Northern peatlands represent a vast wetland complex circulating the globe at northern latitudes and are estimated to account for up to 10% of annual methane emissions to the atmosphere (Charman, 2002). The methane is biogenically produced as a result of anaerobic respiration by archea utilizing carbon as the terminal electron acceptor (the process of methanogenesis). Studies related to CH_4 emissions from peatlands and their response to climate change have increased largely during recent years (Glaser et al., 1981; Kellner et al., 2004). It is now recognized that rapid ebullition releases of free-phase CH_4 from northern peatlands may result in a much greater contribution of methane to the atmosphere than currently estimated (Rosenberry et al., 2006), thereby impacting future climate. These ebullition fluxes are spatially and temporally highly variable and difficult to accurately quantify using direct insertion methods (e.g. gas sampling, TDR, moisture probes) that easily disrupt the in situ gas regime upon insertion and provide poor spatial sampling density.

Ground penetrating radar is very well suited to studying CH_4 dynamics in peatlands. Firstly, in the absence of free-phase CH_4, peat is saturated to within ~0.5 m of the surface and has ϕ on the order of 90%. Secondly, free-phase CH_4 production has been observed to account for up to ~20% changes in θ in peat soils. Considering that successful hydrogeophysical applications of GPR for monitoring vadose-zone moisture dynamics have focused on detecting ~5% changes in θ (e.g. Binley et al., 2002b), these large biogenic-driven θ variations should be readily detectable with GPR techniques. Furthermore, this application of GPR is also encouraged by the fact that saturated peat deposits have a well-constrained ε_r of around 65–70 (Theimer et al., 1994). The spatial variability in ε_r of saturated peat tends to be small such that researchers have found that very precise estimates of peat thickness can be obtained from GPR with an assumed value of ε_r, or better still, a few CMP measurements made in a peatland (Jol and Smith, 1995; Slater and Reeve, 2002).

Comas et al. (2005b) performed a borehole transmission ZOP survey between two boreholes placed 5 m apart and installed at 7 m depth in a peatland in Maine. Using a CRIM [Equation (7.2)] modeling approach (with ε_r of peat fabric measured in the laboratory), they estimated a vertical profile of ϕ_{FPG} varying between 0 and 10%, with two well-defined zones of gas accumulation (a) immediately below

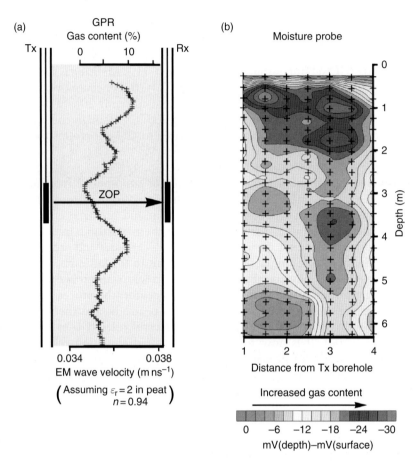

Figure 7.18 (a) Borehole transmission zero-offset profiling (ZOP) survey between two boreholes placed 5 m apart and at 7 m depth at Caribou Bog (Maine). Electromagnetic wave velocities inferred from ground penetrating radar (GPR) are expressed as free-phase gas content using the complex refractive index model (CRIM) model; (b) moisture probe (capacitance) measurements along the same profile show areas of increased gas content that correlated well with GPR results. All data modified from Comas et al. (2005b).

the water table between 0.5 and 2 m, and (b) between 4 and 4.5 m (Figure 7.18a). This vertical distribution of gas content compared well with regions of increased gas content estimated from 175 capacitance probe measurements made in the image plane formed by the two boreholes (see Figure 7.18b), as well as a limited number of samples analyzed for CH_4 with gas chromatography. It has also been suggested that zones of extensive FPG accumulation may be detected as a result of the scattering of the GPR signal (Comas et al., 2005b, 2005c). Backscattering attenuation occurs when small-scale heterogeneities (i.e. smaller than the propagated EM wavelengths) generate weak or undetectable responses but impact the signal passing by (Annan, 2006). Such heterogeneities extract some energy as the EM field passes through the medium, scattering it in all directions and possibly absorbing energy

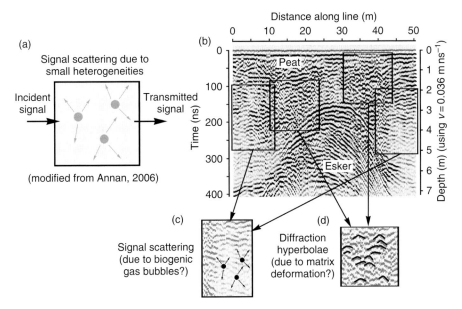

Figure 7.19 (a) Signal scattering in all directions due to small heterogeneities in a medium as an EM field passes though modified from Annan (2006); (b) multiple-offset ground penetrating radar (GPR) profile across a northern peatland (Caribou Bog, Maine), where an esker deposit is underlain by peat soil. Insets show areas of (c) possible signal scattering, attributed to the presence of biogenic gas bubbles and (d) diffraction hyperbolae, attributed to matrix deformation due to gas formation.

through ohmic dissipation. This concept is illustrated graphically in Figure 7.19a (modified from Annan, 2006). Backscattered energy from small fractures, voids, and cavities in massive rock has been exploited as a means of inferring rock quality (Orlando, 2003). Low-reflectivity zones observed in the vadose zone at hydrocarbon-contaminated sites have been attributed to scattering by a hydrocarbon vapor phase (de Castro and Branco, 2003). The size of gas bubbles forming in the peat is unknown but probably 1–2 orders of magnitude smaller than the EM wavelengths in peat soils using antenna frequencies typically deployed in the field (100–500 MHz, wavelengths of ~8–40 cm based on typical ε_r of peat). Zones of EM signal attenuation have been recorded within peatlands where enhanced gas buildup has been suspected from observations of peat surface deformation and episodic outgassing observed in/over pools. An example of such zones of pronounced attenuation is shown in Figure 7.19b for a profile across a portion of Caribou Bog, ME, where hydrological data show overpressurized zones in the peat indicative of extensive FPG accumulation (Comas et al., 2005c). However, the detection of such extensive CH_4 deposits in peatlands from attenuation in the GPR record remains to be fully proven.

Very recent studies have shown that GPR monitoring can be used to capture the temporal buildup and release of FPG in peatlands and thereby contribute to understanding mechanisms and rates of CH_4 buildup and release in peatlands. In a laboratory study, Comas and Slater (2007) measured the transmission time of a high-frequency (1.2 GHz) signal across a peat block during biogenic CH_4 production stimulated by controlled increases in temperature. This concept was extended to field-scale monitoring using surface NMO measurements over a 50-day period during the summer time (Comas et al., 2007) and expanded over almost an entire year to include winter snow and spring melting events (Comas et al., 2008). The NMO-estimated velocities from the mineral soil reflector were converted to estimated FPG content accounting for changes in ϕ due to peat expansion and changes in temperature measured with vertical temperature probes installed in the peat soil. Temporal variations in FPG content were consistent with evidence of methane production from peat surface expansion and were interpreted in terms of ebullition fluxes that showed a clear correspondence with atmospheric pressure variations, suggesting that large, episodic releases occurred during low atmospheric pressures. Figure 7.20 contrasts the variation in GPR-estimated ϕ_{FPG} at a site in Caribou Bog during summer and winter. The ϕ_{FPG} variation is again consistent with other indirect measures of CH_4 accumulation and release (surface deformation and chamber measurements). The GPR monitoring indicates that rapid ebullition-driven buildup and release of FPG occurs during the summer months, whereas a large, sustained reservoir of CH_4 accumulates over the winter due to the freezing of the surface layer acting to confine gas release. This gas is rapidly released upon spring melting. Such studies clearly suggest that GPR has an important contribution to play in the understanding of CH_4 emissions to the atmosphere from northern peatlands. Importantly, the GPR estimates of methane release are considerably higher than estimated in many studies using point-based chamber measurements of emissions. Ground penetrating radar

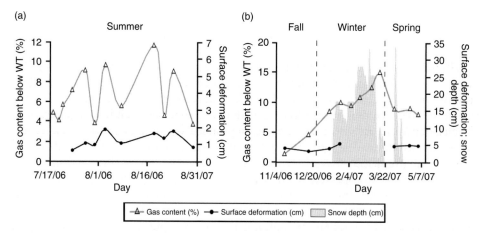

Figure 7.20 Free-phase gas (FPG) content variation estimated from multiple-offset reflection ground penetrating radar (GPR) measurements at a site in a northern peatland during (a) the summer (July–August) and (b) winter–spring (September–May). Ground penetrating radar estimates are well correlated with surface deformation of the peat matrix. Decreases in FPG content during the summer are associated with ebullition events. A consistent accumulation during the winter (e.g. snow season) is followed by a decrease in FPG after the snow/ice melt. All data modified from Comas et al. (2008).

may thus have an important role to play in better quantifying the contribution of peatlands to the atmospheric methane burden and in understanding how CH_4 emissions will change in response to a warming climate.

7.7. CONCLUSIONS

This review demonstrates that GPR is making a growing, valuable contribution to a wide range of water resource-related studies. The fundamental basis of the application of GPR in hydrogeology is the unique sensitivity of the method to θ and the flexibility of the technique in terms of mode of operation to obtain θ at a range of scales from laboratory blocks to the field, and even basin scale. Probably the greatest challenge with future water resource-related developments of the method remains the conversion of the measured ε_r to estimates of hydrogeological properties (i.e. θ, porosity, ϕ_{FPG}). Commonly used petrophysical relations may not be appropriate at every site, depending on the mineralogical characteristics of the soils. An important outstanding issue is the scale dependency of the ε_r–θ relationship that determines how local-scale petrophysical relationships relate to the field-scale relationships needed to interpret GPR data. Mixing models such as CRIM do not account for the geometric distribution of subsurface water, although this distribution is well-known to impact ε_r (Moysey and Knight, 2004). This dependence on geometry means that the petrophysical relationship linking ε_r to θ is scale-dependent where subsurface architecture is complex. Field-scale relationships that capture this geologic heterogeneity and account for the effect of geologic complexity on the ε_r–θ relationship have been predicted using geostatistical characterizations of subsurface structure (Moysey and Knight, 2005). Cassiani and Binley (2005) also showed that accounting for the scale of ZOP borehole measurements was critical to fitting the geophysical data using one-dimensional solute transport modeling.

Further work is also clearly needed to improve field survey techniques in order to optimize subsurface structures resolvable with the GPR method. Ultra high-resolution 3D GPR requires a level of effort typically not invested in a survey but can produce remarkable subsurface information that has illuminated patterns of human land use and laterally variable depositional processes (Grasmueck et al., 2004). More research is also needed to avoid potential pitfalls in processing of hydrogeophysical GPR datasets. Examples include recognition of refracted first arrivals in ZOP data when layering is characterized by high-velocity contrasts, as well as accounting for the effects of borehole antennas on the first arrival from source to receiver in tomographic surveys. Sophisticated data processing methods may also further push the boundaries of GPR in water resource-related studies. Xia et al. (2004) showed how deconvolution of a GPR profile with the source wavelet in air, a process designed to reproduce the reflectivity series generating the observed radar reflections, increased the accuracy of velocity models and the resolution of specific geologic features. Perhaps one of the most exciting recent developments in processing is the evolution of full-waveform modeling and inversion techniques as an alternative to traditional ray-based methods that utilize just first arrival times and first cycle amplitudes (Ernst et al., 2007a, 2007b; Gloaguen et al., 2007). Full-waveform modeling and inversion seeks to use all the information contained in the radar dataset (e.g. radiation patterns, dispersion) and has demonstrated potential to markedly improve resolution of structures within GPR tomograms, even resolving small structures that are a fraction of the size of the dominant wavelength (Ernst et al., 2007a). Full-waveform modeling has also been recently applied to better understand the effects of physical property and structure variations on a scale smaller than the dominant wavelength (Van den Bril et al., 2007).

Despite current limitations, many studies have shown that GPR provides surprisingly accurate θ estimates under a wide range of conditions. The vast volume of literature demonstrating successful applications of GPR for mapping hydrostratigraphy and aquifer architecture, mapping θ, monitoring moisture dynamics, and carbon gas emissions is testimony to the wealth of hydrogeological information available from GPR. Very recent work is increasingly turning to the combined interpretation of GPR and hydrogeological datasets, wherein the spatially rich, indirect information obtained with GPR is an excellent complement to the direct measurement of hydraulic properties using methods that provide poor spatial information. One technological challenge faced by GPR is in the development of autonomous data acquisition protocols. Autonomous data acquisition has recently been applied to resistivity monitoring, enhancing the flexibility of the method as a hydrogeological monitoring tool. The GPR method is inherently more labor-intensive during field data acquisition, and the degree to which the method can be automated for high-resolution spatiotemporal imaging of hydrogeological processes is currently uncertain. Some progress has been made in this direction, with semi-automated acquisition systems developed that use self-tracking laser theodolites with automatic tracking capabilities (Green et al., 2003). Despite the ongoing challenges in the application of GPR to hydrogeophysical studies, the sensitivity of the method to water in the earth will likely ensure that it remains an important hydrogeophysical technique for examining water resource issues.

REFERENCES

al Hagrey, S.A., 2007, Geophysical imaging of root-zone, trunk, and moisture heterogeneity. Journal of Experimental Botany, Vol. 58, No. 4, pp. 839–854.
Al-Qadi, I.L., Lahouar, S., Louizi, A., Elseifi, M.A. and Wilkes, J.A., 2004, Effective approach to improve pavement drainage layers. Journal of Transportation Engineering-Asce, Vol. 130, No. 5, pp. 658–664.
Alumbaugh, D., Chang, P.Y., Paprocki, L., Brainard, J.R., Glass, R.J. and Rautman, C.A., 2002, Estimating moisture contents in the vadose zone using cross-borehole ground penetrating radar: A study of accuracy and repeatability. Water Resources Research, Vol. 38, No. 12.
Annan, A.P., 2005, GPR methods for hydrogeological studies, in Rubin, Y. and Hubbard, S.S.S. (eds), Hydrogeophysics, Springer, The Netherlands, pp. 185–213.
Annan, A.P., 2006, Ground Penetrating Radar: Near Surface Geophysics. D. Butler, Society of Exploration Geophysicists, pp. 357–438.
Bano, M., 2006, Effects of the transition zone above a water table on the reflection of GPR waves. Geophysical Research Letters, Vol. 33, No. 13. doi: 10.1029/2006GL026158.

Beres, M. and Haeni, F.P., 1991, Application of ground-penetrating-radar methods in hydrogeologic studies. Ground Water, Vol. 29, No. 3, pp. 375–387.

Beres, M., Green, A., Huggenberger, P. and Horstmeyer, H. 1995, Mapping the architecture of glaciofluvial sediments with 3-Dimensional georadar. Geology, Vol. 23, No. 12, pp. 1087–1090.

Beres, M., Huggenberger, P., Green, A.G. and Horstmeyer, H., 1999, Using two- and three-dimensional georadar methods to characterize glaciofluvial architecture. Sedimentary Geology, Vol. 129, Nos. 1–2, pp. 1–24.

Bevan, M.J., Endres, A.L., Rudolph, D.L. and Parkin, G., 2003, The non-invasive characterization of pumping-induced dewatering using ground penetrating radar. Journal of Hydrology, Vol. 281, Nos. 1–2, pp. 55–69.

Binley, A., Winship, P., Middleton, R., Pokar, M. and West, J., 2001, High-resolution characterization of vadose zone dynamics using cross-borehole radar. Water Resources Research, Vol. 37, No. 11, pp. 2639–2652.

Binley, A., Cassiani, G., Middleton, R. and Winship, P., 2002a, Vadose zone flow model parameterisation using cross-borehole radar and resistivity imaging. Journal of Hydrology, Vol. 267, Nos. 3–4, pp. 147–159.

Binley, A., Winship, P., West, L.J., Pokar, M. and Middleton, R., 2002b, Seasonal variation of moisture content in unsaturated sandstone inferred from borehole radar and resistivity profiles. Journal of Hydrology, Vol. 267, Nos. 3–4, pp. 160–172.

Blonquist, J.M., Jones, S.B., Lebron, I. and Robinson, D.A., 2006, Microstructural and phase configurational effects determining water content: Dielectric relationships of aggregated porous media. Water Resources Research, Vol. 42, No. 5. doi: 10.1029/2005WR004418.

Bowling, J.C., Rodriguez, A.B., Harry, D.L. and Zheng, C.M., 2005, Delineating alluvial aquifer heterogeneity using resistivity and GPR data. Ground Water, Vol. 43, No. 6, pp. 890–903.

Cardenas, M.B. and Zlotnik, V.A., 2003, Three-dimensional model of modern channel bend deposits. Water Resources Research, Vol. 39, No. 6. doi:10.1029/2002WR001383.

Carreon-Freyre, D., Cerca, M. and Hernandez-Marin, M., 2003, Correlation of near-surface stratigraphy and physical properties of clayey sediments from Chalco Basin, Mexico, using Ground Penetrating Radar. Journal of Applied Geophysics, Vol. 53, Nos. 2–3, pp. 121–136.

Cassiani, G. and Binley, A., 2005, Modeling unsaturated flow in a layered formation under quasi-steady state conditions using geophysical data constraints. Advances in Water Resources, Vol. 28, No. 5, pp. 467–477.

Cassiani, G., Strobbia, C. and Gallotti, L., 2004, Vertical radar profiles for the characterization of deep vadose zones. Vadose Zone Journal, Vol. 3, No. 4, pp. 1093–1105.

Charman, D.J., 2002, Peatland Systems and Environmental Change, John Wiley & Sons, Chichester.

Chen, J., Hubbard, S.S. and Rubin, Y., 2001, Estimating the hydraulic conductivity at the South Oyster Site from geophysical tomographic data using Bayesian techniques based on the normal linear regression model. Water Resources Research, Vol. 37, No. 6, pp. 1603–1613.

Clement, W.P. and Knoll, M.D., 2006, Traveltime inversion of vertical radar profiles. Geophysics, Vol. 71, No. 3, pp. K67–K76.

Clement, W.P., Barrash, W. and Knoll, M.D., 2006, Reflectivity modeling of a ground-penetrating-radar profile of a saturated fluvial formation. Geophysics, Vol. 71, No. 3, pp. K59–K66.

Cognard, A.L., Loumagne, C., Normand, M., Olivier, P., Ottle, C., Vidalmadjar, D., Louahala, S. and Vidal, A., 1995, Evaluation of the Ers-1 synthetic aperture radar capacity to estimate surface soil-moisture – 2-year results over the Naizin watershed. Water Resources Research, Vol. 31, No. 4, pp. 975–982.

Comas, X. and Slater, L., 2007, Evolution of biogenic gases in peat blocks inferred from noninvasive dielectric permittivity measurements. Water Resources Research, Vol. 43, No. 5. doi: 10.1029/2006WR005562.

Comas, X., Slater, L. and Reeve, A., 2005a, Stratigraphic controls on pool formation in a domed bog inferred from ground penetrating radar (GPR). Journal of Hydrology, Vol. 315, Nos. 1–4, pp. 40–51.

Comas, X., Slater, L. and Reeve, A., 2005b, Spatial variability in biogenic gas accumulations in peat soils is revealed by ground penetrating radar (GPR). Geophysical Research Letters, Vol. 32, No. 8. doi: 10.1029/2004GL022297.

Comas, X., Slater, L. and Reeve, A., 2005c, Geophysical and hydrological evaluation of two bog complexes in a northern peatland: Implications for the distribution of biogenic gases at the basin scale. Global Biogeochemical Cycles, Vol. 19, No. 4. doi: 10.1029/2005GB002582.

Comas, X., Slater, L. and Reeve, A. 2007, In situ monitoring of free-phase gas accumulation and release in peatlands using ground penetrating radar (GPR). Geophysical Research Letters, Vol. 34, No. 6. doi: 10.1029/2006GL029014.

Comas, X., Slater, L. and Reeve, A., 2008, Seasonal geophysical monitoring of biogenic gases in a northern peatland: Implications for temporal and spatial variability in free phase gas production rates. Journal of Geophysical Research-Biogeosciences, Vol. 113, p. G01012.

Conant Jr., B., Cherry, J.A. and Gillham, R.W., 2004, A PCE groundwater plume discharging to a river: Influence of the streambed and near-river zone on contaminant distributions. Journal of Contaminant Hydrology, Vol. 73, pp. 249–279.

Copty, N., Rubin, Y. and Mavko, G., 1993, Geophysical-hydrological identification of field permeabilities through Bayesian updating. Water Resources Research, Vol. 29, No. 8, pp. 2813–2825.

Corbeanu, R.M., McMechan, G.A., Szerbiak, R.B. and Soegaardz, K., 2002, Prediction of 3-D fluid permeability and mudstone distributions from ground-penetrating radar (GPR) attributes: Example from the Cretaceous Ferron Sandstone Member, east-central Utah. Geophysics, Vol. 67, No. 5, pp. 1495–1504.

da Silva, C.C.N., de Medeiros, W.E., de Sa, E.F.J. and Neto, P.X., 2004, Resistivity and ground-penetrating radar images of fractures in a crystalline aquifer: A case study in Caicara farm – NE Brazil. Journal of Applied Geophysics, Vol. 56, No. 4, pp. 295–307.

Daniels, J.J., Allred, B., Binley, A., LaBrecque, D. and Alumbaugh, A., 2005, Hydrogeophysical case studies in the vadose zone, in Rubin, Y. and Hubbard, S.S. (eds), Hydrogeophysics, Springer, The Hague, The Netherlands, pp. 413–440.

Davis, J.L. and Annan, A.P., 2002, Ground penetrating radar to measure soil water content, in Dane, J.H. and Topp, G.C. (eds), Methods of Soil Analysis, Part 4, Soil Science Society of America (SSSA), pp. 446–463.

Day-Lewis, F.D. and Lane, Jr., L.W., 2004, Assessing the resolution-dependent utility of tomograms for geostatistics. Geophysical Research Letters, Vol. 31, p. L07503, doi:10.1029/2004GL019617.

Day-Lewis, F.D., Lane, J.W., Harris, J.M. and Gorelick, S.M., 2003, Time-lapse imaging of saline-tracer transport in fractured rock using difference-attenuation radar tomography. Water Resources Research, Vol. 39, No. 10. doi: 10.1029/2002WR001722.

de Castro, D.L. and Branco, R., 2003, 4-D ground penetrating radar monitoring of a hydrocarbon leakage site in Fortaleza (Brazil) during its remediation process: a case history. Journal of Applied Geophysics, Vol. 54, Nos. 1–2, pp. 127–144.

Deiana, R., Cassiani, G., Kemna, A., Villa, A., Bruno, V. and Bagliani, A., 2007, An experiment of non-invasive characterization of the vadose zone via water injection and cross-hole time-lapse geophysical monitoring. Near Surface Geophysics, Vol. 5, No. 3, pp. 183–194.

Dix, C.H., 1955, Seismic velocities from surface measurements. Geophysics, Vol. 20, pp. 68–86.

Doolittle, J.A., Jenkinson, B., Hopkins, D., Ulmer, M. and Tuttle, W., 2006, Hydropedological investigations with ground-penetrating radar (GPR): Estimating water-table depths and local ground-water flow pattern in areas of coarse-textured soils. Geoderma, Vol. 131, Nos. 3–4, pp. 317–329.

Endres, A.L., Clement, W.P. and Rudolph, D.L., 2000, Ground penetrating radar imaging of an aquifer during a pumping test. Ground Water, Vol. 38, No. 4, pp. 566–576.

Eppstein, M.J. and Dougherty, D.E., 1998, Efficient three-dimensional data inversion: Soil characterization and moisture monitoring from cross-well ground-penetrating radar at a Vermont test site. Water Resources Research, Vol. 34, No. 8, pp. 1889–1900.

Ernst, J.R., Maurer, H., Green, A.G. and Holliger, K., 2007a, Full-waveform inversion of crosshole radar data based on 2-D finite-difference time-domain solutions of Maxwell's equations. IEEE Transactions on Geoscience and Remote Sensing, Vol. 45, No. 9, pp. 2807–2828.

Ernst, J.R., Green, A.G., Maurer, H. and Holliger, K., 2007b, Application of a new 2D time-domain full-waveform inversion scheme to crosshole radar data. Geophysics, Vol. 72, pp. J53–J64.

Ezzy, T.R., Cox, M.E., O'Rourke, A.J. and Huftile, G.J., 2006, Groundwater flow modelling within a coastal alluvial plain setting using a high-resolution hydrofacies approach; Bells Creek plain, Australia. Hydrogeology Journal, Vol. 14, No. 5, pp. 675–688.

Galagedara, L.W., Parkin, G.W., Redman, J.D., von Bertoldi, P. and Endres, A.L., 2005, Field studies of the GPR ground wave method for estimating soil water content during irrigation and drainage. Journal of Hydrology, Vol. 301, Nos. 1–4, pp. 182–197.

Gallardo, L.A. and Meju, M.A., 2004, Joint two-dimensional DC resistivity and seismic travel time inversion with cross-gradient constraints. Journal of Geophysical Research, Vol. 109, p. B03311, doi: 10.1029/2003JB002716.

Ghose, R. and Slob, E.C., 2006, Quantitative integration of seismic and GPR reflections to derive unique estimates for water saturation and porosity in subsoil. Geophysical Research Letters, Vol. 33, No. 5. doi: 10.1029/2005GL025376.

Gish, T.J., Walthall, C.L., Daughtry, C.S.T. and Kung, K.J.S., 2005, Using soil moisture and spatial yield patterns to identify subsurface flow pathways. Journal of Environmental Quality, Vol. 34, No. 1, pp. 274–286.

Glaser, P.H., Wheeler, G.A., Gorham, E. and Wright, Jr., H.E., 1981, The patterned mires of the Red Lake peatland, northern Minnesota: Vegetation, water chemistry and landforms. Journal of Ecology, Vol. 69, pp. 575–599.

Gloaguen, E., Giroux, B., Marcotte, D. and Dimitrakopoulos, R., 2007, Pseudo-full-waveform inversion of borehole GPR data using stochastic tomography. Geophysics, Vol. 72, pp. J43–J51.

Grandjean, G. and Gourry, J.C., 1996, GPR data processing for 3D fracture mapping in a marble quarry (Thassos, Greece). Journal of Applied Geophysics, Vol. 36, No. 1, pp. 19–30.

Grasmueck, M., 1996, 3-D ground-penetrating radar applied to fracture imaging in gneiss. Geophysics, Vol. 61, No. 4, pp. 1050–1064.

Grasmueck, M., Weger, R. and Horstmeyer, H., 2004, Three-dimensional ground-penetrating radar imaging of sedimentary structures, fractures, and archeological features at submeter resolution. Geology, Vol. 32, No. 11, pp. 933–936.

Grasmueck, M., Weger, R. and Horstmeyer, H., 2005, Full-resolution 3D GPR imaging. Geophysics, Vol. 70, No. 1, pp. K12–K19.

Greaves, R.J., Lesmes, D.P., Lee, J.M. and Toksoz, M.N., 1996, Velocity variations and water content estimated from multi-offset, ground-penetrating radar. Geophysics, Vol. 61, No. 3, pp. 683–695.

Green, A., Gross, R., Holliger, K., Horstmeyer, H. and Baldwin, J., 2003, Results of 3-D georadar surveying and trenching the San Andreas fault near its northern landward limit. Tectonophysics, Vol. 368, Nos. 1–4, pp. 7–23.

Grote, K., Hubbard, S.S. and Rubin, Y., 2002, GPR monitoring of volumetric water content in soils applied to highway construction and maintenance. The Leading Edge, Vol. 21, pp. 482–485.

Grote, K., Hubbard, S. and Rubin, Y., 2003, Field-scale estimation of volumetric water content using ground-penetrating radar ground wave techniques. Water Resources Research, Vol. 39, No. 11. doi: 10.1029/2003WR002045.

Grote, K., Hubbard, S., Harvey, J. and Rubin, Y., 2005, Evaluation of infiltration in layered pavements using surface GPR reflection techniques. Journal of Applied Geophysics, Vol. 57, pp. 129–153.

Hanafy, S. and al Hagrey, S.A., 2006, Ground-penetrating radar tomography for soil-moisture heterogeneity. Geophysics, Vol. 71, No. 1, pp. K9–K18.

Hubbard, S.S., Peterson, J.E., Majer, E.L., Zawislanski, P.T., Williams, K.H., Roberts, J. and Wobber, F., 1997a, Estimation of permeable pathways and water content using tomographic radar data. The Leading Edge, Vol. 16, pp. 1623–1630.

Hubbard, S.S., Rubin, Y. and Majer, E. 1997b, Ground-penetrating-radar-assisted saturation and permeability estimation in bimodal systems. Water Resources Research, Vol. 33, No. 5, pp. 971–990.

Hubbard, S.S., Chen, J.S., Peterson, J., Majer, E.L., Williams, K.H., Swift, D.J., Mailloux, B. and Rubin, Y., 2001, Hydrogeological characterization of the South Oyster Bacterial Transport Site using geophysical data. Water Resources Research, Vol. 37, No. 10, pp. 2431–2456.

Huisman, J.A., Sperl, C., Bouten, W. and Verstraten, J.M., 2001, Soil water content measurements at different scales: accuracy of time domain reflectometry and ground-penetrating radar. Journal of Hydrology, Vol. 245, Nos. 1–4, pp. 48–58.

Huisman, J.A., Snepvangers, J., Bouten, W. and Heuvelink, G.B.M., 2002, Mapping spatial variation in surface soil water content: Comparison of ground-penetrating radar and time domain reflectometry. Journal of Hydrology, Vol. 269, Nos. 3–4, pp. 194–207.

Huisman, J.A., Hubbard, S.S., Redman, J.D. and Annan, A.P., 2003, Measuring soil water content with ground penetrating radar: A review. Vadose Zone Journal, Vol. 2, pp. 476–491.

Hyndman, D.W. and Gorelick, S.M., 1996, Estimating lithologic and transport properties in three dimensions using seismic and tracer data: The Kesterton aquifer. Water Resources Research, Vol. 32, No. 9, pp. 2659–2670.

Hyndman, D. and Tronicke, J., 2005, Hydrogeophysical case studies at the local scale: the saturated zone, in Rubin, Y. and Hubbard, S.S. (eds), Hydrogeophysics, Springer, The Hague, The Netherlands, pp. 391–412.

Hyndman, D.W., Harris, J.M. and Gorelick, S.M., 1994, Coupled seismic and tracer test inversion for aquifer property characterization. Water Resources Research, Vol. 30, No. 7, pp. 1965–1977.

IPCC (Intergovernmental Panel on Climate Change), 2007, Climate Change 2007: The Physical Science Basis. Summary for Policymakers: Geneva: World Meteorological Organization (WMO) and UN Environment Programme (UNEP); http://ipcc-wg1.ucar.edu/.

Irving, J. and Knight, R., 2005, Effect of antenna length on velocity estimates obtained from crosshole GPR data. Geophysics, Vol. 70, No. 5, pp. K39–K42.

Jacob, R.W. and Hermance, J.F., 2004, Assessing the precision of GPR velocity and vertical two-way travel time estimates. Journal of Environmental and Engineering Geophysics, Vol. 9, No. 3, pp. 143–153.

Jeannin, M., Garambois, S., Gregoire, C. and Jongmans, D., 2006, Multiconfiguration GPR measurements for geometric fracture characterization in limestone cliffs (Alps). Geophysics, Vol. 71, No. 3, pp. B85–B92.

Jol, H.M. and Smith, D.G., 1995, Ground penetrating radar surveys of peatlands for oilfield pipelines in Canada. Journal of Applied Geophysics, Vol. 34, No. 2, pp. 109–123.

Kellner, E., Price, J.S. and Waddington, J.M., 2004, Pressure variations in peat as a result of gas bubble dynamics. Hydrological Processes, Vol. 18, pp. 2599–2605.

Knight, R. and Endres, A.L., 2006, An Introduction to Rock Physics for Near-Surface Applications: Near-Surface Geophysics. D. Butler, Society of Exploration Geophysicists. Volume 1: Concepts and Fundamentals.

Kostic, B., Becht, A. and Aigner, T., 2005, 3-D sedimentary architecture of a Quaternary gravel delta (SW-Germany): Implications for hydrostratigraphy. Sedimentary Geology, Vol. 181, Nos. 3–4, pp. 147–171.

Kowalsky, M.B., Finsterle, S. and Rubin, Y., 2004, Estimating flow parameter distributions using ground-penetrating radar and hydrological measurements during transient flow in the vadose zone. Advances in Water Resources, Vol. 27, pp. 583–599.

Kowalsky, M.B., Chen, J. and Hubbard, S.S., 2006, Joint inversion of geophysical and hydrological data for improved subsurface characterization. The Leading Edge, June, pp. 730–734.

Kunk, J.R., Narbutovskih, S.M., Bergstrom, K.A. and Michell, T.H., 1993, Phase I Summary of Surface Geophysical Studies in the 300-FF-5 Operable Unit, Westinghouse Hanford Company, Richland, WA.

Lai, W.L., Tsang, W.F., Fang, H. and Xiao, D., 2006, Experimental determination of bulk dielectric properties and porosity of porous asphalt and soils using GPR and a cyclic moisture variation technique. Geophysics, Vol. 71, No. 4, pp. K93–K102.

Lambot, S., Antoine, M., van den Bosch, I., Slob, E.C. and Vanclooster, M., 2004a, Electromagnetic inversion of GPR signals and subsequent hydrodynamic inversion to estimate effective vadose zone hydraulic properties. Vadose Zone Journal, Vol. 3, No. 4, pp. 1072–1081.

Lambot, S., Rhebergen, J., van den Bosch, I., Slob, E.C. and Vanclooster, M., 2004b, Measuring the soil water content profile of a sandy soil with an off-ground monostatic ground penetrating radar. Vadose Zone Journal, Vol. 3, No. 4, pp. 1063–1071.

Lambot, S., Antoine, M., Vanclooster, M. and Slob, E.C., 2006a, Effect of soil roughness on the inversion of off-ground monostatic GPR signal for noninvasive quantification of soil properties. Water Resources Research, Vol. 42, No. 3, p. 10.

Lambot, S., Weihermuller, L., Huisman, J.A., Vereecken, H., Vanclooster, M. and Slob, E.C., 2006b, Analysis of air-launched ground-penetrating radar techniques to measure the soil surface water content. Water Resources Research, Vol. 42, No. 11. doi: 10.1029/2006WR005097.

Lambot, S., Slob, E.C., Vanclooster, M. and Vereecken, H., 2006c, Closed loop GPR data inversion for soil hydraulic and electric property determination. Geophysical Research Letters, Vol. 33, No. 21.

Laurens, S., Balayssac, J.P., Rhazi, J., Klysz, G. and Arliguie, G., 2005, Non-destructive evaluation of concrete moisture by GPR: Experimental study and direct modeling. Materials and Structures, Vol. 38, No. 283, pp. 827–832.

Lesmes, D.P. and Friedman, S.P., 2005, Relationships between the electrical and hydrogeological properties of rocks and soils, in Rubin, Y. and Hubbard, S.S. (eds), Hydrogeophysics, Springer, The Hague, The Netherlands, pp. 87–128.

Lesmes, D.P., Decker, S.M. and Roy, D.C., 2002, A multiscale radar-stratigraphic analysis of fluvial aquifer heterogeneity. Geophysics, Vol. 67, No. 5, pp. 1452–1464.

Leucci, G., Cataldo, R. and De Nunzio, G. 2006, Subsurface water-content identification in a crypt using GPR and comparison with microclimatic conditions. Near Surface Geophysics, Vol. 4, No. 4, pp. 207–213.

Linde, N., Chen, J., Kowalsky, M.B. and Hubbard, S., 2006a, Hydrogeophysical parameter estimation approaches for field scale characterization, in Vereecken, H. et al. (eds), Applied Hydrogeophysics, Springer, Dordrecht, pp. 9–44.

Linde, N., Finsterle, S. and Hubbard, S.S., 2006b, Inversion of tracer test data using tomographic constraints. Water Resources Research, Vol. 42, p. W04410, doi:10.1029/2004WR003806.

Linde, N., Binley, A., Tryggvason, A., Pedersen, L. and Revil, A., 2006c, Improved hydrogeophysical characterization using joint inversion of cross-hole electrical resistance and ground-penetrating radar traveltime data. Water Resources Research, Vol. 42, p. W12404, doi:10.1029/2006WR005131.

Loeffler, O. and Bano, M., 2004, Ground penetrating radar measurements in a controlled vadose zone: Influence of the water content. Vadose Zone Journal, Vol. 3, No. 4, pp. 1082–1092.

Lunt, I.A., Hubbard, S.S. and Rubin, Y., 2005, Soil moisture content estimation using ground-penetrating radar reflection data. Journal of Hydrology, Vol. 307, Nos. 1–4, pp. 254–269.

Moysey, S. and Knight, R., 2004, Modeling the field-scale relationship between dielectric constant and water content in heterogeneous systems. Water Resources Research, Vol. 40, No. 3. doi: 10.1029/2003WR002589.

Moysey, S., Caers, J., Knight, R. and Allen-King, R.M., 2003, Stochastic estimation of facies using ground penetrating radar data. Stochastic Environmental Research and Risk Assessment, Vol. 17, No. 5, pp. 306–318.

Moysey, S., Singha, K. and Knight, R., 2005, A framework for inferring field-scale rock physics relationships through numerical simulation. Geophysical Research Letters, Vol. 32, p. L08304, doi:10.1029/2004GL022152.

Nadezhdina, N. and Cermak, J., 2003, Instrumental methods for studies of structure and function of root systems of large trees. Journal of Experimental Botany, Vol. 54, No. 387, pp. 1511–1521.

Nakashima, Y., Zhou, H. and Sato, M., 2001, Estimation of groundwater level by GPR in an area with multiple ambiguous reflections. Journal of Applied Geophysics, Vol. 47, Nos. 3–4, pp. 241–249.

Neal, A., 2004, Ground-penetrating radar and its use in sedimentology: Principles, problems and progress. Earth-Science Reviews, Vol. 66, pp. 261–330.

Oldenborger, G.A., Schincariol, R.A. and Mansinha, L., 2003, Radar determination of the spatial structure of hydraulic conductivity. Ground Water, Vol. 41, No. 1, pp. 24–32.

Orlando, L., 2003, Semiquantitative evaluation of massive rock quality using ground penetrating radar. Journal of Applied Geophysics, Vol. 52, No. 1, pp. 1–9.

Paasche, H., Tronicke, J., Holliger, K., Green, A.G. and Muarer, H., 2006, Integration of diverse physical-property models: Subsurface zonation and petrophysical parameter estimation based on fuzzy c-means cluster analyses. Geophysics, Vol. 71, No. 3, pp. H33–H44.

Parkin, G., Redman, D., von Bertoldi, P. and Zhang, Z., 2000, Measurement of soil water content below a wastewater trench using ground-penetrating radar. Water Resources Research, Vol. 36, No. 8, pp. 2147–2154.

Porsani, J.L., Filho, W.M., Elis, V.R., Shimeles, F., Dourado, J.C. and Moura, H.P., 2004, The use of GPR and VES in delineating a contamination plume in a landfill site: A case study in SE Brazil. Journal of Applied Geophysics, Vol. 55, Nos. 3–4, pp. 199–209.

Porsani, J.L., Elis, V.R. and Hiodo, F.Y., 2005, Geophysical investigations for the characterization of fractured rock aquifers in Itu, SE Brazil. Journal of Applied Geophysics, Vol. 57, No. 2, pp. 119–128.

Porsani, J.L., Sauck, W.A. and Junior, A.O.S., 2006, GPR for mapping fractures and as a guide for the extraction of ornamental granite from a quarry: A case study from southern Brazil. Journal of Applied Geophysics, Vol. 58, No. 3, pp. 177–187.

Quesney, A., Le Hegarat-Mascle, S., Taconet, O., Vidal-Madjar, D., Wigneron, J.P., Loumagne, C. and Normand, M., 2000, Estimation of watershed soil moisture index from ERS/SAR data. Remote Sensing of Environment, Vol. 72, No. 3, pp. 290–303.

Rea, J. and Knight, R., 1998, Geostatistical analysis of ground-penetrating radar data: A means of describing spatial variation in the subsurface. Water Resource Research, Vol. 34, No. 3, pp. 329–339.

Rosenberry, D.O., Glaser, P.H. and Siegel, D.I., 2006, The hydrology of northern peatlands as affected by biogenic gas: Current developments and research needs. Hydrological Processes, Vol. 20, pp. 3601–3610.

Roth, K., Wollschlager, U., Cheng, Z.H. and Zhang, J.B., 2004, Exploring soil layers and water tables with ground-penetrating radar. Pedosphere, Vol. 14, No. 3, pp. 273–282.

Rubin, Y., Mavko, G. and Harris, J., 1992, Mapping permeability in heterogeneous aquifers using hydrologic and seismic data. Water Resources Research, Vol. 28, No. 7, pp. 1809–1816.

Rucker, D.F. and Ferré, P.A., 2003, Near-surface water content estimation with borehole ground penetrating radar using critically refracted waves. Vadoze Zone Journal, Vol. 2, pp. 247–252.

Rucker, D.F. and Ferré, T.P.A., 2004, Correcting water content measurement errors associated with critically refracted first arrivals on zero offset profiling borehole ground penetrating radar profiles. Vadose Zone Journal, Vol. 3, No. 1, pp. 278–287.

Seol, S.J., Kim, J.H., Song, Y. and Chung, S.H., 2001, Finding the strike direction of fractures using GPR. Geophysical Prospecting, Vol. 49, No. 3, pp. 300–308.

Seol, S.J., Kim, J.H., Cho, S.J. and Chung, S.H., 2004, A radar survey at a granite quarry to delineate fractures and estimate fracture density. Journal of Environmental and Engineering. Geophysics, Vol. 9, No. 2, pp. 53–62.

Serbin, G. and Or, D., 2004, Ground-penetrating radar measurement of soil water content dynamics using a suspended horn antenna. IEEE Transactions on Geoscience and Remote Sensing, Vol. 42, No. 8, pp. 1695–1705.

Serbin, G. and Or, D., 2005, Ground-penetrating radar measurement of crop and surface water content dynamics. Remote Sensing of Environment, Vol. 96, No. 1, pp. 119–134.

Skelly, R.L., Bristow, C.S. and Ethridge, F.G., 2003, Architecture of channel-belt deposits in an aggrading shallow sandbed braided river: the lower Niobrara River, northeast Nebraska. Sedimentary Geology, Vol. 158, Nos. 3–4, pp. 249–270.

Slater, L.D. and Reeve, A., 2002, Investigating peatland stratigraphy and hydrogeology using integrated electrical geophysics. Geophysics, Vol. 67, No. 2, pp. 365–378.

Snieder, R., Hubbard, S., Haney, M., Bawden, G., Hatchell, P. and Revil, A., 2007, Advanced noninvasive geophysical monitoring techniques. Annual Review of Earth and Planetary Sciences, Vol. 35, pp. 653–683.

Stevens, K.M., Lodha, G.S., Holloway, A.L. and Soonawala, N.M., 1995, The application of ground-penetrating radar for mapping fractures in plutonic rocks within the whiteshell research area, Pinawa, Manitoba, Canada. Journal of Applied Geophysics, Vol. 33, Nos. 1–3, pp. 125–141.

Stoffregen, H., Yaramanci, U., Zenker, T. and Wessolek, G., 2002, Accuracy of soil water content measurements using ground penetrating radar: Comparison of ground penetrating radar and lysimeter data. Journal of Hydrology, Vol. 267, Nos. 3–4, pp. 201–206.

Szerbiak, R.B., McMechan, G.A., Corbeanu, R.M., Forster, C. and Snelgrove, S.H., 2001, 3-D characterization of a clastic reservoir analog: From 3-D GPR to a 3-D fluid permeability model. Geophysics, Vol. 66, No. 4, pp. 1026–1037.

Talley, J., Baker, G.S., Becker, M.W. and Beyrle, N., 2005, Four dimensional mapping of tracer channelization in subhorizontal bedrock fractures using surface ground penetrating radar. Geophysical Research Letters, Vol. 32, No. 4. doi: 10.1029/2004GL021974.

Tercier, P., Knight, R. and Jol, H., 2000, A comparison of the correlation structure in GPR images of deltaic and barrier-spit depositional environments. Geophysics, Vol. 65, No. 4, pp. 1142–1153.

Theimer, B.D., Nobes, D.C. and Warner, B.G., 1994, A study of the geoelectrical properties of peatlands and their influence on ground-penetrating radar surveying. Geophysical Prospecting, Vol. 42, pp. 179–209.

Theune, U., Rokosh, D., Sacchi, M.D. and Schmitt, D.R., 2006, Mapping fractures with GPR: A case study from turtle mountain. Geophysics, Vol. 71, No. 5, pp. B139–B150.

Tillard, S. and Dubois, J.-C., 1995, Analysis of GPR data: Wave propagation velocity determination. Journal of Applied Geophysics, Vol. 33, Nos. 1–3, pp. 77–91.

Topp, G.C., Davis, J.L. and Annan, A.P., 1980, Electromagnetic determination of soil water content: Measurements in coaxial transmission lines. Water Resources Research, Vol. 16, pp. 574–582.

Tronicke, J., Dietrich, P., Wahlig, U. and Appel, E., 2002, Integrating surface georadar and crosshole radar tomography: A validation experiment in braided stream deposits. Geophysics, Vol. 67, No. 5, pp. 1516–1523.

Truss, S., Grasmueck, M., Vega, S. and Viggiano, D.A., 2007, Imaging rainfall drainage within the Miami oolitic limestone using high-resolution time-lapse ground-penetrating radar. Water Resources Research, Vol. 43, No. 3. doi: 10.1029/2005WR004395.

Tsoflias, G.P. and Hoch, A., 2006, Investigating multi-polarization GPR wave transmission through thin layers: Implications for vertical fracture characterization. Geophysical Research Letters, Vol. 33, No. 20. doi: 10.1029/2006GL027788.

Tsoflias, G.P., Halihan, T. and Sharp, J.M., 2001, Monitoring pumping test response in a fractured aquifer using ground-penetrating radar. Water Resources Research, Vol. 37, No. 5, pp. 1221–1229.

Tsoflias, G.P., Van Gestel, J.P., Stoffa, P.L., Blankenship, D.D. and Sen, M., 2004, Vertical fracture detection by exploiting the polarization properties of ground-penetrating radar signals. Geophysics, Vol. 69, No. 3, pp. 803–810.

United Nations Conference on Environment and Development, 1993, Agenda 21: Programme of action for sustainable development; Rio Declaration on Environment and Development; Statement of forest principles, United Nations, Department of Public Information.

Van Dam, R.L., Van Den Berg, E.H., Van Heteren, S., Kasse, C., Kenter, J.A.M. and Groen, K., 2002, Influence of organic matter on radar-wave reflection: Sedimentological implications. Journal of Sedimentary Research, Vol. 72, No. 3, pp. 341–352.

Van den Bril, K., Gregoire, C., Swennen, R. and Lambot, S., 2007, Ground-penetrating radar as a tool to detect rock heterogeneities (channels, cemented layers and fractures) in the Luxembourg Sandstone Formation (Grand-Duchy of Luxembourg). Sedimentology, Vol. 54, No. 4, pp. 949–967.

van Overmeeren, R.A., 1998, Radar facies of unconsolidated sediments in The Netherlands: A radar stratigraphy interpretation method for hydrogeology. Journal of Applied Geophysics, Vol. 40, Nos. 1–3, pp. 1–18.

van Overmeeren, R.A., Sariowan, S.V. and Gehrels, J.C., 1997, Ground penetrating radar for determining volumetric soil water content; results of comparative measurements at two test sites. Journal of Hydrology, Vol. 197, Nos. 1–4, pp. 316–338.

Wänstedt, S., Carlsten, S. and Tiren, S., 2000, Borehole radar measurements aid structure geological interpretations. Journal of Applied Geophysics, Vol. 43, pp. 227–237.

Weihermuller, L., Huisman, J.A., Lambot, S., Herbst, M. and Vereecken, H., 2007, Mapping the spatial variation of soil water content at the field scale with different ground penetrating radar techniques. Journal of Hydrology, Vol. 340, Nos. 3–4, pp. 205–216.

West, L.J., Handley, K., Huang, Y. and Pokar, M., 2003, Radar frequency dielectric dispersion in sandstone: Implications for determination of moisture and clay content. Water Resources Research, Vol. 39, No. 2. doi: 10.1029/2001WR000923.

Whalen, S.C., 2005, Biogeochemistry of methane exchange between natural wetlands and the atmosphere. Environmental Engineering Science, Vol. 22, pp. 73–94.

WWC (W. W. C.), 2000, Ministerial Declaration: 2nd World Water Forum – The Hague, March 17–22, 2000.

WWC (W. W. C.), 2006, Ministerial Declaration: 4th World Water Forum – Mexico, March 16–22, 2006.

Xia, J.H., Franseen, E.K., Miller, R.D. and Weis, T.V., 2004, Application of deterministic deconvolution of ground-penetrating radar data in a study of carbonate strata. Journal of Applied Geophysics, Vol. 56, No. 3, pp. 213–229.

Yilmaz, O., 1987, Seismic Data Processing. Tulsa, Society of Exploration Geophysicists.

CHAPTER 8

CONTAMINANT MAPPING

J.D. Redman

Contents

8.1. Introduction	247
8.2. Contaminant Types	248
8.3. Electrical Properties of Contaminated Rock and Soil	249
8.3.1. Electrical properties of NAPLs	249
8.3.2. Electrical properties of soil and rock with NAPL contamination	250
8.3.3. Biodegradation effects	253
8.3.4. Inorganics	253
8.4. Typical Distribution of Contaminants	254
8.4.1. DNAPL	254
8.4.2. LNAPL	255
8.4.3. Inorganics	255
8.4.4. Saturated and unsaturated zone	256
8.5. GPR Methodology	256
8.6. Data Processing and Interpretation	257
8.6.1. Visualization	257
8.6.2. Trace attributes	257
8.6.3. Data differencing	257
8.6.4. AVO analysis	258
8.6.5. Detection based on frequency-dependent properties	258
8.6.6. Quantitative estimates of NAPL	258
8.7. Case Studies	259
8.7.1. Controlled DNAPL injection	260
8.7.2. Controlled LNAPL injection	262
8.7.3. Accidental spill sites	262
8.7.4. Leachate and waste disposal site characterization	264
8.8. Summary	265
References	265
Terms for Glossary	269

8.1. INTRODUCTION

Subsurface contamination of soil and groundwater from waste disposal sites, industrial spills, gasoline stations, mine tailings and industrial processes is a serious societal problem. Geophysical methods, including GPR, have an important role to play in the characterization of these sites. Past experience has shown that these methods have been useful in mapping geology, contaminant distribution and in monitoring remediation processes.

The remediation of contaminated sites requires knowledge of the contaminant distribution in the subsurface and of the subsurface geology. Common and necessary methods for characterizing contaminated sites are coring, soil sampling and the installation of monitoring wells for the collection of groundwater samples. These methods are expensive, can in some cases mobilize subsurface contaminants and only provide small localized measurements. For these reasons other non-invasive methods have been sought to extensively characterize contaminated sites and to provide volume averaged properties that support the localized measurements provided by sampling and coring.

Mapping of the subsurface stratigraphy is critical to understanding contaminant migration and distribution. This subject is not discussed because it is addressed in other chapters. The issue of mapping the extent of waste disposal sites or mine tailings is beyond the scope of this chapter. It is not the intent to provide a general purpose methodology for applying GPR to contaminated sites. The methods used to characterize contaminated sites will depend on the specific nature of the site and will involve a suite of tools that may or may not require the use of geophysical techniques. There are many published reports on and guidelines for site characterization (e.g. Canadian Council of Ministers of the Environment (CCME), 1994).

8.2. Contaminant Types

Contaminants may be located within near-surface unconsolidated deposits, within bedrock and within buried wastes. Some contaminants that have strong absorption by soil or have low solubility in water do not necessarily produce groundwater contamination. Other contaminants may have high solubility and mobility resulting in large plumes of contaminated groundwater.

From the perspective of geophysical detection, common subsurface contaminants can be grouped into two main types: inorganics and organics. Inorganic chemical contaminants are typically produced by landfills, mine tailings, chemical spills, sewage lagoons and industrial processes. Common inorganic contaminant types are nitrate, calcium, sulphate, iron and trace metals such as arsenic, chromium, mercury and lead.

Non-aqueous phase liquids (NAPLs) are the most common and important organic contaminants. NAPLs are immiscible in water but have low solubilities that may still be orders of magnitude higher than the acceptable drinking water standards. NAPLs can be further subdivided into those that are denser than water (DNAPLs) and those that are lighter than water (LNAPLs). Tetrachloroethylene, commonly used in dry cleaning, trichloroethylene, an industrial degreaser/cleaner and other chlorinated solvents are common examples of DNAPLs. Hydrocarbon fuels such as gasoline, kerosene and jet fuel are common LNAPL contaminants. The term NAPL distinguishes the immiscible phase of these contaminants from their dissolved phase and vapor phase components. Groundwater is normally contaminated by the dissolved phase, but the immiscible phase is also occasionally found in monitoring and pumping wells. Successful remediation requires removal of the dissolved phase from the groundwater and removal or isolation of the NAPL source zone because it acts as a reservoir for continued contamination through dissolution.

Because of the low solubility and toxicity of NAPLs they have the potential to contaminate large volumes of groundwater if they are not removed from the subsurface. For example, the common contaminant tetrachloroethylene has a solubility of only 200 mg/L, but this is still much higher than the acceptable level in drinking water of 0.005 mg/L (US EPA). At a concentration equivalent to the drinking water standard, 1 L of this contaminant, uniformly distributed within the groundwater, would render 320 thousand liters of water undrinkable. Liquid organic contaminants that have low viscosity are a more serious problem because they have the potential to migrate large distances through the subsurface and contaminate large volumes of rock and soil.

Contaminated sites can be very complicated, having surface and subsurface contamination, buried tanks, metal and other waste. It is common for these sites to contain separate areas of NAPL and inorganic contamination as well as areas in which these two contaminant types are mixed. There are also some important organic contaminants that are miscible with water (e.g. ethanol). These will also have an impact on groundwater but will distribute themselves differently in the subsurface.

8.3. ELECTRICAL PROPERTIES OF CONTAMINATED ROCK AND SOIL

Developing methodologies for subsurface contaminant detection with GPR requires both an understanding of their typical subsurface distribution, as discussed in a following section, and of their affect on the electrical properties of the subsurface. This section discusses the electrical properties of typical contaminants and of soil and rock containing these contaminants.

8.3.1. Electrical properties of NAPLs

The physical properties of common contaminants have been tabulated in Lucius et al. (1992). Table 8.1 illustrates that the relative dielectric permittivities of NAPLs are much lower than water ($\sim$80), and higher than air (1). NAPLs typically have very low electrical conductivities and are non-conducting relative to pore water. For example trichloroethylene has a conductivity of 0.8×10^{-6} S m^{-1} (Lucius et al., 1992) compared with $\sim$0.01–5 S m^{-1} for water. Subsurface contaminants may often not be pure NAPLs, but may contain additives or may be mixed with other substances that modify the electrical properties of the pure NAPL. In particular this may often result in a contaminant mixture that is more conductive than the pure component.

Table 8.1 Electrical properties of common organic contaminants. Crude oil, gasoline and kerosene properties will vary somewhat depending on composition. All organics have very low electrical conductivity (see Lucius et al., 1992)

Contaminant	Type	Density (g/cm^3)	Relative dielectric permittivity	Frequency	Source
Crude Oil	LNAPL	0.70–0.98	2.09	1 GHz	Shen et al., 1985
Gasoline	LNAPL	0.73	2.20	300 MHz	Musil and Zacek, 1986
Kerosene	LNAPL	0.81	2.10	3000 MHz	Von Hippel, 1961
Dichloromethane (DCM)	DNAPL	1.33	9.1	Unknown	Lucius et al., 1992
Trichloroethane (1,1,1 TCA)	DNAPL	1.34	7.6	Unknown	Lucius et al., 1992
Trichloroethylene (TCE)	DNAPL	1.46	3.4	Unknown	Lucius et al., 1992
Carbon Tetrachloride	DNAPL	1.59	2.2	1 MHz	Lucius et al., 1992
Tetrachloro ethylene (PCE)	DNAPL	1.63	2.3	Unknown	Lucius et al., 1992
Ethanol	Miscible	0.79	25	Frequency dependent	Lucius et al., 1992

Based on these properties it is clear that when pure NAPLs displace pore water in rocks and soils they reduce the conductivity and decrease the dielectric permittivity of the medium resulting in an increase in EM wave velocity and decrease in attenuation. If they displace air the effects are smaller, with an increase in permittivity and attenuation. In practice, as will be described later, the effects of NAPLs on the electrical properties of soil or rock are more complicated than this simple explanation would suggest, related to the longer term effects of biodegradation and surface interactions. The dissolved phase components of NAPLs do not directly alter the electrical properties of water but they may indirectly affect water chemistry through biodegradation.

8.3.2. Electrical properties of soil and rock with NAPL contamination

The electrical properties will be described in terms of the dielectric permittivity and electrical conductivity. Both these values can be complex values but the electrical properties can also be completely described in terms of a real permittivity and conductivity. This latter approach has been used in the following discussion. GPR signal attenuation is controlled principally by the electrical conductivity and secondarily by scattering. The wave velocity is determined primarily by the dielectric permittivity.

The dielectric permittivity at typical GPR frequencies (> 100 MHz) is determined by the permittivities of a material's principal components: mineral grains, air, NAPL and water. Typical relative dielectric permittivities are: air 1.0; water 80; soil mineral grains 5; kerosene 2.1; crude oil 2.2; gasoline 2.2.

In general, when NAPLs replace water in the pore space of rock or soil, the relative dielectric permittivity of the media is decreased, thereby increasing the wave velocity. Because attenuation at high frequencies is controlled principally by the electrical conductivity, the presence of NAPLs may increase or decrease attenuation depending on how they change the media conductivity.

The electrical properties of rock and soil, containing mixtures of air, water and NAPL have been estimated using volumetric mixing laws such as the complex refractive index model (CRIM) (Birchak et al., 1974) and effective media theory (Sen et al., 1981; Feng and Sen, 1985; Endres and Redman, 1996). Laboratory measurements of these properties at GPR frequencies have been performed for a variety of NAPLs (Kutrubes, 1986; Olhoeft, 1986; Santamarina and Fam, 1997; Piggott et al., 1998; Piggott et al., 2000). In addition there have been numerous in situ measurements of electrical properties during large-scale controlled NAPL injections at field sites (Redman and Annan, 1992; Schneider and Greenhouse, 1992; Schneider et al., 1993; Redman and DeRyck, 1994) and at accidental spill sites (Atekwana et al., 2000). These measurements at field sites are important because they provide direct in situ data and if measurements have also been performed before a spill, they directly measure the induced changes in electrical properties. These studies have been limited to measurements of the low-frequency electrical conductivity using multiple electrode probes and of the high-frequency permittivity using time domain reflectometry (TDR) probes.

Laboratory measurements of electrical conductivity (Figure 8.1) were acquired during a NAPL (Soltrol 100 oil) imbibition and drainage experiment in an initially

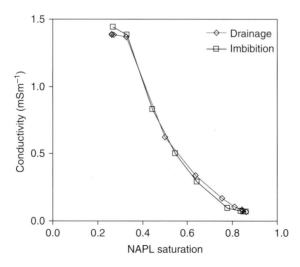

Figure 8.1 Dependence of sand conductivity on NAPL saturation for an initially water saturated Ottawa sand during imbibition and drainage of NAPL. Pore water used was 0.0015 m KCl. (Piggott, 1999).

water-saturated (15.8 mS m^{-1} KCl solution) sand sample (Piggott et al., 1998). The conductivity decreases by over an order of magnitude as the NAPL replaces the water in the pore space. The differences in conductivity at the same NAPL saturation during drainage and imbibition demonstrate the effect of changes in the pore scale distribution of the NAPL and water. Effective media modeling (Endres and Redman, 1996) has also shown the dependence of permittivity on the specific pore distribution, in addition to the dependence on the volumetric fraction of water, air, and NAPL. These results imply that estimates of NAPL concentration based on electrical properties will have inherent inaccuracies related to this lack of uniqueness.

In situ measurements of conductivity (DeRyck et al., 1993) were performed during a controlled injection of 343 L of kerosene into a 3.6-m diameter sand-filled test cell. The conductivity decreased from 5 to 1.3 mS m^{-1} from kerosene replacing water within the saturated zone just above the water table. This experiment will be discussed in greater detail later in this chapter.

The dependence of the relative dielectric permittivity of a water-saturated sandy soil on the NAPL saturation (fraction of pore space filled with NAPL), as determined from the Bruggeman–Hanai–Sen (BHS) mixing model (Sen et al., 1981), is shown in Figure 8.2. The modeled sand had a porosity of 35% and the NAPL (tetrachloroethylene) a relative dielectric permittivity of 2.3. Clearly there are large changes in permittivity induced by the presence of the NAPL that would be detectable with GPR. These induced changes in permittivity have also been observed during monitoring of large-scale controlled spills of NAPLs using multilevel TDR probes.

The effect of NAPLs on the electrical properties of media with relatively low hydraulic conductivity and porosity has not been studied extensively, but there are some generalizations that can be made. NAPLs will not significantly affect the high-frequency electrical properties of low porosity media, such as competent crystalline rocks. Even if the available pore space could be saturated with NAPL, the low

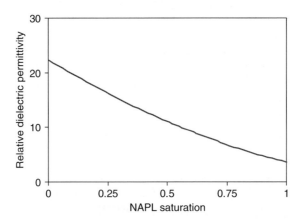

Figure 8.2 Relative dielectric permittivity dependence on NAPL saturation computed using the BHS model for a fully water and NAPL-saturated sample with a porosity of 35%. Grain permittivity is 4.5 and the NAPL permittivity 2.3.

NAPL volumetric content would result in only small affects on the electrical properties. The presence of NAPLs in fracture zones within the rock mass could significantly change the electrical properties of fracture zones if they have sufficient porosity. These effects may be detectable with GPR if the fractures have sufficiently large apertures.

8.3.3. Biodegradation effects

The electrical conductivity of rock and soils is reduced when water in the pore space is displaced by NAPLs. This reduction in conductivity has been measured in laboratory samples by numerous investigators, mixing models predict this behavior and it has been observed during controlled injections of the LNAPL kerosene and the DNAPL tetrachloroethylene (DeRyck et al., 1993; Schneider et al., 1993).

Studies (Nash et al., 1997; Sauck et al., 1998; Atekwana et al., 2000; Sauck, 2000) have shown that there may also be an increase in electrical conductivity associated with the LNAPL at older contamination sites. Atekwana et al. (2004) and Aal et al. (2004) suggest that the production of carbonic and organic acids can enhance the dissolution of soil grains resulting in an increase in the pore water conductivity. In addition the biodegradation process produces biosurfactants resulting in emulsification of LNAPLs. This effect could enhance the conductivity in zones with high LNAPL saturations by dispersing the LNAPL and enhancing conduction pathways through the pooled zones. Based on laboratory measurements in soil columns (Aal et al., 2004), this process requires $\sim$12 weeks to produce a noticeable increase in conductivity. In the LNAPL-controlled injection discussed previously there was a decrease in conductivity that persisted over the 2 month-monitoring period. The time required to change the response from being less conductive to more conductive is likely dependent on local site conditions. The soil grain dissolution process also increases the grain surface area and pore throat geometry potentially changing the frequency dependence of the electrical properties. Recent work investigating the effects of microbial growth in soil columns has shown measurable complex conductivity changes (Davis et al., 2006), and the production of a network of electrically conductive "nanowires" linking bacteria cells to each other and to sand grains potentially increasing the electrical conductivity (Ntarlagiannis et al., 2007). The understanding of biodegradation effects on soil and rock electrical properties has clearly improved but there remain many unanswered questions.

8.3.4. Inorganics

The presence of inorganic contaminants in the subsurface typically results in an increase in electrical conductivity through their dissolution into the groundwater. The resultant effect on subsurface conductivity depends on the solubility of the inorganic contaminant, the mobility of the dissolved ions and the porosity. The effect of the increase in pore water conductivity can be estimated with Archie's Law (Archie, 1942). If there is a specific source zone producing a contaminant plume then the conductivity depends on the distance from the source zone.

8.4. Typical Distribution of Contaminants

Developing methodologies for mapping subsurface contaminants requires an understanding of their typical distribution and concentration in the subsurface. The knowledge base in this field has grown substantially in the past decade and there are numerous publications that discuss the distribution of NAPLs in unconsolidated sediments, rock and fractured rock (Poulson and Kueper, 1992; Kueper et al., 1993).

8.4.1. DNAPL

A DNAPL that has spilled or leaked onto or into the subsurface will move downwards through the unsaturated and saturated zone until it meets any impermeable horizon such as low-permeability bedrock or a clay horizon (Figure 8.3). If sufficient DNAPL is spilled, it will form a pooled zone on this horizon. In a fractured rock environment the DNAPL will penetrate into some of the larger fractures and may move large distances from the source zone (Longino and Kueper, 1999). Along the pathway from the surface the DNAPL will distribute itself as disconnected blobs and ganglia, creating a residual zone in which the DNAPL typically occupies a few percent of the pore space. The residual saturation depends on the initial saturation reached during a spill, resulting in residual saturations that may be up to ~15%. Within the pooled zones the DNAPL saturation could be up

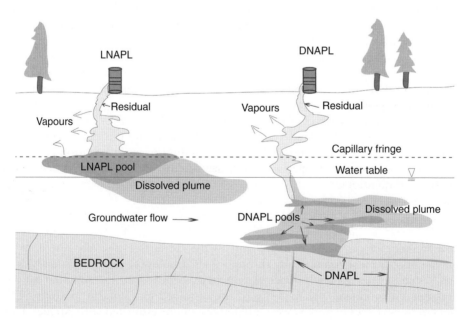

Figure 8.3 Typical distribution of contaminants resulting from an LNAPL and a DNAPL spill at the surface.

to ~90%. Controlled field experiments resulted in pool saturations up to ~30% (Kueper et al., 1993).

In addition to the pooled and residual phases below the water table, a dissolved phase plume will form over time as groundwater flows through the contaminated zones. Above the water table a vapor phase plume will also develop. This plume is important because it can also contaminate groundwater far from the DNAPL source zone.

Although it has been shown that GPR and other geophysical methods can detect sufficient concentrations of the residual and pooled DNAPL, it has not been demonstrated that they can detect the dissolved phase and vapor phase. Considering the small changes in electrical properties that can be attributed to these contaminant phases, it is unlikely that these phases could be detected. It has been shown however that the surface tension of water can be reduced substantially by the presence of dissolved phase organics (Henry and Smith, 2003). Within the unsaturated zone this will result in a change in the water content distribution. Unfortunately this is a secondary effect that will likely be difficult to observe with GPR.

8.4.2. LNAPL

LNAPLs, being lighter than water, behave differently from DNAPLs in the subsurface. An LNAPL that has leaked into the subsurface from a storage tank or surface spill will move downwards through the unsaturated zone until it encounters an impermeable horizon or a water saturated zone such as the water table (Figure 8.3). It will then spread laterally on this horizon and form a zone of the connected immiscible phase liquid that is referred to as the "LNAPL pool." A residual phase of up to 10–15% may remain trapped within the unsaturated zone. Within the pooled zone most of the water may be displaced resulting in LNAPL saturations of up to 80–90%. A dissolved phase plume will form within the groundwater flow regime below the water table. Fluctuations in the water table will often smear the LNAPL over greater depth intervals potentially resulting in the residual phase being trapped below the water table.

8.4.3. Inorganics

Inorganic contaminants typically result from the leaching of a contaminated source zone, such as waste disposal and mine-tailing sites, into both surface and groundwater. The contaminant from these sites will move in the groundwater flow direction creating a plume of dissolved phase-contaminated water. Remediation of these sites may require characterization of both the source zone and the plume.

Contaminant plumes can be up to many kilometers in length and of varying width depending on the groundwater flow velocity and flow direction, and the age of the plume. The plume shape can be quite variable depending on the hydraulic conductivity distribution in the subsurface and the groundwater flow regime. There are many examples in the literature of typical plume geometries, contaminant concentration and evolution over time (MacFarlane et al., 1983; van der

Kamp et al., 1994). These are based on modeling exercises, field measurements of groundwater contaminants and terrain conductivity surveys of existing contaminant plumes (e.g. CCME Report, 1994). Source zones can be variable in their areal extent, from a local spill a few tens of meters in size, to a waste disposal site or a mine-tailing site that may be kilometers in size.

8.4.4. Saturated and unsaturated zone

DNAPLs and LNAPLs have roughly similar electrical properties. Residual zones of these two contaminants within the unsaturated zone would have similar electrical properties. They may present quite different scenarios in terms of geophysical detection because of the way they distribute themselves in the subsurface (Figure 8.3). DNAPL pools are usually below the water table in contrast to LNAPL pools that are situated near the water table, at the interface between the unsaturated and the saturated zones. This makes the geophysical detection of LNAPLs, with electrical techniques, more difficult because this interface is usually complicated and variable depending on the infiltration history and movement of the water table.

8.5. GPR Methodology

The specific GPR methodology used will clearly depend on the nature of the site investigation or study being performed. Choosing the correct GPR center frequency, to provide sufficient sampling depth and spatial resolution, and providing sufficiently high spatial sampling will be required to properly characterize the site. These issues and others related to survey design are covered in Chapter 1.

The use of mixing formulae to predict the electrical properties of the contaminated subsurface and numerical modeling to determine the GPR response for a typical contaminant distribution can be very helpful in understanding a characteristic response for a contaminant, and the results of this process will likely impact the survey design. In many cases GPR will not be appropriate because of insufficient sampling depth due to high attenuation, lack of contrast in electrical properties between the contaminated and the uncontaminated media or insufficient spatial resolution.

GPR applied to contaminant detection is more successful if it can be used in a monitoring mode where changes with time resulting from remediation or contaminant migration can be observed. In this case it is easier to differentiate between responses from natural geological heterogeneity or a contaminated zone. During remediation of contaminant sites this approach can be used to monitor the removal of a contaminant or to monitor the effectiveness of remediation processes (Tomlinson et al., 2003).

Because contaminant distribution can often be very complicated, the use of 3D survey methods is recommended if possible. GPR surveys are normally performed in the common offset mode but multi-offset surveys when feasible can provide both improved imaging capability and allow for amplitude versus offset (AVO) analysis (Bradford, 2003, 2004; Jordan and Baker, 2004).

8.6. DATA PROCESSING AND INTERPRETATION

The objective of GPR is to determine the subsurface electrical property distribution from the acquired data. Based on the measured subsurface electrical property variation it is possible to extract information on both the subsurface geology and the contaminant distribution. There is no simple means of achieving this mapping objective. This section describes practical methods that have been used in contaminant applications to provide visual aids to the interpreter, to identify anomalous zones and to provide an approximate quantification of the contaminant concentrations.

8.6.1. Visualization

GPR transects and 3D GPR data sets contain a wealth of detailed information that is often best presented in some visual form. For example, quickly cycling through a series of migrated depth slices from 3D data or GPR sections acquired on the same transect at different times provides a powerful visual aid for interpretation. Gain functions should be chosen to enhance the attributes of interest. If the character of the section is important then AGC gains are useful. If the relative reflectivity of horizons is important then a spherical exponential compensation (SEC) gain is best.

8.6.2. Trace attributes

Trace attributes can be used to extract useful information from the GPR data. For example in cases where reflection amplitude can be used to estimate NAPL saturation (Brewster and Annan, 1994; Hwang et al., 2008) or in cases where estimates of relative attenuation are required, the envelope or instantaneous amplitude attribute of a trace provides a robust method of estimating amplitude that is insensitive to phase changes in the reflected pulse. In addition instantaneous amplitude, phase and frequency was used by Orlando (2002) to characterize an LNAPL spill.

8.6.3. Data differencing

It would be useful to be able to invert the GPR data, collected before and after a spill, into electrical properties data and then difference the electrical properties to allow visualization of changes. This can be done in borehole GPR tomography to visualize subsurface changes in velocity or attenuation (Sander et al., 1992; Redman et al., 2000; Tomlinson et al., 2003; Johnson et al., 2007). Currently this is not practical for GPR surface reflection data. Simple differencing of GPR sections or 3D data sets can sometimes be used to visualize changes over time; however, if there are significant velocity changes this method is ineffective. Hwang et al. (2008) successfully differenced GPR cross-sections collected before and following a DNAPL injection to monitor its dissolution over a 66-month period. An alternative method that is more reliable is to difference an attribute of the data. Versteeg and Birken (2001) have used this approach by computing the average energy in a sliding time window. The average energy values for a background 3D data set were then subtracted from each subsequent data set collected following an oil injection. This process was used to successfully image the development of the injected oil pool over time.

8.6.4. AVO analysis

The AVO method has been used as method for determining the electrical properties at subsurface horizons and thus in principle allowing differentiation of NAPL pools and residual zones from natural stratigraphic horizons (Bradford, 2003; Jordan and Baker, 2004; Jordan et al., 2004). The method relies on measuring the dependence of a reflected wavelet's amplitude on the angle of incidence. This dependence is controlled by the electrical properties above and below the horizon from which the measured wavelet is reflected. Data for AVO analysis are acquired by performing multi-offset or common mid-point surveys (CMP). The AVO method determines the layer properties from the reflection characteristics at the top of the layer, unlike velocity analysis which requires that energy penetrate through the layer and produce a detectable reflection event from a lower horizon.

Although the technique appears to provide some useful information on NAPL distribution there are many factors that make AVO analysis difficult. These include accounting for the antenna radiation pattern, incorporating geometrical spreading, and dealing with the effects of heterogeneities and variable attenuation above the reflection horizon.

8.6.5. Detection based on frequency-dependent properties

The presence of NAPLs in rock and soil can result in frequency-dependent properties at frequencies up to $\sim$200 MHz. It has not been demonstrated that this effect can be observed with GPR in normal practice but recent work (Bradford, 2007) has shown that it is possible to measure a frequency-dependent attenuation that is correlated with the presence of NAPLs. In practice the frequency dependence of the reflection coefficient and of the attenuation for scattering losses will not allow this process to be definitive, but this method may provide a useful relative attribute for mapping these contaminants.

8.6.6. Quantitative estimates of NAPL

Providing quantitative measurements of NAPL saturation (fraction of pore space filled with NAPL) based on borehole GPR measurements may be possible in some situations. Using surface GPR in the reflection mode for this task is more difficult because this method normally requires determining an empirical relationship between NAPL saturation and reflection amplitude, as described in this section.

In the following discussion it is assumed that the media contaminated with the NAPL is fully water saturated. In the water-saturated zone the NAPL must displace water in the pore space while in the unsaturated zone it may displace water or air making the problem of estimating NAPL saturation impossible with the relatively simple methodology described in the following discussion.

Determining NAPL saturation requires that the electrical properties be measured and that the NAPL saturations be estimated from these properties. Both of these tasks are quite difficult. The AVO method, CMP velocity analysis and fitting of trace data with a 1D model (Powers and Olhoeft, 1994) have all been used to estimate electrical properties. Relating the NAPL saturation to the electrical properties could in principle be done using numerical modeling if the grain permittivity, grain shape and porosity are known or if the relationship could be established with laboratory measurements. In practice this process has not worked because of lack of information about these properties and because the subsurface is generally too heterogeneous.

If the subsurface permittivity is measured before an NAPL injection then it is possible to provide estimates of NAPL saturation. This has been done with in situ TDR probes (Redman and Annan, 1992).

If a strong subsurface reflection event exists in the GPR section then comparisons of the travel time to this event, measured before and after a NAPL injection, provide an average permittivity measurement to the reflecting horizon. From these measurements the average water content to this depth can be estimated using a mixing formula such as the Topp relationship (Topp et al., 1980) or the CRIM model. In the case of media that is water saturated, the difference in water content provides an estimate of the NAPL saturation.

Measurements of the reflection amplitudes from subsurface layers of pooled or residual NAPL can also be used to estimate the product of the NAPL saturation and the pool thickness. This saturation-thickness product can provide estimates of the total amount of NAPL pooled on a horizon. The relationship between the saturation-thickness product and the reflection amplitude is determined from an empirical relationship between measured NAPL saturations from core samples and the measure reflection amplitudes from the same location and NAPL layer. This procedure, described by Brewster et al. (1995), was used by these authors as well as by Hwang et al. (2008) to provide quantitative estimates of DNAPL distribution and its evolution in time during and following controlled injections.

8.7. CASE STUDIES

GPR and other geophysical techniques have been used successfully at many contaminated sites to characterize subsurface geology, to locate inorganic contaminant plumes, and to find buried barrels, pipes and storage tanks. As well, these methods have in some cases been able to delineate NAPL contaminants and to monitor the remediation processes applied to these contaminants. Although GPR and other geophysical methods have been helpful in mapping NAPL distribution at some sites, there are currently no standard accepted geophysical techniques that can reliably map NAPL distribution at contaminated sites.

At many sites the contamination may be deeper than the penetration available from GPR, or as commonly occurs, the natural geological heterogeneity can mask the response from the contaminant. GPR contaminant mapping is performed indirectly by detecting the electrical property variation induced by the presence of the contaminant. Geological heterogeneity can, in many cases, produce similar changes in electrical properties related to changes in density, porosity, water content, water conductivity or mineralogy. For example, a high clay content zone in the subsurface could increase conductivity in a similar fashion to an inorganic plume and changes in subsurface porosity within the water-saturated zone could produce the same response as a DNAPL zone. Attribution of anomalous responses to contaminants is often constrained by other information such as direct measurements of contaminants in core or surface samples, core logs, other geophysical survey results and geological knowledge of the site. In addition there is usually an understanding of the contaminant's typical distribution.

If GPR surveys can be performed before the contaminant is present then comparisons of results before and after will sometimes allow mapping of the contaminant. It has been clearly demonstrated that many methods, including GPR, can successfully monitor and map the distribution of NAPLs during controlled experimental injections and in some cases following accidental spills (Redman et al., 1991; Brewster et al., 1992, 1995; DeRyck et al., 1993; Greenhouse et al., 1993; Grumman and Daniels, 1995; Bermejo et al., 1997; Sauck et al., 1998; Kim et al., 2000; Versteeg and Birken, 2001; Lopes de Castro and Castelo Branco, 2003; Bradford, 2004; Hwang et al., 2008).

The case studies presented in this section have been performed mostly as research projects to evaluate the methodology. Most of the studies were controlled injections of contaminants in which GPR surveys were performed before, during and following the injections. The injection environments in these cases were either large test cells packed with the test media by the researchers or were an undisturbed natural environment, about which the researcher had considerable knowledge from coring and sampling performed before or after the spill.

8.7.1. Controlled DNAPL injection

Experiments performed at CFB Borden, Ontario, Canada, have demonstrated that geophysical techniques can successfully monitor the DNAPL distribution within a natural saturated sandy aquifer during and following a DNAPL injection. Test cells were used in most of these experiments. A section of the natural undisturbed aquifer was isolated within sheet pile walls and underlain by a thick clay layer. Performing the injections in a natural aquifer is important because it is extremely difficult, if not impossible, to pack test cells with soil to simulate a typical natural aquifer with the subtle changes in hydraulic properties that control DNAPL migration. The DNAPL used was tetrachloroethylene, an effectively non-conducting and low relative dielectric permittivity (2.3) liquid.

In the first experiment, 231 L were injected into a $3 \, m \times 3 \, m \times 3 \, m$ deep test cell. Multilevel TDR probes and GPR monitored the induced changes in dielectric permittivity (Redman et al., 1991; Kueper et al., 1993). In a more extensive experiment 770 L of tetrachloroethylene was injected into a $9 \, m \times 9 \, m \times 3 \, m$ deep test cell over a period of 70 h. In this experiment, the 3D distribution of the pooled and residual DNAPL zones was mapped by interpreting reflection events observed in GPR sections (Brewster and Annan, 1994; Brewster et al., 1995). Multilevel TDR probes were used to measure the dielectric stratigraphy and its evolution during and following the injection (Redman and Annan, 1992).

Example GPR sections (Figure 8.4) collected before the DNAPL injection and at three times during and following the injection clearly show the development of DNAPL horizons over time at depths of ~0.8 m at early times and at ~2.8 m 178 h after the start of the injection. The reflection event from the aquitard at 3.4 m is pulled up in time for later sections because of the significant amounts of the higher velocity DNAPL replacing the water within the section above the aquitard. The reflection amplitude of the aquitard is also reduced at later times from energy reflected back to the surface by zones of high DNAPL saturation. The dipping events at the edges of the sections are diffraction events from the locations where the steel sheet pile walls intersect reflective soil horizons.

This case study pointed out the importance of using GPR antennas with a range of center frequencies. The data presented in Figure 8.4 were acquired using a centre frequency of 200 MHz. GPR data were also acquired at 500 and 900 MHz (Sander et al., 1992), providing higher resolution images of the DNAPL pool development

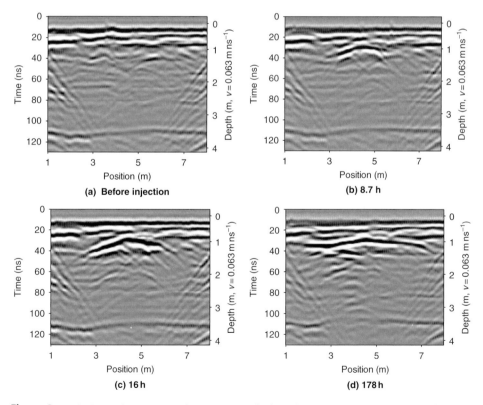

Figure 8.4 GPR sections acquired at 200 MHz, before the DNAPL injection (a), and at 8.7 h (b), at 16 h (c) and 178 h (d) after the start of the injection.

but the 900 MHz system only penetrated to ~1.5 m while the 500 MHz system was limited to about 2.5 m. The 200 MHz antennas were able to clearly see the aquitard event at 3.3 m.

Another injection of 50 L of tetrachloroethylene into the Borden aquifer adjacent to the previous test cell was performed without sheet pile wall containment to monitor the effect of natural ground water flow on the DNAPL (Hwang et al., 2008). This experiment demonstrated the ability of GPR to see the effects of smaller amounts of DNAPL and to monitor the dissolution of the pooled DNAPL over a 66-month period. The authors show an excellent example of the reduction in the areal extent of the pool and in the total DNAPL mass within the pool during this period.

These experiments demonstrated that having background measurements available before the DNAPL injection made it possible to identify the location of the pools and residual zones with GPR. These techniques were successful principally because background measurements were performed before the injection, the cells were saturated to the surface and the aquifer was relatively homogeneous but still sufficiently inhomogeneous to produce a non-uniform distribution of DNAPL.

These large-scale spill experiments into natural aquifers were important because they provided many unexpected results. Although the aquifer is characterized as a relatively uniform medium to fine-grained sand, it had subtle changes in hydraulic conductivity that controlled the DNAPL distribution, resulting in relatively thin pools perched on indistinct soil horizons. These large-scale experiments in natural media are unlikely to be repeated because they are expensive and difficult to perform and there is the potential for accidental groundwater contamination.

8.7.2. Controlled LNAPL injection

In another experiment, 343 L of the LNAPL (kerosene) were injected into a 3.6 m diameter by 1.7 m deep cylindrical test cell (DeRyck et al., 1993; DeRyck, 1994). A medium to fine sand was packed around monitoring instrumentation. The kerosene was injected in a series of separate injections and monitoring was performed following each injection. Permittivity data were acquired from a high-resolution multilevel TDR probe (Figure 8.5). These data clearly demonstrate the induced changes in dielectric permittivity that occurred at the top of the water table as the kerosene replaced water within the capillary-saturated zone. The interface is shifted to a greater depth and becomes somewhat more gradational. This effect would result in a GPR reflection event with reduced amplitude occurring later in time. Modeling also shows a similar effect but the results from the GPR sections results were less conclusive (Redman et al., 1994).

8.7.3. Accidental spill sites

In addition to controlled injections of LNAPLs there have been numerous GPR studies of accidental spill sites (Bermejo et al., 1997; Sauck et al., 1998; Lopes de Castro and Castelo Branco, 2003). As described previously, in cases of older

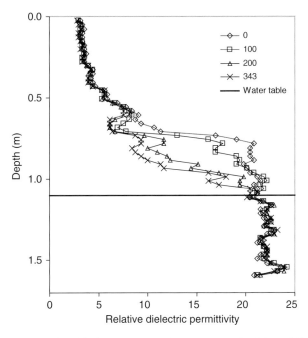

Figure 8.5 Vertical profile of dielectric permittivity measured with multilevel TDR probe before and following kerosene injections in a large test cell filled with a sandy soil. The total amount of kerosene injected is shown. The water table was maintained at a depth of 1.1 m.

hydrocarbon fuel spills it has been observed in some cases that biodegradation can cause an increase in conductivity associated with the presence of the LNAPL. An example GPR section from a former fire-training facility at the Wurtsmith AFB (Oscoda, Michigan, USA) is shown in Figure 8.6. The site, consisting of 20 m of uniform sands with a water table at ∼5 m, has been studied extensively (Sauck et al., 1998; Smart et al., 2004). In the GPR section a strong water table reflection event can be observed at ∼90 ns. As well in the region from 170 to 210 m the reflection

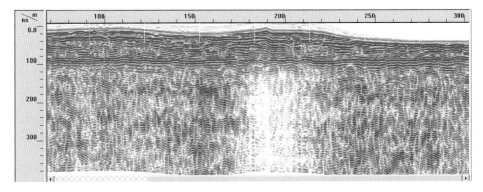

Figure 8.6 GPR section from an LNAPL spill showing a "shadow zone" related to high conductivity associated with biodegradation of hydrocarbon fuels.

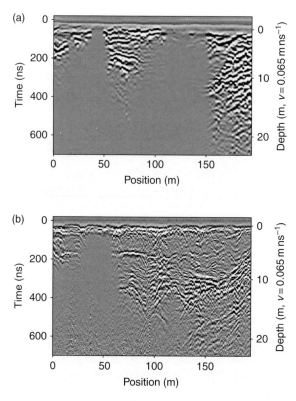

Figure 8.7 GPR sections acquired over a leachate plume from a waste disposal site (a) before and (b) following remediation.

events from below the water table are more subdued or not visible in the section (Figure 8.6). This zone has often been referred to as "the shadow zone" and is attributed to the high conductivity related to the biodegradation of LNAPL as described by Sauck (2000). The source of the LNAPL contamination in this case is from bi-weekly fire-training exercises.

8.7.4. Leachate and waste disposal site characterization

GPR can be a useful tool to provide relatively fast and high resolution definition of the lateral extent of conductive zones such as leachate plumes and waste disposal sites. Calabrese et al. (2004) used GPR to delineate the boundaries of an industrial waste disposal site for an industrial mud-containing oil products, metal dust and other waste. Electrical resistivity tomography profiles were used to characterize the depth and approximate lateral extent of the landfill.

Although EM conductivity and DC resistivity surveys are the preferred methods for mapping leachate plumes, GPR can sometimes be useful for these applications. GPR has higher spatial resolution and is sensitive to small changes in conductivity but often has insufficient penetration for these applications. An example of a GPR transect collected over a leachate plume in glacial/fluvial sands shows the results of an original survey and one conducted 5 years later after remediation (Figure 8.7). The waste disposal site is off to the right of the GPR section. In areas affected by the conductive leachate plume the natural stratigraphy is not visible because of the high attenuation.

8.8. SUMMARY

GPR can provide important information on contaminant distribution. It has the advantage over other geophysical methods of providing relatively high spatial resolution, but GPR penetration depth will in many cases be insufficient because of attenuation related to clays or high pore water conductivity.

Research studies focused on contaminant migration and remediation processes often require the use of non-invasive methods to monitor contaminant distribution over long time periods. These studies are often performed in packed test cells or in relatively simple natural environments. Because these applications often allow for measurements before the contaminant is present, the issue of natural heterogeneity is less important, resulting in a higher probability of being able to map the contaminant distribution with GPR.

Recently, work on AVO analysis has shown that it may be useful for direct detection of NAPLs, but there is still further work needed to demonstrate the practical applicability of the method.

GPR is not a general purpose tool for delineation of subsurface contaminants at accidental spill sites, but it can be very useful in specific cases were the subsurface conditions are suitable and when other direct sampling or core data are available to constrain the GPR interpretation. It is also an excellent tool in research applications when experiments can be designed to take advantage of its capabilities.

REFERENCES

Aal, G.A., Atekwana, E.A., Slater, L.D. and Atekwana, E.A., 2004, Effects of microbial processes on electrolytic and interfacial electrical properties of unconsolidated sediments. Geophysical Research Letters, Vol. 31, p. L12505, doi:10.1029/2004GL020030.

Archie, G.E., 1942, The electrical resistivity log as an aid in determining some reservoir characteristics. Transactions of the American Institute of Mining and Metallurgical Engineers, Vol. 146, pp. 54–67.

Atekwana, E.A., Sauck, W.A. and Werkema, D.D., 2000, Investigations of geoelectrical signatures at a hydrocarbon contaminated site. Journal of Applied Geophysics, Vol. 44, pp. 167–180.

Atekwana, E.A., Werkema, D.D., Duris, J.W., Rossbach, S., Atekwana, E.A., Sauckzz, W.A., Cassidy, D.P., Means, J. and Legall, F.D., 2004, In-situ apparent conductivity measurements and microbial population distribution at a hydrocarbon-contaminated site. Geophysics, Vol. 69, pp. 56–63.

Bermejo, J.L., Sauck, W.A. and Atekwana, E.A., 1997, Geophysical discovery of a new LNAPL plume at the former Wurtsmith AFB, Oscoda, Michigan. Ground Water Monitoring Remediation, Vol. 17, pp. 131–137.

Birchak, J.R., Gardner, C.G., Hipp, J.E. and Victor, J.M., 1974, High dielectric constant microwave probes for sensing soil moisture. Proc. IEEE, Vol. 67, pp. 93–98.

Bradford, J.H., 2003, GPR offset dependent reflectivity analysis for characterization of a high-conductivity LNAPL plume. Proceedings, Symposium on the Application of Geophysics to Engineering and Environmental Problems, San Antonio, Texas, Paper CON08, p. 15.

Bradford, J.H., 2004, 3D multi-offset, multi-polarization acquisition and processing of GPR data: a controlled DNAPL spill experiment. Proceedings, Symposium on the Application of Geophysics to Engineering and Environmental Problems, Colorado Springs, Colorado, Paper DNA08, p. 14.

Bradford, J., 2007, Frequency-dependent attenuation analysis of ground-penetrating radar data. Geophysics, Vol. 72, No. 3, pp. J7–J16.

Brewster, M.L. and Annan, A.P., 1994, Ground penetrating radar monitoring of a controlled DNAPL release: 200 MHz radar. Geophysics, Vol. 59, pp. 1211–1221.

Brewster, M.L., Redman, J.D. and Annan, A.P., 1992, Monitoring a controlled injection of perchloroethylene in a sandy aquifer with ground penetrating radar and time domain reflectometry. Proceedings, Symposium on the Application of Geophysics to Engineering and Environmental Problems, Oakbrook, Illinois, pp. 611–618.

Brewster, M.L., Annan, A.P., Greenhouse, J.P., Kueper, B.H., Olhoeft, G.R., Redman, J.D. and Sander, K.A., 1995, Observed migration of a controlled DNAPL release by geophysical methods. Ground Water, Vol. 33, pp. 977–987.

Calabrese, M., Zanzi, L. and Lualdi, M., 2004, Mapping an industrial landfill area from penetration of GPR data. Proceedings, Symposium on the Application of Geophysics to Engineering and Environmental Problems, Colorado Springs, Colorado, Paper ENG02, 14p.

Canadian Council of Ministers of the Environment (CCME), 1994, Subsurface Assessment Handbook for Contaminated Sites. Report: CCME-EPC-NCSRP-48E prepared by Waterloo Centre for Groundwater Research, 293p.

Davis, C.A., Atekwana, E., Slater, L.D., Rossbach, S. and Mormile, M.R. 2006, Microbial growth and biofilm formation in geological media is detected with complex conductivity measurements. Geophysical Research Letters, Vol. 33, p. L18403, doi:10.1029/2006GL027312.

DeRyck, S.M., 1994, Monitoring a controlled LNAPL spill. MSc Thesis, University of Waterloo, Waterloo, Ontario, Canada, 238p.

DeRyck, S.M., Redman, J.D. and Annan, A.P., 1993, Geophysical monitoring of a controlled kerosene spill. Proceedings, Symposium on the Application of Geophysics to Engineering and Environmental Problems, April, 1993, San Diego, California, pp. 5–20.

Endres, A. and Redman, J.D., 1996, Modelling the electrical properties of porous rocks and soils containing immiscible contaminants. Journal of Environmental Engineering Geophysics, Vol. 0, No. 2, pp. 105–112.

Feng, S. and Sen, P.N., 1985, Geometrical model of conductive and dielectric properties of partially saturated rocks. Journal of Applied Physics, Vol. 58, pp. 3236–3243.

Greenhouse, J.P., Brewster, M.L., Schneider, G.W., Redman, J.D., Annan, A.P., Olhoeft, G.R., Sander, K. and Mazzella, A., 1993, Geophysics and solvents: The Borden experiments. The Leading Edge, Vol. 12, pp. 261–267.

Grumman, D.L. and Daniels, J.J., 1995, Experiments on the detection of organic contaminants in the vadose zone. Journal of Environmental Engineering Geophysics, Vol. 0, pp. 31–38.

Henry, E.J. and Smith, J.E., 2003, Surfactant-induced flow phenomena in the vadose zone: A review of data and numerical modeling. Vadose Zone Journal, Vol. 2, pp. 154–167.

Hwang, Y.K., Endres, A.L., Piggott, S.D. and Parker, B.L., 2008, Long-term ground penetrating radar monitoring of a small volume DNAPL release in a natural groundwater flow field. Journal of Contaminant Hydrology, Vol. 97, pp. 1–12.

Johnson, T.C., Routh, P.S., Barrash, W. and Knoll, M.D., 2007, A field comparison of Fresnel zone and ray-based GPR attenuation-difference tomography for time-lapse imaging of electrically anomalous tracer or contaminant plumes. Geophysics, Vol. 72, No. 3, pp. J7–J16.

Jordan, T.E. and Baker, S.B., 2004, Reprocessing GPR data from the CFB Borden experiment using APVO/GPR techniques. Proceedings, Symposium on the Application of Geophysics to Engineering and Environmental Problems, Colorado Springs, Colorado, Paper DNA07, p. 8.

Jordan, T.E., Baker, S.B., Henn, K. and Messier, J., 2004, Using amplitude variation with offset and normalized residual polarization analysis of ground penetrating radar data to differentiate an NAPL release from stratigraphic changes. Journal of Applied Geophysics, Vol. 56, pp. 41–58.

Kim, C., Daniels, J.J., Guy, E.D., Radzevicius, S.J. and Holt, J., 2000, Residual hydrocarbons in a water saturated medium: A detection strategy using ground penetrating radar. Environmental Geoscience, Vol. 7, pp. 169–176.

Kueper, B.H., Redman, J.D., Starr, R.C., Reitsma, S. and Mah, M., 1993, A field experiment to study the behaviour of tetrachloroethylene below the watertable: Spatial distribution of residual and pooled DNAPL. Journal of Ground Water, Vol. 31, No. 5, pp. 756–766.

Kutrubes, D.L., 1986, Dielectric permittivity measurements of soils saturated with hazardous fluids. MSc. Thesis, Colorado School of Mines, Golden, Colorado, USA.

Longino, B.L. and Kueper, B.H., 1999, Non-wetting phase retention and mobilization in rock fractures. Water Resources Research, Vol. 35, No. 7, pp. 2085–2093.

Lopes de Castro, D. and Castelo Branco, R.M.G., 2003, 4-D ground penetrating radar monitoring of a hydrocarbon leakage site in Fortaleza (Brazil) during its remediation process: A case history. Journal of Applied Geophysics, Vol. 54, pp. 127–144.

Lucius, J.E., Olhoeft, G.R., Hill, P.L. and Duke, S.K., 1992, Properties and hazards of 108 selected substances -1992 Edition, U.S. Geological Survey Open-File Report 92-527.

MacFarlane, D.S., Cherry, J.A., Gillham, R.W. and Sudicky, E.A., 1983, Migration of contaminants in groundwater at a landfill: A case study, 1. Groundwater flow and contaminant distribution. Journal of Hydrology, Vol. 63, pp. 1–29.

Musil, J. and Zacek, F., 1986, Microwave Measurements of Complex Permittivity by Free Space Methods and Their Applications, Elsevier, Amsterdam, 275p.

Nash, M.S., Atekwana, E. and Sauck, W.A., 1997, Geophysical investigation of anomalous conductivity at a hydrocarbon contaminated site. Proceedings, Symposium on the Application of Geophysics to Engineering and Environmental Problems, Reno Nevada, pp. 675–683.

Ntarlagiannis, D., Atekwana, E.A., Hill, E.A. and Gorby, Y., 2007, Microbial nanowires: Is the subsurface "hardwired"? Geophysical Research Letters, Vol. 34, p. L17305, doi:10.1029/2007GL030426.

Olhoeft, G.R., 1986, Direct detection of hydrocarbon and organic chemicals with ground penetrating radar and complex resistivity. Proceedings, NWWA/API Conference on Petroleum Hydrocarbons and Organic Chemicals in Ground Water – Prevention Detection and Restoration, Natl.Water Well Assoc., pp. 284–305.

Orlando, L., 2002, Detection and analysis of LNAPL using the instantaneous amplitude and frequency of ground-penetrating radar data. Geophysical Prospecting, Vol. 50, pp. 27–41.

Piggott, S.D., Redman, J.D. and Endres, A.L., 1998, Hysteresis in the electrical conductivity-saturation relationship of sand. Proceedings, Symposium on the Application of Geophysics to Engineering and Environmental Problems, March 1998, Chicago, Illinois, pp. 939–946.

Piggott, S.D., 1999, Saturation history effects on the electrical properties of Ottawa sans during water-air and water-LNAPL drainage-imbibition experiments, MSc Thesis, Univ. of Waterloo, Waterloo, Ont., Canada, p. 331.

Piggott, S.D., Redman, J.D. and Endres, A.L., 2000, Frequency dependence in the complex resistivity of Ottawa sand containing water-air and water-NAPL mixtures. Proceedings, Symposium on the Application of Geophysics to Engineering and Environmental Problems, February, 2000, Arlington, Virginia, pp. 945–954.

Poulson, M.M. and Kueper, B.H., 1992, A field experiment to study the behavior of tetrachloroethylene in unsaturated porous media. Environmental Science and Technology, Vol. 26, pp. 889–895.

Powers, M.H. and Olhoeft, G.R., 1994, Modeling dispersive ground penetrating radar data. Proceedings, Fifth International Conference on Ground Penetrating Radar, June 1994, Kitchener, Ontario, pp. 173–183.

Redman, J.D., Kueper, B.H. and Annan, A.P, 1991, Dielectric stratigraphy of a DNAPL spill and implications for detection with ground penetrating radar. Proceedings, Fifth National Outdoor

Action Conference on Aquifer Restoration, Ground Water Monitoring and Geophysical Methods, May 13–16, 1991, Las Vegas, Nevada, pp. 1017–1030.

Redman, J.D. and Annan, A.P., 1992, Dielectric permittivity monitoring in a sandy aquifer following the controlled release of a DNAPL. Proceedings of Fourth International Conference on Ground Penetrating Radar, June 8–13, 1992, Rovaniemi, Finland, Geological of Survey of Finland, Special Paper 16, pp. 191–196.

Redman, J.D. and DeRyck, S.M., 1994, Monitoring non-aqueous phase liquids in the subsurface with multilevel time domain reflectometry probes. Proceedings, Symposium on Time Domain Reflectometry in Environmental, Infrastructure and Mining Applications, September 1994, Northwestern University, Evanston Illinois, pp. 207–214.

Redman, J.D., DeRyck, S.M. and Annan, A.P., 1994, Detection of LNAPL pools with GPR: Theoretical modelling and surveys of a controlled spill. Proceedings, Fifth International Conference on Ground Penetrating Radar, June 1994, Kitchener, Ontario, pp.1283–1294.

Redman, J.D., Parkin, G.P. and Annan, A.P., 2000, Borehole GPR measurement of soil water content during an infiltration experiment. Proceedings, 8th International Conference on Ground Penetrating Radar, May 2000, Queenland, Australia, pp. 501–505.

Sander, K.A., Olhoeft, G.R. and Lucius, J.E., 1992, Surface and borehole radar monitoring of a DNAPL spill in 3D versus frequency, look angle and time. Proceedings, Symposium on the Application of Geophysics to Engineering and Environmental Problems, Oakbrook, Illinois, pp. 375–382.

Santamarina, J.C. and Fam, M., 1997, Dielectric permittivity of soils mixed with organic and inorganic fluids (0.02 GHz to 1.3 GHz). Journal of Environmental and Engineering Geophysics, Vol. 2, pp. 37–52.

Sauck, W.A., 2000, A model for the resistivity structure of LNAPL plumes and their environs in sandy sediments. Jounal of Applied Geophysics, Vol. 44, pp. 167–180.

Sauck, W.A., Atekwana, E.A. and Nash, M.S., 1998, High conductivities associated with an LNAPL plume imaged by integrated geophysical techniques. Journal of Environmental and Engineering Geophysics, Vol. 2, No. 3, pp. 203–212.

Schneider, G.W. and Greenhouse, J.P., 1992, Geophysical detection of perchloroethylene in a sandy aquifer using resistivity and nuclear logging techniques. Proceedings, Symposium on the Application of Geophysics to Engineering and Environmental Problems, Chicago, Illinois, pp. 619–628.

Schneider, G.W., DeRyck, S.M. and Ferre, P.A., 1993, The application of automated high resolution resistivity in monitoring hydrogeological field experiments. Proceedings of the Symposium on the Application of Geophysics to Engineering and Environmental Problems, San Diego, California, pp. 145–162.

Sen, P.N., Scala, C. and Cohen, M., 1981, A self similar model for sedimentary rocks with application to the dielectric constant of fused glass beads. Geophysics, Vol. 46, pp. 781–795.

Shen, L.C., Savre, W.C., Price, J.M. and Athavale, K., 1985, Dielectric properties of reservoir rocks at ultra-high frequencies. Geophysics, Vol. 50, pp. 692–704.

Smart, L.A., Nash, M. and Sauck, W.A., 2004, Wurtsmith air force base revisited. Proceedings, Symposium on the Application of Geophysics to Engineering and Environmental Problems, Colorado Springs, Colorado, Paper CON06, p. 12.

Tomlinson, D.W., Thomson, N.R., Johnson, R.L. and Redman, J.D., 2003, Air distribution in the Borden aquifer during in situ air sparging. Journal of Contaminant Hydrology, Vol. 67, pp. 113–132.

Topp, G.C., Davis, J.L. and Annan, A.P., 1980, Electromagnetic determination of soil water content: Measurements in coaxial transmission lines. Water Resources Research, Vol. 16, No. 3, pp. 574–582.

van der Kamp, G., Luba, L.D., Cherry, J.A. and Maathuis, H., 1994, Field study of a long and very narrow contaminant plume. Ground Water, Vol. 32, No. 6, pp. 1008–1016.

Versteeg, R. and Birken, R., 2001, Imaging fluid flow and relative hydraulic conductivity using ground penetrating radar in a controlled setting. Proceedings, Symposium on the Application of Geophysics to Engineering and Environmental Problems, Paper GP1-6, 16p.

Von Hippel, A.R., 1961, Dielectric Materials and Applications, MIT Press, Cambridge, MA, 438p.

TERMS FOR GLOSSARY

Dense non-aqueous phase liquid (DNAPL) – Class of liquid organic contaminants that have higher densities than water and have low solubility in water resulting in a separate immiscible phase. Their high densities allow them to sink below the water table and cause serious groundwater contamination. The chlorinated solvents are a common contaminant in this class.

Light non-aqueous phase liquid (LNAPL) – Class of liquid organic contaminants that have lower densities than water and have low solubility in water resulting in a separate immiscible phase. The liquids will move through the unsaturated zone and because they are less dense than water they will pool at the water table. Hydrocarbon fuels (e.g. gasoline, kerosene) are a common contaminant in this class.

Non-aqueous phase liquid (NAPL) – Class of liquid organic contaminants that have low solubility in water resulting in a separate immiscible phase.

PART III

EARTH SCIENCE APPLICATIONS

CHAPTER 9

Ground Penetrating Radar in Aeolian Dune Sands

Charlie Bristow

Contents

9.1. Introduction	274
9.2. Sand Dunes	274
9.3. Survey Design	277
9.3.1. Line spacing	277
9.3.2. Step size	277
9.3.3. Orientation	278
9.3.4. Survey direction	278
9.3.5. Vertical resolution	278
9.4. Topography	279
9.4.1. Topographic surveys	280
9.4.2. Topographic correction	281
9.4.3. Apparent dip	281
9.5. Imaging Sedimentary Structures and Dune Stratigraphy	281
9.6. Radar Facies	282
9.7. Radar Stratigraphy and Bounding Surfaces	283
9.8. Aeolian Bounding Surfaces	285
9.8.1. Reactivation surfaces	285
9.8.2. Superposition surfaces	285
9.8.3. Interdune surfaces	286
9.9. Dune Age and Migration	288
9.10. Stratigraphic Analysis	288
9.11. Ancient Aeolian Sandstones	290
9.12. Three-Dimensional Images	290
9.13. Pedogenic Alteration and Early Diagenesis	291
9.13.1. Evaporites	291
9.13.2. Environmental noise	291
9.13.3. Diffractions	293
9.13.4. The water table	293
9.13.5. Multiples	293
9.14. Conclusions	294
Acknowledgments	294
References	294

9.1. INTRODUCTION

Before the advent of GPR the internal structure of sand dunes was investigated from trenches dug by hand (e.g. McKee and Tibbitts, 1964), or mechanical excavation (McKee, 1966, 1979), making observations of partially deflated dune surfaces (Hunter, 1977; Tsoar, 1982) or numerical modeling (Rubin and Hunter, 1985; Rubin, 1987). Digging or mechanical excavation of trenches through dunes are effective but destructive, and the depth of investigation is limited by the stability of dry or slightly damp sand. Furthermore, the excavation of trenches in unconsolidated sands can be dangerous due to the risk of the trench walls collapsing. In contrast, GPR is a non-invasive technique which can be used to image the internal structure of dunes leaving no more than a set of footprints on the dune surface. GPR offers a fast and efficient method for the collection of high resolution, almost continuous images of the shallow subsurface structure in aeolian sands. In addition, GPR can be used to produce 3D visualizations of the sedimentary structures within sand dunes, and from these structures it is possible to reconstruct the history of dune development and migration (e.g. Bristow et al., 2000a, 2005, 2007a). Dune sands are suitable targets for GPR surveys because they usually have low conductivity and low magnetic permeability allowing good depths of penetration (low attenuation). The depth of penetration of GPR in dune sands can exceed 25 m which is deeper than most trenches and a recent survey of sand dunes in Antarctica reached depths of around 75–80 m (unpublished data).

This paper includes a brief introduction to aeolian sand dunes and then outlines GPR survey methodology with particular reference to surveys over aeolian sand dunes. It explains the imaging of primary sedimentary structures within dune sands with examples of both 2D and 3D GPR data sets. The paper also includes a brief review of recent developments in determining dune chronology and dune stratigraphy with the aid of GPR and geochronology based on recent case studies.

9.2. SAND DUNES

Sand dunes are hills of sand built up and shaped by the wind. They range in length from 1 m to tens of kilometres and in height from a few tens of centimeters to over 300 m. Sand dunes come in a variety of forms, and there are several classification schemes for dunes. The classification of McKee (1979) is followed here, with minor modifications following Pye and Tsoar (1990). McKee (1979) distinguished between dunes that are simple, compound or complex. Simple dunes are classified based on their overall form and the number of slipfaces (see below). Compound dunes consist of two or more dunes of the same type, while complex dunes consist of two or more different types of dune. Types of dunes classified by the number of slipfaces and morphology are illustrated in Figure 9.1. Dome dunes are circular or elliptical in outline and have no slipface, zibar are coarse grained and also lack a slipface. Dunes that have one slipface on the downwind side include crescent-shaped barchan dunes, coalesced barchans referred to as barchanoid ridges

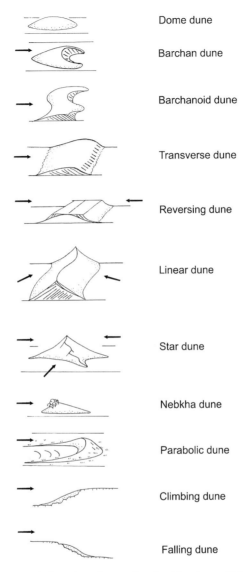

Figure 9.1 Sand dune morphology based upon the classifications of McKee (1979) and Pye and Tsoar (1990). These represent the main types of dunes but there are many variations on the forms. In each case the arrows indicate the direction of the main formative winds.

and essentially straight crested transverse dunes. Dunes with two slipfaces include reversing dunes, which as the name implies, are formed under a reversing wind regime, and linear dunes that are believed to form from bimodal winds. Dunes with multiple slipfaces with a high central peak and three or more arms extending radially are called star dunes. In addition to these dunes that Pye and Tsoar (1990) class as autogenic dunes, that is dunes whose form is related to the interaction of the wind and the sand, there are two other classes of sand dunes which are influenced by vegetation (phytogenic), and those influenced by topography. Dunes formed by the accumulation of sand related to vegetation include parabolic dunes, vegetated linear dunes, nebkha and coastal foredunes. Parabolic dunes have a U or V shape with the trailing arms anchored by vegetation and pointing upwind and a nose that migrates downwind. Nebkha are wind-aligned sand accumulations around vegetation and include coppice and shrub dunes. Coastal foredunes are shore parallel accumulations of sand that are usually vegetated. The vegetation serves the dual role of stabilizing the dune and trapping sand allowing the dunes to accrete vertically and accrete "upwind" toward the beach because of sand capture by the vegetation. Topographically controlled dunes include climbing dunes on the upwind side of a topographic obstacle and falling dunes on the downwind side.

Sand is transported by wind in three ways: reptation, saltation and suspension. In addition grainflows of sand avalanche down the lee-side slipface of sand. These different deposition processes produce primary sedimentary structures with slight differences in depositional texture. Reptation is the movement of sand along the sediment surface with grains rolling along the surface, the term is derived from the latin *reptare* to crawl. Saltation, from the latin *saltare* to leap, describes grains that bounce over the surface. As grains impact the bed they can either bounce back up into the air again or eject other grains into the air setting further sand grains in motion. The rolling and saltation of sand along the surface of the sediment creates wind ripples and the migration of wind ripples forms laminae. Wind ripple laminae are commonly fine grained and form inverse graded laminae commonly less than 1 cm thickness that are inclined from horizontal up to the angle of repose around 32–34°. Wind ripple laminae have inverse grading (coarsen-up) because the sand grains on crest of wind ripples are coarser than the troughs. Less common are aeolian plane bed laminae that are fine grained and thinly laminated. Fine-grained sand, silt and clay-sized particles can be transported in suspension by turbulence and where this fine-grained sediment is deposited it forms air-fall laminae. Air-fall laminae are fine grained and very thin commonly less than 1 mm in thickness. Failure of the steep, lee side of dunes results in avalanching where sand grains flow down the dune slipface in a grainflow. Grainflow strata from avalanches on the lee side slipface of sand dunes form relatively coarse-grained, inverse-graded, strata inclined at around 32–34° that can be mm to a few cm in thickness. Beds of aeolian cross-strata deposited by sand dunes can be up to tens of meters in thickness, e.g. Mountney et al. (1999).

Wind is highly selective in entraining sand grains and as a result aeolian sands are usually well sorted. In addition, sand grains being transported by air have a relatively high-impact velocity, and this can destroy weak grains leading to a concentration of more resistant sand grains. As a consequence aeolian sands tend to be both well sorted and mineralogically mature. These physical characteristics, together with the suite of primary sedimentary structures aid their identification in the sedimentary record. Furthermore, mature aeolian sands have a low-conductivity and low-magnetic permeability which gives good depths of penetration for GPR. Indeed, the thickness of preserved sets of aeolian cross-strata means that they can be readily imaged by GPR. The different types of sedimentary structures, wind ripple lamination, air fall lamination and grainflow deposits have different grainsize and bedding characteristics which enables them to be detected using GPR because of the associated changes in water content Bristow et al 2006, van Dam 2001.

For further information on sand dune morphology and dynamics readers are referred to books by Pye and Tsoar (1990) and Lancaster (1995) which provide much more detailed information in a readily accessible format.

9.3. Survey Design

Choosing an appropriate survey design helps to optimize field time and collection of useful data. A 3D survey can provide the best visualization of subsurface sedimentary structures, but they are time consuming to collect and process and may not be the most efficient survey design for larger areas. As a consequence most 3D GPR surveys to date have covered an area of a few tens of meters. Collecting 3D GPR data over large areas (km^2) are technically possible but not yet widely applied in sedimentary studies. A well-designed survey grid can be almost as effective and a lot quicker to collect. However, it is essential to plan the grid and select an appropriate line spacing and orientation.

9.3.1. Line spacing

The selection of line spacing is a compromise between the detail required from a survey and the size of the survey area. In order to effectively cover a large area the number of survey lines will increase with a consequent increase in survey time and cost. In order to increase the efficiency of a survey the number and total length of survey lines can be reduced by increasing the spacing between GPR profiles. However, it is important to consider the dimensions of subsurface targets. If the target is an extensive stratigraphic horizon or the water table then a line spacing of 100 m may suffice. For example, in their survey of coastal dunes stratigraphy in Denmark Pedersen and Clemmensen (2005) used a line spacing approaching 0.5 km which was constrained by paths through the dunes. In this case the spacing was sufficient to resolve the dune stratigraphy. In contrast smaller targets require much closer line spacing. For example, sets of cross-stratification 5 m wide cannot be resolved with GPR profiles spaced 10 m apart. Thus if the target is 5 m across, the spacing between lines needs to be less than 5 m and in order to avoid spatial aliasing (where a target is not correctly resolved) the line spacing needs to be around one-fourth of the extent of the sedimentary structures (Jol and Bristow, 2003). In 3D surveys Grasmueck and Weger (2002) suggest that the line spacing should be one-quarter of a wavelength of the signal in the ground to provide successful imaging of submeter-scale sedimentary structures.

9.3.2. Step size

Step size is the distance between each point where a measurement is made along a GPR profile. GPR equipment manufacturers usually suggest a minimum step size for each antenna frequency, e.g. 0.25 m for 100 MHz and 0.1 m for 200 MHz antennas based upon the Nyquist sampling interval which is one-quarter of the wavelength in the ground. Grasmueck and Weger (2002) found that a similar relationship pertains to 3D surveys of sedimentary structures where a horizontal grid-spacing approaching a quarter of a wavelength provides successful imaging of submeter-scale sedimentary structures. In principle it is better to over-sample than under-sample and follow the manufacturers' instructions. A small step size is required in order to avoid spatial aliasing which occurs when a subsurface feature is under-sampled. Spatial aliasing in sediments is explained by Jol and Bristow (2003) and illustrated with examples in Woodward et al. (2003).

9.3.3. Orientation

As well as designing with an appropriate line spacing it is important to consider the line orientation. The most appropriate orientation for survey lines is parallel and perpendicular to the sedimentary dip direction. Lines parallel to the dip direction show the dip of foresets, the thickness of sets of cross-strata, and the dip of set-bounding surfaces. Lines perpendicular to the sedimentary dip direction show the width of sets of cross-strata and bounding surfaces. In practice, trough cross-strata are curved and locally change orientation and dip direction. As a consequence the dip of cross-strata seen on GPR profiles are commonly apparent dips rather than true dips. Intersecting cross-lines or 3D surveys are required for the reconstruction of the true dip angle and dip direction. In addition, dipping reflections need to be corrected to restore the position of dipping reflectors using a migration package to process the data. Migration has the effect of restoring apparent dips to their true dip. This can be done manually using simple trigonometry. If the apparent dip on the GPR profile is β then the actual dip α can be calculated using the equation $\alpha = \sin^{-1}(\tan \beta)$ (Ulriksen, 1982).

9.3.4. Survey direction

Field experience has shown that images of sedimentary structures are improved if the GPR survey is run in the up–dip direction. That is starting at the downwind end and working up-wind. This is probably due to the orientation of the dipping reflectors with respect to the positions of the transmitter and receiver.

9.3.5. Vertical resolution

Vertical resolution is generally considered to be around one-quarter of the wavelength of the radar signal in the ground (Reynolds, 1997). Wavelength is a function of both antenna frequency and the velocity of the radar signal through the ground. A table of wavelengths for given antenna frequencies and typical velocities in wet, damp and dry sands is included as Table 9.1. Low-frequency antennas (50 MHz) have longer wavelengths and higher frequency antennas (400 MHz) have short wavelengths. For a given frequency the wavelength is longer if the velocity is high and shorter if the velocity is low (Table 9.1). Thus a high-frequency signal in a

Table 9.1 Theoretical values for GPR resolution in typical wet, damp, and dry sands depend on the wavelength of the GPR which is a function of the pulse length and the velocity. The theoretical resolution is around one-quarter of the wavelength (Reynolds, 1997). The wavelength is calculated using the following equation: $\lambda = v/f$, where λ is wavelength, v is velocity, and f is frequency

Antenna central frequency (MHz)	Theoretical resolution saturated sand (0.06 m ns^{-1})	Theoretical resolution damp sand (0.1 m ns^{-1})	Theoretical resolution dry sand (0.15 m ns^{-1})
50	0.3 m	0.5 m	0.75 m
100	0.15 m	0.25 m	0.375 m
200	0.075 m	0.125 m	0.1875 m
400	0.0375 m	0.0625 m	0.09375 m

Modified from Jol and Bristow (2003)

low-velocity medium has the shortest wavelength and highest resolution, while a low-frequency signal in a high-velocity medium has the lowest resolution. For example a 100-MHz transmitter has a pulse period of 10 ns and in sand saturated with fresh water where the velocity is 0.06 m ns^{-1} the wavelength will be 0.6 m and the resolution will be one quarter of the wavelength or 0.15 m (0.6 m × 0.25 = 0.15 m). In a dry sand with a velocity of 0.15 m ns^{-1} the same 100 MHz transmitter produces a wavelength of 1.5 m and the resolution will be 0.325 m (1.5 m × 0.25 = 0.325 m).

Resolution is important because it determines the ability of the radar to "see" or image strata in sediments. Sets of cross-strata with a thickness close to the limits of GPR resolution may be imaged as a reflection but the cross-strata within the bed will not be imaged. Individual cross-strata may only be 1 cm thick and therefore below the resolution of some high-frequency antennae (Table 9.1). However, while individual foresets or laminae are beneath the radar resolution, sets of cross-stratification are usually imaged so long as the bed of cross-stratified sand is thick enough. For example, a GPR survey with 100 MHz antennas across an aeolian sand with a typical velocity of 0.15 m ns^{-1} will have a wavelength of 1.5 m and therefore a resolution of 0.75–0.375 m (between one-half to one-quarter of the wavelength). A bed of cross-stratified sand 0.5 m thick may be imaged as a subsurface reflection. However, the cross-stratification within the bed is unlikely to be imaged because it is below the resolution of the GPR. Thus a bed of cross-stratified sand can appear as a single reflection without any cross-stratification which can lead to misinterpretation.

9.4. TOPOGRAPHY

Aeolian dunes are piles of sand built up by the wind in a variety of morphologies including barchan, transverse, linear, star, and parabolic forms. Active dunes commonly have a steep lee-side slip face where sand is at the angle of repose, around 32–34° (Allen, 1970). This presents two problems: firstly, physically climbing a steep, unstable sand dune whilst collecting GPR data; secondly, the GPR profiles across the dune need to be corrected for topography.

When collecting GPR profiles field experience shows that it is better to work up-slope. There are two reasons or this. Firstly, working up-slope on slip faces should give a suitable antennae configuration with respect to the dipping foresets within the dune (See *Survey Direction* above). Secondly, it is easier to hold location when working up-slope. Working down an avalanche slope might appear to be easier but usually results in operator and equipment sliding down-slope, sometimes in an uncontrolled manner, making it difficult to hold station while collecting data. As a consequence, the geometry between the antennae and the reflecting horizons can change during measurements with a consequent decrease in data quality, the step size between measurements can be variable, and the line may appear shorted than it really is.

9.4.1. Topographic surveys

Topographic data should be collected at the same time as, just before, or immediately after collection of GPR data. This is especially important on active dunes where the topography can vary from one day to the next depending on dune migration, and crest reversal in response to the wind. Temporary markers left in the field to mark the location of surveys can be buried by wind-blown sand, blown over, or create scour so it is not a good idea to leave marks in the field expecting to be able to relocate them later. On dunes where there is a lot of relief, with frequent breaks of slope, topographic measurements need to be closely spaced along the profile. A spacing of measurements every 5 m measured from a tape-measure laid along the profile and at breaks of slope, points where the surface topography changes abruptly, has been found to work well. It is worth noting that making topographic measurements across a dune can take almost as long as collecting the GPR data.

Four common methods for measuring surface topography in the field are: staff and level, laser level, total station, DGPS. Staff and level is accurate but slow. The main drawback is the need to move the level frequently across the varied relief on a dune field where differences in elevation exceed the height of the staff. A laser level is much faster than a staff and level and works well in areas with low relief but in areas with high relief such as dune fields this method suffers from the same drawbacks as staff and level. A total station is the preferred method because it is possible to obtain elevations on steep inclines without having to relocate the instrument so long as there is line of sight between the instrument and the target. However, on long profiles across dune fields the total station may also need to be relocated and requires foresight and backsight measurements between stations to ensure continuity within the topographic survey data. Differential global positioning systems (DGPS) have high accuracy but unless running real-time kinematic (RTK) they require additional processing. They can suffer from loss of lock if satellites disappear over the horizon at the base of steep slopes or in hollows on dunes. DGPS is a fast and cost-effective method for collecting topographic data across large areas with irregular relief. However, total station is my preferred method because I know that the data are safely stored in my notebook and require minimal post-processing.

9.4.2. Topographic correction

Topographic correction can be performed by elevation static corrections that reposition the time zero in the vertical axis and adjust reflections in the vertical axis accordingly. Elevation static corrections assume that the ray-path between reflectors and the surface is perpendicular. This is the case when both the surface and the subsurface stratigraphy is horizontal. However, in aeolian sands reflectors are usually inclined with respect to the surface. Due to the spherical expansion of the signal the reflection returns perpendicular to the reflector and not perpendicular to the surface. As a consequence, the reflections will be offset both vertically and laterally. The standard procedure to correct for this effect is to migrate the data so that dipping reflections are restored to the correct dip, and then to apply the static correction for topography to restore the depth to reflections beneath the surface. However, as explained by Lehmann and Green (2000), these procedures do not take full account of the changes in geometry of ray-paths between an undulating acquisition surface and an irregular subsurface reflector. They describe an algorithm which allows GPR data to be migrated directly from gently or highly irregular acquisition surfaces. Lehmann and Green (2000) suggest that this correction should be applied when the land surface slope exceeds 10% or 6°, as is the case on sand dunes. On dunes there are many steep slopes and abrupt changes in slope where the configuration of the antennae change which results in misrepresentation of subsurface reflectors. Topographic effects are most pronounced at the dune crest or at the brink of a slipface. Another problem at a steep dune crest is the presence of a reflection from the interface between the dune and the air on the opposing face which forms a reflection that could be mistaken for a dipping reflection within the dune.

9.4.3. Apparent dip

Inclined reflections on GPR profiles through sand dunes usually originate from primary sedimentary structures within the dunes. The dip angle and dip direction of the inclined reflections is essentially an apparent dip, because one cannot be certain that the GPR profile is collected exactly perpendicular to the depositional dip of the sedimentary structures. In order to determine the true dip angle and direction requires either a 3D survey or a grid of intersecting profiles from which the true dip can be reconstructed. Neal and Roberts (2001) used a grid with lines spaced between 5 and 10 m to reconstruct the dip angle and direction of aeolian cross-strata using apparent dip from the intersecting profiles and a trigonometric method (Simpson, 1968) to correct the apparent dips to dip angle and direction.

9.5. IMAGING SEDIMENTARY STRUCTURES AND DUNE STRATIGRAPHY

Early investigations of dune structure by McKee (1966, 1979) used trenches to expose the internal structure of sand dunes. This is clearly destructive and rarely possible in the 21st century when there is much greater concern for the environment. In addition, digging trenches through unconsolidated sand can be dangerous with a risk of being buried if the trench walls collapse. On the other hand, trenches allow direct examination of the dune structure whereas GPR profiles are effectively a form of remote sensing. Early GPR surveys of sand dunes (Bristow et al., 1996) compared the results of a GPR survey of dunes in the Liwa area of Abu Dhabi with a trench section cut by a bulldozer to establish ground truth. They found inclined reflections that they interpreted as cross-strata and bounding surfaces. Bristow et al. (1996) suggested that the reflections were caused by the contracts in relative permittivity between dry and slightly damp sand associated with changes in the grainsize in the cross-strata. Subsequently, van Dam et al. (2002a, 2002b, 2003) conducted a series of tests to determine the causes of reflections in sands including aeolian dune sands. Reflections on GPR profiles come from interfaces where there is an abrupt change in the dielectric properties in the subsurface. Within aeolian dune sands, van Dam and Schlager (2000), and van Dam et al. (2002a, 2002b, 2003) conducted a series of tests to determine the relative importance of iron, organic matter and water content in producing changes in relative permittivity. Their results indicate that water content is the most important factor affecting the electromagnetic properties of sediments. Consequently, the ability of sediment to hold water governs the GPR reflections (van Dam, 2001). Within sand, water can be held within clay or organic rich layers and fine-grained sand. Fine sands have smaller pore spaces and pore throats than coarse sands and as a result more water can be held by capillary effects in the unsaturated vadose zone above the ground water surface. Coarse sands with larger pore spaces and higher permeability are more free draining and do not retain so much water in the unsaturated vadose zone. However, if coarse sands have a higher porosity than fine sands then they will have a higher water content within the saturated zone beneath the ground water surface. Changes in grainsize within aeolian sands are commonly associated with primary sedimentary structures, and it is the associated changes in the water content that is primarily responsible for the changes in relative permittivity that causes reflections on GPR profiles across sand dunes. An intriguing issue is that the cross-strata within sand dunes are a few millimeters or a few centimeters thick whereas the wavelength of the GPR used to image these structures are usually decimetres (Table 9.1), much greater than the thickness of the laminae or cross-strata within the sand dunes. And yet, sets of cross-strata and bounding surfaces are imaged by GPR. The apparent contradiction between the theoretical resolution and the practical results may in part be explained by the spectrum of the signal produced by a GPR transmitter. Although the central frequency is quoted as, say 100 MHz, in practice the frequency has a bandwidth including both higher and lower frequencies. The higher frequency component of the signal will have a shorter wavelength and therefore can achieve higher resolution than would be predicted from the central frequency. In addition, constructive and destructive interference at the top and bottom of thin layers can influence reflections (Hollender and Tillard 1998).

9.6. Radar Facies

A radar facies can be defined as a mappable, 3D sedimentary unit composed of GPR reflections whose parameters differ from adjacent units (definition modified from Mitchum (1977), by Jol and Bristow (2003)). Within aeolian sands different authors have identified a range of radar facies; van Overmeeren (1998) identified combinations of inclined, parallel continuous, and undulating reflections within aeolian dune sand. The most common is inclined reflections from cross-strata (e.g. van Overmeeren, 1998), although inclined tangential reflections are also common (e.g. Bristow et al., 2005). In vegetated dunes, hummocky discontinuous reflections, trough-shaped reflections, and low-angle inclined reflections have been recognized (Clemmensen et al., 1996, 2007; Bristow et al., 2000b), whereas van Heteren et al. (1998) recognized a so called bounding surface facies which is poorly defined "individual reflections may dip in either direction on any section, showing occasional bounding surfaces cf. Schenk et al. (1993)" (van Heteren et al., 1998, p. 191), which they interpret as dune sands including palaeosols. They also interpret reflection-free areas as homogenous dune sand which is unusual because aeolian sands usually show inclined reflections from sets of cross-stratification. Pedersen and Clemmensen (2005) recognized five different aeolian radar facies, including: organic horizons, aeolian sand sheet strata, aeolian cross-strata, blowout structures, and bounding surfaces. They also identified a reflection-free facies which they interpreted as lake sand strata. Hugenholtz et al. (2007) recognize four radar facies: high-angle planar, high-angle oblique tangential, horizontal subparallel, and moderate to low-angle convex-up, which they relate directly to primary sedimentary structures within a parabolic dune.

There are limitations to this approach because, firstly; radar facies are not unique to a specific sedimentary structure (Jol and Bristow, 2003), and secondly; a 3D object like a set of trough cross-strata can appear as a different-shaped reflection on a GPR profile depending on orientation of the cross-strata and direction of the GPR profile. Inclined reflections in the dip direction can be trough shaped in the strike direction (see Figures 3, 5 and 6 in Bristow et al., 1996). This reinforces the requirement that radar facies should be defined as 3D objects and not just 2D reflection patterns as is commonly the case.

9.7. Radar Stratigraphy and Bounding Surfaces

The concepts of radar stratigraphy were introduced by Beres and Haeni (1991) and Jol and Smith (1991). The terms radar sequence and radar sequence boundary (were introduced by Gawthorpe et al. (1993)) are based on the terminology developed for seismic interpretation by Mitchum et al., (1977). Radar stratigraphy is similar to seismic stratigraphy but it is at a much higher resolution (10s of cm instead of 10s of m). Following seismic stratigraphic interpretation principles, it is necessary to identify reflection terminations to identify radar sequence (Gawthorpe et al., 1993). The identification of reflection terminations is the basis for constructing a relative chronology because reflections are regarded as isochronous surfaces (Vail et al., 1977), and terminations or truncations of reflections represent breaks in time – chronostratigraphic gaps. Successive radar sequences can be used to construct a relative chronology following the laws of superposition and cross-cutting relations. This interpretation methodology has been applied to radar profiles across dunes in the Namib Sand Sea (Figures 9.2 and 9.3) resulting in a reconstruction of dune migration (Bristow et al., 2005).

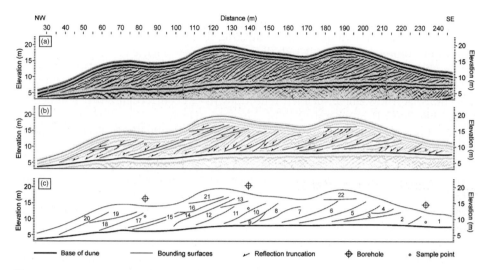

Figure 9.2 (a) GPR profile across the southern end of a linear dune in the Namib Sand Sea shows inclined reflections and low-angle inclined reflections interpreted as sets of cross-strata and reactivation surfaces. Net migration is from East to West but a bi-modal wind regime produced the reactivation/redefinition surfaces. (b) The truncation of reflections are picked as radar sequence boundaries which indicate breaks in deposition and dune migration, (c) on the basis of the cross cutting relationships a relative chronology for the dune deposits was established (from Bristow et al., 2005).

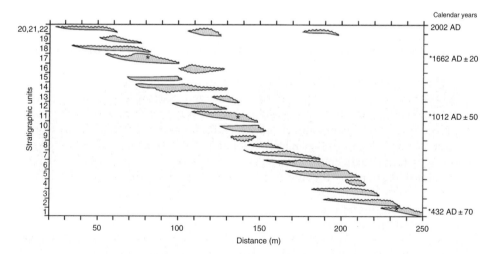

Figure 9.3 A Wheeler diagram constructed on the basis of the stratigraphic interpretation of the GPR profile in Figure 9.2 showing depositional packages bounded by erosion surfaces at the radar sequence boundaries (from Bristow et al., 2005).

9.8. AEOLIAN BOUNDING SURFACES

The cross-strata produced by a migrating dune are said to be simple if they do not contain internal bounding surfaces, and compound if they do. Bounding surfaces are erosional surfaces found within or between sets of cross-strata (Kocurek, 1996, p. 133). Kocurek identified three different types of bounding surfaces, interdune surfaces, superposition surfaces and reactivation or redefinition surfaces. These are discussed below.

9.8.1. Reactivation surfaces

The reactivation or redefinition surfaces are formed if the lee face of a dune is eroded as a result of changes in dune morphology in response to fluctuations in wind direction or wind velocity. These surfaces are very common in dunes because natural winds are almost never steady, continuous and unidirectional. On GPR profiles reactivation or redefinition surfaces appear as angular discordances between dipping reflections where reflections are truncated by or downlapped by an overlying dipping reflection (see Figure 9.2).

9.8.2. Superposition surfaces

Superposition surfaces form by migration of dunes superimposed on a larger bedform, for example transverse dunes superimposed on the flanks of linear dunes in the Namib Sand Sea (Figure 9.4).

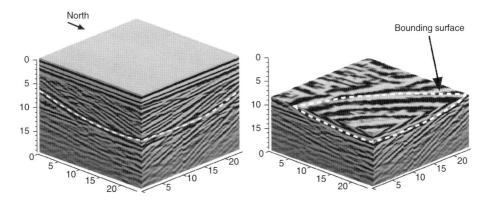

Figure 9.4 A 3D data cube from the flanks of a linear dune in the Namib Sand Sea where transverse dunes migrate along the flank of a linear dune. The low-angle bounding surfaces between sets of cross-strata are superposition surfaces. In the right side image the data cube has been sliced at a depth of 8 m revealing the curved trough cross-strata formed by the superimposed dune. The superposition surface is picked out by dashed white lines.

9.8.3. Interdune surfaces

Interdune surfaces are formed in the trough between migrating dunes. They originate with erosion on the stoss slope and progress to the depth of the scour defined by the interdune trough (Kocurek, 1996). Interdune surfaces are downlapped by the strata from overlying dunes and interdune surfaces can be associated with interdune deposits including inland sabkha sediments, or fluvial sediments deposited in interdune areas. Their preservation is usually associated with changes in the water table which controls the depth of scour in interdune areas. On GPR profiles, interdune surfaces should appear as continuous sub-horizontal reflections that are downlapped by overlying reflections from sets of cross-strata. The interdune surfaces should also truncate underlying dipping reflections. The example shown in Figure 9.5 is from the Jurassic Navajo Sandstone. In this case large transverse dunes migrated across wet interdune surfaces.

Identification of the type of bounding surface in the field is not a trivial exercise (Kocurek, 1996), and the same applies to the interpretation of bounding surfaces on GPR profiles. Although, as a general rule, reactivation surfaces can be distinguished from superposition surfaces because the former are sub-parallel to the foresets, whereas the latter have a different mean dip direction between the bounding surfaces and the foresets as shown in computer simulation models of Rubin (1987). Identifying a change in dip direction can be readily identified on a 3D survey (Figure 9.6) but will not be possible on a 2D profile.

Figure 9.5 Outcrop photograph of the Navajo Sandstone, a Jurassic aeolian sandstone at Zion Canyon National Park, Utah, USA. The outcrop shows sets of cross-stratified sandstone separated by horizontal layers that were deposited on wet interdune surfaces. Figure 9.6 shows a GPR profile of the same section for comparison.

Ground Penetrating Radar in Aeolian Dune Sands

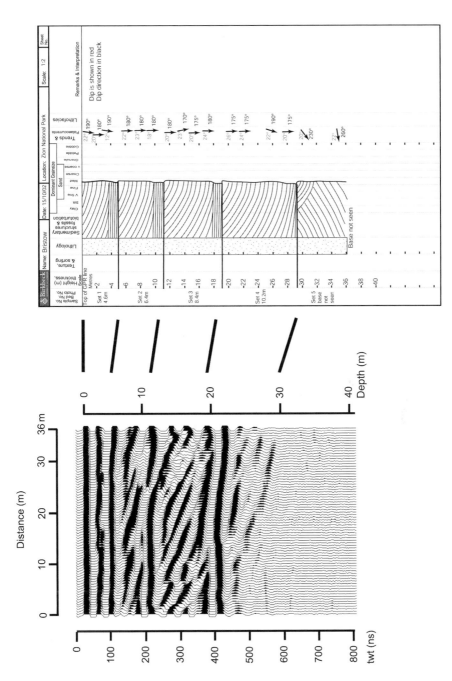

Figure 9.6 Comparison between a GPR profile collected with 25 MHz antennas (left) and a measured section through the Jurassic Navajo Sandstone outcrop illustrated in Figure 9.5. In this example, the uppermost set of cross strata is too thin to be resolved by the relatively long wavelength of the 25 MHz antennas.

9.9. Dune Age and Migration

GPR images of stratification within dunes include sets of cross-strata and their bounding surfaces. Sets of cross-stratification record phases of deposition while bounding surfaces are largely because of erosion (Kocurek, 1996). Between them they record changes in dune morphology and migration. Reflection terminations at bounding surfaces and cross-cutting relationships between radar reflections can be used to generate a relative chronology (Bristow et al., 2005). The relative chronology can be used to select sample points for dating. One of the best techniques for dating aeolian sand is optical dating (see Duller (2004) for reviews). OSL measurements of quartz grains extracted from sand dunes, combined with measurements of the annual dose rate, yield the time since sediment was last exposed to daylight. In most dunes this equates to the time since the sand was deposited on the surface of the dune. OSL signals can be obtained from even very young samples, allowing rates of deposition to be determined over periods from decades to millennia (e.g. Bailey et al., 2001). Previous work on linear dunes from Namibia (Bristow et al., 2005, 2007a) has demonstrated that the method has the resolution and the accuracy required to provide an absolute chronology for the structures imaged by GPR. Alternatively, organic rich layers within dune sands can be dated using radiocarbon dating, and this has been used successfully to date soil horizons within coastal dunes (e.g. Clemmensen et al., 1996, 2001; Pedersen and Clemmensen, 2005).

By combining GPR profiles with OSL dating it is possible to constrain the time of deposition and thus the age of the dune sand. Determining the age of either end of a dune by OSL can be used to derive an end-point migration rate. More detailed chronology of dune evolution can be derived by selecting samples from either side of an erosional bounding surface defined by the relative chronology from GPR profiles within a dune (Bristow et al., 2005). Alternatively, sampling from sets of cross-strata can be used to identify the timing of deposition and dune migration. Using his approach it is possible to achieve a chronology of dune migration and evolution. Bailey et al. (2001) used GPR to define the relative chronology of coastal dune sands at Aberffraw in north Wales and then dated the phases of aeolian dune deposition using OSL. Bristow et al. (2007a) used GPR to define the relative chronology of dune evolution from bounding surfaces on GPR profiles, which, used in combination with optical dating were used to deduce the aeolian dune response to climate change in the Namib Sand Sea. Sample points for OSL dating were selected on the basis of the GPR profile to constrain the chronology of dune migration and accumulation.

9.10. Stratigraphic Analysis

In dry desert environments dunes can migrate across a surface leaving little record of their passage (Kocurek, 1999). The preservation of aeolian dune stratigraphy generally requires some special circumstances, usually a rising water table, because of either a marine transgression on a coastal plain or an increased humidity within a continental basin (Kocurek and Havholm, 1993). A rising water table helps to stabilize the dune sediments so that they are not reworked and eroded by the wind. For example, if migrating dunes encounter standing bodies of water at the downwind end of their path their migration is halted and sand accumulates along the edge of the lake basin. The migration of dunes into a lake has been used by Bailey et al. (2001) to investigate the timing of dune activity in a coastal dune field at Aberffraw in Anglesey. In the examples studied by Bailey et al. (2001) GPR was used to determine the dune stratigraphy with auger samples used to collect sand samples for OSL dating which shows that the dune sands had accumulated within the past 700 years.

In South Africa, Botha et al. (2003) collected GPR profiles across a coastal plain dominated by vegetated dunes in order to assess the depositional history of pedogenically altered aeolian dune sands. They found stacked aeolian sand strata separated by bounding surfaces indicating polyphase vertical accretion on some parabolic dunes. Buried sand units identified on the GPR profiles were sampled using hand augers for infrared-stimulated luminescence dating. The results indicate that some dunes were stabilized around 6–7 thousand years ago, whereas others were active 15–11 thousand years ago, and the oldest dune sand sampled gave an age of 35.8 ka, suggesting intermittent and localized aeolian activity through the Late Pleistocene and Holocene.

In North America, Loope et al. (2004) used GPR to investigate the Late Quaternary and Holocene fill of a valley on the southern shores of Lake Superior. The GPR profiles imaged a buried valley with buried soil horizons overlain by aeolian sands which were interpreted to have impounded a lake. Stratigraphic analysis of the GPR profiles combined with radiocarbon dating of organic rich soil horizons and luminescence (OSL) dating of the aeolian sands reveals the chronology of dune building and drowned forests around the dune dammed lake.

A study of linear dunes in Australia by Bristow et al. (2007b) showed poor resolution of sedimentary structures and only 5 m depth of penetration on GPR profiles. The low penetration and poor resolution of sedimentary structures is attributed to the high mud content of the dunes and pedogenic modification of the dunes. However, palaeosol horizons were identified and these helped to constrain phases of dune accumulation. Results of thermoluminescence dating of the dune sands show that the linear dunes are locally up to 23.7 ± 1.9 ka, and that the linear dunes have extended by up to 3000 m within the last 10 000 years (Bristow et al., 2007b).

An extensive GPR survey in the form of a loose grid was used by Pedersen and Clemmensen (2005) to map the stratigraphy within coastal dunes at Thy in Denmark. They identified packages of aeolian dune sand separated by organic rich horizons that form continuous, medium- to high-amplitude sub-horizontal reflections. Mapping the depth, thickness and extent of these horizons, Pedersen and Clemmensen (2005) were able to produce isopach maps of sediment thickness and reconstruct the palaeotopography. Radiocarbon dating of the organic rich horizons allowed Pedersen and Clemmensen (2005) to reconstruct the dune field evolution over the past 4000 years. More recently, Clemmensen et al. (2007) used GPR and dating to investigate dune mobility in response to storminess and deforestation on the island of Anholt, Denmark.

In Washington and Oregon on the Pacific coast of North America GPR profiling of beach retreat scarps and corresponding OSL and radio carbon dating of the beach deposit shells and drift wood has been used to correlate shoreline response to earthquakes (Peterson et al. 2008).

9.11. Ancient Aeolian Sandstones

There are very few published examples of GPR profiles through ancient aeolian sandstones. The Jurassic, Navajo Sandstone at Zion Canyon National Park has been surveyed by Jol et al. (2003). Their results show sets of inclined reflections separated by continuous horizontal reflections (Figure 9.6). The inclined reflections are interpreted as reflections from cross-stratified sandstones, and the horizontal reflections are interpreted as interdune deposits. The depth of penetration and resolution of different frequency antennae have been tested on this outcrop (Jol et al., 2003). Comparison against a measured stratigraphic section for the outcrop (Figure 9.6) shows excellent resolution of ancient aeolian sedimentary structures and stratigraphy at depths of up to 30 m.

9.12. Three-Dimensional Images

GPR is ideally suited for making 3D images of the shallow subsurface with the rapid acquisition of high-resolution digital data. Three-dimensional interpretations of aeolian sands have been gained from grids of 2D profiles, e.g. Bristow et al. (1996), van Dam (2002), Pedersen and Clemmensen (2005). Bristow et al. (1996) used a grid of GPR profiles spaced at 10 m intervals to image sets of cross-strata and bounding surfaces in the Liwa dunes of Abu Dhabi. van Dam (2002) used an 8-m grid and interpolated between the lines to make a 3D interpretation of the stratigraphy of a river dune in Holland. Pedersen and Clemmensen (2005) used stratigraphic picks on continuous reflections from organic rich horizons within the coastal sand dunes at Thy in Denmark, and interpolation to produce isopach maps and topographic subsurface maps of the dune stratigraphy. These three papers (Bristow et al., 1996; van Dam, 2002; Pedersen and Clemmensen, 2005) all use grids of GPR data to make 3D reconstructions. True 3D data sets should be collected with equal spacing along and between the GPR profiles. Collecting such data sets is time consuming but can produce exceptional visualization of the shallow subsurface. A 3D image of cross-strata in the Jurassic Navajo Sandstone is featured on the front cover of the GPR in Sediments book edited by Bristow and Jol (2003). Bristow et al. (2007a) collected 3D data cuboids on the flanks of a linear dune in Namibia and a similar data set is shown in Figure 9.4 with sets of cross-strata separated by bounding surfaces interpreted as superposition surfaces. The faces of the cuboid show the apparent dip of the cross-strata whereas the horizontal slice through the cuboid reveals the true dip direction of the foresets.

9.13. PEDOGENIC ALTERATION AND EARLY DIAGENESIS

Aeolian dune sands are usually composed of well-sorted, quartz arenites which have low conductivity and low magnetic permeability allowing good depths of penetration (low attenuation). GPR can also achieve good depths of penetration and resolution in carbonate sands, and GPR profiles across carbonate dune sands show good results (Figure 9.7). Problems occur on older aeolian sandstones as a result of early diagenesis, especially dissolution and precipitation of carbonates which alters the primary depositional texture and can lead to the formation of calcretes. Calcretes, or caliches, can cause high-amplitude reflections in GPR profiles. When calcrete horizons occur at or near the surface they can significantly reduce the depth of penetration. Lack of GPR penetration of aeolian sands in the Lander Sandhills has been attributed to pedogenic silt, clay, carbonates and iron oxides (Havholm et al., 2004).

Clay minerals can be incorporated into dune sands either by infiltration of aeolian dust, or illuviation, or from sand-sized clay pellets blown into dunes at the time of formation, or diagenetic alteration of labile minerals. Some dunes in the Simpson Desert, central Australia, have clay contents up to 35%. GPR surveys across these dunes show limited depths of penetration to around 5 m, and little or no resolution of cross-strata (Bristow et al., 2007b). The lack of penetration is attributed to the high clay content of the sands. The lack of cross-strata is attributed to pedogenic alteration, especially the destruction of primary depositional structures by burrowing insects including ants and termites. Another source of biological disturbance is from plant roots. However, vegetation on desert dunes is usually sparse and vegetated coastal dunes do not appear to be significantly altered by roots (e.g. Bristow et al., 2000, Havholm et al., 2004), but soils and palaeosols in dune sands can be imaged on GPR profiles (e.g. Bristow et al., 2007).

9.13.1. Evaporites

Some desert dunes around playa lakes contain evaporite minerals such as gypsum, e.g., sand dunes around Chott Djerid in Tunisia and White Sands National Monument in New Mexico. The presence of gypsum ($CaSO_4\ 2H_2O$) within a dune sand should not make much difference to the performance of GPR, although this remains to be tested. However, gypsum dunes and siliciclastic dunes around evaporite basins can include salt-crusts and evaporite minerals in solution. The electrolytes increase conductivity, attenuate the GPR signal, and reduce the depth of penetration. Saline groundwaters in coastal areas have a similar effect. However, some sensitivity to salt water makes GPR a useful tool for imaging subsurface saltwater intrusions in coastal sand dune aquifers (Peterson et al., 2007).

9.13.2. Environmental noise

High-frequency electromagnetic waves from communications equipment produces high-frequency noise on GPR profiles. Even desert areas are not free from high-frequency interference which can be directional, stronger on one side of a dune and weaker on the other. This noise usually increases toward the crest of a dune and in some

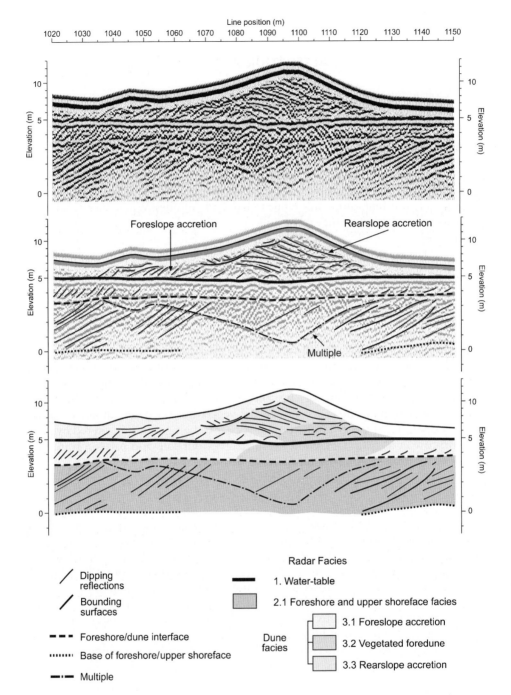

Figure 9.7 Interpreted GPR profile across a vegetated coastal foredune ridge at Guichen Bay in South Australia (from Bristow and Pucillo, 2006). The profile shows examples of radar facies analysis using reflection patterns to distinguish between foreslope accretion (seaward dipping reflections) and rearslope accretion (landward dipping reflections) within the foredune ridge. Beneath the dune sediments the water table forms a continuous high-amplitude sub-horizontal reflection that cuts across the inclined reflections from sedimentary structures. Another feature of this profile is a multiple reflection from the water table giving a mirror image of the surface topography beneath the water table.

9.13.3. Diffractions

Buried objects with a strong contrast in dielectric properties, or voids in sediments, can produce diffraction hyperbolae. These hyperbolae are useful for the recognition of buried objects such as sub-surface pipes or other utilities. They can also be used to calculate the velocity of radar waves using curve matching. Diffraction hyperbolae can also come from objects above ground such as trees or fences.

Diffraction hyperbolae can be collapsed by migration. However, care has to be taken in applying migration because it can introduce artifacts especially where there are changes in velocity. Performing migration with an airwave velocity to collapse hyperbolae from trees will overmigrate other reflections producing U shaped "smiles" that could be mistaken for trough cross-strata.

9.13.4. The water table

The water-table forms a high-amplitude, continuous, sub-horizontal reflection on many GPR profiles because there is a big contrast in relative permittivity between the dry sand above the water table and the saturated sand beneath the water table (Figure 9.7). There is also a change in velocity between dry, damp and saturated sand, with typical velocities for dry sand $0.15\,\text{m ns}^{-1}$, slightly damp sand $0.12\,\text{m ns}^{-1}$, and saturated sand $0.06\,\text{m ns}^{-1}$. The change in velocity at the water table is important because the time/depth scale has to be adjusted. The lower velocities beneath the water table means that the depth scale is expanded. In addition, the resolution increases beneath the water table because the wavelength decreases at a lower velocity. The reflection that appears to be the water table on some GPR profiles can come from the top of the capillary fringe above the saturated zone. This is because the contrast in relative permittivity between dry to slightly damp sand is greater than the change in relative permittivity from damp to saturated sand. In an investigation of the ground water surface beneath coastal dunes Peterson et al. (2007) found that they could image the phreatic ground water surface with lower frequency antennas (50–100 MHz), whereas higher frequency antennas produce reflections from the capillary fringe.

Many desert dunes are well above the water table but still hold small amounts of water by capillary pressure. This water is crucial in causing the changes in dielectric properties that produce reflections on GPR profiles through dune sands. A gradual increase in water content with depth will result in a gradual reduction in the GPR velocity with depth. As a result, buried objects may not be as deep as they appear on the GPR profile if a single velocity for dry sand is used on the depth scale.

9.13.5. Multiples

Multiples are produced when the transmitted signal is reflected back and forth between the surface and a reflecting horizon giving the impression of two or more reflections. The example shown in Figure 9.7 includes a horizontal, high-amplitude reflection from the water table with a dipping reflection beneath the water table that is a mirror image of the surface topography. This dipping reflection is a multiple. The presence of multiples, either as repeated horizontal reflections or dipping reflections like those shown in Figure 9.7 have to be considered during interpretation.

9.14. Conclusions

Dune sands are suitable targets for GPR surveys because they are usually have low conductivity and low-magnetic permeability allowing good depths of penetration (low attenuation). They commonly contain sets of cross-strata which are readily imaged by GPR. The reflections are due to changes in moisture content between fine and coarse-grained laminae. The fine-grained laminae hold more water than the coarse-grained laminae because of capillary effects. This effect can be seen in trenched sections where fine sands which are slightly damp are more competent than the coarser laminae. Other changes in composition such as organic content, iron, or heavy mineral lags are usually less significant than changes in moisture. GPR can be used to investigate the stratigraphy of aeolian sands and develop a relative chronology of dune deposits which can then be dated using OSL. The combination of GPR and luminescence dating can be used to investigate rates of dune migration and longer term controls on coastal and desert dune stratigraphy and the effects of forcing mechanisms such as climate change (Bristow et al., 2007a) and the effects of deforestation and increased storminess (Clemmensen et al., 2007). Depths of penetration can be on the order of 20–40 m with resolution less than 1 m. No other geophysical technique offers the same high-resolution profiles for subsurface investigation of sand body geometry and sedimentary structures.

Acknowledgments

The author thanks all the people who have provided assistance in collecting GPR profiles across sand dunes, especially Simon Armitage, Paul Augustinus, Simon Bailey, Greg Botha, Dana Derickson, Geoff Duller, Ryan Ewing, Harry Jol, Gary Kocurek, Nick Lancaster, David Upton and Irene Wallis. The manuscript has been improved following constructive criticism from John Bridge and Curt Petersen.

REFERENCES

Allen, J.R.L., 1970, The avalanching of granular solids on dune and similar slopes. The Journal of Geology, Vol. 78, pp. 326–351.

Bailey, S.D., Wintle, A.G., Duller, G.A.T. and Bristow, C.S., 2001, Sand deposition during the last millennium at Aberffraw Anglesey, North Wales as determined by OSL dating of quartz. Quaternary Science Reviews, Vol. 20, pp. 701–704.

Beres, M. and Haeni, F.P., 1991, Application of ground-penetrating-radar methods in hydrogeologic studies. Ground Water, Vol. 29, pp. 375–386.

Botha, G.A., Bristow, C.S., Porat, N., Duller, G.A.T., Armitage, S.J., Roberts, H.M., Clarke, B.M., Kota, M.W. and Schoeman, P., 2003, Evidence for dune reactivation from GPR profiles on the Maputaland coastal plain, South Africa, in Bristow, C.S. and Jol, H.M. (eds), Ground Penetrating Radar in Sediments, Geological Society Special Publication 211, pp. 29–46.

Bristow, C.S. and Jol, H.M., 2003, Ground Penetrating Radar in Sediments. Geological Society Special Publication 211, p. 330.

Bristow, C.S. and Pucillo, K., 2006, Quantifying rates of coastal progradation from sediment volume using GPR and OSL: The Holocene fill of Guichen Bay, southeast South Australia. Sedimentology, Vol. 53, pp. 769–788.

Bristow, C.S., Bailey, S.D. and Lancaster, N., 2000, The sedimentary structure of linear sand dunes. Nature, Vol. 406, pp. 56–59.

Bristow, C.S., Chroston, P.N. and Bailey, S.D., 2000, The structure and development of foredunes on a locally prograding coast: Insights from ground penetrating radar surveys, Norfolk, England. Sedimentology, Vol. 47, pp. 923–944.

Bristow, C.S., Duller, G.A.T. and Lancaster, N., 2007a, Age and dynamics of linear dunes in the Namib desert. Geology, Vol. 35, pp. 555–558.

Bristow, C.S., Jones, B.G., Nanson, G.C., Hollands, C., Coleman, M. and Price, D.M., 2007b, GPR surveys of vegetated linear dune stratigraphy in central Australia: Evidence for linear dune extension with vertical and lateral migration, in Baker, G.S. and Jol, H.M. (eds), Stratigraphic Analysis Using GPR, Geological Society of America Special Paper 432, pp. 19–34.

Bristow, C.S., Lancaster, N. and Duller, G.A.T., 2005, Combining ground penetrating radar surveys and optical dating to determine dune migration in Namibia. Journal of the Geological Society, Vol. 162, Part 2, pp. 315–322.

Bristow, C.S., Pugh, J. and Goodall, T., 1996, Internal structure of aeolian dunes in Abu Dhabi revealed using ground penetrating radar. Sedimentology, Vol. 43, pp. 995–1003.

Clemmensen, L.B., Andreasen, F., Nielsen, S.T. and Sten, E., 1996, The late Holocene coastal dunefield at Vejers, Denmark: Characteristics, sand budget and depositional dynamics. Geomorphology, Vol. 17, pp. 79–98.

Clemmensen, L.B., Pye, K., Murray, A. and Heinemeir, J., 2001, Sedimentology, stratigraphy and landscape evolution of a Holocene dune system, Lodbjerg, NW Jutland, Denmark. Sedimentology, Vol. 48, pp. 3–27.

Clemmensen, L.B., Bjornsen, M., Murray, A. and Pedersen, K., 2007, Formation of aeolian dunes on Anholt, Denmark since AD 1560: A record of deforestation and increased storminess. Sedimentary Geology, Vol. 199, pp. 171–187.

Duller, G.A.T., 2004, Luminescence dating of Quaternary sediments: Recent advances. Journal of Quaternary Science, Vol. 19, pp. 183–192.

Gawthorpe, R.L., Collier, R.E.Ll., Alexander, J., Bridge, J.S. and Leeder, M.R., 1993, Ground penetrating radar: Application to sandbody geometry and heterogeneity studies, in North, C.P. and Prosser, D.J. (eds), Characterisation of Fluvial and Aeolian Reservoirs, Geological Society Special Publication 73, pp. 421–432.

Grasmueck, M. and Weger, R., 2002, 3D GPR reveals complex internal structure of Pleistocene oolitic sandbar. The Leading Edge, Vol. 21, No. 7, pp. 634–639.

Havholm, K.G., Ames, D.V., Whittecar, G.R., Wenell, B.A., Riggs, S.R., Jol, H.M., Berger, G.W. and Holmes, M.A., 2004, Stratigraphy of back-barrier coastal dunes, northern North Carolina and southern Virginia. Journal of Coastal Research, Vol. 20, pp. 980–999.

Havholm, K.G., Bergstrom, N.D., Jol, H.M., and Running, G.L. IV., 2003, GPR survey of a Holocene aeolian/fluvial/lacustrine succession, Lauder Sandhills, Manitoba, Canada. In Bristow, C.S., and Jol, H.M., (ed.) Ground Penetrating Radar in Sediments. Geological Society Special Publication 211, p. 47–54.

Hugenholtz, C.H., Moorman, B.J. and Wolfe, S.A., 2007, Ground penetrating radar (GPR) imaging of the internal structure of an active parabolic sand dune, in Baker, G.S. and Jol, H.M. (eds), Stratigraphic Analysis Using GPR, Geological Society of America Special Paper 432, pp. 19–34.

Hunter, R.E., 1977, Basic types of stratification in small eolian dunes. Sedimentology, Vol. 24, pp. 361–387.

Jol, H.M. and Bristow, C.S., 2003, GPR in sediments: Advice on data collection, basic processing and interpretation, a good practice guide, in Bristow, C.S. and Jol, H.M. (eds), Ground Penetrating Radar in Sediments, Geological Society, London, Special Publication 211, pp. 9–27.

Jol, H.M., Bristow, C.S., Smith, D.G., Junck, M.B. and Putnam, P., 2003, Stratigraphic imaging of the Navajo Sandstone using ground-penetrating radar. The Leading Edge, September, pp. 882–887.

Jol, H.M., and Smith, D.G., 1991, Ground penetrating radar of northern lacustrine deltas. Canadian Journal of Earth Sciences, Vol. 28, pp.1939–1947.

Kocurek, G.A., 1996, Desert aeolian systems, in Reading, H.G. (ed.), Sedimentary Environments: Processes, Facies and Stratigraphy, Blackwell Science, Oxford, pp. 125–153.

Kocurek, G., 1999, The aeolian rock record (Yes, Virginia, it exists, but it really is rather special to create one), in Goudie, A. and Livingstone, I. (eds), Aeolian Environments Sediments and Landforms, John Wiley and Sons Ltd, Chichester, pp. 239–259.

Kocurek, G. and Havholm, K.G., 1993, Eolian sequence stratigraphy – a conceptual framework, in Weimer, P. and Posamentier, H. (eds), in Recent Advances in the Application of Sliciclastic Sequence Stratigraphy, Memoir of the American Association of Petroleum Geologists 58, pp. 393–409.

Lancaster, N., 1995, Geomorphology of Desert Dunes, Routledge, London and New York, p. 290.

Lehmann, F. and Green, A.G., 2000, Topographic migration of georadar data: Implications for acquisition and processing. Geophysics, Vol. 65, pp. 836–848.

Loope, W.L., Fisher, T.G., Jol, H.M., Goble, R.J., Anderton, J.B. and Blewett, W.L., 2004, A Holocene history of dune mediated landscape change along the southeastern shore of Lake Superior. Geomorphology, Vol. 61, pp. 303–322.

McKee, E.D., 1966, Structures of dunes at White Sands National Monument, New Mexico (and a comparison with structures of dunes from other selected areas). Sedimentology Vol. 7, pp. 1–69.

McKee, E.D., 1979, A study of Global Sand Seas. US Geological Survey Professional Paper 1052.

McKee, E. and Tibbitts, G.C., Jr., 1964, Primary structures of a seif dune and associated deposits in Libya. Journal of Sedimentary Petrology, Vol. 34, pp. 5–17.

Mitchum, R.M. Jr., Vail. P.R. and Thompson, S. III., 1977, Seismic stratigraphy and global changes in sea level, Part 2: The depositional sequence as a basic unit for stratigraphic analysis. In Payton, C.E., (ed.) Seismic Stratigraphy – applications to hydrocarbon exploration. American Association of Petroleum Geologists Memoir 26, pp. 53–62.

Mountney, N., Howell, J., Flint, S. and Jerram, D., 1999, Relating eolian bounding surface geometries to the bed forms that generated them: Etjo Formation, Cretaceous, Namibia. Geology, Vol. 27, pp. 159–162.

Neal, A. and Roberts, C.L., 2001, Internal structure of a trough blowout determined from migrated ground-penetrating radar profiles. Sedimentology, Vol. 48, pp. 791–810.

Pye, K. and Tsoar, H., 1990, Aeolian Sand and Sand Dunes, Unwin Hyman, London.

Pedersen, K. and Clemmensen, L.B., 2005, Unveiling past aeolian landscapes: A ground penetrating radar survey of a Holocene coastal dunefield system, Thy, Denmark. Sedimentary Geology, Vol. 177, pp. 57–86.

Peterson, C.D., Jol, H.M., Percy, D. and Nielsen, E.L., 2007, Groundwater surface trends from ground penetrating radar (GPR) profiles taken across Late Holocene barriers and beach plains of the Columbia littoral system, Pacific Northwest Coast, USA, in Baker, G.S. and Jol, H.M. (eds), Stratigraphic Analysis Using GPR, Geological Society of America Special Paper 432, pp. 59–76.

Peterson, C.D., Jol, H.M., Vanderburgh, S., Phipps, J.B., Percy, D. and Gelfenbaum, G., 2008, Dating of late Holocene beach shoreline positions by regional correlation of coseismic retreat events in the Columbia River littoral cell, USA. Marine Geology, p.

Reynolds, J.M., 1997, An Introduction to Applied and Environmental Geophysics, John Wiley and Sons, Chichester.

Rubin, D.M., 1987, Cross-bedding, bedforms, and palaeocurrents. Society of Economic Paleontologists and Mineralogists Concepts in Sedimentology and Paleontology, Volume 1, Society of Economic Paleontologists and Mineralogists, Tulsa, Oklahoma.

Rubin, D.M. and Hunter, R.E., 1985, Why deposits of longitudinal dunes are rarely recognised in the geologic record. Sedimentology, Vol. 32, pp. 147–157.

Schenk, C.J., Gautier, D.L., Olhoeft, G.R. and Lucius, J.E., 1993, Internal structure of an aeolian dune using ground-penetrating radar, in Pye, K. and Lancaster, N. (eds), Aeolian Sediments Ancient and Modern. Spec. Pub. Int. Ass. Sediment 16, pp. 61–69.

Simpson, B., 1968, Geological Maps, Pergamon Press, Oxford.

Tsoar, H., 1982, Internal structure and surface geometry of longitudinal seif dunes. J Sed. Pet., Vol. 52, pp. 823–831.

Ulriksen, C.P.F., 1982, Application of impulse radar to civil engineering. PhD Thesis, Lund University of Technology, Lund, Sweden.

Vail, P.R., Todd, R.G. and Sangree, J.B., 1977, Seismic stratigraphy and global changes in sea level, part 5: Chronostratigraphic significance of seismic reflections, in Payton, C.E. (ed.), Seismic Stratigraphy – Applications to Hydrocarbon Exploration AAPG Memoir 26, pp. 99–116.

van Dam, R.L., 2001. Causes of ground-penetrating radar reflections in sediment. PhD Thesis, Vrije Universiteit Amsterdam, Netherlands, p. 110.

van Dam, R.L., 2002, Internal structure and development of an aeolian river dune in the Netherlands, using 3-D interpretation of ground-penetrating radar data. Netherlands Journal of Geosciences/Geologie en Mijnbouw, Vol. 81, pp. 27–37.

van Dam, R.L. and Schlager, W., 2000, Identifying causes of ground-penetrating radar reflections using time-domain reflectometry and sedimentological analyses. Sedimentology, Vol. 47, pp. 435–449.

van Dam, R.L., Schlager, W., Dekkers, M.J. and Huisman, J.A., 2002a, Iron oxides as a cause of GPR reflections. Geophysics, Vol. 67, pp. 536–545.

van Dam, R.L., Schlager, W., Dekkers, M.J. and Huisman, J.A., 2002b. Influence of organic matter in soils on radar-wave reflection: Sedimentological implications. Journal of Sedimentary Research, Vol. 72, pp. 341–352.

van Dam, R.L., Van Den Berg, E.H., Schaap, M.G., Broekema, L.H. and Schlager, W., 2003, Radar reflections from sedimentary structures in the vadose zone, in Bristow, C.S. and Jol, H.M. (eds), GPR in Sediments, Geological Society, London, Special Publication 211, pp. 257–273.

van Heteren, S., Fitzgerald, D.M., McKinley, P.A. and Buynevich, I.V., 1998, Radar facies of paraglacial barrier systems: Coastal New England, USA, Sedimentology Vol. 45, pp. 181–200.

van Overmeeren, R.A., 1998, Radar facies of unconsolidated sediments in The Netherlands: A radar stratigraphy interpretation method for hydrogeology. Journal of Applied Geophysics, Vol. 40, pp. 1–18.

Woodward, J., Ashworth, P.J., Best, J.L., Sambrook Smith, G.H. and Simpson, C.J., 2003, The use and application of GPR in sandy fluvial environments: Methodological considerations, in Bristow, C.S. and Jol, H.M. (eds), Ground Penetrating Radar in Sediments, Geological Society Special Publication 211, pp. 127–142.

CHAPTER 10

COASTAL ENVIRONMENTS

Ilya V. Buynevich, Harry M. Jol *and* Duncan M. FitzGerald

Contents

10.1. Introduction	299
10.2. Methodology	301
10.3. Ground Penetrating Radar Strengths in Coastal Environments	303
10.4. Ground Penetrating Radar Limitations in Coastal Environments	304
10.5. Ground Penetrating Radar Studies in Coastal Environments	305
10.6. Examples of Ground Penetrating Radar Images from Coastal Environments	305
10.6.1. Record of coastal progradation	306
10.6.2. Signatures of coastal erosion	307
10.6.3. Coastal Paleochannels	308
10.6.4. Ground penetrating radar signal response to lithological anomalies in coastal dunes	310
10.6.5. Deltas	312
10.6.6. Reservoir characterization – hydrocarbon and hydrogeology	313
10.7. Summary	314
Acknowledgments	315
References	315

10.1. INTRODUCTION

Depositional coastal landforms, in both marine and lacustrine settings, contain various accretionary and erosional elements that are created through the interaction of wind, wave, tidal, riverine, and/or other geologic processes. These regions are characterized by beaches, dunes, barriers, deltas, strandplains, backbarrier marshes, lagoons, and tidal flats. In extreme climates, such as along the Arctic coast, features are influenced by ice processes such as the patterned ground and ice-push barriers. In equatorial regions, geochemical processes often dominate sedimentation processes forming deposits such as oolite shoals, beachrock, and sabkhas. The resulting sedimentary packages represent a range of facies, which depend upon the sediment source and the environment of deposition. The facies assemblage comprising coastal deposits consists of stacked sedimentary sequences (packages) that differ in lithology, geometry, sedimentary structures, and bounding surfaces due to varying physical, biological, and chemical processes that operate under different climatic and sea level conditions, energy regimes, sediment supplies, and basement controls. For example, a transgressive barrier sequence may overlie different mainland sedimentary deposits including riverine, estuarine, aeolian, or other sediments. The transgressive barrier lithosome consists of amalgamated washover, beach, and aeolian deposits, which may be interlayered with lagoon, tidal inlet fill, and flood–tidal delta sediments. Due to the similarity in sedimentologic character of many of these facies, a simple coring study yielding grain size data, sedimentary structures, and other small-scale sedimentary characteristics may be insufficient to identify a transgressive barrier deposit or to distinguish individual facies within the barrier lithosome.

Traditional methods for obtaining sedimentologic and stratigraphic information such as coring programs, logging exposures, and surface morphology mapping are rarely capable of capturing the complexity of coastal sedimentary deposits or ascertaining the extent of a particular lithosome. In response to this problem, ground penetrating radar (GPR), a high-resolution geophysical technique, has increasingly been used during the past two decades to study coastal systems (Figure 10.1). Ground penetrating radar helps visualize the subsurface geology and produces an image of the major erosional and depositional surfaces (Figure 10.2). This noninvasive geophysical survey technique allows for the rapid mapping of lithofacies and is an effective tool for planning a coring program to ground-truth the sedimentary surfaces that correspond to the GPR reflections (Bristow and Jol, 2003; Baker and Jol, 2007).

The primary objective of this chapter is to provide an overview of how GPR has been used along coasts in both marine and lacustrine settings. In addition, examples will be provided from various coastal depositional environments to demonstrate that GPR is an effective tool for the investigation of subsurface features and stratigraphy in coastal settings that exhibit different sediment compositions

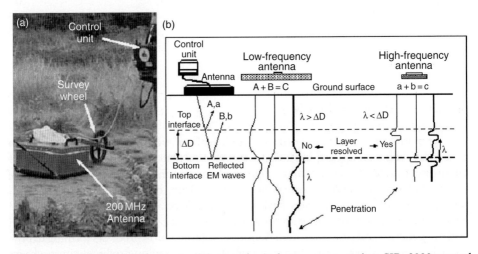

Figure 10.1 (A) Main components of the geophysical survey systems inc. SIR-2000 ground penetrating radar (GPR) with a 200-MHz monostatic antenna (photo courtesy: James O'Connell). (B) Resolution and penetration depth of the low- and high-frequency GPR antennae. Low-frequency wave (A) defines the top interface, but the distance between the interfaces (ΔD) is less than the wavelength (B). The relatively long wavelength (λ) of the resulting electromagnetic signal (C) will not resolve both interfaces. A high-frequency EM wave defines both the top (a) and bottom (b) interfaces. The resulting wave (c) is shorter than the distance between the two interfaces. The penetration depth in this case, however, is less than that with a low-frequency antenna (modified after Conyers and Goodman, 1997).

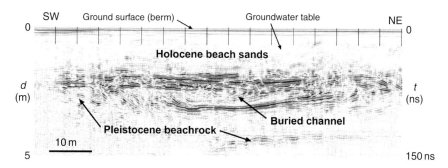

Figure 10.2 A shore-parallel ground penetrating radar (GPR) profile over a berm, Myrtle Beach, South Carolina, USA. The record shows a buried channel fill over lithified Pleistocene beach rock, which can be resolved due to its high dielectric contrast. The subhorizontal reflections at the top are typical of shore-parallel (strike) sections along the upper beach (these reflections dip seaward in a shore-normal section). Due to presence of fresh groundwater, it is sometimes possible to collect GPR images of the shallow subsurface just above the mean high tide elevation. Unless otherwise noted, all records were collected using a digital GSSI SIR-2000 system with a 200-MHz monostatic antenna [d – approximate depth (m) based on coring data or empirical EM velocities; t – two-way travel time in nanoseconds (ns)].

(sand, shell fragments, gravel/boulder, and organic). Note that aeolian environments are discussed in a separate chapter of this volume and only partly treated herein.

10.2. METHODOLOGY

Ground penetrating radar profiles are similar in appearance to seismic-reflection profiles, except that GPR data are acquired using transient electromagnetic (EM) energy reflection instead of acoustic energy and thus provide greater resolution. A short pulse of high-frequency EM energy (~10–1000 MHz) is transmitted into the ground. When the signal encounters a contrast in material properties, some of the energy is reflected back to the surface due to a change in the bulk electrical properties of different subsurface lithologies (see Figure 10.1). The interface between these two layers may be characterized by bedrock contact, organic-rich sediments, groundwater table, and changes in sediment grain size, mineralogy, and packing (Davis and Annan, 1989). A change in the dielectric constant (relative permittivity) of the sediment also affects the rate of attenuation of energy passing through the ground. The resolution and the penetration depth in most coastal environments are based on the frequency and pulser voltage of the initial GPR signal and the material properties it travels through. Resolution ranges from subdecimeter to greater than a meter and depths range from less than a meter to tens of meters (Davis and Annan, 1989; Jol, 1995; Smith and Jol, 1995).

The best survey results are obtained using a network of intersecting transects that capture the extent and variability of GPR reflections. One significant limitation of the GPR system setup is penetration depth and resolution. Both of these attributes depend on the choice of the antennae, which is governed by the research objectives (Jol, 1995; Smith and Jol, 1995). High-frequency antennae (500–1000 MHz) are smaller in size and provide high resolution at the expense of relatively limited penetration. Antennae with frequency range of 12.5–50 MHz have poor resolution but allow for a maximum probable penetration of 45–65 m (Smith and Jol, 1995). The most common antennae frequencies used in coastal stratigraphic research are 100 and 120 MHz, which allow a good penetration depth (∼10–20 m; deeper in unsaturated sequences), while still providing high-resolution images. Data collected in coastal settings are commonly collected using either step mode or continuous mode. In step mode, the transmitter and the receiver are placed on the ground, and after the reading is obtained, it is moved to the next location. For imaging detailed stratigraphy in coastal sedimentary environments, a step mode is most often preferred (Jol and Bristow, 2003; Jol et al., 2006). Continuous mode involves the unidirectional movement of both transmitter and receiver (or a single transceiver) antennae, which are dragged across the study site providing rapid collection of continuous traces over large areas. On relatively smooth substrates, the antennae are often used in conjunction with an odometer wheel, which triggers the start and termination of a recording and provides consistent distance marks. This method has been shown to be an effective means for collecting subsurface data, particularly during reconnaissance surveys. Neal and Roberts (2000) and Jol and Bristow (2003) provide a background on GPR system setup, survey design, and basic processing within coastal sedimentary systems.

Whenever GPR profiles are collected in areas having relief, detailed topographic surveys should be conducted and applied to the datasets. This need arises from the fact that the ground surface on the time-record output is represented as a horizontal surface. For example, on a profile taken over a coastal dune, the dune surface will be depicted as a horizontal reflection and the (sub-) horizontal layers of the underlying sequence (e.g., water table, dune/beach contact, washover deposit, and peat horizon) will appear as concave-upward reflections (i.e., the mirror images of dune topography). The latter can be misinterpreted as shallow paleochannels if no topographic measurements or notes were made. Commonly, in areas where surface relief exceeds vertical resolution of the GPR or where detailed measurements of dip angles are required, it is necessary to collect topographic data for the proper surface normalization.

Radar stratigraphic analysis provides the framework for assessing both lateral and vertical geometries and the stratification of coastal deposits (Beres and Haeni, 1991; Jol and Smith, 1991; Gawthorpe et al., 1993; Huggenberger, 1993). The analysis is based on well-developed principles of seismic stratigraphy (e.g., geometry of bounding surfaces, internal reflections; Mitchum et al., 1977) and provides a systematic methodology to objectively describe and interpret GPR reflection profiles for geologic applications. The development of radar stratigraphic analysis has allowed for the delineation and mapping of genetically related stratigraphic units within sedimentary deposits (van Heteren et al., 1998). In addition to stratigraphic applications, recent studies demonstrate the use of reflection geometry as a sea level indicator (e.g., upper contacts between oblique beach/shoreface and dune reflections; van Heteren and van de Plassche, 1997; van Heteren et al., 2000; Rodriguez and Meyer, 2006; Storms and Kroonenberg, 2007).

Widely spaced lines that intersect over coastal features tie transects together so that fence diagrams can be constructed. Fence diagrams allow associated facies to be mapped over a broad distances. For smaller areas (e.g., 25 m × 25 m), a grid of closely spaced, parallel lines (and cross lines) are collected. A 3D image, or cube, constructed from high-resolution GPR datasets (often 2D lines) can provide a unique representation of the subsurface. Such 3D images have been utilized in interpreting various geomorphic environments, including marine and lacustrine coastal features (Smith and Jol, 1992a, 1922b; Jol et al., 1994; Beres et al., 1995; Jol et al., 1996a, 1996b, 2002; Grasmueck and Weger, 2002; Grasmueck et al., 2004, 2005). The processing and display of these 3D datasets require additional software packages but results can provide a detailed 3D image of the subsurface from which one can view the data from any angle, plan view (time slices), or individual reflections. These visualizations often lead to a better understanding of sedimentary deposits and the relationships among facies (e.g., Daniels et al., 1988; Beres et al., 1995; Olsen and Andreasen, 1995; Thompson et al., 1995).

10.3. Ground Penetrating Radar Strengths in Coastal Environments

Ground penetrating radar is well suited for investigating coastal landforms due to the well-developed GPR reflections that are produced within coastal deposits. Sediments comprising these environments are formed by various depositional processes and possess a range of grain sizes and mineralogies. Among other factors, strong GPR reflections are produced from abrupt changes in grain size/mineralogy that are commonly found in storm deposits, washovers, and channel fill sequences. Coastal settings are also regions where shell, heavy minerals, and mica are common (Jol et al., 1996a). High concentrations of heavy minerals (e.g., magnetite, ilmenite, garnet, tourmaline, and other minerals with densities exceeding 2.9 $g\,cm^{-3}$) that form as storm-lag deposits (or erosional remnants/scarps) on beaches often produce prominent reflections on GPR records (Topp et al., 1980; Meyers et al., 1996; van Heteren et al., 1998; Smith et al., 1999). With the exception of areas dominated by thick glaciolacustrine or glaciomarine clays, GPR works particularly well along paraglacial coasts due to the highly variable sediments that occur in these regions. These deposits commonly consist of intercalated layers of sand and gravel and are ideal for imaging using GPR (Smith and Jol, 1997). In these environments, GPR surveying is likely to be most successful where ground conductivities and resulting attenuation are low, thus allowing significant penetration of the GPR signal. Consequently, coastal deposits dominated by sands, gravels, and peats that are either unsaturated or below a fresh groundwater table are likely to be the most amenable to GPR reflection profiling.

Georadar (GPR) can be used to infer stratigraphic trends such as directions of progradation and/or aggradation, delineation of sedimentary facies, and assessment of depth to the freshwater–brackish water interface in shallow freshwater conditions. For example, internal structure of most coastal systems of sand, broken shell fragments, and/or gravel–boulder often shows either shore-parallel horizontal reflections or seaward-dipping reflections of paleobeach surfaces ranging from 1° to 26°. The original beachface slope is a function of incident wave energy, as well as grain size of sediments (1–2° in fine sand, 3–7° in medium-coarse sand to granules of broken shells, and up to 26° in gravel/boulders). Some coastal environments are dominated by peat deposits. Since these organic deposits are electrically resistive, GPR is a good tool for imaging their thickness and lateral extent as well as the coastal deposits below (Jol and Smith, 1991, 1995; Mellett, 1995).

In general, unless the subsurface sediments of a study site are well known, ground truthing (e.g., sediment cores) should be conducted to confirm the major reflections. In turn, the GPR images of dipping subsurface horizons (e.g., sloping bedrock surface and tidal inlet channel) may be used to maximize the coring effort by planning a core site in the area where the depth to a target reflection is minimal and can provide a detailed stratigraphic context for the core's interpretation.

10.4. Ground Penetrating Radar Limitations in Coastal Environments

Ground penetrating radar does not work well in fine-grained (silt and clay) and saline/brackish coastal environments. These electrically conductive deposits result in signal dissipation and loss. As a result, sections of many profiles extending below the high tidewater line on either ocean-facing beach or landward margin of a barrier, along with deeper profile sections affected by saltwater intrusion, are often reflection-free (van Heteren et al., 1998). However, for delineating the freshwater–brackish water contact, GPR becomes an excellent mapping tool (Figure 10.3).

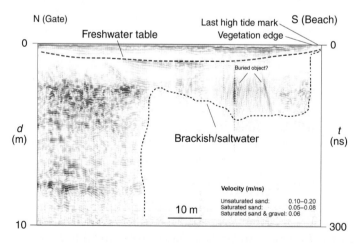

Figure 10.3 An example of signal attenuation by shallow saltwater. The shore-perpendicular ground penetrating radar (GPR) profile was collected across the beach at the head of Waquoit Bay, Massachusetts. Note that even though saltwater may penetrate only the outwash deposits as a narrow tongue, it attenuates the EM signal over the entire depth range. The landward extent of saltwater intrusion is delineated by the sharp vertical contact and appearance of the full record on the left side of the image.

In some cases, running a GPR survey following a freshet and/or a major rainstorm allows for a better opportunity to collect GPR data in these problematic settings. Along many marine settings, the barrier is wide enough so that the groundwater is fresh and thus most of the stratigraphy can be imaged by GPR without attenuation. Similarly, attenuation due to thick, fine-grained deposits can be used for mapping the extent of the offshore or backbarrier facies (e.g., intervening paleolagoonal deposits within sand-dominated, prograded strandplains; Buynevich et al., 2005).

In some cases, lithological anomalies that contain high concentrations of heavy minerals, as well as thick, iron-stained horizons (e.g., soils), may have high-enough magnetic permeability values such that they attenuate the magnetic portion of the EM signal and preclude or reduce penetration (Topp et al., 1980). Similar problems can occur when postdepositional processes (i.e., caliches, cementation) change the dielectric properties within the sediments. As discussed above and illustrated below, many lithological causes of signal attenuation can be useful in other aspects of research where such anomalies have geologic significance.

10.5. Ground Penetrating Radar Studies in Coastal Environments

Initial GPR investigations into coastal environments were first conducted in the mid-to-late 1980s by Ulriksen (1982), Leatherman (1987), and Truman et al. (1988), with subsequent studies in the early 1990s confirming its potential (Baker, 1991; Jol and Smith, 1991; FitzGerald et al., 1992; Jol and Smith, 1992; Smith and Jol, 1992a; Jol, 1993; Bristow et al., 1995a, 1995b; van Overmeeren, 1998). Based on published studies, much of the coastal GPR research has concentrated in North America and northern Europe, with fewer studies outside these regions. Neal and Roberts (2000) noted that since this research is published in both the geologic and geophysical literature, it is often difficult for the nonspecialist to find publications that aid in determining whether GPR will provide the desired information at a particular coastal study site or how to properly interpret the collected datasets. This may be one of the reasons why there have been a relatively limited number of published coastal GPR studies, although the numbers have increased substantially in recent years.

10.6. Examples of Ground Penetrating Radar Images from Coastal Environments

The application of GPR to coastal stratigraphic, geomorphic, and environmental problems has shown many positive results (Leatherman, 1987; Beres and Haeni, 1991; Jol and Smith, 1991; FitzGerald et al., 1992; Smith and Jol, 1992b; Barnhardt et al., 2002: Buynevich, 2003; Moller and Anthony, 2003; Bode and Jol, 2006; Buynevich et al., 2004, 2007c). A number of GPR studies have been conducted along the many coastlines of North America. These examples include research carried out in various settings along the Atlantic Coast (Leatherman, 1987; Truman et al., 1988; FitzGerald et al., 1992; van Heteren et al., 1994, 1996, 1998; Jol et al., 1996a; Wyatt and Temples, 1996; Buynevich and FitzGerald, 2001, 2002, 2003; O'Neal and Dunn, 2003; Buynevich, 2006), Pacific Coast (Jol et al., 1994, 1996a, 1996c, 1998a, 1998b; 2002; Meyers, 1994; Meyers et al., 1994, 1996; Moore et al., 2003a, 2003b, 2004; Roberts and Jol, 2002; Roberts et al., 2003; Peterson et al., 2007) and the Gulf of Mexico coast (Zenero et al., 1995; Jol et al., 1996a). A range of lacustrine and ancient coastal environments were investigated as well (Jol and Smith, 1991, 1992; Smith and Jol 1992a, 1992b, 1995, 1997; Jol et al., 1994, 1996a, 1996b, 1996c; Tercier et al., 2000; Buynevich and FitzGerald, 2003; Kruse and Jol, 2003; Smith et al., 2003, 2005; Johnston et al., 2007; Wilkins and Clement, 2007).

In this section, we present examples of GPR studies from a number of coastal environments that exhibit a range of sediment compositions. In order to demonstrate various types of geomorphological and sedimentologic data that can be obtained from GPR surveys, complete datasets and full interpretations are not furnished. Rather our examples will demonstrate the nature and range of information that can be obtained and its potential contribution to future coastal research.

10.6.1. Record of coastal progradation

Along coastal areas that have experienced progradation (seaward growth), the geomorphic expression of this process, such as consecutive beach ridges, is often used to determine the origin, magnitude, orientation, and chronology of ridge sets or individual ridges (Tanner, 1995; Bristow and Pucillo, 2006). However, in many cases, dense vegetation, dune migration, and human development have modified or obscured the surface expression of barrier growth. In such areas, subsurface records, complemented with sediment cores, may provide the only means of analyzing the erosional–depositional history of a barrier. Recent studies have shown the value of continuous, shore-normal GPR profiles in visualizing the style of coastal progradation (Figure 10.4; Buynevich and FitzGerald, 2001) and addressing the volumes and progradation rates of coastal lithosomes (Jol et al., 1996a; van Heteren et al., 1996; Buynevich et al., 2005; Bristow and Pucillo, 2006; FitzGerald et al., 2007).

As mentioned earlier, given a particular texture, the equilibrium slope of a beachface is a function of incident wave energy (Bascom, 1951; Masselink and Puleo, 2006). Therefore, paleogradients determined from GPR profiles (see Figure 10.4), combined with granulometric information from sediment cores, may be used to assess the past wave regime. For instance, without substantial changes in sediment texture or storm frequency, a measurable decrease in the angle of GPR reflections in a seaward (younging) direction can be interpreted as a response of a beach to an increase in wave energy through time. This, in turn, can be linked to changes in sea level (accommodation space) or sediment supply.

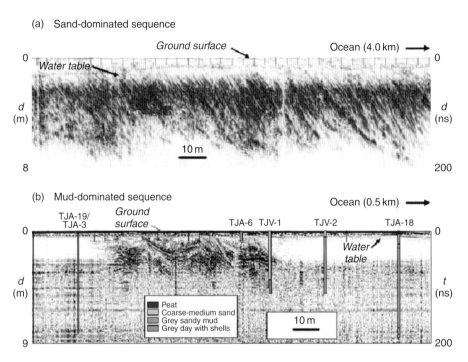

Figure 10.4 Contrasting shore-normal ground penetrating radar (GPR) records collected north of Tijucas River, Santa Catarina, Brazil (modified from Buynevich et al., 2005; Fitzgerald et al., 2007). (A) The older part of the coastal plain is dominated by steep seaward-dipping reflections produced by deposition of coarse-medium sands in a relatively high-energy setting; (B) subsurface image from the younger part of the plain reveals a different, mud-dominated regime, with sand deposited in narrow, low ridges (TJA – Eijkelkamp auger core; TJV – vibracore). Presently, the mud suspended in the nearshore exerts a major control on dampening the incident waves.

10.6.2. Signatures of coastal erosion

Episodes of erosion, retreat, and flooding of low-lying areas punctuate the evolution of sandy coasts. They may have diverse origins, such as tropical cyclones (hurricanes, typhoons), El Niño events, extratropical storms, and tsunamis. Primarily due to limited documentation and instrumental records, the return periods of large-magnitude erosional events are poorly understood. Prograding sequences seldom exhibit geomorphic evidence of an erosional event, except occasional washovers or dune and berm scarps confined to the youngest portion of the barrier (Figure 10.5a). These features, particularly in the earlier constructional history of the barrier, are often preserved as buried accumulations of coarse-grained sediments or heavy-mineral concentrations (HMCs; Figure 10.5a) and are often missed in isolated cores. Such lithological anomalies may be observed in sediment cores, but their geometry and continuity can be confirmed only in geophysical records (Meyers et al., 1996; Moore et al., 2003a, 2003b, 2004; Nichol, 2002; Buynevich et al., 2004; Costas et al., 2006; Jol and Peterson, 2006).

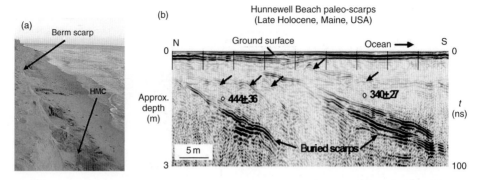

Figure 10.5 Erosional features and their geologic record: (a) extensive heavy-mineral concentration (HMC) at the base of a berm scarp at Madaket, Nantucket Island, Massachusetts; (b) a shore-normal image from Hunnewell Beach, Maine, reveals buried erosional scarps as a series of prominent, steeply dipping reflections within a progradational barrier sequence. In sediment cores, these structures coincide with HMCs and represent upper-beach erosion surfaces produced by large storms. Two of the four optical dates (in years before 2003) help reconstruct the history of erosion (modified after Buynevich et al., 2007c).

These HMCs are found in many parts of the world and can be used as indicators of erosion. Their formation is due, primarily, to selective density sorting during the waning stages of storms (Komar and Wang, 1984). Similar HMCs preserved within Willapa barrier, Washington, were attributed to temporary coseismic subsidence and subsequent severe wave erosion (Meyers et al., 1996). Regardless of their origin, reworking and concentration of heavy-mineral fractions ensures their recognition in the stratigraphic record, which is particularly important where surface expression of past events has been masked by subsequent deposition. For example, the GPR profile in Figure 10.5b illustrates a series of prominent reflections, which represent buried erosional berm scarps, which when combined with optically stimulated luminescence (OSL) dating of overlying sands provide a 1500-year record of erosional events along the coast of Maine (Buynevich et al., 2007c). When GPR images are combined with trenches and sediment cores, these relict marker horizons provide information, which cannot be obtained by other means; specifically they provide the (1) landward extent of erosion or inundation; (2) longshore extent of storm impact; (3) thickness of eroded sediment; and (4) intensity of the erosional episode. In addition, the attributes of individual HMCs (sedimentary structures, grain size, sorting, and density) can be used to calculate threshold conditions of sediment transport and deposition.

10.6.3. Coastal Paleochannels

Channel-fill sequences of tidal inlets along microtidal coastlines may comprise a significant portion of the barrier lithosome (Moslow and Heron, 1978), and in some instances, the locations and dimensions of former inlet channels can be detected with GPR (Figures 10.2 and 10.6). Mixed-sediment barriers are ideal for the recognition of inlet-fill structures. Due to large contrasts between the

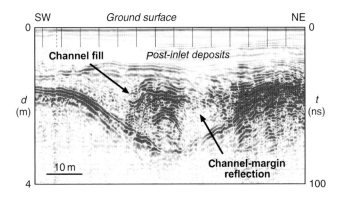

Figure 10.6 Shore-parallel ground penetrating radar (GPR) profile over the buried inlet channel beneath a mixed-sediment baymouth barrier fronting Oyster Pond, Cape Cod, Massachusetts. Note the strong, channel-margin reflection and the internal stratification of the inlet fill.

coarse-grained, channel-lag and finer-grained, channel-fill deposits, the outline of the channel commonly appears as a prominent concave-upward reflection. Figure 10.6 shows a paleoinlet channel that has migrated along a retrograding, sand-and-gravel barrier as evidenced by a series of northward-dipping reflections. Eventually, inlet stabilization and subsequent infilling by sediment from a seaward source is recorded as subhorizontal reflections within the paleochannel. Using GPR profiling, the locations of the former inlets can be mapped and compared to historical maps, where available. In addition, such elements of inlet channel geometry as depth, width, and approximate length (using a series of records) can be determined. The thalweg depth of a paleoinlet channel relative to the present sea level can also be estimated.

In recent years, GPR images, ground-truthed with sediment cores, have confirmed the positions of several historical inlets and revealed a number of previously undocumented tidal inlet and storm–breach channels along the US Atlantic Coast (FitzGerald et al., 2001; Daly et al., 2002; Buynevich, 2003; Havholm et al., 2004; Buynevich and Donnelly, 2006). Along a retrograding mixed-sediment barrier at Duxbury, Massachusetts, at least 18 buried channels were imaged along one continuous, shore-parallel GPR line (FitzGerald et al., 2001). Although most of these channels were ephemeral storm breachways, there is evidence for longshore migration of the channels in several cases. The earliest historical charts of the area (1774–1777) depict only a few of the openings that were imaged during the survey period. Often, subsurface signatures of nonmigrating inlets or breachways lack lateral migration surfaces but still show the internal stratification of the channel fill (see Figure 10.6).

Paraglacial coasts or coastal segments with fluvial bed load sources are particularly suitable for subsurface imaging due to the high contrasts in dielectric properties of the coastal lithofacies (high degree of textural and compositional heterogeneity; van Heteren et al., 1998; Buynevich and FitzGerald, 2001; FitzGerald et al., 2001; Neal et al., 2002, 2003). In the sand-dominated barrier lithosomes of the Atlantic Coastal

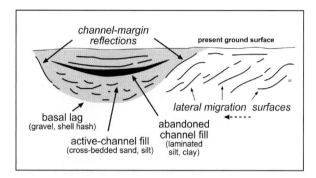

Figure 10.7 Schematic cross section of a buried channel (modified after Buynevich et al., 2003). The basal-lag deposit of the inlet floor often contrasts with the sediments above and below produces prominent reflections in geophysical records. A set of inward-dipping, channel-margin reflections are often the only subsurface signature of the final channel position (grey). In tidally influenced settings, fine-grained, abandoned channel fill may be present (Moslow and Tye, 1985). Many ephemeral or stabilized channels will lack longshore-dipping reflections diagnostic of spit elongation and channel migration.

Plain, subsurface GPR reflections may correspond to more subtle lithological transitions, such as slight textural and compositional variations. In such settings, mud plugs are diagnostic of tidally influenced abandoned channels (Moslow and Heron, 1978). Thick shallow muds will likely attenuate the GPR signal, whereas shell-rich, inlet-throat deposits have greater potential to be resolved in the shallow subsurface (Figure 10.7). Often, even where the reflection of the thalweg of the channel is attenuated by saltwater, a set of inward-dipping, channel-margin (contour) reflections are reliable indicators of the final channel position (see Figures 10.6 and 10.7). Where space permits, a grid of closely spaced, shore-parallel and shore-normal profiles over the buried channel may be used to construct a 3D image of the subsurface.

10.6.4. Ground penetrating radar signal response to lithological anomalies in coastal dunes

Dune fields of various sizes and origins are common landforms along many coasts. Their stratigraphy often reveals important clues about the sensitivity of aeolian landforms to major climatic and geologic forces (Hesp and Thom, 1990; Wilson et al., 2001; Clarke et al., 2002). In recent years, continuous, high-resolution GPR images have helped reconstruct landscape change, particularly where older periods of aeolian activity are in question (Clemmensen et al., 1996, 2001; Botha et al., 2003; van Dam et al., 2003; Barnhardt et al., 2004; Havholm et al., 2004). Being relatively homogeneous compared to other settings, dune lithosomes may still exhibit dielectrically distinct textural variations resulting from changes in sediment source and wind-flow patterns and intensity. Some sedimentologic changes (e.g., grain fabric, packing, grading, and water retention) are sufficient enough to produce distinct signal responses but may be too subtle to be resolved by standard analyses of sediment cores and outcrops (van Dam and Schlager, 2000, 2002).

Ground penetrating radar has been used successfully to identify and map bounding surfaces within dune sequences (Schenk et al., 1993; Jol et al., 1996a; Bristow et al., 2005). Several recent studies have also addressed the origin and geologic significance of individual reflections, which is important for accurate correlation of geophysical records with sedimentologic features observed in cores or outcrops (Guha et al., 2005; Buynevich et al., 2007a).

For correlating GPR reflections with specific sedimentary features, outcrops or trenches are typically preferable to point-source information from sediment cores (van Dam et al., 2000, 2002). Using ground exposures of migration surfaces (slipfaces) of a relict Holocene coastal dune along the southeastern Baltic Sea, Buynevich et al. (2007a) have investigated the causes of prominent reflections on geophysical profiles. High-amplitude reflections on GPR images correlate well with major lithological anomalies: paleosols developed on dune slipfaces and slipfaces consisting of HMCs (Figure 10.8). Paleosols are indicators of dune stability, which represent datable chronostratigraphic surfaces and help reconstruct dune paleomorphology. Heavy-mineral concentrations have noticeably higher magnetic susceptibility values than background quartz-rich sands and in some instances can be used for spatial correlation. Based on their occurrence at the study site, these enriched horizons likely represent periods of increased wind activity (storminess) along the southeast Baltic Coast (Buynevich et al., 2007a, 2007b) and demonstrate the value of an integrated sedimentologic and geophysical approach to reconstructing aeolian dynamics in other coastal regions.

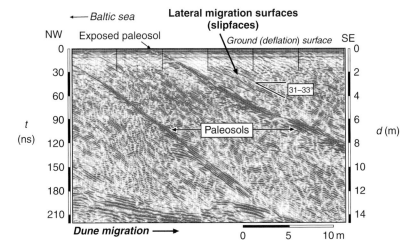

Figure 10.8 Ground penetrating radar (GPR) transect across a relict coastal dune, northern Curonian Spit, Lithuania. The geometry and the extent of two paleosols exposed on the surface are visible on the image. The numerous prominent reflections in the surrounding aeolian sands coincide with coarser laminae and heavy-mineral concentrations (HMCs), which are also exposed on the deflation surface and confirmed in trenches and sediment cores (modified from Buynevich et al., 2007a).

10.6.5. Deltas

Deltas are an important component of many marine and lacustrine coastal settings. Although large parts of marine deltas may not be suitable for GPR imaging due to signal attenuation by thick muds or pore-water salinity, a number of proglacial and lacustrine deltas have been successfully imaged. Like inland glacial or fluvial deposits, given favorable ground conditions, the combination of freshwater conditions and lithological heterogeneity makes these sequences ideal for GPR investigations. In particular, fan-foreset, wave-influenced and braided deltas reveal distinct radar facies (Jol and Smith, 1991; Jol and Smith, 1992; Smith and Jol, 1992a; Jol, 1993; Roberts et al., 2000; Smith et al., 2005; Tary et al., 2001; 2007).

A number of large sandplains in coastal Maine (Sanford and Brunswick sandplains, Pineo Ridge complex) exhibit diagnostic clinoforms (offlap; Figure 10.9) and have been interpreted as proglacial and glacial–marine deltas (Crider et al., 1997; Tary et al., 2001, 2007). These studies have revised previous interpretations of some of the Maine's sandplains, largely due to extensive GPR coverage complemented with outcrops and sediment cores. Regardless of their origin, images of clinoform geometry of proglacial or deglacial deltas provide mesoscale analogs of large deltaic sequences in the rock record.

A number of studies of relict and active lacustrine deltas demonstrate the capability of GPR to resolve key bounding surfaces and internal stratigraphy, while achieving considerable penetration (Figure 10.10; Jol and Smith, 1992; Smith and Jol, 1992a; Jol et al., 1996b; Smith et al., 2005; Stevens and Robinson, 2007). Similar to proglacial and glacial–marine deltas, reconstruction of lacustrine delta architecture can provide important insight into facies

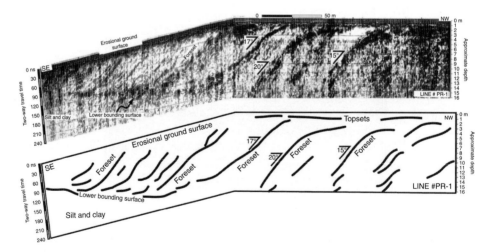

Figure 10.9 Ground penetrating radar (GPR) transect over a section of Pineo Ridge sandplain, Maine (modified from Tary et al., 2007). Variations in the amplitude of seaward-dipping reflections (clinoforms) are likely the result of lithological changes related to sediment flux. This record was collected using a GSSI SIR-3 system with a 120-MHz monostatic antenna.

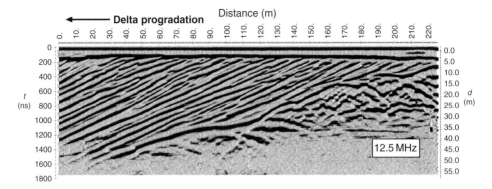

Figure 10.10 Ground penetrating radar (GPR) transect collected on an active delta of Llewellyn Glacier, northwestern British Columbia, Canada. The profile shows a thick (55–60 m) fan-foreset delta, which is prograding into Atlin Lake. The data were collected with a pulseEKKO 100A GPR system with 12.5-MHz antennae.

relationships and assess depositional styles based on clinoform geometry (Adams and Schlager, 2000). The clinoform geometry has long been the focus of seismic–stratigraphic analyses (Mitchum et al., 1977; Jol, 1988; Bhattacharya and Willis, 2001; Jol and Roberts, 1992), and the superior resolution of the GPR can provide a more detailed view of the depositional elements and key sequence-stratigraphic surfaces. Furthermore, the boundary between foreset and topset beds often resolved in subsurface records provides a useful indicator of basinal water level changes (Smith and Jol, 1997).

10.6.6. Reservoir characterization – hydrocarbon and hydrogeology

Geophysical studies in reservoir characterization of coastal settings continue to provide models for hydrocarbon and hydrogeology applications. This interest has led to GPR imaging of both modern and ancient coastal environments that aids in an improved definition and understanding of internal reservoir structure, as well as adds new insights regarding the direction of fluid migration (e.g., Figure 10.11, Knight et al., 1997). Such analysis could indicate optimum zones of reservoir permeability, which may enhance fluid recovery of hydrocarbons and also aid in the determination of volume estimates. In addition, 3D datasets can enhance our ability to interpret the depositional framework of these coastal environments and provide a more detailed perspective of the geometry of individual facies units within reservoirs (e.g., Thompson et al., 1995; Jol et al., 2002). Recently, rapid urbanization of coastal environments has initiated further research so that coastal communities can better plan for and mitigate sustainable freshwater supplies, sewage and sanitary-fill disposal, and long-term erosion and depositional problems, all of which can lead to large financial expenditures (e.g., Peterson et al., 2007).

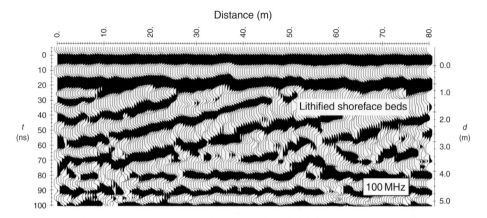

Figure 10.11 Ground Penetrating Radar (GPR) profile collected along an exposed ridge of the Panther Sandstone Tongue (Cretaceous) near Helper, Utah, USA. The profile shows continuous dipping (1–2°) reflections, which is similar to outcrop sections, and demonstrates that GPR is useful in investigating lithified coastal sedimentary rocks. Depth of penetration was limited to approximately 4–5 m, likely due to the interbedded layers of finer-grained material and cementation in the rock, both contributing to an increased attenuation of the EM signal. The data were collected with a pulseEKKO 100A GPR system with 100-MHz antennae.

10.7. Summary

Ground penetrating radar is an effective tool for imaging and interpreting modern and ancient coastal depositional environments composed of electrically resistive materials such as clean, coarse sediments (little to no clay or silt) and peats. These are important requirements since the GPR method has severe signal loss problems in fine-grained and brackish/saline groundwater conditions. For subsurface investigations of coastal sedimentary environments, GPR is presently the most promising land-based geophysical device available that provides high resolution, time and cost-effective datasets of shallow stratigraphy. Radar facies have been shown to compare well with sedimentary facies and can also provide 3D views of the subsurface. From the interpretation of GPR datasets, we can now infer sedimentary facies, directions of paleodeposition, and patterns of progradation and aggradation of sediments. With these and other benefits, GPR will continue to be used effectively in coastal geomorphic environments.

Ground penetrating radar has proven to be a valuable tool for high-resolution imaging of antecedent geology, stratigraphy, and hydrogeology of coastal systems. Although saltwater attenuation presents a significant limitation in coastal lowlands, areas with moderate to high rainfall and relatively good sediment permeability often contain considerable freshwater lenses (5–20 m), which ensure good penetration of electromagnetic signal. Varying degrees of textural and compositional heterogeneity of sediments in many coastal sequences produce the lithological contrast necessary to generate subsurface reflections. These systems provide excellent natural laboratories for effective and detailed stratigraphic analysis using GPR profiling supplemented with sediment cores. Such studies have already significantly improved our knowledge of coastal development over a wide range of temporal (years to millennia) and spatial (centimeter to tens of kilometers) scales and served to emphasize the complexity of coastal processes and resulting stratigraphic records.

The examples in this chapter indicate that GPR can accurately delineate the stratigraphy and internal sedimentary structure of coastal barriers, spits and strandplains, both above and below a freshwater–groundwater table. Sand- and/or gravel-dominated beach–dune systems on moderate to high wave energy, macrotidal coasts appear to provide optimum settings for GPR deployment. Like GPR technology, data processing software and data interpretation techniques continue to develop; the technique is likely to become a routine reconnaissance and primary data collection tool in many coastal sedimentary and geomorphic studies. Future coastal-stratigraphic research will benefit from an integrated approach that will combine GPR surveys with nearshore geophysical data and complementary sedimentologic and chronological datasets in order to enhance our understanding of coastal sedimentary systems.

ACKNOWLEDGMENTS

This chapter is based on a number of studies by the authors, with funding provided by the Geological Society of America Grant 6398-99, American Chemical Society contract 32527-AC8, Ocean and Climate Change Institute, Woods Hole Oceanographic Institution, and the Natural Sciences and Engineering Research Council of Canada. We are grateful to numerous colleagues and students for their assistance in the field. We thank the reviewers for their critical comments that greatly improved the manuscript.

REFERENCES

The following list of publications referenced in the text is not meant to be comprehensive but rather to provide a reader with a basis from where to start in the literature. GPR papers are published in a wide interdisciplinary range of journals and publications.

Adams, E.W. and Schlager, W., 2000, Basic types of submarine slope curvature. Journal of Sedimentary Research, Vol. 70, pp. 814–828.
Baker, P.L., 1991, Response of ground-penetrating radar to bounding surfaces and lithofacies variations in sand barrier sequences. Exploration Geophysics, Vol. 22, pp. 19–22.
Baker, G.S. and Jol, H.M., 2007, Stratigraphic analyses using GPR. Geological Society of American Special Publication 432, 181 p.
Barnhardt, W.A., Gonzalez, R., Kelley, J.T., Neal, W.J., Pilkey, O.H., Jr., Jose, H. and Dias, J.A., 2002, Geologic evidence for incorporation of flood tidal deltas at Tavira Island, southern Portugal. Journal of Coastal Research, Special Issue 36, pp. 28–36.
Barnhardt, W.A., Jaffe, B.E., Kayen, R.E. and Cochrane, G.R., 2004, Influence of near-surface stratigraphy on coastal landslides at Sleeping Bear Dunes National Lakeshore, Lake Michigan, USA. Journal of Coastal Research, Vol. 20, pp. 510–522.

Bascom, W.H., 1951, The relationship between sand size and beach face slope. Transactions, American Geophysical Union, Vol. 32, pp. 866–874.

Beres, M., Jr. and Haeni, F.P., 1991, Application of ground-penetrating radar methods in hydrogeologic studies. Ground Water, Vol. 29, pp. 19–22.

Beres, M., Green, A., Huggenberger, P. and Horstmeyer, H., 1995, Mapping the architecture of glaciofluvial sediments with three-dimensional georadar. Geology, Vol. 23, pp. 1087–1090.

Bhattacharya, J.P. and Willis, B.J., 2001, Lowstand deltas in the Frontier Formation, Wyoming Powder River basin, Wyoming, U.S.A.: Implications for sequence stratigraphic models. AAPG Bulletin, Vol. 85, pp. 261–294.

Bode, J.A. and Jol, H.M., 2006, GPR investigation of Hapuna Beach, Hawaii: Coastal and fluvial deposits. Proceedings of the Eleventh International Conference on Ground Penetrating Radar (GPR 2006), June 19–22, The Ohio State University, Columbus, Ohio, USA, Papers on CD-ROM.

Botha, G.A., Bristow, C.S., Porat, N., Duller, G., Armitage, S.J., Roberts, H.M., Clarke, B.M., Kota, M.W. and Schoeman, P., 2003, Evidence for dune reactivation from GPR profiles on the Maputuland coastal plain, South Africa, in Bristow, C.S. and Jol, H.M. (eds), Ground Penetrating Radar in Sediments, Geological Society of London, Special Publication 211, pp. 29–46.

Bristow, C., 1995a, Facies analysis in the Lower Greensand using ground-penetrating radar. Journal of the Geological Society, London, Vol. 152, pp. 591–598.

Bristow, C., 1995b, Internal geometry of ancient tidal bedforms revealed using ground penetrating radar, in Flemming, B.W. and Bartholoma, A. (eds), Tidal Signatures in Modern and Ancient Sediments, International Association of Sedimentologists, Special Publication, Vol. 24, pp. 313–328.

Bristow, C.S. and Pucillo, K., 2006, Quantifying rates of coastal progradation from sediment volume using GPR and OSL: The Holocene fill of Guichen Bay, south-east South Australia. Sedimentology, Vol. 53, pp. 769–788.

Bristow, C.S. and Jol, H.M. (eds), 2003, Ground penetrating radar in sediments. Geological Society of London, Special Publication 211, 366 p.

Bristow, C.S., Lancaster, N. and Duller, G.A., 2005, Combining ground penetrating radar surveys and optical dating to determine dune migration in Namibia. Journal of the Geological Society, London, Vol. 162, pp. 315–321.

Buynevich, I.V., 2003, Subsurface evidence of a pre-1846 breach across Menauhant Barrier, Cape Cod, Massachusetts. Shore and Beach, Vol. 71, pp. 3–6.

Buynevich, I.V., 2006, Coastal environmental changes revealed in geophysical images of Nantucket Island, Massachusetts, USA. Environmental and Engineering Geoscience, Vol. 12, pp. 227–234.

Buynevich, I.V. and Donnelly, J.P., 2006, Geological signatures of barrier breaching and overwash, southern Massachusetts, U.S.A. Journal of Coastal Research, Special Issue 39, pp. 112–116.

Buynevich, I.V. and FitzGerald, D.M., 2001, Styles of coastal progradation revealed in subsurface records of paraglacial barriers, New England, USA. Journal of Coastal Research, Special Issue 34, pp. 194–208.

Buynevich, I.V. and FitzGerald, D.M., 2002, Organic-rich facies in paraglacial barrier lithosomes of northern New England: Preservation and paleoenvironmental significance. Journal of Coastal Research, Special Issue 36, pp. 109–117.

Buynevich, I.V. and FitzGerald, D.M., 2003, High-resolution subsurface (GPR) imaging and sedimentology of coastal ponds, Maine, U.S.A.: Implications for Holocene back-barrier evolution. Journal of Sedimentary Research, Vol. 73, pp. 550–571.

Buynevich, I.V., Evans, R.L. and FitzGerald, D.M., 2003, High-resolution geophysical imaging of buried inlet channels. Proceedings of the 6th International Conference on Coastal Sediments, May 18–23, 2003, CD-ROM. Published by World Scientific Publishing Corp and East Meets West Productions, Inc., Corpus Christi, TX, USA, ISBN 981-238-422-7.

Buynevich, I.V., FitzGerald, D.M. and van Heteren, S., 2004, Sedimentary records of intense storms in Holocene barrier sequences, Maine, USA. Marine Geology, Vol. 210, pp. 135–148.

Buynevich, I.V., Asp, N.E., FitzGerald, D.M., Cleary, W.J., Klein, A.H.F., Siegle, E. and Angulo, R., 2005, Mud in the surf: Nature at work in a Brazilian bay. Eos Transactions, AGU, Vol. 86, pp. 301 and 304.

Buynevich, I.V., Bitinas, A. and Pupienis, D., 2007a, Lithological anomalies in a relict coastal dune: geophysical and paleoenvironmental markers. Geophysical Research Letters, Vol. 34, L09707, doi:10.1029/2007GL029767.

Buynevich, I.V., Bitinas, A. and Pupienis, D., 2007b, Reactivation of coastal dunes documented by subsurface imaging of the Great Dune Ridge, Lithuania. Journal of Coastal Research, Special Issue 50, pp. 226–230.

Buynevich, I.V., FitzGerald, D.M. and Goble, R.J., 2007c. A 1,500-year record of North Atlantic storm activity based on optically dated relict beach scarps. Geology, Vol. 35, pp. 543–546.

Clarke, M.L., Rendell, H.M., Tastet, J.-P., Clavé, B. and Massé, L., 2002, Late-Holocene sand invasion and North Atlantic storminess along the Aquitaine Coast, southwest France. Holocene, Vol. 12, pp. 231–238.

Clemmensen, L.B., Andreasen, F., Nielsen, S.T. and Sten, E., 1996, The late Holocene coastal dunefield at Vejers, Denmark: Characteristics, sand budget and depositional dynamics. Geomorphology, Vol. 17, pp. 79–98.

Clemmensen, L.B., Andreasen, F., Heinemeier, J. and Murray, A., 2001, A Holocene coastal aeolian system, Vejers, Denmark: Landscape evolution and sequence stratigraphy. Terra Nova, Vol. 13, pp. 129–134.

Conyers, L.B. and Goodman, D., 1997, Ground-penetrating Radar: An Introduction to Archaeologists, Walnut Creek, Altamira Press, 232 p.

Costas, S., Alejo, I., Rial, F., Lorenzo, H. and Nombela, M.A., 2006, Cyclical evolution of a modern transgressive sand barrier in northwestern Spain elucidated by GPR and aerial photos. Journal of Sedimentary Research, Vol. 76, pp. 1077–1092.

Crider, H.B., FitzGerald, D.M., Buynevich, I.V. and Weddle, T.K., 1997, Evidence for late Pleistocene drainage of Androscoggin River through Thomas Bay, Maine: GSA Northeastern Section Abstracts with Programs, King of Prussia, Pennsylvania, Vol. 29, p. 38.

Daly, J., McGeary, S. and Krantz, D.E., 2002, Ground-penetrating radar investigation of a late Holocene spit complex: Cape Henlopen, Delaware. Journal of Coastal Research, Vol. 18, pp. 274–286.

Daniels, D.J., Gunton, D.J. and Scott, H.F., 1988, Introduction to subsurface radar.IEEE Proceedings, Vol. 135, part F, No. 4, pp. 277–320.

Davis, J.L. and Annan, A.P., 1989, Ground penetrating radar for high resolution mapping of soil and rock stratigraphy. Geophysical Prospecting, Vol. 37, pp. 531–551.

FitzGerald, D.M., Baldwin, C.T., Ibrahim, N.A. and Humphries, S.M., 1992, Sedimentologic and morphologic evolution of a beach-ridge barrier along an indented coast: Buzzards Bay, Massachusetts, in Fletcher, C.H., III and Wehmiller, J.F. (eds), Quaternary Coasts of the United States: Marine and Lacustrine Systems, SEPM-IGCP Special Publication 48, pp. 65–75.

FitzGerald, D.M., Buynevich, I.V. and Rosen, P.S., 2001, Geological evidence of former tidal inlets along a retrograding barrier: Duxbury Beach, Massachusetts, USA. Journal of Coastal Research, Special Issue 34, pp. 437–448.

FitzGerald, D.M., Cleary, W.J., Buynevich, I.V., Hein, C., Klein, A.H.F., Asp, N.E. and Angulo, R., 2007, Strandplain evolution along the southern coast of Santa Catarina, Brazil. Journal of Coastal Research, Special Issue 50, pp. 152–156.

Gawthorpe, R.L., Collier, R.E.Ll., Alexander, J., Leeder, M. and Bridge, J.S., 1993, Ground penetrating radar: Application to sandbody geometry and heterogeneity studies, in North, C.P. and Prosser, D.J. (eds), Characteristaion of Fluvial and Aeolian reservoirs, Geological Society of London, Special Publications, Vol. 73, pp. 421–432.

Grasmueck, M. and Weger, R., 2002, 3D GPR reveals complex internal structure of Pleistocene oolitic sandbar: Leading Edge of Exploration. Society of Exploration Geophysics, Vol. 21, pp. 634–639.

Grasmueck, M., Weger, R. and Horstmeyer, H., 2004, Three-dimensional ground-penetrating radar imaging of sedimentary structures, fractures, and archeological features at submeter resolution. Geology, Vol. 32, pp. 933–936.

Grasmueck, M., Weger, R. and Horstmeyer, H., 2005, Full-resolution 3-D GPR imaging. Geophysics, Vol. 70, pp. K12–K10.

Guha, S., Kruse, S.E., Wright, E.E. and Kruse, U.E., 2005, Spectral analysis of ground penetrating radar response to thin sedimentary layers. Geophysical Research Letters, Vol. 32, p. L23304, doi:10.1029/2005GL023933.

Havholm, K.G., Ames, D.V., Whittecar, G.R., Wenell, B.A., Riggs, S.R., Jol., H.M., Berger, G.W. and Holmes, M.A., 2004, Stratigraphy of back-barrier coastal dunes, northern North Carolina and southern Virginia. Journal of Coastal Research, Vol. 20, pp. 980–999.

Hesp, P.A. and Thom, B.G., 1990, Geomorphology and evolution of transgressive dunefields, in Nortdstrom, K., Psuty, N. and Carter, R.W.G. (eds), Coastal Dunes: Processes and Morphology, John Wiley and Sons, Chichester, pp. 253–288.

Huggenberger, P., 1993, Radar facies: Recognition of characteristic braided river structures of the Pleistocene Rhine gravel (NE part of Switzerland), in Best, J. and Bristow, C.S. (eds), Braided Rivers, Geological Society of London, Special Publications, Vol. 75, pp. 163–175.

Johnston, J.W., Thompson, T.A. and Baedke, S.J., 2007, Systematic pattern of beach-ridge development and preservation: conceptual model and evidence from ground penetrating radar, in Baker, G.S. and Jol, H.M., Stratigraphic Analysis Using GPR. Geological Society of America Special Paper 432, pp. 47–58.

Jol, H.M., 1995, Ground penetrating radar antennae frequencies and transmitter powers compared for penetration depth, resolution and reflection continuity. Geophysical Prospecting, Vol. 43, pp. 693–709.

Jol, H.M., 1993, Ground penetrating radar (GPR): A new geophysical methodology used to investigate the internal structure of sedimentary deposits (field experiments on lacustrine deltas) [Ph.D. Thesis], Calgary, Alberta, University of Calgary, 135 p.

Jol, H.M., 1988, Seismic stratigraphic analysis of the southeastern Fraser river delta, British Columbia [M.S. Thesis], Burnaby, British Columbia, Simon Fraser University, 176 p.

Jol, H.M. and Bristow, C.S., 2003, GPR in sediments: advice on data collection, basic processing and interpretation, a good practice guide, in Bristow, C.S. and Jol, H.M. (eds), GPR in Sediments, Geological Society of London, Special Publication 211, pp. 9–28.

Jol, H.M. and Roberts, M.C., 1992, The seismic facies of a tidally influenced Holocene delta: Boundary Bay, Fraser River delta, B.C. Sedimentary Geology, Vol. 77, pp. 173–183.

Jol, H.M. and Peterson, C.D., 2006, Imaging earthquake scarps and tsunami deposits in the Pacific Northwest, USA, in Proceedings of the Symposium on the Application of Geophysics to Engineering and Environmental Problems (SAGEEP), 19th Annual Meeting, Seattle, Washington, April 2–6, Papers on CD-ROM, pp. 217–229.

Jol, H.M. and Smith, D.G., 1995, Ground penetrating radar of peatlands for oil field pipelines in Canada. Journal of Applied Geophysics, Vol. 34, pp. 109–123.

Jol, H.M. and Smith, D.G., 1992, Geometry and structure of deltas in large lakes: A ground penetrating radar overview, in Hǒnninen, P. and Autio, S. (eds), Fourth International Conference on Ground Penetrating Radar, June 8–13, Rovaniemi, Finland, Geological Survey of Finland, Special Paper 16, pp. 159–168.

Jol, H.M. and Smith, D.G., 1991, Ground penetrating radar of northern Lacustrine deltas. Canadian Journal of Earth Sciences, pp. 28, 1939–1947.

Jol, H.M., Peterson, C.D., Vanderburgh, S. and Phipps, J., 1998a, GPR as a regional geomorphic tool: shoreline accretion/erosion along the Columbia river littoral cell. Proceedings of the Seventh International Conference on Ground Penetrating Radar (GPR'98), May 27-30, University of Kansas, Lawrence, KS, USA, Vol. 1, pp. 257–262.

Jol, H.M., Smith, D.G. and Meyers, R.A., 1996a. Digital ground penetrating radar (GPR): An improved and very effective geophysical tool for studying modern coastal barriers (examples for the Atlantic, Gulf and Pacific coasts, U.S.A.). Journal of Coastal Research, Vol. 12, pp. 960–968.

Jol, H.M., Smith, D.G. and Meyers, R.A., 1996b, Three dimensional GPR imaging: of a fan-foreset delta: An example from Brigham City, Utah, U.S.A. Proceedings of the Sixth International Conference on Ground Penetrating Radar (GPR'96), September 30–October 3, Sendai, Japan, pp. 33–37.

Jol, H.M., Smith, D.G. and Meyers, R.A., 1994, Ground penetrating radar of lakeshore spits in northwestern Saskatchewan, Canada: Variable internal structure and inferred depositional process. Proceeding of the Fifth International Conference on Ground Penetrating Radar, Kitchener, Ontario, June 12–16, Vol. 2, pp. 817–830.

Jol, H.M., Smith, D.G., Meyers, R.A. and Lawton, D.C., 1996c, Ground penetrating radar: High resolution stratigraphic analysis of coastal and fluvial environments, in Pacht, J.A., Sheriff, R.E. and Perkins, B.F. (eds), Stratigraphic Analysis using Advanced Geophysical, Wireline and Borehole Technology for Petroleum Exploration and Production, Gulf Coast Section Society of Economic Paleontologists and Mineralogists Foundation 17th Annual Research Conference, pp. 153–163.

Jol, H.M., Vanderburgh, S. and Havholm, K.G., 1998b, GPR studies of coastal aeolian (foredune and crescentic) environments: Examples from Oregon and North Carolina, U.S.A. Proceedings of the Seventh International Conference on Ground Penetrating Radar (GPR'98), May 27–30, University of Kansas, Lawrence, KS, USA, Vol. 2, pp. 681–686.

Jol, H.M., Lawton, D. and Smith, D.G., 2002, Ground penetrating radar: 2D and 3D examples from Long Beach Washington. Geomorphology, Vol. 53, pp. 165–181.

Jol, H.M., Grote, K.R., Smith, D.G. and Smith, N.D., 2006, Investigation of data quality problems in portions of the William River Delta, Saskatchewan, Canada, in Proceedings of the Eleventh International Conference on Ground Penetrating Radar (GPR 2006), June 19–22, The Ohio State University, Columbus, Ohio, USA, Papers on CD-ROM.

Knight, R. Tercier, P. and Jol, H., 1997, The role of ground penetrating radar and geostatistics in reservoir description. The Leading Edge, Vol. 16, pp. 1576–1582.

Komar, P.D. and Wang, C., 1984, Processes of selective grain transport and the formation of placers on beaches. Journal of Geology, Vol. 92, pp. 637–655.

Kruse, S.E. and Jol, H.M., 2003, Amplitude and waveform analysis of repetitive GPR reflections: A Lake Bonneville Delta, Utah, in: Bristow, C.S. and Jol, H.M. (eds), GPR in Sediments, Geological Society of London, Special Publication 211, pp. 287–298.

Leatherman, S.P., 1987, Coastal geomorphological applications of ground penetrating radar. Journal of Coastal Research, Vol. 3, pp. 397–399.

Masselink, G. and Puleo, J.A., 2006, Swash-zone morphodynamics. Continental Shelf Research, Vol. 26, pp. 661–680.

Mellett, J.S., 1995, Profiling of ponds and bogs using ground-penetrating radar. Journal of Paleolimnology, Vol. 14, pp. 233–240.

Meyers, R., Smith, D.G., Jol, H.M. and Peterson, C.R., 1996, Evidence for eight great earthquake-subsidence events detected with ground-penetrating radar, Willapa barrier, Washington. Geology, pp. 24, 99–102.

Meyers, R.A., Smith, D.G., Jol, H.M. and Hay, M.B. 1994, Internal structure of Pacific Coast barrier spits using ground penetrating radar. Proceeding of the Fifth International Conference on Ground Penetrating Radar, Kitchener, Ontario, June 12–16, Vol. 2, pp. 831–842.

Meyers, R.A., 1994, Willapa barrier spit of S.W. Washington State: Depositional processes inferred from ground penetrating radar [M.S. Thesis], University of Calgary, Calgary, Alberta, 101 p.

Mitchum, R.M., Jr., Vail, P.R. and Sangree, J.B., 1977, Stratigraphic interpretation of seismic reflection patterns in depositional sequences, in Payton, C.E., (ed.), Seismic Stratigraphy – Application to Hydrocarbon Exploration, AAPG Memoir 26, pp. 117–133.

Moller, I. and Anthony, D., 2003, GPR study of sedimentary structure within a transgressive coastal barrier along the Danish North Sea Coast, in Bristow, C.S. and Jol, H.M. (eds), GPR in Sediments, Geological Society of London, Special Publication 211, pp. 55–65.

Moore, L.J., Jol, H.M., Kaminsky, G.M. and Kruse, S., 2003a, Severe winter storm effects revealed in stratigraphy of prograding coastal barrier, Southwest Washington, USA. Proceedings of the 6th International Conference on Coastal Sediments 2003, May 18–23, CD-ROM. Published by World Scientific Publishing Corp and East Meets West Productions, Inc., Corpus Christi, TX, USA, ISBN 981-238-422-7.

Moore, L.J., Kaminsky, G.M. and Jol, H.M., 2003b, Linkages between shoreline progradation and El Nino southern oscillations, Southwest Washington, USA. Geophysical Research Letters, Vol. 39, p. 1448, doi:10.1029/2002GL016147.

Moore, L.J., Jol, H.M., Kruse, S., Vanderburgh, S. and Kaminsky, G.M., 2004, Annual layers revealed in the subsurface of a prograding coastal barrier. Journal of Sedimentary Research, Vol. 74, pp. 690–696.

Moslow, T.F. and Heron, S.D., Jr., 1978, Relict inlets: Preservation and occurrence in the Holocene stratigraphy of southern Core Banks, North Carolina. Journal of Sedimentary Petrology, Vol. 48, pp. 1275–1286.

Moslow, T.F. and Tye, R.S., 1985, Recognition and characterization of Holocene tidal inlet sequences. Marine Geology, Vol. 63, pp. 129–152.

Neal, A., Pontee, N.G., Pye, K. and Richards, J., 2002, Internal structure of mixed-sand-and-gravel beach deposits revealed using ground-penetrating radar. Sedimentology, Vol. 49, pp. 789–804.

Neal, A., Richards, J. and Pye, K., 2003, Sedimentology of coarse-clastic beach-ridge deposits, Essex, southeast England. Sedimentary Geology, Vol. 162, pp. 167–198.

Neal, A. and Roberts, C.L., 2000, Applications of ground penetrating radar (GPR) to sedimentological, geomorphological and geoarchaeological studies in coastal environments, in Pye, K. and Allen, J.R.L. (eds), Coastal and Estuarine Environments: Sedimentology, Geomorphology and Geoarchaeology, Geological Society of London, Special Publications, Vol. 175, pp. 139–171.

Nichol, S.L., 2002, Morphology, stratigraphy and origin of Last Interglacial beach ridges at Bream Bay, New Zealand. Journal of Coastal Research, Vol. 18, pp. 149–159.

Olsen, H. and Andreasen, F., 1995, Sedimentology and ground-penetrating radar characteristics of a Pleistocene sandur deposit. Sedimentary Geology, Vol. 99, pp. 1–15.

O'Neal, M.L. and Dunn, R.K., 2003, GPR investigation of multiple stage-5 sea-level fluctuations on a siliciclastic estuarine shoreline, Delaware Bay, southern New Jersey, USA, in Bristow, C.S. and Jol, H.M. (eds), GPR in Sediments, Geological Society of London, Special Publication 211, pp. 67–77.

Peterson, C.D., Jol, H.M., Percy, D. and Nielsen, E.L., 2007, Groundwater surface trends from ground penetrating radar (GPR) profiles taken across Late Holocene barriers and beach plains of the Columbia River Littoral system, Pacific Northwest Coast, in Baker, G.S. and Jol, H.M., Stratigraphic Analysis Using GPR, Geological Society of America Special Paper 432, pp. 59–76.

Roberts, M.C. and Jol, H.M., 2000, The sedimentary architecture and geomorphology of a cuspate spit: Tsawwassen, British Columbia, in Völkel, J. and Barth, H.J. (eds), Regensburger Geographische Schriften, Beiträge zur Quartärforschung, Vol. 33, pp. 141–156.

Roberts, M.C., Niller, H.P. and Helmstetter, N., 2003, Sedimentary architecture and radar facies of fan delta, Cypress Creek, West Vancouver, British Columbia, in Bristow, C.S. and Jol, H.M. (eds), GPR in Sediments, Geological Society of London, Special Publication 211, pp. 11–126.

Roberts, M.C., Vanderburgh, S. and Jol, H., 2000, Radar facies and geomorphology of the seepage face of the Brookswood aquifer, Fraser Lowland, in Ricketts, B.D. (ed.), Mapping, Geophysics, and Groundwater Modeling in Aquifer Delineation, Fraser Lowland and Delta, British Columbia, Geological Survey of Canada Bulletin, Vol. 552, pp. 95–102.

Rodriguez, A.B. and Meyer, C.T., 2006, Sea-level variation during the Holocene deduced from the morphologic and stratigraphic evolution of Morgan Peninsula, Alabama, U.S.A. Journal of Sedimentary Research, Vol. 76, pp. 257–269.

Schenk, C.J., Gautier, D.L., Olhoeft, G.R. and Lucius, J.E., 1993, Internal structure of an aeolian dune using ground-penetrating radar, in Pye, K. and Lancaster, N. (eds), Aeolian Sediments Ancient and Modern, International Association of Sedimentologists, Special Publications, Vol. 16, pp. 61–69.

Smith, D.G. and Jol, H.M., 1997, Ground penetrating radar investigation of the Peyto Delta. Sedimentary Geology, Vol. 113, pp. 195–209.

Smith, D.G. and Jol, H.M., 1995. Ground penetrating radar: Antennae frequencies and maximum probable depths of penetration in Quaternary sediments. Journal of Applied Geophysics, Vol. 33, pp. 93–100.

Smith, D.G. and Jol, H.M., 1992a, Ground penetrating radar investigation of a Lake Bonneville delta, Provo level, Brigham City, Utah. Geology, Vol. 20, pp. 1083–1086.

Smith, D.G. and Jol, H.M., 1992b, GPR results used to infer depositional processes of coastal spits in large lakes, in Hänninen, P. and Autio, S. (eds), Fourth International Conference on Ground Penetrating Radar, June 8–13, Rovaniemi, Finland, Geological Survey of Finland, Special Paper 16, pp. 169–177.

Smith, D.G., Meyers, R.A. and Jol, H.M., 1999, Sedimentology of an upper-mesotidal (3.7 m) Holocene barrier, Willapa Bay, SW Washington, U.S.A. Journal of Sedimentary Research, Vol. 69, pp. 1290–1296.

Smith, D.G., Simpson, C.J., Jol, H.M., Meyers, R.A. and Currey, D.R., 2003, Radar stratigraphy used to infer transgressive or regressive deposition and internal structure of a barrier and spit, Lake

Bonneville, Stockton, Utah. USA, in Bristow, C.S. and Jol, H.M. (eds), GPR in Sediments, Geological Society of London, Special Publication 211, pp. 79–86.

Smith, D.G., Jol, H.M., Smith, N.D., Kostachuk, R.A. and Pearce, C.M., 2005, The wave-dominated William River Delta, Lake Athabasca, Canada: its morphology, radar stratigraphy, and history, in Giosan, L. and Bhattacharya, J.P., River Deltas – Concepts, Models and Examples, SEPM Special Publication #83, pp. 295–318.

Stevens, C.W. and Robinson, S.D., 2007, The internal structure of relict lacustrine deltas, northern New York, in Baker, G.S. and Jol, H.M., Stratigraphic Analysis Using GPR, Geological Society of America Special Paper 432, pp. 93–102.

Storms, J.E.A. and Kroonenberg, S.B., 2007, The impact of rapid sea level changes in recent Azerbaijan beach ridges. Journal of Coastal Research, Vol. 23, pp. 521–527.

Tanner, W.F., 1995, Origin of beach ridges and swales. Marine Geology, Vol. 129, pp. 149–161.

Tary, A.K., FitzGerald, D.M. and Buynevich, I.V., 2001, Late Quaternary morphogenesis of a marine-limit delta plain, southwest Maine, in Weddle, T.K and Retelle, M.J. (eds), Deglacial history and relative sea-level changes, northern New England and adjacent Canada, Geological Society of America Special Paper 351, pp. 125–149.

Tary, A.K., FitzGerald, D.M. and Weddle, T.K., 2007, A ground penetrating radar investigation of a glacial-marine ice-contact delta, Pineo Ridge, eastern coastal Maine, in Baker, G.S. and Jol, H.M., Stratigraphic Analysis Using GPR, Geological Society of America Special Paper 432, pp. 77–92.

Tercier, P., Knight, R. and Jol, H., 2000, A comparison of the correlation structure in GPR images of deltaic and barrier spit depositional environments. Geophysics, Vol. 65, pp. 1142–1153.

Thompson, C., McMechan, C., Szerbiak, R. and Gaynor, N. 1995, Three-dimensional GPR imaging of complex stratigraphy within the Ferron sandstone, Castle Valley, Utah. Proceedings of the Symposium on the Application of Engineering and Environmental Problems, pp. 435–443.

Topp, G.C., Davis, J.L. and Annan, A.P., 1980, Electromagnetic determination of soil water content: Measurements in coaxial transmission lines. Water Resources Research, Vol. 16, pp. 574–582.

Truman, C.C., Perkins, H.F., Asmussen, L.E. and Allison, H.D., 1988, Some applications of ground-penetrating radar in the southern coastal plains region of Georgia: The Georgia Agricultural Experiment Stations, College of Agriculture, The University of Georgia, 27 p.

Ulriksen, C.P.F., 1982, Application of impulse radar to civil engineering [Ph.D. Thesis]: Lund, Sweden: Lund University of Technology, (republished by Geophysical Survey Systems Inc., Hudson, Hew Hampshire), 175 p.

Van Dam, R.L. and Schlager, W., 2000, Identifying causes of ground-penetrating radar reflections using time-domain reflectometry and sedimentological analyses. Sedimentology, Vol. 47, pp. 435–449.

Van Dam, R.L., Van Den Berg, E.H., Van Heteren, S., Kasse, C., Kenter, J.A.M. and Groen, K., 2002, Influence of organic matter on radar-wave reflection: Sedimentological implications. Journal of Sedimentary Research, Vol. 72, pp. 341–352.

van Dam, R.L., Nichol, S.L., Augustinus, P.C., Parnell, K.E., Hosking, P.L. and McLean, R.F., 2003, GPR stratigraphy of a large active dune on Parengarenga Sandspit, New Zealand. The Leading Edge, Vol. 22, pp. 865–870.

van Heteren, S. and van de Plassche, O., 1997, Influence of relative sea-level change and tidal-inlet development on barrier-spit stratigraphy, Sandy Neck, Massachusetts. Journal of Sedimentary Research, Vol. 67, pp. 350–363.

van Heteren, S., FitzGerald, D.M., Barber, D.C., Kelley, J.T. and Belknap, D.F., 1996, Volumetric analysis of a New England barrier system using ground-penetrating-radar and coring techniques. The Journal of Geology, Vol. 104, pp. 471–483.

van Heteren, S., FitzGerald, McKinaly, P.A. and Buynevich, I.V., 1998, Radar facies of paraglacial barrier systems: Coastal New England, USA. Sedimentology, Vol. 45, p. 181–200.

van Heteren, S., FitzGerald, D.M. and McKinlay, P.A., 1994, Application of ground penetrating radar in coastal stratigraphic studies. Proceedings of the Fifth International Conference on Ground Penetrating Radar, Vol. 2, pp. 869–881.

van Heteren, S., Huntley, D.J., van der Plassche, O. and Lubberts, R.K., 2000, Optical dating of dune sand for the study of sea-level change. Geology, Vol. 104, pp. 411–414.

van Overmeeren, R.A., 1998, Radar facies of unconsolidated sediments in The Netherlands – a radar stratigraphy interpretation method of hydrogeology. Journal of Applied Geophysics, Vol. 40, pp. 1–18.
Wilkins, D.E. and Clement, W.P., 2007, Palaeolake shoreline sequencing using ground penetrating radar: Lake Alvord, Oregon and Nevada, in Baker, G.S. and Jol, H.M., Stratigraphic Analysis Using GPR, Geological Society of America Special Paper 432, pp. 103–110.
Wilson, P., Orford, J.D., Knight, J., Braley, S.M. and Wintle, A.G., 2001, Late-Holocene (post-4000 years BP) coastal dune development in Northumberland, northeast England. The Holocene, Vol. 11, pp. 215–229.
Wyatt, D.E. and Temples, T.J., 1996, Ground-penetrating radar detection of small-scale channels, joints and faults in the unconsolidated sediments of the Atlantic coasts plain. Environmental Geology, Vol. 27, pp. 219–225.
Zenero, R.R., Seng, D.L., Byrnes, R.R. and McBride, R.A., 1995, Geophysical techniques for evaluating the internal structure of cheniers, southwestern Louisiana. Transactions of the Gulf Coast Association of Geological Societies, Vol. 45, pp. 611–620.

CHAPTER 11

ADVANCES IN FLUVIAL SEDIMENTOLOGY USING GPR

John Bridge

Contents

11.1. Introduction	323
11.2. Scales of Fluvial Deposits and GPR Resolution	324
11.3. Examples of Use of GPR in Fluvial Sedimentology	327
11.3.1. South Esk, Scotland	327
11.3.2. Calamus, Nebraska	329
11.3.3. Brahmaputra (Jamuna), Bangladesh	331
11.3.4. Niobrara, Nebraska	336
11.3.5. South Saskatchewan, Canada	340
11.3.6. Sagavanirktok, northern Alaska	343
11.3.7. Fraser and Squamish Rivers, Canada	349
11.3.8. Pleistocene outwash deposits in Europe	350
11.3.9. Mesozoic deposits of SW USA	353
11.4. Concluding Discussion	354
Acknowledgments	355
References	355

11.1. INTRODUCTION

Understanding of river deposits has improved significantly over the past two decades, leading to the development of new and improved fluvial depositional models (reviews by Bridge, 2003, 2006; Bridge and Lunt, 2006). This progress has been facilitated in no small measure by the use of ground penetrating radar (GPR) in combination with cores, trenches, natural outcrops and quarry faces to describe fluvial deposits in detail and in 3D. GPR has been particularly useful for describing unconsolidated recent and Quaternary fluvial deposits both above and below the water table (e.g., Jol and Smith, 1991; Gawthorpe et al., 1993; Huggenberger, 1993; Huggenberger et al., 1994; Olsen and Andreasen, 1995; Bridge et al., 1995, 1998; Beres et al., 1995, 1999; Roberts et al., 1997; Asprion and Aigner, 1997, 1999; Van Overmeeren, 1998; Bristow et al., 1999; Vandenberghe and Van Overmeeren, 1999; Regli et al., 2002; Best et al., 2003; Heinz and Aigner, 2003; Skelly et al., 2003; Woodward et al., 2003; Lunt and Bridge, 2004; Lunt et al., 2004a, 2004b; Wooldridge and Hickin, 2005; Sambrook Smith et al., 2005, 2006; Best et al., 2006; Bakker et al.,

2007; Kostic and Aigner, 2007). Also, GPR has been used for imaging near-surface rocks linked to large nearby outcrops (e.g. Stephens, 1994; McMechan et al., 1997; Szerbiak et al., 2001; Corbeanu et al., 2001, 2004). In general, GPR allows high-resolution, 2D and 3D imaging (vertical resolution up to the order of a decimeter) of the upper 10 m of deposits. GPR data can be acquired in real time, and subsequent processing is commonly minimal. The character of radar reflections can be linked closely to the character of the sedimentary strata.

The organization of this paper is (1) review of different scales of fluvial deposits and ability of GPR to resolve them; (2) examples of how GPR (plus supplementary information) has facilitated description of fluvial deposits, and development of depositional models; (3) discussion of benefits and potential problems in the use of GPR, and future applications.

11.2. Scales of Fluvial Deposits and GPR Resolution

Different scales of strataset occur in river deposits, depending on the scale of the associated topographic feature (e.g. bar, dune), the time and spatial extent over which deposition occurred, and the degree of preservation. Typical scales of strataset (Figure 11.1) that are relevant here are (1) a complete channel belt (e.g. a sand-gravel body); (2) deposits of individual compound channel bars and channel fills (sets of compound large-scale inclined strata, also known as storeys); (3) deposits of unit bars that comprise compound bars and channel fills (sets of simple large-scale inclined strata); (4) depositional increments (large-scale inclined strata) on channel bars and in channel fills formed during distinct floods; and (5) depositional increments associated with the passage of bed forms such as dunes, ripples and bed-load sheets (sets of medium-scale and small-scale cross strata and planar strata). The geometry of a particular scale of strataset is related to the geometry and migration of the associated bedform. In particular, the length-to-thickness ratio of stratasets is similar to the wavelength-to-height ratio of associated bedforms (Figure 11.2; Bridge and Lunt, 2006). Furthermore, the wavelength and height of bedforms such as dunes and bars are related to channel depth and width. Therefore, the thickness of a particular scale of strataset (i.e. medium-scale cross sets and large-scale sets of inclined strata) will vary with river dimensions.

GPR reflections are due to changes in dielectric permittivity (and less commonly because of magnetic permeability), related to the amount and type of pore-filling material, sediment texture and composition (Van Dam and Schlager, 2000; Kowalsky et al., 2001, 2004; Van Dam et al., 2002; Lunt et al., 2004a, 2004b). As reflections are primarily caused by changes in pore-water saturation or volume, which are closely related to sediment texture and composition, reflections give a record of sedimentary strata. The amplitude of radar reflections depends on contrasts in radar velocity between adjacent strata and on stratal thickness compared with the wavelength of the transmitted pulse (Kowalsky et al., 2001). A particular reflection amplitude can be caused by different combinations of sediment texture (e.g. sand adjacent to sandy gravel, open-framework gravel adjacent to sandy

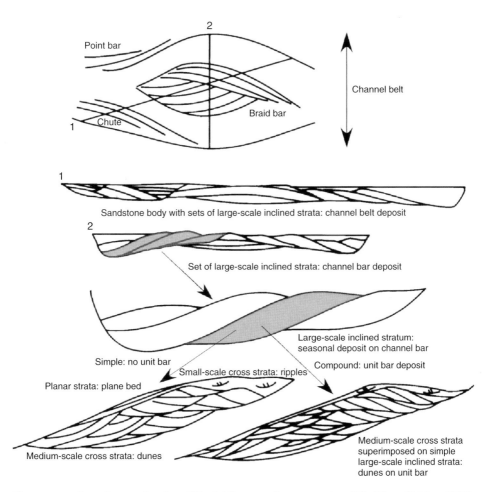

Figure 11.1 Superimposed scales of fluvial forms and stratasets (modified from Bridge, 2003). Cross sections (1) and (2) through an idealized braided channel belt show several sets of large-scale inclined strata formed by deposition on channel bars. Each large-scale inclined stratum can be simple (deposited during a single flood) or compound (deposited as a unit bar over one or more floods). Large-scale inclined strata contain smaller-scale stratasets associated with passage of ripples, dunes and bedload sheets over bars.

gravel). Above the water table, reflections are primarily related to variations in water saturation. Below the water table, reflections are related mainly to variations in porosity and sediment composition. In permafrost, reflections are related to porosity and relative proportions of water and ice in the pores.

Central frequencies of antennae commonly used in sedimentological studies are 50–1000 MHz. For a given subsurface sediment type, the depth of penetration decreases, and the resolution of sedimentary strata increases, with increasing antenna frequency (Jol, 1995; Woodward et al., 2003). For example, 100 MHz antennae may have a penetration depth on the order of 10 m and be able to resolve

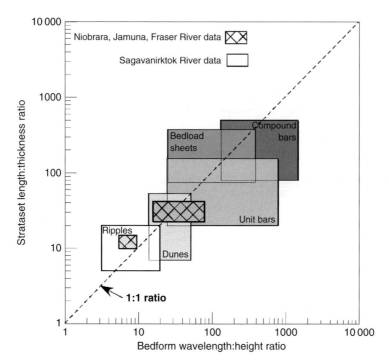

Figure 11.2 Bedform length/height plotted against corresponding strataset length/thickness (from Bridge and Lunt, 2006). Bedform geometry was taken from aerial photos or descriptions, and strataset dimensions were taken from GPR profiles and descriptions. Some bedform dimensions were measured at low flow stage, and may not represent formative bedform dimensions.

strata 0.3 m thick. 450-MHz antenna may have depth of penetration of several meters and be able to resolve strata 0.1 m thick. In order to distinguish all scales of stratification, it is desirable to use a range of antenna frequencies (50–1000 MHz). As an example, cross strata that are centimeters thick within cross sets on the order of decimeters thick can be discerned using high-frequency antennae. With low-frequency antennae, only the set boundaries would be represented. If the set thickness of cross strata is less than a decimeter or so, even the high-frequency antennae will only pick up the set boundaries. As a result, GPR has been most useful for imaging the larger scales of strata in channel belts. Furthermore, radar facies cannot be clearly related to sedimentary facies unless their 3D geometry is investigated using a range of antenna frequencies, the antenna frequency is stated, and reasons for variation in reflection amplitude are understood. For example, concave-upward reflections in radar profiles may be associated with confluence scours, main channel fills, small cross-bar channels adjacent to unit bars, scours upstream of obstacles such as logs or ice blocks, or bases of trough cross strata. Interpretation of the origin of such reflections will depend on their 3D geometry and an understanding of adjacent strata. The central antenna frequency and vertical exaggeration of GPR profiles need to be considered when making sedimentological interpretations.

11.3. Examples of Use of GPR in Fluvial Sedimentology

GPR studies of modern river deposits have been conducted on the Madison River, Montana (Gawthorpe et al., 1993; Alexander et al., 1994); River South Esk, Scotland (Bridge et al., 1995); Rhone River in France (Roberts et al., 1997); floodplains and alluvial fans in British Columbia (Leclerc and Hickin, 1997; Ekes and Hickin, 2001; Ekes and Friele, 2003); Calamus River in Nebraska, USA (Bridge et al., 1998); Niobrara River in Nebraska, USA (Bristow et al., 1999; Skelly et al., 2003); Brahmaputra River in Bangladesh (Best et al., 2003); Sagavanirktok River on the North Slope of Alaska (Lunt and Bridge, 2004; Lunt et al., 2004a, 2004b); South Saskatchewan River in Canada (Woodward et al., 2003; Sambrook Smith et al., 2005, 2006; Best et al., 2006); Squamish and Fraser Rivers in Canada (Wooldridge, 2002; Wooldridge and Hickin, 2005). With modern river studies, it is possible to relate deposits to the associated bedforms and their migration. GPR studies of ancient deposits include Holocene–Pleistocene fluvial deposits in the Netherlands (Van Overmeeren, 1998; Vandenberghe and Van Overmeeren, 1999; Bakker et al., 2007); Pleistocene braided outwash in Denmark, Switzerland and Germany (Huggenberger, 1993; Siegenthaler and Huggenberger, 1993; Olsen and Andreasen, 1995; Beres et al., 1995, 1999; Asprion and Aigner, 1997, 1999; Heinz and Aigner, 2003; Kostic and Aigner, 2007); Mesozoic rocks in southwestern USA (Stephens, 1994; McMechan et al., 1997; Szerbiak et al., 2001; Corbeanu et al., 2001, 2004). With studies of ancient deposits, we cannot be sure of the interpretations of the relationship between deposits and topographic features. The main points that came from some of these studies are discussed below.

11.3.1. South Esk, Scotland

The River South Esk in Glen Clova, Scotland, is a meandering river with sandy and gravelly point-bar deposits (Figure 11.3). Previous detailed studies of channel geometry, flow and sediment transport, and the nature of channel migration and point bar deposition were undertaken by Bridge and Jarvis (1976, 1977, 1982). The deposits were described using GPR (300 MHz antennae), vibracores and trenches (Bridge et al., 1995). Channel migration is by bend expansion and downstream migration, producing point-bar sequences that fine upwards and downstream. Radar facies were linked closely to sedimentary facies. Lower point-bar deposits are associated with low-amplitude, relatively discontinuous (across-stream length up to a meter, and along-stream length of meters) curved GPR reflections, corresponding to the bases of medium-scale trough cross strata in coarse to very coarse sand (formed by dune migration). Upper point-bar deposits are associated with moderate- to high-amplitude, relatively continuous (across-stream length up to 10 m, and along-stream length of tens of meters) GPR reflections, which correspond to the boundaries between small-scale cross-stratified medium to fine sand (formed by ripple migration) and vegetation-rich silty sand. Sets of inclined reflections record seasonal accretion on the point bar, in places as unit bars. Discordances and changes in inclination of the reflections are due to the formation of unit bars, lower bar platforms and swale fills (Figure 11.4).

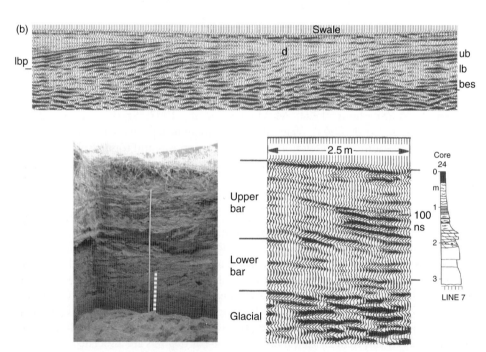

Figure 11.3 (a) Photo of South Esk study area. (b) GPR section of South Esk point-bar deposits showing upper-bar (ub) and lower-bar (lb) deposits, basal erosion surface (bes), discontinuity (d) in inclination of large-scale inclined strata below swale, and lower-bar platform (lbp) above lower (unit) bar deposits (modified from Bridge et al., 1995). Profile is oriented normal to channel direction, and bar migration was right to left. Distance between vertical dashes at top of GPR sections is approximately 2 m, and thickness of GPR section is approximately 4 m. Lower pictures show detail of lower-bar and upper-bar deposits.

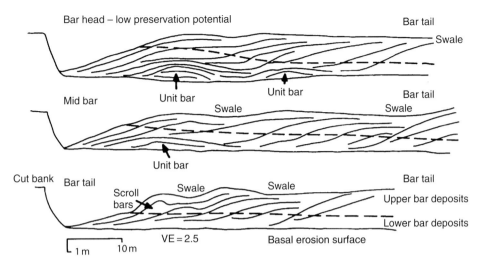

Figure 11.4 South Esk depositional model showing upper-bar and lower-bar deposits, discontinuities in inclination of large-scale inclined strata, lower-bar platforms above unit bar deposits, and discordances below swales (from Bridge, 2003). Point bar migration from right to left.

11.3.2. Calamus, Nebraska

The Calamus River in central Nebraska is a sandy, low-sinuosity, slightly braided river containing point bars and some braid bars (Figure 11.5). The evolution of the channel geometry was determined in the study by Bridge et al. (1986). Braid bars form either from chute cutoff of point bars or from amalgamation of lobate unit bars in mid channel, as is the case in most braided rivers. Growth of braid bars is mainly by increase in width and downstream length (i.e. lateral and downstream accretion). Lateral accretion is not necessarily symmetrically distributed on each side of the braid bar. The upstream ends of braid bars may experience erosion or deposition. A curved channel around a braid bar may become abandoned as a side bar or point bar grows into its entrance. During the early stages of filling, the channel contains relatively small unit bars, especially at the upstream end.

The interaction between channel geometry, water flow, sediment transport and deposition around a compound braid bar on the sandy Calamus River was studied using measurements made over a large discharge range from catwalk bridges (Bridge and Gabel, 1992; Gabel, 1993). In the curved channels on either side of the braid bar, the patterns of flow velocity, depth, water surface topography, bed shear stress, and rate and mean grain size of bedload transport rate are similar to those in single channel bends. A theoretical model of bed topography, flow and bedload transport in bends (Bridge, 1992) agreed well with Calamus River data, and was subsequently used as the basis for simple theoretical models for deposition in braided channels (Bridge, 1993).

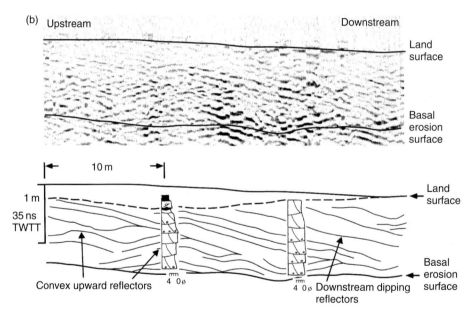

Figure 11.5 (a) Calamus River study site in lower part of photo (see bridges). (b) Example of 500 MHz GPR profile (along-stream view) and core logs through Calamus compound braid bar deposit. Reflections are due to bases of medium-scale cross strata (associated with dunes). Convex upward patterns of reflections are associated with unit bars. Downstream inclination of reflections indicates dominantly downstream accretion of the braid bar. Modified from Bridge et al. (1998). (c) Example of 500 MHz GPR profile (across stream view) through Calamus compound braid bar and adjacent fill. Reflections are due to bases of medium-scale cross strata (associated with dunes). Convex upward patterns of reflections are associated with both unit bars and the compound braid bar. Modified from Bridge et al. (1998).

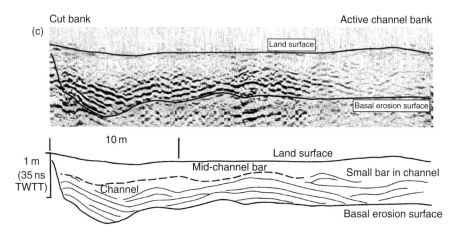

Figure 11.5 (Continued)

Extensive vibracoring of channel-bar and channel-fill deposits and box coring of the bed by Bridge et al. (1986) revealed the nature of the deposits. Bridge et al. (1998) further studied the large-scale structure of the deposits in 3D using GPR (500 MHz antennae). Vibracores and GPR profiles show that channel-bar deposits reflect (1) the distribution of grain size and bedform on the streambed during high flow stages, and (2) the mode of bar growth and migration. The geometry and orientation of large-scale inclined strata reflect mainly lateral and downstream migration of the compound bars, and accretion in the form of lobate unit bars and scroll bars (Figure 11.5). Within the large-scale strata are mainly medium-scale cross strata associated with the migration of dunes, which cover most of the bed at high flow stages. Small-scale cross strata from ripple migration occur in shallow water near banks. Point-bar and braid-bar sequences have an erosional base and generally fine upwards except for those near the bar head which show little vertical variation in mean grain size. Channel-fill deposits are similar to channel-bar deposits, except that the large-scale inclined strata are concave upward in cross-channel view, and the deposits are generally finer grained in the downstream part of the channel fill (Figure 11.5). Figure 11.6 is a model of the channel geometry, mode of channel migration, and deposits of the Calamus River.

11.3.3. Brahmaputra (Jamuna), Bangladesh

The Brahmaputra River in Bangladesh is a large sandy braided and anastomosing river (Figure 11.7). There have been many studies of the geometry, flow and sedimentary processes, channel migration and deposits of the Brahmaputra River (e.g., Coleman, 1969; Bristow, 1987, 1993; Bristow et al., 1993; Thorne et al., 1993; Mosselman et al., 1995; Richardson et al., 1996; Best and Ashworth, 1997; Richardson and Thorne, 1998; McLelland et al., 1999; Ashworth et al., 2000; Best

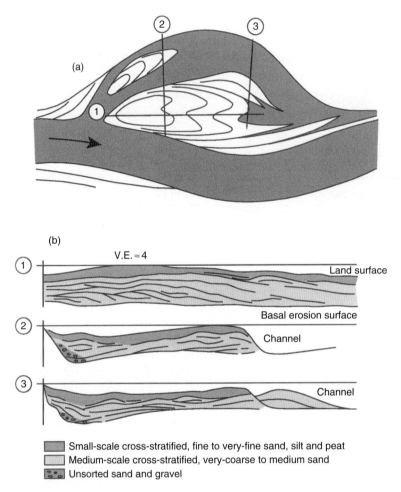

Figure 11.6 Model of channel geometry, mode of channel migration and deposits of the Calamus River (modified from Bridge et al., 1998). The map (a) shows accretion topography on a compound braid bar, suggesting bar growth mainly by incremental lateral and downstream accretion. Relatively small unit bars occur within the filling channel. The cross sections (b) show large-scale inclined strata associated with the accretion of unit bars (convex upward patterns), the lateral and downstream accretion of the compound braid bar and channel filling. Lower-bar deposits are composed mainly of medium-scale cross strata because of dune migration. Upper-bar deposits and channel fills are composed mainly of small-scale cross strata (due to ripples), plus bioturbated silt and peat. The overall sedimentary sequence generally fines upward from very coarse or coarse sand to fine or very fine sand.

et al., 2003). Compound braid bars in the Brahmaputra originate from the amalgamation of mid-channel lobate unit bars and by chute cutoff of point bars, as seen in other braided rivers (Bristow, 1987; Ashworth et al., 2000). Once formed, braid bars grow episodically by lateral and downstream accretion. The upstream parts of braid bars may be sites of accretion or erosion, depending on flow stage and geometry. The braid bars are typically up to 12 m high, 1.5 to 3 km long, and 0.5 to 1 km wide. The amount of lateral and downstream accretion during a monsoonal flood can be on the order of kilometers. The accretion on braid bars is normally

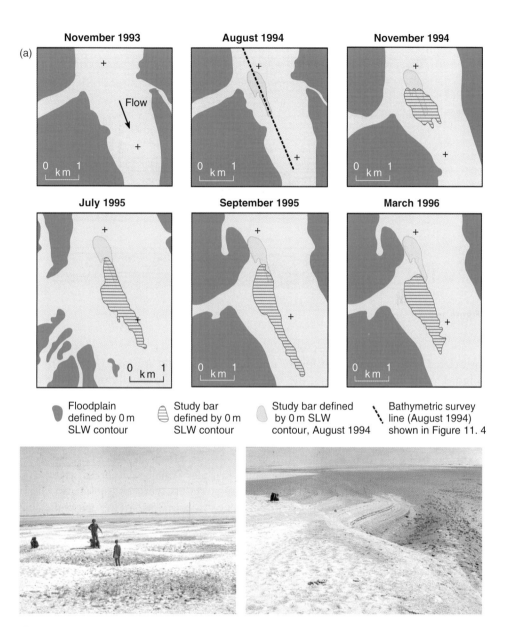

Figure 11.7 (a) Evolution of the Brahmaputra River channel bar studied in detail by Ashworth et al. (2000) and Best et al. (2003). Photos below are typical Brahmaputra River compound bars with dunes (left) and a unit bar margin (right). (b) Examples of 100 MHz GPR profiles through Brahmaputra compound braid bar deposits (modified from Best et al., 2003). Vertical exaggeration is 1.58. Profile 6, across stream through the downstream part of the braid bar (location on Figure 11.8), shows large-scale inclined strata associated with unit bar migration (1). The spatially variable dip angles (up to angle-of-repose) are typical of unit-bar deposits (2). Profile BH, alongstream through the upstream part of the bar, shows concave upward reflections (4) that were ascribed by Best et al. to migration of large sinuous-crested dunes. However, trough cross sets that are up to 2 m thick cannot be formed by migration of dunes that are only about a meter high. Lower in this profile, 3 m thick inclined strata (3) are ascribed to downstream accretion of a bar head.

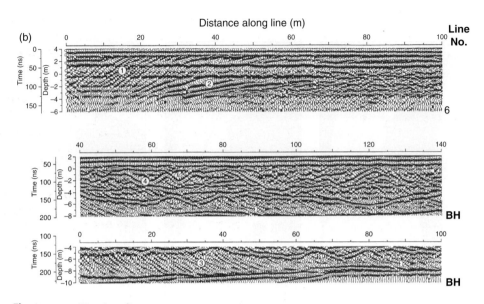

Figure 11.7 (Continued)

associated with migration of lobate unit bars. These bars are up to several meters high. During high flow stages, the bed is covered with dunes, although there may be restricted areas of upper-stage plane beds and ripples near bar tops. In the deeper parts of the channels, dunes are typically 3 to 4 m high, but are 0.5 to 1 m high near bar tops.

Bristow (1993) described the deposits in the upper 4 m of channel bars exposed in long cut banks during low flow stage. He described the deposits of sinuous-crested dunes (trough cross strata), ripples (small-scale cross strata) and upper stage plane beds (planar strata). These stratasets were arranged into larger scale sets of strata indicative of seasonal deposition on bars. The orientation of these flood-generated stratasets (large-scale inclined strata) indicated bar migration by upstream, downstream and lateral accretion. Bristow cites evidence for migration of small dunes over larger dunes at various stages of a monsoonal flood; however, it is likely that some of the larger, meters-high bedforms are actually unit bars. The fills of cross-bar channels are common in these upper-bar deposits.

Best et al. (2003) studied the deposits of a single compound braid bar using GPR (mainly 100 MHz antennae), vibracores and shallow trenches (Figure 11.7). They recognized the following types of cross (inclined) strata: (1) large-scale (sets up to 8 m thick, cross strata dipping at the angle of repose or less), due to deposition associated with unit bars on compound bar margins; (2) medium-scale (1 to 4 m thick, cross strata dipping at the angle of repose or less), associated with migration of large dunes, and; (3) small-scale (sets 0.5 to 2 m thick) due to dunes migrating over compound bar flanks. Set thickness of small-scale cross strata decreases upwards. Cross strata associated with ripples occur locally at the top of the bar, but they were not called small-scale cross strata by Best et al. (2003). This terminology of large, medium and small is different from that used by other workers. Furthermore, the ranges of set thickness in these categories overlap, so that it is difficult to know how to classify any specific cross set. Best et al. (2003) did not explicitly recognize the distinction between unit bars and compound bars, and the different scales of deposits associated with them. Cross sets associated with dunes that have a mean height of 3 m are likely to have a mean set thickness on the order of 1 m (Leclair and Bridge, 2001). Therefore, most of the medium-scale cross sets (with thickness of 1–4 m) are likely to be formed by unit bars rather than dunes.

Figure 11.8 indicates that vertical-accretion deposits on the bar top pass laterally into both upstream- and lateral-accretion deposits, and that both bar-margin slipface deposits and vertical-accretion deposits in the channel pass laterally into downstream/oblique-accretion deposits. In fact, if vertical deposition occurs on any mound-like form (i.e. a braid bar), 2D sections will appear to show components of accretion in the upstream, lateral and downstream directions. Also, as most braid bars experience both lateral and downstream growth simultaneously, a lateral-accretion deposit will also be a downstream-accretion deposit.

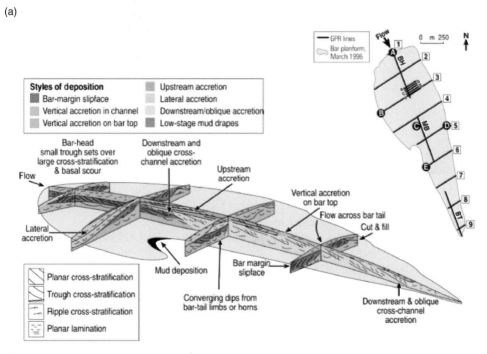

Figure 11.8 Best et al. (2003) depositional model for a Brahmaputra braid bar. (a) Three-dimensional diagram of the principal styles of deposition. The bar is 3 km long, 1 km wide and 12–15 m high. (b) Schematic sedimentary logs at five localities within the braid bar (see inset in (a) for location), illustrating the characteristic sedimentary structures, large-scale bedding surfaces and styles of deposition. Arrows depict approximate flow directions for sedimentary structures, with flow down the page indicating flow parallel to the mean flow direction. See text for discussion of this model.

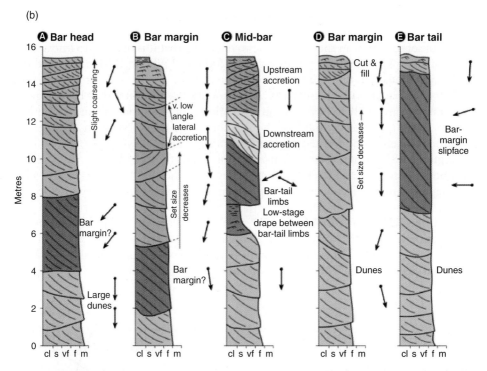

Figure 11.8 (Continued)

11.3.4. Niobrara, Nebraska

The Niobrara River in northern Nebraska is a sandy braided river. The part of the river studied by Bristow et al. (1999) and Skelly et al. (2003) has recently experienced channel aggradation, crevasse splay formation and avulsion associated with construction of a dam downstream.

Bristow et al. (1999) described the form, evolution and deposits of two sandy crevasse splays, using topographic surveys, cores and GPR profiles (Figure 11.9). Two hundred megahertz GPR antennae were used, having a maximum vertical resolution of 0.15 m. The crevasse splays formed between 1993 and 1997, but unfortunately the details of their development were not observed. The crevasse splays are on the order of hundreds of meters long and wide, and up to 2 m thick. Crevasse channels are up to 30 m wide and 1–2 m in maximum depth. Sinuous crested dunes migrated in the channels, and ripples migrated in shallow parts of channels and at the margins of the splays where channels were greatly reduced in size. Avalanche faces up to 1 m high occurred at the margins of splays that prograded into standing water. Channel migration and switching apparently occurred commonly during splay formation, but no details were given by Bristow et al. (1999).

Advances in Fluvial Sedimentology using GPR 337

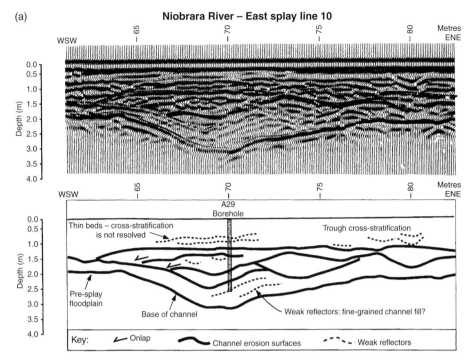

Figure 11.9 Niobrara crevasse splay data (from Bristow et al., 1999) (a) Alongstream GPR profile from downstream part of a crevasse splay. Fine-grained channel fill at the base of the section is interpreted to predate the crevasse splay. Overlying strata are interpreted as crevasse channels filled with medium-scale trough cross-stratified sand (formed by dunes). (b) Alongstream GPR profile (flow left to right) through a crevasse splay different from that in (a). GPR profile shows low-angle reflections (large-scale inclined strata) interpreted as due to progradation of the front of the splay. In places, these reflections approach the angle-of-repose, and are interpreted as avalanche faces at the distal margin of the splay. Internal texture and structure are indicated in core logs. Log 19 shows three sedimentation units in a splay sequence that coarsens upward. In ascending order, these units contain wavy laminae, planar cross strata and trough cross strata overlain by ripple lamination. Log 20 contains a coarsening upward sequence with four sedimentation units that are composed, from base to top of wavy laminae; ripple laminae; planar laminae capped by deformed laminae; and trough cross strata.

Cores show that splay deposits are represented by 1–2 m thick, sharp-based sequences that coarsen upwards or coarsen upwards then fine upwards at the top (Figure 11.9). Commonly, 2–3 dm-thick fining-upward sequences occur within the total thickness of a crevasse splay. These were related either to distinct seasonal floods or to discrete episodes of channel switching during a single flood. Internal structures within these sequences were medium-scale trough cross strata (sets 0.2–0.5 m thick) formed by dunes, small-scale cross strata formed by ripples and planar strata formed on plane beds.

GPR profiles enabled recognition of the base of a crevasse splay, overlapping crevasse channel fills within a splay (in cross-stream views), the boundaries of the

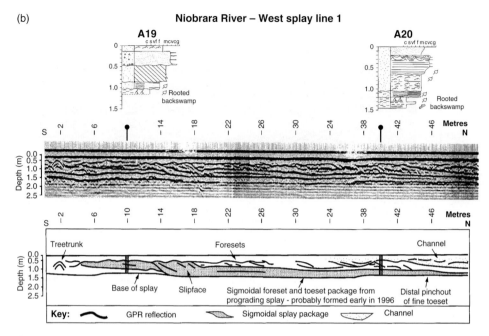

Figure 11.9 (Continued)

decimeter-thick fining upward units, bases of some medium-scale trough cross sets and some thick angle-of-repose cross strata at the margin of a splay (Figure 11.9). In alongstream views, the decimeter-thick fining upward units are inclined at up to 5° in the downstream direction, indicating downstream progradation of the splay. In places, these units pass laterally into angle-of-repose cross strata (splay margin deposits) or are truncated by crevasse channels (Figure 11.10).

Skelly et al. (2003) interpreted the following sedimentary facies from radar facies using 100 and 200 MHz GPR antennae: trough cross beds associated with small 3D dunes; planar cross beds associated with small 2D dunes; horizontal to low-angle planar strata or cross-set boundaries associated with large 3D dunes (or linguoid bars); channel-shaped erosion surfaces associated with secondary channels (and filled with deposits of 2D and 3D dunes); sigmoidal strata associated with accretion on braid bar margins. They also interpreted superimposed sequences, each sequence 1–1.5 m thick, as due to stacking of braid bars. Most bar sequences fined upward. These depositional patterns were associated with high-stage bar construction and upstream, lateral and downstream bar accretion, and low-stage dissection of the bars by small channels. Unfortunately, the uppermost parts of the radar profiles were not linked to the geometry and migration of specific extant channels and bars. The distinction between unit bars and compound bars was not made. However, Figure 11.11 (from Skelly et al., figure 10) shows an example of a lateral transition from low-angle strata to angle-of-repose strata to low-angle strata, as observed in unit-bar deposits elsewhere. Skelly et al. (2003) interpreted these deposits incorrectly as formed by small 2D dunes. These unit-bar deposits accreted onto the western margin of a compound

Advances in Fluvial Sedimentology using GPR 339

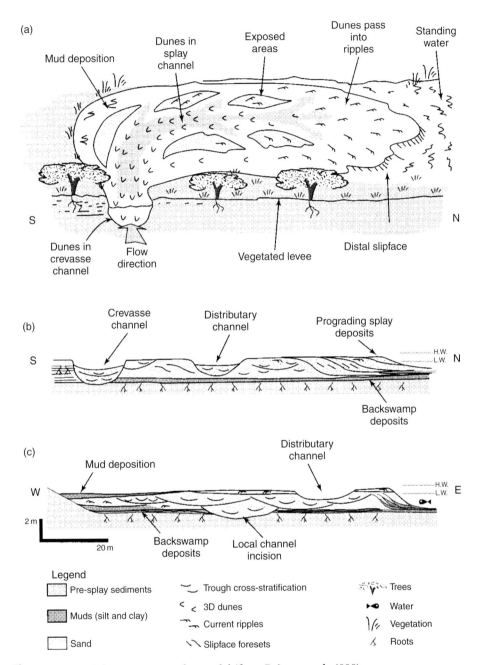

Figure 11.10 Niobrara crevasse splay model (from Bristow et al., 1999).

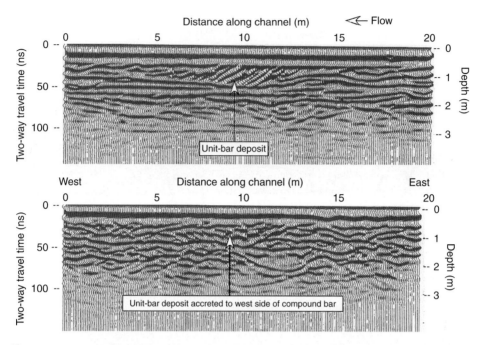

Figure 11.11 Examples of 200 MHz GPR profiles from the Niobrara River (Skelly et al., 2003, figures 10 and 11) with reinterpretation.

bar deposit (from Skelly et al., Figure 11). The dimensions of these unit-bar deposits are consistent with those in other rivers (Figure 11.2).

11.3.5. South Saskatchewan, Canada

The sandy, braided South Saskatchewan River near Outlook in Canada has been studied intensively from 2000 to 2007 using GPR, coring, topographic surveying and sequential aerial photos. Preliminary results from this study have been published (e.g. Woodward et al., 2003; Sambrook Smith et al., 2005, 2006; Best et al., 2006) but the bulk of this work (from 2004 to 2007) awaits publication. The river contains compound braid bars that form, expand laterally and migrate downstream by the sequential amalgamation of lobate unit bars (Cant, 1978; Cant and Walker, 1978). These unit bars are for the most part covered in sinuous-crested dunes, except in shallow, slow-moving water where they are covered in ripples. The lee sides of unit bars have slopes that are mainly less than the angle-of-repose, and this is related to the presence of superimposed bedforms (Reesink and Bridge, 2007). However, if the height of superimposed bedforms is less than 0.2 of the height of the unit bars, the lee sides of unit bars can reach the angle of repose, locally and episodically. Main channels can become abandoned and filled, and unit bars occur in the channel fills also.

The compound-bar and channel-fill deposits are composed of one main radar facies, and two subordinate radar facies (Figure 11.12). The main radar

Advances in Fluvial Sedimentology using GPR 341

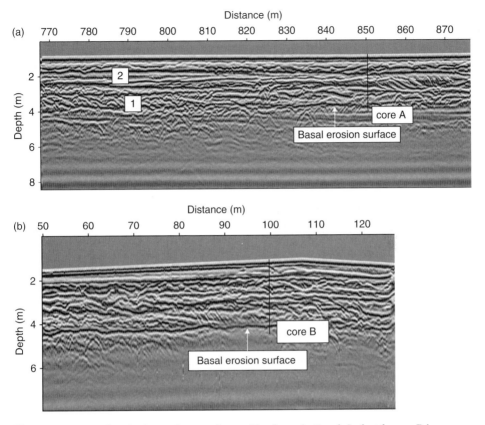

Figure 11.12 Two hundred megahertz radar profiles from the South Saskatchewan River, near Outlook, Canada. Flow is from left to right in all profiles, and vertical exaggeration is about 4.5. In (a) radar facies 1 is inclined reflections (in places undulatory) dipping to the right resulting from downstream migration of unit bars with superimposed dunes. To the right of core A, they approach the angle of repose. Radar facies 2 is low-angle (less than 2°) planar reflections resulting from migration of ripples over low-relief unit bars. The basal erosion surface is marked by diffraction hyperbolae. In (b) all of the profile is composed of radar facies 1. Isolated channel fills occur in (c). Unit bar deposits are indicated in the cores (d).

facies is inclined reflections that range in inclination for near horizontal to (rarely) the angle of repose. The inclined reflections may be undulatory in places. The inclined reflections occur in sets with sharp, near-horizontal bases composed of high-amplitude, laterally continuous (tens of meters) reflections. Each set is mainly in the range of 0.4–1 m thick. In cores, this facies is composed of sharp-based units that normally fine upward, but also show little vertical variation in grain size (Figure 11.12). Internally, these units are composed of medium-scale cross strata (set thickness normally 0.05–0.1 m) formed by dunes. Therefore, this facies is due to downstream and lateral migration of unit bars on which dunes were migrating. The angle of the inclined reflections reflects the slope of the unit bars over which the dunes were migrating. The

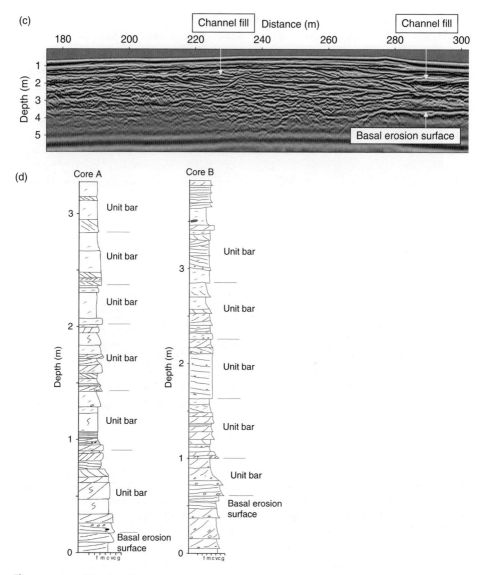

Figure 11.12 (Continued)

undulatory nature of some of the inclined reflections is due to the presence of medium-scale trough cross strata. Normally, four to six unit bar sets make up the thickness of a compound-bar deposit (Figure 11.12). The compound-bar deposits normally fine upwards from a basal erosion surface, from gravelly very coarse to coarse sand up to fine sand (Figure 11.12), and this fining upwards reflects the downstream migration of the compound bars. The basal erosion surface is commonly marked by diffractions related to the boulders scattered over this surface (Figure 11.12).

One of the subordinate facies is low-angle inclined (up to about 2°) planar reflections, which occurs in the upper parts of channel bars and in channel fills (Figure 11.12). Cores through this radar facies indicate that the internal structure is mainly small-scale cross strata (due to ripples) with subordinate thin medium-scale cross sets (due to low dunes), with a grain size of fine sand. The low angle of the reflections is due to the migration of ripples and low dunes over low-relief unit bars. Such unit bars occur in the shallow water of compound-bar tops, and in channel fills. The other subordinate facies is isolated concave-upward reflections that define channel shapes that are up to about 30 m wide and 1 m deep (Figure 11.12). This facies represents the fills of cross-bar channels and bar-top hollows (Best et al., 2006), and occurs only in the upper parts of compound-bar deposits. These deposits have counterparts in the deposits of the sandy braided Brahmaputra and Niobrara rivers discussed above.

11.3.6. Sagavanirktok, northern Alaska

The Sagavanirktok River on the North Slope of Alaska is gravelly, braided and anastomosing. The deposits were described using cores, trenches and GPR (110 and 450 MHz) profiles (Lunt and Bridge, 2004; Lunt et al., 2004a, 2004b). The origin of the deposits was inferred from (1) interpretation of channel and bar formation and migration, and channel filling, using annual aerial photographs; (2) observations of water flow and sediment transport during floods and (3) observations of bed topography and sediment texture at low-flow stage.

The Sagavanirktok River contains compound braid bars and point (side) bars (Figure 11.13). Compound braid bars originate by amalgamation of lobate unit bars and by chute cutoff of point bars. Compound bars migrate by downstream and lateral accretion of successive unit bars. The upstream ends of compound bars may be sites of erosion or accretion. During floods, most of the active riverbed is covered with sinuous-crested dunes, with minor proportions of bed-load sheets. Transverse ribs and ripples occur rarely in very shallow water. A channel segment may be abandoned if its entrance becomes blocked by a channel bar. Channels that are becoming abandoned and filled contain lobate unit bars.

Channel-belt deposits are composed of deposits of compound braid bars and point bars, and large channel fills (Figure 11.14). Channel-belt deposits are up to 7 m thick and 2.4 km wide across-stream, and are composed mainly of gravels, with minor sands and sandy silts. Episodic migration of compound bars within channel belts produces sets of compound large-scale inclined strata. Each compound large-scale stratum is a set of simple large-scale inclined strata, formed by migration of a unit bar. Small channel fills (cross-bar channels) occur at the top of compound bar deposits. Individual compound-bar deposits are hundreds of meters long and wide, meters thick, have basal erosion surfaces and terminate laterally in large channel fills. Thickness of compound-bar deposits, and vertical trends in grain size within them, depends on the bed geometry and surface grain size of the compound bars and the nature of bar migration. Relatively thick fining-upward sequences form as bar-tail regions migrate downstream into a curved channel or confluence scour. Grain size may increase towards the top of a thick fining-upward sequence where bar-head

Figure 11.13 Compound braid bar in the Sagavanirktok river, composed of unit bars (ub). Bars are normally covered with dunes (d) at high flow stage. Note cross-bar channels. Lower photos show dunes (left) and bedload sheets (right) on bar surfaces.

lobes migrate over the bar-tail. Relatively thin compound sets of large-scale strata with no vertical grain size trend are found in riffle regions. Compound-bar deposits are composed mainly of sandy gravel, but open-framework gravel is common near their base (Figure 11.14). A complete understanding of the evolution of a compound-bar deposit is only possible with a combination of frequent aerial photography, orthogonal GPR profiles and cores.

Large channel fills are also composed mainly of sets of simple large-scale inclined strata because of unit bars (Figures 11.14 and 11.15), but are capped with sandy strata containing small- and medium-scale cross sets, and planar strata. These deposits are also generally sandier in downstream parts of a channel fill.

Sets of simple large-scale inclined strata (Figures 11.14 and 11.15), formed by unit bar migration, are decimeters to meters thick, tens of meters long and wide, and generally fine upwards, although they may show no grain size trend. Between 3 and 7 unit bar deposits make up the thickness of a compound bar deposit. Open-framework gravels are common at the base of a unit-bar deposit. The large-scale strata are generally inclined at less than 10°, but may reach the angle of repose at the margin of the unit bar. Each simple large-scale stratum generally contains medium-scale cross strata or planar strata (see below).

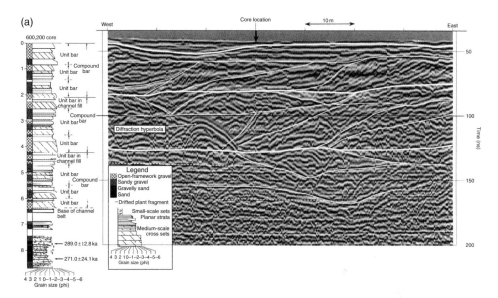

Figure 11.14 Sagavanirktok channel deposits (a) Example of 110 MHz GPR profile (across stream view) and core log through the channel belt. Core log shows three compound-bar deposits each comprising 3 or 4 unit-bar deposits. Internal structure of unit-bar deposits is medium-scale cross strata (from dunes) and planar strata (from bedload sheets). The base of the active channel-belt (not visible on GPR profile) is 6.4 m below the surface and overlies burrowed and rooted sands containing small-scale and medium-scale cross strata. The GPR profile has a vertical exaggeration of 5 and shows the 2D geometry of simple sets of large-scale inclined strata because of unit bars (set bases marked by thin white lines) and compound sets of large-scale inclined strata because of compound bars (set bases marked with thick white lines). From Lunt et al. (2004). (b) Example of 110 MHz GPR profile (across stream view) showing compound large-scale inclined strata associated with lateral accretion of a braid bar (upper part). The basal erosion surface and the top of the braid bar are marked by thick white lines, and unit-bar deposits and a channel fill are bordered by thin white lines. The lower part of the profile shows accretion of side bars (bordered by thin white lines) towards a central fill of a confluence scour zone. The base and top of the confluence scour zone are marked by thick white lines. Modified from Bridge (2003). (c) Example of 110 MHz GPR profile (across stream view) showing channel filling with unit bar deposits (bases marked by small arrows and thin white lines) that accreted onto the western margin of a compound bar. Basal erosion surface, channel margin and top of channel fill marked by large arrows and thick white lines. From Bridge (2003). (d) Trench photomosaic and 450 MHz GPR profile through two superimposed unit bar deposits, cut oblique to flow direction. Large-scale inclined strata dipping to the east are formed by migration of unit bars. These strata vary in inclination laterally and reach the angle of repose in places. The large-scale inclined strata are composed internally of sets of medium-scale cross strata (formed by dunes) or planar strata (formed by bedload sheets). From Bridge (2003).

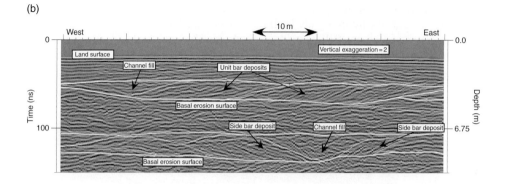

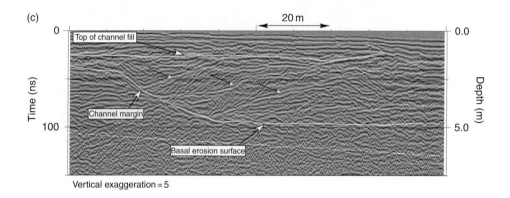

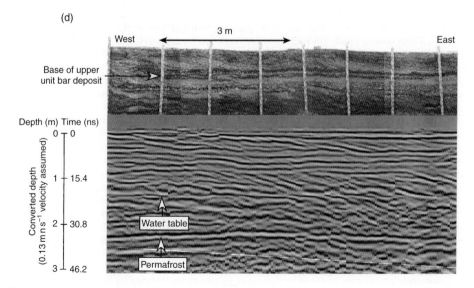

Figure 11.14 (Continued)

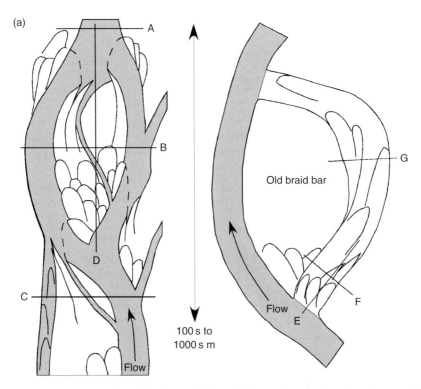

Figure 11.15 Depositional model of gravelly braided river deposits based on Sagavanirktok data (from Lunt et al. 2004). (a) Map showing idealized channels, compound bars and simple (unit) bars in active and abandoned channels. Cross sections A–D correspond to those shown in (b). Cross sections E–G correspond to those shown in (C); (b) Cross sections showing large-scale inclined strata (associated with compound and unit bars) from deposits in the active part of the channel belt. Vertical exaggeration is 2. Thin lines represent large-scale strata, medium weight lines represent bases of large-scale sets and thick lines represent bases of compound sets. Large-scale strata generally dip at less than 12°, but may be up to the angle of repose; (c) Cross sections showing large-scale inclined strata (associated with compound and unit bars) from deposits in the abandoned part of the channel belt; (d) Vertical logs of typical sequences through different parts of compound bar deposits and channel fills.

Small channel fills (cross-bar channels) are made up of small- and medium-scale cross sets, and planar strata (Figure 11.15). They are tens of meters long, meters wide and decimeters thick, and occur where simple large-scale sets occur at the top of compound large-scale sets.

Dune migration forms medium-scale sets of cross strata. These sets are decimeters thick and wide, up to 3 m long, and contain isolated open-framework gravel cross strata and sandy trough drapes. The thickness of medium-scale cross sets decreases upward in compound-bar deposits. Planar strata are formed by bedload sheets and may be made up of sandy gravel, open-framework gravel or sand. Individual strata are centimeters thick and decimeters to meters long and wide. Imbricated pebbles and pebble clusters are commonly found at the base of planar strata.

(b)

(A) Across stream view of compound side bars adjacent to a confluence scour

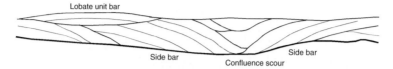

(B) Across-stream view of compound braid bar that migrated over a confluence

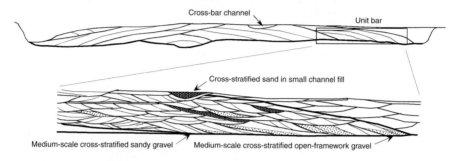

(C) Across-stream view of compound point bar that accreted laterally

(D) Along-stream view through compound bar that migrated laterally and downstream

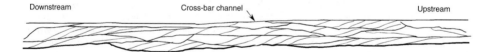

Figure 11.15 (Continued)

Small-scale sets of cross strata, formed by ripples, generally occur in channel fills, as trough drapes, and as overbank deposits. Small-scale cross sets are centimeters thick and long, are always composed of sand, and may contain organic remains, root traces or burrows where they occur in channel fills or overbank deposits. These smaller scale stratasets cannot be discerned on GPR profiles.

The GPR and trench data allowed the geometry of the different scales of strataset in the Sagavanirktok River to be related to the geometry and migration of their formative bed forms, in particular, the wavelength; height ratio of bedforms is similar to the length; thickness ratio of their associated deposits (Figure 11.2). Furthermore, the wavelength and height of bed forms like dunes and bars are related to channel depth and width. Therefore, the thickness of a particular scale of strata set (i.e. medium-scale cross sets and large-scale sets of inclined strata) will vary

Advances in Fluvial Sedimentology using GPR   349

(c)

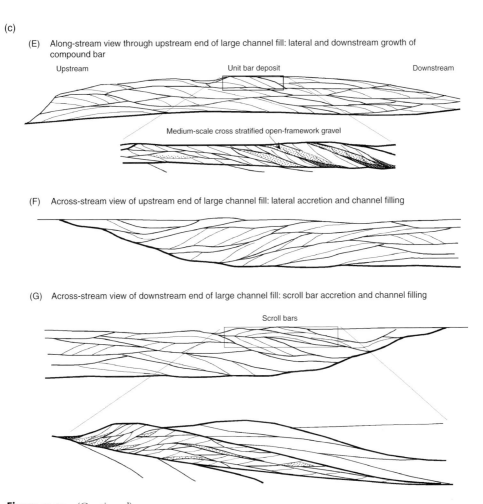

Figure 11.15 (Continued)

with the scale of the paleoriver. These relationships between the dimensions of stratasets, bedforms and channels mean that the depositional model developed from the Sagavanirktok River can be applied to other gravelly fluvial deposits.

11.3.7. Fraser and Squamish Rivers, Canada

Wooldridge and Hickin (2005) described the evolution of a compound bar in the Fraser River and one in the Squamish River, both in British Columbia, Canada. They referred to these rivers as wandering, but in fact they are really braided rivers with meandering reaches. Limited GPR surveys were conducted on the upstream tips of two compound bars. They defined radar facies as seen in 2D sections, and thus failed to recognize that several different radar facies are really different views of the same 3D structure. Partly as a result of this, interpretations of the radar facies are

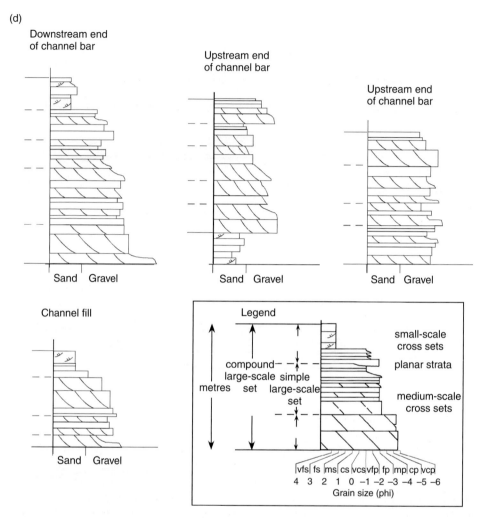

Figure 11.15 (Continued)

suspect. The two limited radar surveys were used to develop a model for the entire channel deposits of these two rivers, which of course cannot be justified. Despite the shortcomings of this study, the radar data are of high quality and show some interesting features, and several profiles are reproduced in Figure 11.16 with my reinterpretations.

11.3.8. Pleistocene outwash deposits in Europe

Sedimentary structures and textures from Pleistocene gravel outcrops, and associated GPR profiles, in Switzerland and Germany have been described and interpreted as braided river deposits by Huggenberger (1993), Siegenthaler and Huggenberger (1993), Beres et al. (1995, 1999), Asprion and Aigner (1997, 1999), Regli et al.

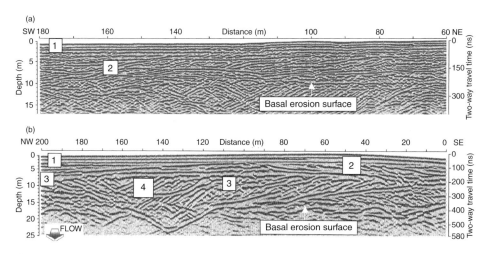

Figure 11.16 GPR profiles from the Squamish (a) and Fraser (b) Rivers (from Wooldridge and Hickin, 2005). Profile a is 100 MHz and is parallel to the flow direction (right to left). Radar facies 1 is near horizontal planar reflections because of migration of bedload sheets or low-relief dunes. Radar facies 2 is inclined reflections because of migration of unit bars with superimposed dunes or bedload sheets. Profile b is 50 MHz. Radar facies 1 and 2 are as for a. Radar facies 3 are inclined reflections that dip towards the central channel fill (4). Radar facies 3 and 4 represent a confluence scour fill (4) with side bars on either side (3).

(2002), and Heinz and Aigner (2003). These studies are potentially useful in that they link deposits seen in large quarries with high-quality GPR data (3D, antenna frequencies of 100 MHz). Sedimentary facies have been related to radar facies and to hydrofacies. However, the depositional environment has been inferred rather than known. Two main sedimentary and radar facies are described: (1) scour fills formed in confluence scours and channel bends, and (2) horizontally bedded gravel sheets formed in channels. The scour fills have inclined strata dipping towards the deepest part of the scour, and these strata are commonly composed of alternations of open framework and sandy gravel. These gravel couplets have been interpreted in terms of flow separation over dunes, following Carling and Glaister (1987). Scour fills are commonly overlain by horizontal gravel sheets. The different types and scales of strataset described above (e.g., Figure 11.1) are not recognized by these workers, nor are the deposits of unit bars, compound bars and channel fills.

These interpretations cannot be justified in general. The scour-fill deposits described by these workers apparently do not have the characteristic side-bar deposits adjacent to the confluence scour fills, as shown in numerous studies of modern rivers and predicted in Bridge's (1993) model. Alternative explanations for some of their scour-fill deposits might be swales adjacent to unit-bar deposits, small channel fills, log-jam/ice-block scour fills or even trough cross strata formed by dunes. These various types of scour fill may well look similar on GPR profiles obtained with radar waves of different frequency, depending on the size of the original river. Apparently horizontal strata may be produced by bedload sheets or may in fact contain medium-scale cross strata, formed by dunes. Furthermore, cross sets may appear as horizontal strata on some GPR profiles because of strong reflections from set boundaries, and use of GPR antennae with inadequate resolution to detect cross strata.

The radar facies models of Beres et al. (1999), which are linked to sedimentological interpretations, are misleading for several reasons (Figure 11.17). Sedimentary strata imaged on GPR profiles will look different depending on the scale of the strata relative to the antenna frequency. For example, thin planar strata imaged by high-frequency antennae will look similar to thick planar strata imaged

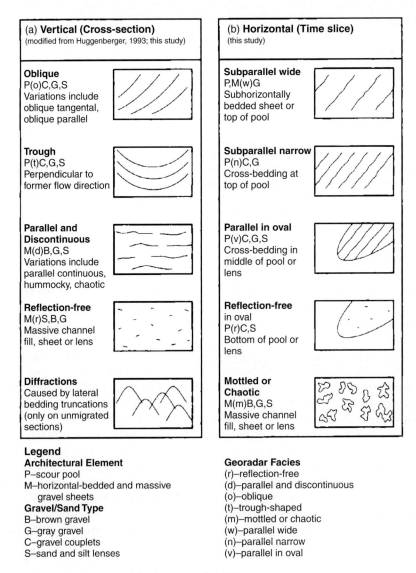

Figure 11.17 Radar facies model of Beres et al. (1999).

by low-frequency antennae. Also, cross strata imaged with high-frequency antennae may appear planar when imaged using low-frequency antennae. Thus, planar strata, as imaged by GPR using different antenna frequencies, could be formed by migration of bedload sheets, or dunes or unit bars. Trough-shaped cross strata could be formed in association with either sinuous-crested dunes, scours around obstacles, channel fills, or confluence scours. The range of stratal patterns proposed by Beres et al. (1999) is too restricted. It is necessary to recognize large-scale inclined strata resulting from the migration of bars. Channel fills must also be recognized. Finally, it is necessary to define radar facies in 3D, because the same radar (and sedimentary) facies may look different in each of the three orthogonal orientations.

Kostic and Aigner (2007) described a 3-m thick sequence of Quaternary gravelly deposits from SW Germany using GPR and quarry exposures, and interpreted them as meandering river deposits overlying braided river deposits. They defined a number of depositional elements as seen in 2D sections, and named them using interpretive terms such as gravel dune, chute channel, lateral accretion and confluence scour fill. In fact, depositional elements should be defined in 3D and given descriptive names before interpretation. Their confluence-scour element does not resemble confluence scour deposits, which must have a central channel fill bounded laterally by side-bar deposits (see above and Bridge, 1993). The cross strata in the gravel-dune element are not at the angle of repose and resemble the deposits of unit bars. Unit bars are not interpreted by Kostic and Aigner, although they are ubiquitous in all rivers. The small channel fills in the upper parts of the deposits are probably cross-bar channel fills, although this term is not used. Some of the lateral-accretion deposits may be upper point-bar deposits (Bridge et al., 1995), but this term is also not used. There was no attempt to relate the thickness of the sedimentary sequence to the depth of the meandering channels. However, it is clear from the size of the preserved meander loops that the maximum channel depth was probably greater than the thickness of deposits studied. In this case, all of the deposits are meandering river deposits, and the braided river interpretation for the lower deposits should be questioned.

11.3.9. Mesozoic deposits of SW USA

Stephens (1994) studied the Jurassic Kayenta Formation in southwestern Colorado using descriptions of two large outcrops and seven nearby GPR lines. Fifty megahertz antennae were used, which allowed depth of penetration of around 10 m and vertical resolution of reflections of about 1 m. Only one GPR profile was shown, although interpretations from all seven profiles were shown. Stephens (1994) does not discuss exactly what is being imaged by the GPR, and radar facies were not defined. However, it appears that the reflections marked on his interpreted sections correspond to basal erosion surfaces of channel bars and channel fills, and also boundaries of large-scale inclined strata resulting from bar migration.

The Cretaceous Ferron Sandstone in Utah has been studied extensively using 3D GPR, cores and outcrops, at least partly because it is considered to be a hydrocarbon reservoir analog (McMechan et al., 1997; Szerbiak et al., 2001; Corbeanu et al., 2001, 2004). One hundred megahertz antennae were used, and were capable of penetrating to a depth of about 15 m. The resolution of reflections was about 0.5 m, sufficient to image large-scale inclined strata formed by episodic accretion on a channel bar. The increased resolution of 200 MHz antennae enabled imaging of the bases of trough cross sets, as was the case in studies of modern sandy fluvial deposits.

11.4. Concluding Discussion

The GPR and associated data reviewed above have resulted in greatly improved understanding of fluvial deposits, which has facilitated the building of generalized depositional models (e.g., Bridge, 2006; Bridge and Lunt, 2006). GPR in combination with coring and trenching has been particularly useful for describing the larger scale structure of fluvial deposits, associated with migration of bars and dunes. Interpretation of such deposits has been facilitated by detailed studies of flow and sedimentary processes, bed form geometry and migration. There is remarkable similarity in the flow and sedimentary processes, bed form geometry and migration, and deposits of rivers of different channel size, plan geometry and sediment size: e.g. unit bars and dunes at high flow stage; bar geometry; evolution and migration (see also Sambrook Smith et al., 2005). Differences between gravel-bed and sand-bed rivers are more bed-load sheets and associated planar strata in gravel-bed rivers, increasing as mean grain size increases; more abundant ripples (small-scale cross strata) and upper stage plane beds (planar strata) in sand-bed rivers, increasing as mean grain size decreases. It is clear from these data that there is no truth to the common claim that braided rivers transport relatively coarse sediment as bed-load and have unstable banks, whereas meandering rivers transport relatively fine sediment as suspended load and have stable banks.

Meandering and braided rivers cannot be distinguished based on spatial variations in mean grain size and smaller scales of sedimentary structures. Unit bars and supposedly associated sets of planar cross strata have been specifically associated with braided rivers (e.g. Collinson, 1970; Smith, 1971, 1972, 1974; Bluck, 1976,1979; Cant and Walker, 1976, 1978; Cant, 1978; Blodgett and Stanley, 1980; Crowley, 1983). However, unit bars occur in meandering rivers also (e.g. McGowen and Garner, 1970; Bluck, 1971; Jackson, 1976; Levey, 1978; Bridge et al., 1995). Furthermore, it has become increasingly apparent that most of the internal structure of unit bars is not planar cross strata, but is due to the bedforms (dunes, ripples, bed-load sheets) migrating over them (Collinson, 1970; Jackson, 1976; Nanson, 1980; Bridge et al., 1986, 1995, 1998; Ashworth et al., 2000; Best et al., 2003; Lunt et al., 2004a, 2004b). It appears that angle-of-repose (planar) cross strata in unit bars only occur when the height of superimposed bedforms (especially dunes) is less than 20% of the height of the unit bar, in other words for the highest unit bars (Reesink and Bridge, 2007). It is only the largest scales of strataset within channel belts that allow meandering and braided rivers to be distinguished: cross sections through confluence scours with side bars; cross sections through braid bars with adjacent coeval channels (Bridge and Tye, 2000; Bridge, 2003). Interestingly, Lunt and Bridge

(2004) documented an example where a point bar was transformed into a braid bar by chute cutoff. The deposits seen in GPR profiles and cores could not have revealed such an origin.

It is expected that GPR will be increasingly used to help describe near-surface fluvial deposits. It is desirable to explore in much more detail the link between the geometry of topographic features (e.g. dunes, bars) and their associated stratasets. Description of floodplain deposits using GPR is really only just beginning. In order to avoid misinterpretation of GPR data, it is necessary to define radar facies as they appear in 3D rather than 2D; recognize all scales of strataset and how resolution of these stratasets depends on antenna frequency; gain as much knowledge as possible about the geometry, flow and sedimentary processes in rivers and floodplains; and avoid confusing terminology.

ACKNOWLEDGMENTS

The South Saskatchewan GPR and core data were collected by a research team that included Phil Ashworth, Jim Best, John Bridge, Ian Lunt, Greg Sambrook-Smith, Chris Simpson and Arjan Reesink.

REFERENCES

Alexander, J., Bridge, J.S., Leeder, M.R., Collier, R.E.Ll. and Gawthorpe, R.L., 1994, Holocene meander belt evolution in an active extensional basin, southwestern Montana. Journal of Sedimentary Research, Vol. B64, pp. 542–559.

Ashworth, P.J., Best, J.L., Roden, J.E., Bristow, C.S. and Klaassen, G.J., 2000, Morphological evolution and dynamics of a large, sand braid-bar, Jamuna River, Bangladesh. Sedimentology, Vol. 47, pp. 533–555.

Asprion, U. and Aigner, T., 1997, Aquifer architecture analysis using ground-penetrating radar: Triassic and quaternary examples. Environmental Geology, Vol. 31, pp. 66–75.

Asprion, U. and Aigner, T., 1999, Towards realistic aquifer models; three-dimensional georadar surveys of Quaternary gravel deltas (Singen Basin, SW Germany). Sedimentary Geology, Vol. 129, pp. 281–297.

Bakker, M.A.J., Maljers, D. and Weerts, H.J.T., 2007, Ground-penetrating radar profiling on embanked floodplains. Netherlands Journal of Geosciences, Vol. 86, pp. 55–61.

Beres, M., Green, A.G., Huggenberger, P. and Horstmeyer, H., 1995, Mapping the architecture of glaciofluvial sediments with three-dimensional georadar. Geology, Vol. 23, pp. 1087–1090.

Beres, M., Huggenberger, P., Green, A.G. and Horstmeyer, H., 1999, Using two- and three-dimensional georadar methods to characterize glaciofluvial architecture. Sedimentary Geology, Vol. 129, pp. 1–24.

Best, J.L. and Ashworth, P.J., 1997, Scour in large braided rivers and the recognition of sequence stratigraphic boundaries. Nature, Vol. 387, pp. 275–277.

Best, J.L., Ashworth, P.J., Bristow, C. and Roden, J., 2003, Three-dimensional sedimentary architecture of a large, mid-channel sand braid bar, Jamuna River, Bangladesh. Journal of Sedimentary Research, Vol. 73, pp. 516–530.

Best, J., Woodward, J., Ashworth, P., Sambrook Smith, G. and Simpson, C., 2006, Bar-top hollows: A new element in the architecture of sandy braided rivers. Sedimentary Geology, Vol. 190, pp. 241–255.

Blodgett, R.H. and Stanley, K.O., 1980, Stratification, bedforms and discharge relations of the Platte River system, Nebraska. Journal of Sedimentary Petrology, Vol. 50, pp. 139–148.

Bluck, B.J., 1971, Sedimentation in the meandering River Endrick. Scottish Journal of Geology, Vol. 7, pp. 93–138.

Bluck, B.J., 1976, Sedimentation in some Scottish rivers of low sinuosity. Transactions of the Royal Society of Edinburgh, Vol. 69, pp. 425–456.

Bluck, B.J., 1979, Structure of coarse grained braided stream alluvium. Transactions of the Royal Society of Edinburgh, Vol. 70, pp. 181–221.

Bridge, J.S., 1992, A revised model for water flow, sediment transport, bed topography and grain size sorting in natural river bends. Water Resources Research, Vol. 28, pp. 999–1023.

Bridge, J.S., 1993, The interaction between channel geometry, water flow, sediment transport and deposition in braided rivers, in Best, J.L. and Bristow, C.S. (eds), Braided Rivers, Geological Society of London Special Publication, Vol. 75, pp. 13–72.

Bridge, J.S., 2003, Rivers and Floodplains, Blackwells, Oxford, 491 p.

Bridge, J.S., 2006, Fluvial facies models: Recent developments, in Posamentier H.W. and Walker R.G. (eds), Facies Models Revisited, SEPM Special Publication, Vol. 84, pp. 85–170.

Bridge, J.S. and Gabel, S.L., 1992, Flow and sediment dynamics in a low sinuosity, braided river: Calamus River, Nebraska Sandhills. Sedimentology, Vol. 39, pp. 125–142.

Bridge, J.S. and Jarvis, J., 1976, Flow and sedimentary processes in the meandering river South Esk, Glen Clova, Scotland. Earth Surface Processes, Vol. 1, pp. 303–336.

Bridge, J.S. and Jarvis, J., 1977, Velocity profiles and bed shear stress over various bed configurations in a river bend. Earth Surface Processes, Vol. 2, pp. 281–294.

Bridge, J.S. and Jarvis, J., 1982, The dynamics of a river bend: A study in flow and sedimentary processes. Sedimentology, Vol. 29, pp. 499–541.

Bridge, J.S. and Lunt, I.A., 2006, Depositional models of braided rivers, in Sambrook Smith, G.H., Best, J.L., Bristow, C.S. and Petts, G. (eds), Braided Rivers II, IAS Special Publication 36, pp. 11–50.

Bridge, J.S. and Tye, R.S., 2000, Interpreting the dimensions of ancient fluvial channel bars, channels, and channel belts from wireline-logs and cores. American Association of Petroleum Geologists Bulletin, 84, 1205–1228.

Bridge, J.S., Smith, N.D., Trent, F., Gabel, S.L. and Bernstein, P., 1986, Sedimentology and morphology of a low-sinuosity river: Calamus River, Nebraska Sand Hills. Sedimentology, Vol. 33, pp. 851–870.

Bridge, J.S., Alexander, J., Collier, R.E.L., Gawthorpe, R.L. and Jarvis, J., 1995, Ground-penetrating radar and coring used to document the large-scale structure of point-bar deposits in 3-D. Sedimentology, Vol. 42, pp. 839–852.

Bridge, J.S., Collier, R.E.Ll. and Alexander, J., 1998, Large-scale structure of Calamus river deposits revealed using ground-penetrating radar. Sedimentology, Vol. 45, pp. 977–985.

Bristow, C.S., 1987, Brahmaputra River: Channel migration and deposition, in Ethridge, F.G., Flores, R. M. and Harvey, M.D. (eds), Recent Developments in Fluvial Sedimentology, SEPM Special Publication 39, pp. 63–74.

Bristow, C.S., 1993, Sedimentary structures exposed in bar tops in the Brahmaputra River, Bangladesh, in Best, J.L. and Bristow, C.S. (eds), Braided Rivers, Geological Society of London Special Publication 75, pp. 277–289.

Bristow, C.S., Best, J.L. and Roy, A.G., 1993, Morphology and facies models of channel confluences, in Marzo, M. and Puigdefabregas, C. (eds), Alluvial sedimentation, IAS Special Publication 17, pp. 91–100.

Bristow, C.S., Skelly, R.L. and Ethridge, F.G., 1999, Crevasse splays from the rapidly aggrading, sand-bed, braided Niobrara River, Nebraska: Effect of base-level rise. Sedimentology, Vol. 46, pp. 1029–1047.

Cant, D.J., 1978, Bedforms and bar types in the South Saskatchewan River. Journal of Sedimentary Petrology, Vol. 48, pp. 1321–1330.

Cant, D.J. and Walker, R.G., 1976, Development of a braided fluvial facies model for the Devonian Battery Point Sandstone, Quebec, Canada. Canadian Journal of Earth Sciences, Vol. 13, pp. 102–119.

Cant, D.J. and Walker, R.G., 1978, Fluvial processes and facies sequences in the sandy braided South Saskatchewan River, Canada. Sedimentology, Vol. 25, pp. 625–648.

Carling, P.A. and Glaister, M.S., 1987, Rapid deposition of sand and gravel mixtures downstream of a negative step: The role of matrix-infilling and particle-overpassing in the process of bar-front accretion. Journal of the Geological Society of London, Vol. 144, pp. 543–551.

Coleman, J.M., 1969, Brahmaputra River: Channel processes and sedimentation. Sedimentary Geology, Vol. 3, pp. 129–239.

Collinson, J.D., 1970, Bedforms of the Tana River, Norway. Geografiska Annaler, Vol. 52A, pp. 31–56.

Corbeanu, R.M., Soegaard, K., Szerbiak, R.B., Thurmond, J.B., McMechan, G.A., Wang, D., Snelgrove, S.H., Forster, C.B. and Menitove, A., 2001, Detailed internal architecture of fluvial channel sandstone determined from outcrop and 3-D ground penetrating radar: Example from mid-Cretaceous Ferron Sandstone, east-central Utah. American Association of Petroleum Geologists Bulletin, Vol. 85, pp. 1583–1608.

Corbeanu, R.M., Wizevich, M.C., Bhattacharya, J.P., Zeng, X. and McMechan, G.A., 2004, Three-dimensional architecture of ancient lower delta-plain point bars using ground-penetrating radar, Cretaceous Ferron Sandstone, Utah, in Chidsey, T.C. Jr., Adams, R.D. and Morris, T.H. (eds), Regional to Wellbore Analog for Fluvial-deltaic Reservoir Modeling: The Ferron Sandstone of Utah, AAPG Studies in Geology No. 50, pp. 427–449.

Crowley, K.D., 1983, Large-scale bed configurations (macroforms), Platte River Basin, Colorado and Nebraska: Primary structures and formative processes. Geological Society of America Bulletin, Vol. 94, pp. 117–133.

Ekes, C. and Friele, P., 2003, Sedimentary architecture and post-glacial evolution of Cheekeye fan, southwestern British Columbia, Canada, in Bristow, C.S. and Jol, H.M. (eds), Ground Penetrating Radar in Sediments. Geological Society of London Special Publication 211, pp. 87–98.

Ekes, C. and Hickin, E.J., 2001, Ground penetrating radar facies of the paraglacial Cheekye Fan, southwestern British Columbia. Sedimentary Geology, Vol. 143, pp. 199–217.

Gabel, S.L., 1993, Geometry and kinematics of dunes during steady and unsteady flows in the Calamus River, Nebraska, USA. Sedimentology, Vol. 40, pp. 237–269.

Gawthorpe, R.L., Collier, R.E.Ll., Alexander, J., Bridge, J.S. and Leeder, M.R., 1993, Ground penetrating radar: Application to sandbody geometry and heterogeneity studies, in North, C.J. and Prosser, D.J. (eds), Characterization of Fluvial and Aeolian Reservoirs. Geological Society of London Special Publication 73, pp. 421–432.

Heinz, J. and Aigner, T., 2003, Three-dimensional GPR analysis of various Quaternary gravel-bed braided river deposits (southwestern Germany), in Bristow, C.S. and Jol, H.M. (eds), Ground Penetrating Radar in Sediments. Geological Society of London Special Publication 211, pp. 99–110.

Huggenberger, P., 1993, Radar facies: Recognition of characteristic braided river structures of the Pleistocene Rhine gravel (NE part of Switzerland), in Best, J.L. and Bristow, C.S. (eds), Braided Rivers, Geological Society of London Special Publication, Vol. 75, pp. 163–176.

Huggenberger, P., Meier, E. and Pugin, A., 1994, Ground-probing radar as a tool for heterogeneity estimation in gravel deposits: Advances in data-processing and facies analysis. Journal of Applied Geophysics, Vol. 31, pp. 131–184.

Jackson, R.G., 1976, Large-scale ripples of the Lower Wabash River. Sedimentology, Vol. 23, pp. 593–623.

Jol, H.M., 1995, Ground penetrating radar antennae frequencies and transmitter powers compared for penetration depth, resolution and reflection continuity. Geophysical Prospecting, Vol. 43, pp. 693–709.

Jol, H.M. and Smith, D.G., 1991, Ground penetrating radar of northern lacustrine deltas. Canadian Journal of Earth Sciences, Vol. 28, pp. 1939–1947.

Kostic, B. and Aigner, T., 2007, Sedimentary architecture and 3D ground-penetrating radar analysis of gravelly meandering river deposits (Neckar Valley, SW Germany). Sedimentology, Vol. 54, pp. 789–808.

Kowalsky, M.B., Dietrich, P., Teutsch, G. and Rubin, Y., 2001, Forward modeling of ground-penetrating radar data using digitized outcrop images and multiple scenarios of water saturation. Water Resources Research, Vol. 37, pp. 1615–1625.

Kowalsky, M.B., Rubin, Y. and Dietrich, P., 2004, The use of ground-penetrating radar for characterizing sediments under transient conditions, in Bridge, J.S. and Hyndman, D. (eds), Aquifer Characterization, SEPM Special Publication 80, pp. 107–127.

Leclair, S.F. and Bridge, J.S., 2001, Quantitative interpretation of sedimentary structures formed by river dunes. Journal of Sedimentary Research, Vol. 71, pp. 13–716.
Leclerc, R.F. and Hickin, E.J., 1997, The internal structure of scrolled floodplain deposits based on ground-penetrating radar, North Thompson River, British Columbia. Geomorphology, Vol. 21, pp. 17–38.
Levey, R.A., 1978, Bedform distribution and internal stratification of coarse-grained point bars, Upper Congaree River, South Carolina, in Miall, A.D. (ed.), Fluvial Sedimentology, Canadian Society of Petroleum Geologists Memoir 5, pp. 105–127.
Lunt, I.A. and Bridge, J.S., 2004, Evolution and deposits of a gravelly braid bar and a channel fill, Sagavanirktok River, Alaska. Sedimentology, Vol. 51, pp. 415–432.
Lunt, I.A., Bridge, J.S. and Tye, R.S., 2004a, Development of a 3-D depositional model of braided river gravels and sands to improve aquifer characterization, in Bridge, J.S. and Hyndman, D. (eds), Aquifer characterization, SEPM Special Publication 80, pp. 139–169.
Lunt, I.A., Bridge, J.S. and Tye, R.S., 2004b, A quantitative, three-dimensional depositional model of gravelly braided rivers. Sedimentology, Vol. 51, pp. 377–414.
McGowen, J.H. and Garner, L.E., 1970, Physiographic features and stratification types of coarse grained point bars: Modern and ancient examples. Sedimentology, Vol. 14, pp. 77–111.
McLelland, S.J., Ashworth, P.J., Best, J.L., Roden, J. and Klaasen, G.J., 1999, Flow structure and transport of sand-grade suspended sediment around an evolving braid bar, Jamuna River, Bangladesh, in Smith, N.D. and Rogers, J. (eds), Fluvial Sedimentology VI, IAS Special Publication 28, pp. 43–57.
McMechan, G.A., Gaynor, G.C. and Szerbiak, R.B., 1997, Use of ground-penetrating radar for 3-D sedimentological characterization of clastic reservoir analogs. Geophysics, Vol. 62, pp. 786–796.
Mosselman, E., Koomen, M.H.E. and Seijmonsbergen, A.C., 1995, Morphological changes in a large braided sand-bed river, in Hickin, E.J. (ed.), River Geomorphology, Wiley, Chichester, UK, pp. 235–249.
Nanson, G.C., 1980, Point bar and floodplain formation of the meandering Beatton River, northeastern British Columbia, Canada, Sedimentology, Vol. 27, pp. 3–29.
Olsen, H. and Andreasen, F., 1995, Sedimentology and ground-penetrating radar characteristics of a Pleistocene sandur deposit. Sedimentary Geology, Vol. 99, pp. 1–15.
Reesink, A.J.H. and Bridge, J.S., 2007, Influence of superimposed bedforms and flow unsteadiness on formation of cross strata in dunes and unit bars. Sedimentary Geology, Vol. 202, pp. 281–296.
Regli, C., Huggenberger, P. and Rauber, M., 2002, Interpretation of drill core and georadar data of coarse gravel deposits. Journal of Hydrology, Vol. 255, pp. 234–252.
Richardson, W.R.R. and Thorne, C.R., 1998, Secondary currents and channel changes around a braid bar in the Brahmaputra River, Bangladesh. Journal of Hydraulic Engineering, ASCE, Vol. 124, pp. 325–328.
Richardson, W.R.R., Thorne, C.R. and Mahmood, S., 1996, Secondary flow and channel changes around a bar in the Brahmaputra River, Bangladesh, in Ashworth, P.J., Bennett, S.J., Best, J.L. and McLelland, S.J. (eds), Coherent flow structures in open channels. Wiley, Chichester, pp. 519–543.
Roberts, M.C., Bravard, J.-P. and Jol, H.M., 1997, Radar signatures and structure of an avulsed channel: Rhone River, Aoste, France. Journal of Quaternary Science, Vol. 12, pp. 35–42.
Sambrook Smith, G.H., Ashworth, P.J., Best, J.L., Woodward, J. and Simpson, C.J., 2005, The morphology and facies of sandy braided rivers: Some considerations of spatial and temporal scale invariance, in Blum, M.D., Marriott S.B. and Leclair, S.F. (eds), Fluvial Sedimentology VII, IAS Special Publication 35, pp. 145–158.
Sambrook Smith, G.H., Ashworth, P.J., Best, J.L., Woodward, J. and Simpson, C.J., 2006, The sedimentology and alluvial architecture of the sandy braided South Saskatchewan River, Canada. Sedimentology, Vol. 53, pp. 413–434.
Siegenthaler, C. and Huggenberger, P., 1993, Pleistocene Rhine gravel: Deposits of a braided river system with dominant pool preservation, in Best, J.L. and Bristow, C.S. (eds), Braided Rivers: Geological Society of London Special Publication 75, pp. 147–162.
Skelly, R.L., Bristow, C.S. and Ethridge, F.G., 2003, Architecture of channel-belt deposits in an aggrading shallow sandbed braided river: The lower Niobrara River, northeast Nebraska. Sedimentary Geology, Vol. 158, pp. 249–270.

Smith, N.D., 1971, Transverse bars and braiding in the Lower Platte River, Nebraska. Geological Society of America Bulletin, Vol. 82, pp. 3407–3420.

Smith, N.D., 1972, Some sedimentological aspects of planar cross-stratification in a sandy braided river. Journal of Sedimentary Petrology, Vol. 42, pp. 624–634.

Smith, N.D., 1974, Sedimentology and bar formation in the upper Kicking Horse River, a braided outwash stream. Journal of Geology, Vol. 81, pp. 205–223.

Stephens, M., 1994, Architectural element analysis within the Kayenta formation (Lower Jurassic) using ground-probing radar and sedimentological profiling, southwestern Colorado. Sedimentary Geology, Vol. 90, pp. 179–211.

Szerbiak, R.B., McMechan, G.A., Corbeanu, R., Forster, C. and Snelgrove, S.H., 2001, 3-D characterization of a clastic reservoir analog: From 3-D GPR data to a 3-D fluid permeability model. Geophysics, Vol. 66, pp. 1026–1037.

Thorne, C.R., Russell, A.P.G. and Alam, M.K., 1993, Platform pattern and channel evolution of Brahmaputra River, Bangladesh, in Best, J.L. and Bristow, C.S. (eds), Braided Rivers, Geological Society of London Special Publication 75, pp. 257–276.

Van Dam, R.L. and Schlager, W., 2000, Identifying the causes of ground-penetrating radar reflections using time-domain reflectometry and sedimentological analyses. Sedimentology, Vol. 47, pp. 435–449.

Van Dam, R.L., Van den Berg, E.H., van Heteren, S., Kasse, C., Kenter, J.A.M. and Groen, K., 2002, Influence of organic matter in soils on radar-wave reflection: Sedimentological implications. Journal of Sedimentary Research, Vol. 72, pp. 341–352.

Vandenberghe, J. and Van Overmeeren, R.A., 1999, Ground-penetrating images of selected fluvial deposits in the Netherlands. Sedimentary Geology, Vol. 128, pp. 245–270.

Van Overmeeren, R.A., 1998, Radar facies of unconsolidated sediments in The Netherlands: A radar stratigraphy interpretation method for hydrogeology. Applied Geophysics, Vol. 40, pp. 1–18.

Woodward, J., Ashworth, P.J., Best, J.L., Sambrook Smith, G.H. and Simpson, C.J., 2003, The use and application of GPR in sandy fluvial environments: Methodological considerations, in Bristow, C.S. and Jol, H.M. (eds), Ground Penetrating Radar in Sediments, Geological Society of London Special Publication 11, pp. 127–142.

Wooldridge, C.L., 2002, Channel bar radar architecture and evolution in the wandering gravel-bed Fraser and Squamish Rivers, British Columbia, Canada. Unpublished MSc thesis, Department of Geography, Simon Fraser University, Burnaby, BC, Canada, 122 p.

Wooldridge, C.L. and Hickin, E.J., 2005, Radar architecture and evolution of channel bars in wandering gravel-bed rivers: Fraser and Squamish Rivers, British Columbia, Canada. Journal of Sedimentary Research, Vol. 75, pp. 844–860.

CHAPTER 12

GLACIERS AND ICE SHEETS

Steven A. Arcone

Contents

12.1. Introduction	361
12.2. Antarctica	363
12.2.1. Alpine glaciers: Dry Valleys	365
12.2.2. Polar firn: West Antarctica	367
12.2.3. Englacial stratigraphy: West Antarctica	371
12.2.4. Ice shelf: McMurdo Sound	373
12.2.5. Crevasses: Ross Ice Shelf	376
12.3. Alaska	379
12.3.1. Temperate valley glacier: Matanuska Glacier	380
12.3.2. Temperate valley glacier: Gulkana Glacier	382
12.3.3. Temperate firn: Bagley Ice Field, Alaska	384
12.3.4. Temperate hydrology: Black Rapids Glacier	385
12.4. Summary	388
References	389

12.1. INTRODUCTION

Dry snow and ice generally provide the best propagation medium of all geologic materials for GPR pulses with dominant frequency above about 1 MHz. The extremely low attenuation rates result from low conductivity (σ) and the absence of any dielectric (relative permittivity $= \varepsilon$) or magnetic relaxation processes centered above this frequency. Airborne, 50- to 150-MHz radars routinely profile Antarctic or Greenland stratification to 3–4 km depth. Even temperate glaciers exhibit good penetration characteristics for GPR signals. At 0°C, or slightly below, depending on the pressure, water can exist along grain boundaries, or in pockets and conduits. In the grain boundary case, such as candling ice (lake ice, in which melting occurs along the crystal boundaries), spring snow, or soaked firn (old glacier snow undergoing metamorphism to ice during burial), the conductivity is usually low ($<0.01\,\text{S m}^{-1}$) because the water content is rarely more than a few percent, and the ionic pathways are not well connected. In addition, the strong 0°C, water dielectric relaxation centered at 9 GHz is still way above GPR bandwidths and so does not affect loss significantly. In the case of pockets and conduits, such as in temperate englacial ice, scattering losses limit penetration.

Snow and ice are also ideal GPR media because their stratification presents reflecting horizons with great continuity and interesting configurations. Polar ice shelves may contain layers of basal sea ice a few to hundreds of meters thick (Blindow, 1994), while temperate terrestrial glaciers may contain stratified, debris-rich basal ice tens of meters thick (Arcone et al., 1995; Lawson, 1998). Englacial polar sheet ice contains layers of ions at concentrations ranging to about 250 ppb for volcanic sulfate deposits (LeGrand and Mayewski, 1997), layers of dust (e.g., tephra), and layers of morainal or bottom debris. There is even evidence that some of the layers profiled in the englacial regime are caused by changes in ice fabric (Fujita et al., 1999). Within the overlying firn, and within any snow cover, density contrasts dominate the causes of radar reflectivity, and mainly result from the formation of depth hoar (large-grained, sublimating buried snow of low density), refreezing of hoar vapor into ice "crusts," and wind packing (which causes low density but very small grain size). Some of these contrasts persist for hundreds of kilometer within polar regions (Arcone et al., 2004). Within the right frequency bandwidth, nearly all stratification is detectable with GPR even though the electrical properties associated with the chemical or physical changes often provide contrasts of less than a percent with those of the ice matrix. For example, a 1-mm layer of ice is easily detectable with a 16-bit radar operating in the virtually noise-free environment of Antarctica even though it provides a reflection coefficient of only −60 dB to a 400-MHz signal. Whether the change is caused by a contrast in permittivity or conductivity, both the concentration and the thickness of the anomalous layer help to determine its reflectivity.

The examples in this chapter are restricted to cases that used a pulsed GPR with a waveform consisting of about 1.5 cycles (Figure 12.1). The dominant frequency of GPR pulses commonly used ranges from 2 to 1500 MHz, and this rarely varies over dry snow and firn because the low relative dielectric permittivities of these materials do not offer strong impedance loading (reactive electromagnetic forces) upon the current propagating along the antennas. GPR has always been the system of choice for ground-based applications, but there are a few examples of its airborne use in the open literature (e.g., Arcone, 2002b). The classical airborne radar method is radio-echo sounding (RES), for which the pulse generally is a Gaussian modulated carrier of 10–12 cycles at 35–60 MHz (Bogorodsky et al., 1985). RES has been used to penetrate over 4 km into the East Antarctic ice sheet (EAIS), which far exceeds any possibility with GPR because of its limited power, antenna gain, and trace sampling rate. Recently, airborne chirp systems (Gogineni et al., 1998; Fahenstock et al., 2001) operating near 150 MHz have shown great success. Spread spectrum systems such as FMCW (Yankielun et al., 1992; Arcone and Yankielun, 2000), FM step frequency (Hamran et al., 1995; Richardson et al., 1997; Kanagaratnam et al., 2001) and phase code modulation (Nicollin and Kofman, 1994) have also been tried successfully.

The features in this chapter are divided into particular glacial examples, rather than by features (e.g. strata or depth). All examples are relevant to typical glacial regimes. They are taken from Antarctica and Alaska because of the amount of data available to the author from these regions. GPR profiles also exist from Greenland (Hempel and Thyssen, 1993; Hempel et al., 2000), Svalbard (Bjornsson et al., 1996; Murray et al., 1997; Odegard et al., 1997), Iceland (Murray et al., 2000), and Scandinavia (Kohler et al., 1997; Maijala et al., 1998).

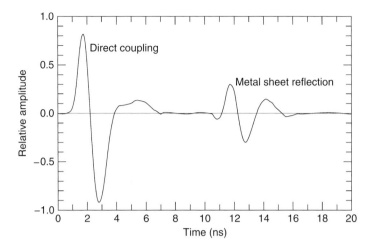

Figure 12.1 The pulsed GPR waveform from a 400-MHz dominant frequency antenna within a commercially available, shielded transducer. The transmitter and receiver antennas are spaced only 15 cm apart. The metal sheet was buried 1.4 m deep in snow of $\varepsilon = 1.7$, and the exercise was performed on the polar plateau of West Antarctica. Consequently, the reflection is an inversion of the transmitted waveform. Depending on surface properties, the relative amplitude and phases of the various frequency components will change. The result is usually a decrease in the dominant frequency as permittivity increases, and a change in the relative amplitudes of the various half-cycles. The relative amplitudes of the direct coupling and a metal sheet reflection have been used as a reference to calculate the reflectivity of deeper reflections (Arcone et al., 2004).

Some of the profiles in this chapter have not been published because there is insufficient positional or ground-truth information to verify the features. Consequently, some interpretative suggestions are offered to show the possible scientific applications of GPR in glacial areas. A few examples have been reviewed previously (Daniels, 2004).

12.2. ANTARCTICA

The glacial features of Antarctica (Figure 12.2) include broad ice sheets, ice streams, valley and alpine glaciers, and ice shelves. The East (EAIS) and West Antarctic Ice Sheet (WAIS) are generally 2–4 km thick and are snow accumulation areas (areas with a positive "mass balance"). Most Antarctic ice is lost by the breakage of small masses into the ocean (calving) or of large icebergs. A small percentage is lost to ablation by wind or natural sublimation. Accumulation can be as little as 2–4 cm yr^{-1} water equivalent within the EAIS, or nearly 100 cm yr^{-1} in some locations along the WAIS coast. Ice velocity is negligible near the ice divides, whereas away from the divides it can speed up to 1000 m yr^{-1} within the several major ice streams that drain the WAIS (Alley and Bindschadler, 2001) into the Ross Ice Shelf.

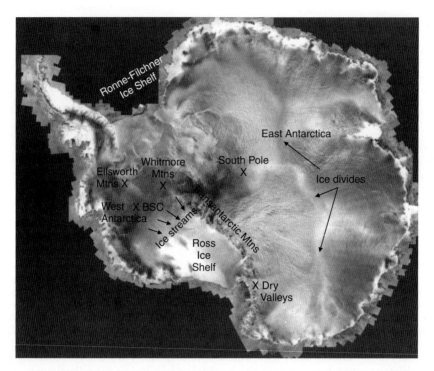

Figure 12.2 Radarsat composite image of Antarctica, with some major features labeled. BSC is Byrd surface camp.

Almost all the valley glaciers reside in the Transantarctic Mountains, where they drain the EAIS into the Ross Ice Shelf at more southerly latitudes, and into the Ross Sea at the more northerly latitudes. The Ross Ice Shelf, amalgamates several ice streams from West Antarctica, and valley glaciers from the Transantarctic Mountains. The larger, including the Byrd, Beardmore, Shackleton, Scott, and Reedy Glaciers, are greater than 2 km deep along their main channels. In turn, these large glaciers amalgamate the flow of smaller, tributary glaciers, many of which are generally 300 m deep or less. Near the coasts, these shelves are relatively stronger accumulation areas and may be moving at up to 300 m yr^{-1}.

Alpine glaciers originate in the mountains. The most well known reside in the Dry Valleys area near McMurdo Sound and no longer contribute to any of the larger glaciers. Most are several kilometers long, originate at less than 1 km above sea level, and terminate in a steep, free surface just above or bordering the floor of one of the valleys. They are frozen to the bed and move at speeds on the order of 1 m yr^{-1}. They are interesting sites for ice core studies of past climate because of their slow movement and low rates of snow accumulation.

12.2.1. Alpine glaciers: Dry Valleys

Alpine glaciers, as with any valley glacier, accumulate snow mainly within either a major zone in the upper reaches of the main trunk or within several smaller, tributary glaciers at higher elevations, each of which is an accumulation zone in itself. A single, main accumulation zone can present an uninteresting GPR or physical profile of nearly flat layers, examples of which are better discussed below in the context of the ice divides. The more interesting cases are where tributaries merge, such as the upper basin of the Commonwealth Glacier, above lower Taylor Valley (Figure 12.3). Here, there are two main tributaries, along with steep ridges that suggest a basin of several hundred meters depth. There is no airborne photography or satellite imagery that reveals the flow patterns in this area. The area is also an ecological sanctuary where use of snowmobiles is sometimes not allowed. Consequently, the following profiles were obtained by very slow hand-towing at a speed of about 0.5 m s^{-1}.

Figure 12.4 (top) shows the upper section of a 100-MHz profile of the 1.3 km cross-sectional north–south transect seen in Figure 12.3. The profile was recorded at a range of 4000 ns, with 4096 samples/trace and 16 bits/sample. The profile is not corrected for topographic elevation because the changes are minor compared with the depths profiled. No layering beneath about 70 m is apparent in the profile, partly because the dips of the folds are too steep to follow. The profile was recorded at an effective rate of only 3 traces/s because a dynamic, running stack

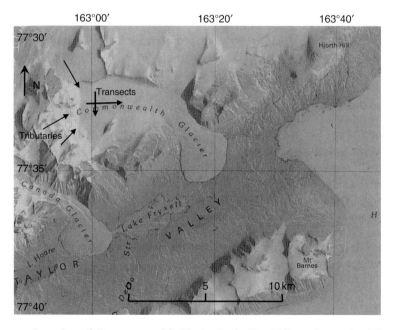

Figure 12.3 Location of Commonwealth Glacier in the Dry Valleys, Antarctica. The crossed arrows indicate the approximate location of the N–S 100-MHz and the W–E 800-MHz profiles in the upper accumulation zone. The other small arrows indicate tributary glaciers, which compressed the ice.

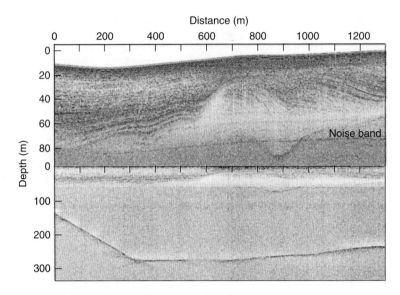

Figure 12.4 Top: 100-MHz profile across the flow in upper Commonwealth Glacier, Antarctica. The convergence of tributary glaciers has folded the ice. The depth scale is based on an estimated, average $\varepsilon = 2.4$ for firn. Bottom: The profile of the whole glacier. The loss of stratigraphy at depth is partly caused by a lack of density contrasts, partly by the inability of the radar to follow the steep folds, and insensitivity to weak contrasts in ionic concentrations. The depth scale is based on an average $\varepsilon = 3.0$ for the firn plus ice.

was needed to lower the signal to noise ratio and make the stratigraphy visible. There appears to be an upper stratigraphic regime with anticlinal folding, and a lower one with synclinal folding. A possible interpretation is that the anticlinal regime is within firn that has accumulated only in this main basin, whereas the synclinal regime shows ice that was previously folded before entering the basin, and has since undergone further compression.

Figure 12.4 (bottom) shows the bottom section of the profile. The signal strength indicates that an ice bottom at twice this depth could have been recorded. There appears to be little correlation between bottom features and the upper folding. There are only a few diffractions visible along the bottom in this display. A closer look at the rising part of the horizon (detail not shown) reveals that it is composed almost entirely of diffractions, which appears typical of any glacier margin we have profiled (e.g., Arcone et al., 2000; Moran et al., 2000; Arcone, 2002b). These diffractions may indicate far more active erosion (i.e., a rough surface) than the generally smoother and more diffraction-free profiles of the bottom of the basin.

A natural assumption regarding profiles of glacier stratigraphy is that higher frequencies should produce better resolution. This is not necessarily so because at some point the layer roughness, or uneven deposition can produce a poor stratigraphic image. As will be seen later, a 400-MHz pulse appears about optimal for resolving stratigraphy in polar regions. However, on the Commonwealth, we recorded an 800-MHz profile (Figure 12.5) along the longitudinal profile shown

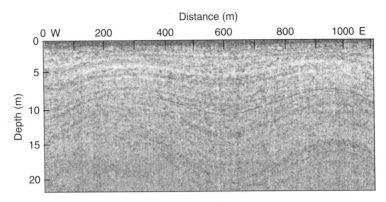

Figure 12.5 An 800-MHz profile along the glacier axis. Only some layers appear well resolved. The use of higher frequencies would probably make this image even less clear. At this frequency, the radar is sensitive to density, and not conductivity contrasts.

in Figure 12.3 with fairly decent results. The antenna unit rode smoothly with elevation changes over hard snow patches of less than about 20 cm. The profile shows layering to about 25 m depth or just about 250 ns of time range. The generally dark nature of the profile appears to result from the scattering component within the reflections themselves.

12.2.2. Polar firn: West Antarctica

The firn regime is the top 50–100 m in which snow compresses, and metamorphoses to ice. The transition, by convention, generally occurs when firn reaches a density of 830 kg m^{-3}, at which point the firn is no longer permeable, yet still porous (solid ice density = 917 kg m^{-3}). In temperate regions, where melting and refreezing is common within firn, this process may occur within about 20 m depth. In interior West Antarctica, where summer temperatures rarely climb above freezing, the transition takes place at about 60–70 m depth, whereas in East Antarctica it may occur at 100 m. 100 MHz GPR profiles of brine infiltration in Antarctic firn nearer the coast in the McMurdo Sound area were first presented by Kovacs and Gow (1975). Later, Arcone (1996) discussed profiles and these brine reflections at 400 and near 800 MHz. Vaughn et al. (1999) looked at firn on the West Antarctica plateau at 100 MHz to show interesting stratigraphic features. In this section we discuss some more recent 400-MHz profiles from which we have interpreted the nature of the density change that causes the reflections and the depositional processes that cause stratigraphic variations over tens of kilometers.

Figure 12.6 locates the several GPR transects of the 1999–2002 U.S. International Transantarctic Scientific expedition (ITASE; Mayewski, 2003) program, from which small sections of the thousands of kilometers of profiles recorded are discussed here. The profiles were recorded to select sites for shallow core drilling, but the immense volume of data permitted us to draw other conclusions. Figure 12.7 shows an example of firn stratigraphy recorded at 400 MHz near an ice divide close to the

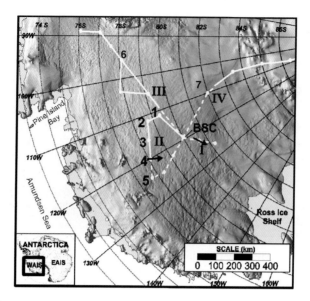

Figure 12.6 The U.S. ITASE transects in West Antarctica. Roman numerals refer to the four transects conducted from 1999 to 2002, all starting at Byrd Surface Camp (BSC). The numbers refer to some of the ice core locations. The arrows indicate ice flow directions. They are also the approximate directions of the mean wind. The dashed portions indicate transects along which prominent reflections have been tracked. Locations 2 and 3 are on an ice divide and 6, north of the Ellsworth Mountains, is near one. Location 7 is the Whitmore Mountains.

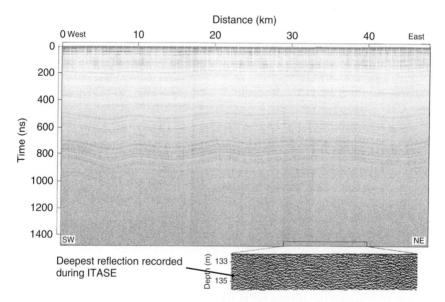

Figure 12.7 A 400-MHz profile of stratigraphy near an ice divide near the Ellsworth Mountains. In the main profile faint strata are seen to about 1000 ns (about 93 m depth). With signal processing, deeper reflections can be seen on a computer screen, but need to be highlighted for a printed display. Even deeper reflections, from the bottom of an ice shelf, with a 400-MHz antenna will be seen later. The highlighted wavelet in the detail at bottom has a negative–positive–negative phase polarity sequence for the half-cycles.

Ellsworth Mountains (Figure 12.6; location 6). The profile was recorded at a time range of 1500 ns (Arcone, 2002a), 8192 samples/trace and with a running stack of 32-fold, which gives an effective recording rate of only one trace every 15 m for the traverse speed of 12 kph. The vertical compression precludes the detail of the stratigraphy available within the 8192 samples. The display is of the Hilbert magnitude transform, which captures the amplitude envelope of the horizon wavelets. This rate of recording combined with the Fresnel zone width provides a horizontal integration length of about 22 m by 100 m depth. The amplitude is corrected for geometric spreading losses and the topographic variation was insignificant at this location. The profile contains the deepest 400-MHz reflections recorded for the entire U.S. ITASE program.

The low dielectric constant of the surface snow ensured that the 400-MHz center frequency of the wavelet and intended by the manufacturer, was indeed, very close to 400 MHz, as shown in Figure 12.1, and not the usual, near 300-MHz value characteristic of terrestrial applications. As Figure 12.1 shows, there are three major half-cycles of the 400-MHz wavelet and the 1.5 cycle wavelet is then about 63 cm long in ice, which provides a vertical resolution of about 32 cm. In firn of $\varepsilon = 2$, the resolution is then about 40 cm. This is far greater than the yearly snow accumulation rates over most of Antarctica, yet all the profiles show horizons comprising this type of well-resolved wavelet. In addition, almost all of the horizons have a particular three-banded phase polarity sequence opposite to that shown in Figure 12.1. Mostly, we find a negative–positive–negative polarity, as seen in the detail of Figure 12.7, and in Figure 12.8. Such a basic wavelet response is characteristic of a single interface between denser firn above less-dense firn below, a thin layer of less dense firn less than about 10 cm thick, or a cluster of such low-density thin layers spanning less than 10 cm, and all with interfaces between similar density contrasts. Reflection horizons with these types of wavelets have

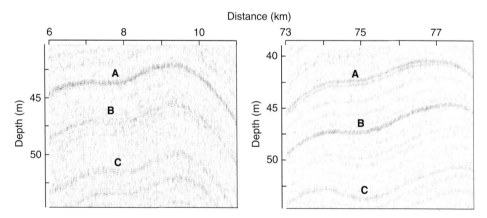

Figure 12.8 Detail of some of the layer reflections seen in the profile of Figure 12.9. Horizons A and B are continuously visible along the dashed portions of the transects in Figure 12.6. Close examination of their banding reveals that they are often blue (negative)- red (positive)- blue (negative) in phase structure.

been followed along transects that span ranges of nearly 600 km (Figure 12.6; dashed lines). These horizons appear to have been caused by widespread clusters of intense hoar growth (Arcone et al., 2004, 2005a) because all ice cores show as many as 360 such layers of hoar within a 60 m length.

Ice accelerates and thins as it spreads from a divide, and so increases the sensitivity of the surface topography to that of the bottom. The wind blowing over the hills and valleys, as gentle in slope as they may be, causes differential accumulation. Leeward slopes, where wind speeds up as it moves downglacier, receive relatively less deposition; windward slopes receive relatively more (Black and Budd, 1964; Arcone et al., 2005b). The consequences of this process can be seen in Figure 12.9, which shows as much as 30 m change in the depths of individual firn strata along a 400-MHz profile recorded while traveling west from Byrd Surface Camp in interior West Antarctica (Figure 12.6, transect I) and over a series of hills (Arcone et al., 2005b). In some places, the fold features, such as synclinal hinges, line up almost vertically, rather than moving deeper and progressively downglacier (to the right in the figure). This can be simply explained by the occurrence of accumulation anomalies, whose extra or less time delay translated vertically downward through successive layers to create the appearance of an artificial fold feature in the GPR profile. However, much of the distortion is a consequence of the superposition of time delays through consecutive layers whose

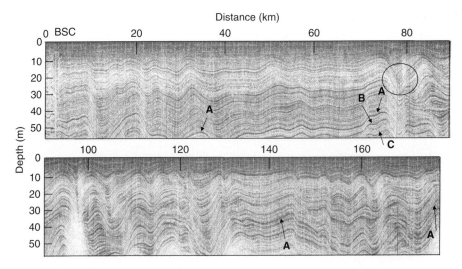

Figure 12.9 Hilbert-transformed 400-MHz profile of firn stratigraphy in a hilly area in interior West Antarctica. The profile is nearly parallel with the directions of the wind and ice flow (left to right). Differential accumulation between leeward and windward slopes causes the apparent stratigraphic deformation. Horizon A, at about 25 m depth at the right side end, falls below 56 m depth near 30 and 40 km. The circle delineates an accumulation anomaly, whose extra time delay translates vertically down and gives the appearance of a fold hinge. However, much of the distortion also comes from the superposition of variable accumulation surfaces. Detail of horizons A, B, and C are seen in the Figure 12.8.

Glaciers and Ice Sheets

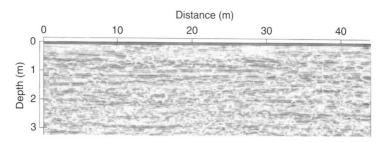

Figure 12.10 A 1200-MHz profile of stratigraphy recorded 87 km west of Byrd Surface Camp. At this scale stratigraphy is still present, but continuity is not evident.

thicknesses vary in a sinusoidal fashion with distance, and progressively move downglacier with depth (Arcone et al., 2005b). The sinusoidal type variation in thickness is a direct consequence of snow deposition within hilly topography and alignment between wind direction and ice movement. A result of this process is that fold hinge loci within areas of periodic accumulation move downglacier at only half the ice speed. When a depth whose age equals the ratio of accumulation wavelength (in kilometers) divided by ice speed (in km/yr), the fold hinge locus may reverse its direction.

The continuity of the strata imaged in the profiles of Figures 12.7 and 12.8 raises the question as to the degree of the continuity of reflection horizons on a small scale, and whether the features thought to produce it could be correlated with observations taken in snow pits or ice cores. Adjacent U.S. ITASE ice cores do not show obvious physical correlations (they are spaced 100 km apart along the U.S. ITASE routes; a further difficulty). However, the question may in part be answered by a 1200-MHz profile (Figure 12.10) recorded at the 87 km distance of Figure 12.9. Despite the fact that the 400-MHz profile shows continuity over kilometer lengths (Figure 12.7, and along the dashed lines in Figure 12.6), and the 1200-MHz profile shows layering to a depth of 3 m, the higher frequency profile shows no particular layer that is continuous throughout all 44 m. Therefore, the larger scale integration of the 400 MHz profile parameters may be necessary to find continuity of density layering in firn. If so, the physical stratigraphy within individual snow pits might never be able verify this continuity.

12.2.3. Englacial stratigraphy: West Antarctica

Classically, ground-based GPR profiles in Antarctica have concentrated on depth soundings at dominant frequencies centered from about 2–5 MHz (Jacobel et al., 1993; Nereson et al., 2000). The U.S. ITASE program included stratigraphic and bottom profiling with an improved version of these earlier systems, operating at about 3 MHz (Welch and Jacobel, 2003). The 3-MHz dominant frequency produces a 1.5 cycle pulse about 84 m long in ice. Therefore, the interface resolution is about 42 m. The low frequency allows front-end digitization of the real time signal so that a running stack of 1024-fold or more can be applied to the data, which greatly enhances the signal to noise ratio.

The dipoles are approximately 40 m long and separated by a distance of 125 m center-to-center to reduce the ringing that accompanies the directly coupled air wave propagating between them. They are dragged in a co-linear orientation with the GPR operator riding with the receiver and between the two lengths of the receiver antenna. The free-running transmitter trails behind and its powerful air wave triggers the receiver to record. There is no time variable gain applied to the 14-bit data. The dipole length places the start of the far field directivity pattern at about 56 m depth within the ice.

A 56-km long profile recorded near the ice divide in the vicinity of the Whitmore Mountains (Figure 12.6) is shown in Figure 12.11. The ringing obscures reflections for a few hundred meters, below which stratification emerges and then the bottom. The profile appears to have been successfully migrated because most of the bottom diffractions have collapsed. Stratification is visible throughout the ice depth over the shallower bottom topography. Over the deeper ice the stratification fades near the bottom, but it appears that this is entirely a function of system sensitivity; there is no time variable gain. Recent field work by Barwick et al. (2005) shows extremely low attenuation rates at 380 MHz, which implies very low rates at 3 MHz because the attenuation mechanism must be the same.

At these low frequencies for sure, and probably at much higher (Kanagaratnam et al., 2001), reflections are believed to be caused by contrasts in electrical conductivity, the stronger of which are associated with the more acidic, sulfate layers (as done by comparing reflections depths with ice core conductance profiles, e.g., Hempel et al., 2000). We can estimate the reflectivity of these contrasts from a simplified expression for the absolute value of the Fresnel reflection coefficient, Γ, for a simple conductive interface (a more exact expression for an anomalous layer is given by Kanagaratnam et al., 2001). For normal incidence $|\Gamma| = \Delta\sigma/(4\omega\varepsilon_o\varepsilon)$, where ω is the radian frequency, ε_o is the permittivity of free space and $\Delta\sigma$ is the conductivity contrast across the detected

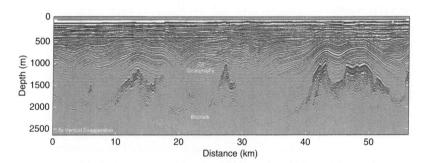

Figure 12.11 A 3-MHz migrated profile of internal stratigraphy and bottom topography recorded near the Whitmore Mountains, Antarctica. The flow is probably nearly perpendicular with the profile because the transect was nearly parallel with an ice divide. A fault is visible at 30 km. The stratigraphy is caused by layers of anomalous conductivity, which are visible to the bottom in several sections. The profile was recorded and provided by Brian Welch and Robert Jacobel of St. Olaf College, Northfield, Minnesota.

interface (there is no contrast in ice relative permittivity, ε). The Antarctic conductivity soundings of Shabtaie and Bentley (1995) gave background values of 1×10^{-6} S m^{-1} for firn, and many acidic layers show up at ten times the concentration of the background level. Therefore, we estimate a maximum $\Delta\sigma$ to be on the order of 1×10^{-5} S m^{-1}, for which $\Gamma = -47$ dB at 3 MHz. This value, when added to the approximately −35 dB geometric spreading losses over 3 km (referenced to the 56 m far field distance), places the total propagation loss well below the estimated 120–150 dB performance figure of the radar, and leaves much sensitivity to image weaker conductivity contrasts.

We have attributed radar horizons to density contrasts in firn and to conductivity contrasts in englacial ice. This simple model precludes the sensitivity of higher frequency (e.g., 400 MHz) signals to conductivity contrasts because of the inverse frequency dependence of reflectivity to conductivity. However, the wideband high-frequency profiles discussed by Kanagaratnam et al. (2001, 2004) and Paden et al. (2005) show sensitivity to ice below firn and deep within englacial regimes, respectively. Similarly, the profile shown in Figure 12.7 also shows a 400-MHz sensitivity to layering well beneath the firn–ice transition where density contrasts should be insignificant. Consequently, there is much room for research into the exact mechanisms responsible for these reflection horizons.

12.2.4. Ice shelf: McMurdo Sound

Polar ice shelves provide ideal environments for profiling ice depths because the bottom reflectivity is so high; for either sea water or frozen-on sea ice the magnitude of the reflectivity is nearly unity. The only obstacle preventing its detection is brine infiltration within firn, such as occurs in the McMurdo Ice Shelf because permeable firn is in contact with sea water just north of McMurdo Station (Figure 12.12). The delineation of stratification within the englacial zone, however, may not be straightforward because ice shelves are composed of many glacial streams, the merging of which can cause folding.

Ice shelf radar profiles have been acquired with both airborne, RES systems (Neal, 1979) and with GPR (80 MHz: Jezek et al., 1979; 40 MHz: Blindow, 1994), whereas profiles primarily of the brine layer and the firn stratigraphy have been acquired with a variety of GPR frequencies (20 MHz: Morse and Waddington, 1994; 100 MHz: Kovacs and Gow, 1975; 400 and 800 MHz: Arcone, 1996). With the availability of trace sampling rates as high as 8192 samples/trace, it is possible to use 400 MHz to profile depths of about 250 m, the time range for which (3000 ns) would still allow adequate sampling of the GPR waveform. However, when using 8192 samples/trace, the recording rate slows down and so a lower frequency may be necessary to acquire good continuity in the stratigraphic detail.

This reversal in the traditional roles of low and high frequency are illustrated in Figure 12.13, which shows the upper stratigraphy of the McMurdo Ice Shelf at a lower frequency, and the corresponding bottom topography at a higher

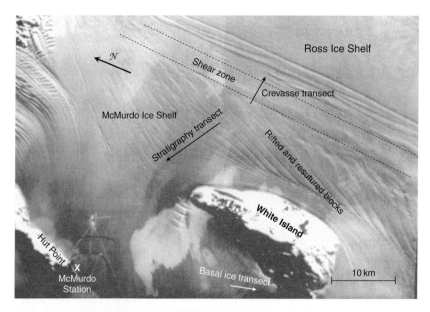

Figure 12.12 The McMurdo Ice Shelf and part of the Ross Ice Shelf regions in Antarctica, and the approximate locations of our transects.

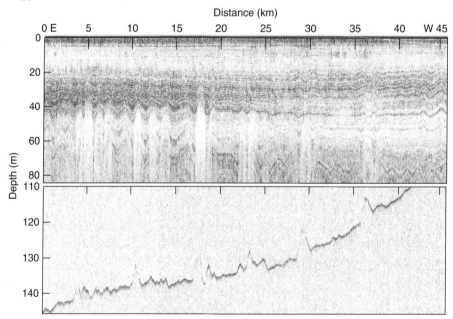

Figure 12.13 Top: 100-MHz near-surface stratigraphy along the stratigraphy transect in Figure 12.12. The folding is caused by compression of the large, rifted blocks seen in Figure 12.12. The 400-MHz profile would not resolve this stratigraphy any deeper than about 40 m because the recording rate was low in order to get a large amount of samples per trace to record the bottom reflection. Bottom: The 400-MHz profile of the bottom topography corresponding with the profile at top.

frequency. The profiles were recorded in December 2003. The 100-MHz profile reveals two regimes of folding, separated around 60–80 m depth and unsynchronized. Arcone and Laatsch (2004) interpret the upper regime to be within the firn, and the lower within older shelf ice that has underwent compression from resutured blocks that previously rifted (Figure 12.12). The higher frequency profile reveals uneven bottom topography. Although this profile has not been matched with the surface features seen in Figure 12.12, it seems likely that the notches along the bottom, most of which extend about 1 km, may be suture zones between blocks that once rifted apart. The advantage of this frequency for bottom profiling is seen from a bottom section of the 100-MHz profile, the faint diffractions of which crudely suggest a crevasse centered at 18 km (Figure 12.14).

The 400-MHz profile of Figure 12.13 suggests that the contact between ice and seawater is complicated in geometry although simple in the materials present, i.e., ice over water. However, sea ice is known to accrete sometimes under ice shelves (Blindow, 1994) because the ice at the bottom can be much colder than the sea water. This phenomenon may explain the events in a 100-MHz profile (Figure 12.15) recorded in 1995 on the McMurdo Ice Shelf along an approach to the passage between White and Black Islands (Figure 12.12). This profile, recorded just beyond where the brine reflection terminated (at left), shows at least three different interfaces, and evidence of buckling and crushing. A higher frequency profile would have revealed far more detail on the processes happening here.

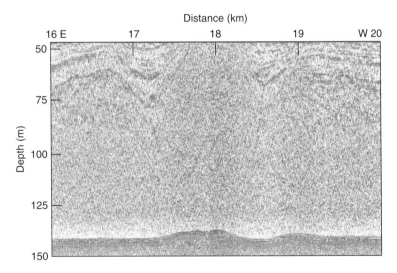

Figure 12.14 A segment of the 100-MHz profile that shows the compressive folding, most likely above a suture zone. Compare the detail of the bottom with that seen in the 400-MHz profile in Figure 12.13.

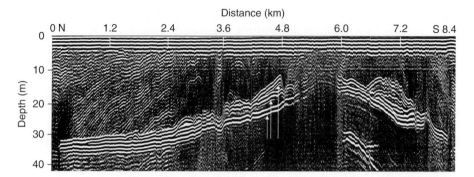

Figure 12.15 Ice bottom reflections recorded along the basal ice transect shown in Figure 12.12. The arrows indicate reflections from three interfaces, which appear to have been buckled as they lead to the unstratified section between 4.8 and 6.0 km where ice may be crushed. The dipping firn stratigraphy at left is also evidence of compressive forces.

12.2.5. Crevasses: Ross Ice Shelf

Shear zones are areas of ice where velocity gradients occur transverse to the flow direction. Consequently, there is always a component of tension within the horizontal plane, and it causes crevassing at a 45° angle to the flow direction. Shear zones are mainly associated with the margins of glaciers and ice streams, and one also occurs where the Ross Ice Shelf, moving at about $300\,\text{m}\,\text{yr}^{-1}$ to the north, is very close to the McMurdo Ice Shelf, moving about $5\,\text{m}\,\text{yr}^{-1}$. In between is a 5-km wide zone of large (up to 15 m wide) and frequent (as many as 26 along a transverse transect) crevassing, as revealed in the RADARSAT satellite imagery of Figure 12.16. The anomalies in the satellite imagery appear too wide to be individual crevasses, and so

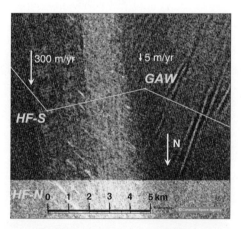

Figure 12.16 Radarsat image of the shear zone, with part of a 2002 traverse from McMurdo Station to South Pole Station traverse (Figure 12.12) superimposed. Grid Area West (GAW) and Home Free South (HF-S) were the start and end, respectively, of the traverse across this zone. The traverse was profiled to detect, and then remediate, the crevasses to establish a route for heavy equipment.

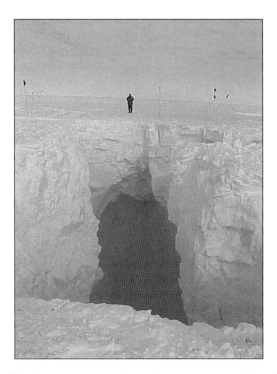

Figure 12.17 One of the largest crevasses along the transect exposed by dynamite. Widths in excess of 10 m, and snow bridge thicknesses in excess of 7 m are common in this shear zone. The person standing is 198 cm (6' 6" in boots) tall.

may actually represent swarms of crevasses, most likely occurring in en echelon formation. Some of these crevasses, exposed by dynamite, are huge (Figure 12.17).

The sensitivity of GPR to crevasses has long been known (Jezek et al., 1979; Glover and Rees, 1992; Clarke and Bentley, 1994). GPR has been used extensively in the shear zone since 1995 (Arcone and Delaney, 2000) with a 400-MHz antenna unit placed in a truck tire tube and pushed about 5 m ahead of a tracked vehicle (Figure 12.18). The tire tube easily glides over the surface and the boom gives an extra 5 m of warning. A crevasse is detected in advance by monitoring backscatter from the crevasse discontinuities (Figure 12.19) or the onset of the dipping strata within the crevasse snow bridge. Crevasses of only a few centimeters width still give strong responses. Ideally, as the antenna passes over the crevasse cavity, a reflectionless image appears, in contrast with the stratified firn (Figure 12.20). The nearly ideal cavity image occurs frequently, and often reveals a vaulted arch within the snow bridge, which must add to the snow bridge strength. The vaulted shape may result from the sagging of the snow bridge and simultaneous growth of ice between the crevasse walls and the edges of the bridge. Sometimes the image shows no cavity but instead, is filled with diffractions. Exploration has shown such images to correspond with exfoliated sheets of ice running through the crevasses.

Figure 12.18 Pisten Bully vehicle pushing a tire tube with a 400-MHz antenna inside (inset). The arrows represent the many directions of the radiation; the heavy arrow indicates downwards. Radiation forward of the antenna detects signals backscattered from a crevasse. The short, double-sided arrows in the inset indicate the direction of the antennas inside their housing.

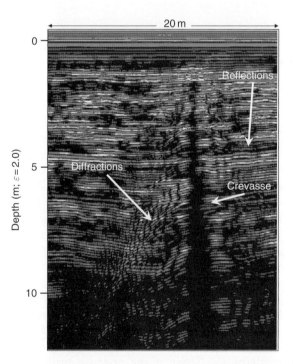

Figure 12.19 A 400-MHz radar image of a simple crevasse. The diffractions originate from discontinuities in the crevasse walls, most likely related to the intersections of the density horizons with the wall, and from roughness features on the wall. The depth scale is based on an average dielectric constant of 2.0, based on firn density measurements.

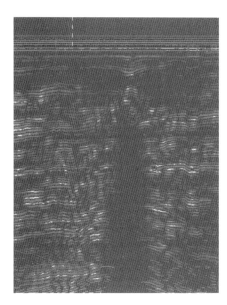 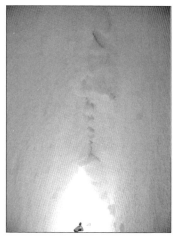

Figure 12.20 A 400-MHz crevasse image (left) that shows a convergence toward the crevasse roof. Many crevasse roofs show this arched structure (right).

12.3. ALASKA

The dominant glaciers of Alaska are the large piedmont glaciers of the Chugach Range and coastal mountains such as the Columbia, Bering-Bagley system, Malaspina, Hubbard and Taku, and the large alpine glaciers of the interior Alaska Range such as the Ruth, Muldrow, and Kahiltna on Mt. McKinley. The depths or englacial strata of any of these have not been studied with GPR or any other radar system because of difficult access, poor weather, and crevasses. However, there are much data to suggest depth soundings should be possible with lower frequency GPR. The Taku has been attempted with GPR and measured at 1500 m depth by seismic soundings (Nolan et al., 1995). The other glaciers may exceed 1000 m in some places, but most are of unknown depth. There are no reported uses of GPR on the glaciers of the Brooks Range, but they are not likely to be more than a few hundred meters deep.

Most glaciers in the Alaska and southerly mountain ranges are temperate. This means that the ice is at its pressure melting point, which is always close to 0°C. Consequently, water can exist within snow, firn, and ice. In the otherwise solid ice, water may exist along grain boundaries, and within filled or partially filled conduits and cavities. The conduits may be on the order of a millimeter or meters in diameter. This water will cause propagation loss, mainly through scattering from the larger conduits and cavities (Watts and England, 1976). The combination of a hydraulic network (Lawson, 1993) and the presence of water make radar sounding of a deep temperate glacier in spring, summer, or early fall a formidable task. Therefore, it has long been assumed that the GPR dominant frequencies should not exceed about 5 MHz where depths in excess of about 100 m need to profiled. Small glaciers present little problem (Welch et al., 1998) for low frequencies.

Almost all glacier GPR depth-sounding profiles in Alaska have been performed on moderate- to small-sized glaciers that are accessible in summer. Some examples are the Matanuska, Gulkana and Muir (Arcone et al., 2000; Lawson et al., 1998), Black Rapids (Arcone and Yankielun, 2000), Worthington (Welch et al., 1998), and Variegated (Jacobel and Anderson, 1987). Most of these investigations, with the exception of the Gulkana, have taken place in the ablation zone and so the depths are not maximal. Studies of the firn regime are available for the Bagley Ice Field (Arcone, 2002b), part of which is discussed here.

12.3.1. Temperate valley glacier: Matanuska Glacier

Figure 12.21 identifies the location on the Matanuska Glacier of the Chugach Mountains where a depth profile was recorded only about 2–3 km from the terminus (Arcone et al., 2000). The Matanuska is about 46 km long and about 8 km wide at its terminus, where significant thicknesses of basal ice have been observed and profiled with GPR (Arcone et al., 1995) within the "study area" indicated on the map. Basal ice in temperate glaciers may reach tens of meters

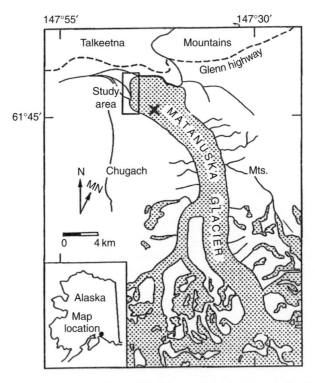

Figure 12.21 Location of a 1998 profile (marked by an X) on the Matanuska Glacier, located in the Chugach Mountains of south central Alaska. The box labeled "study area," locates where studies have been conducted of basal ice (Arcone et al., 1995; Lawson et al., 1998).

in thickness and may be associated with unfrozen but supercooled water on the bottom (Lawson et al., 1998). The elevation of the profile discussed here was about 1000 m above sea level, and there was marginal snow cover at the time of profiling. The profile (Figure 12.22) was recorded at a pulse dominant frequency of 30 MHz and at an antenna separation of 2.6 m. The apparent bottom reflection appears to indicate a smooth bottom. However, there are several apparent, sub-bottom reflections near the south end. The maximum apparent depth is about 270 m, but a hot water drill hole at 0.3 km found the ice thickness there to be 336 m. Therefore, the apparent bottom horizon is most likely from the western confining wall, and the deeper reflections are responses from complex bottom topography. These off-axis reflections are known as side-swipe and are common in radioglaciology. They occur because of the wide directivity of the antenna (Arcone, 1995), whether in the direction parallel or perpendicular to the antenna axis.

The profile also shows englacial stratification for which there are several possible explanations. We doubt the strata are caused by refrozen meltwater in crevasses because there should be no significant contrast between the permittivity of refrozen water and that of the englacial ablation zone ice matrix. A second explanation is that it is entrained basal debris dragged from the bottom, possibly as part of an

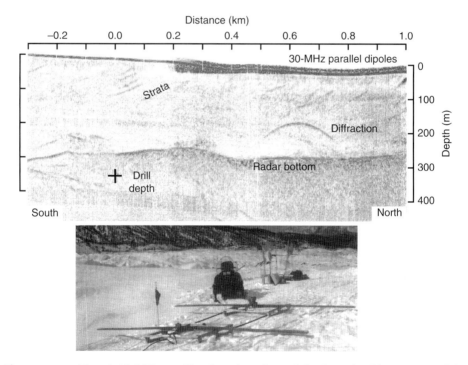

Figure 12.22 Top: A 30-MHz profile of stratigraphy and depth in the ablation zone of the Matanuska Glacier. Bottom: the 30-MHz antenna system. The dip in the strata may be caused by upthrusting, as is common in ablation zones. The diffraction centered just beyond 0.6 km may be a response from a conduit.

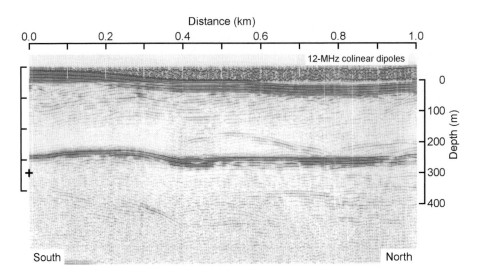

Figure 12.23 A 12-MHz version of the last kilometer of the profile in Figure 12.22. The wide spacing of the antennas makes the englacial strata appear conformable with the surface. The prominent diffraction centered near 0.63 km seen at 30 MHz is not apparent.

excavation process known as overdeepening (Lawson et al., 2000). For this to occur, the basal layers would have to be thrusted within the clear englacial ice. A third explanation is that they are dust bands or concentrations of chemical impurities existing within old layers. The updip of the layers is the orientation of old ice as it surfaces in ablation zones of any valley glacier.

A 12-MHz version of the final kilometer is shown in Figure 12.23. For this profile the antennas were co-linear, about 10 m long and with a center-to-center spacing of about 15 m. There is obvious smoothing of the bottom profile, and of the response to the englacial stratigraphy. However, the stratigraphy appears greatly distorted relative to its appearance at 30 MHz because of the wide spacing of the antennas.

12.3.2. Temperate valley glacier: Gulkana Glacier

The Gulkana Glacier of the central Alaska Range is about 8 km long and about 0.6 km at its widest. Moran et al. (2000) have discussed 50-MHz array processing to delineate the bed, and Arcone et al. (2000) have discussed 12-MHz surveys to profile the depth along single transects. One of these transects (Figure 12.24) started at the top of the main trunk and its elevation corrected profile is shown in Figure 12.25. The deepest return is from 295 m. However, it is not certain that any reflection horizon is from beneath the antenna because there are multiple bottom events, seen either in the complex bottom horizons of the full profile or in the complexity of the wavelets, as seen in the traces in Figure 12.25. The multiple oscillations of the bottom reflections could be caused by an additional reflection from a valley wall, which must be steep because the ice depth is half the width of the valley. Alternatively, the bottom wavelets

Figure 12.24 Photograph of the Gulkana Glacier, Alaska Range, with profile transects superimposed. The one discussed is along the dashed, axial line. The solid line leading to the upper transect from the left locates an array study of the bottom topography (Moran et al., 2000). The straight, dashed line across the top is the upper limit of the glacier. Features seen beyond this line are across a valley.

could be formed by a thick layer of basal ice or of wet sediments. The existence of basal ice beneath the accumulation zone has not been observed.

The strength of the bottom reflection, relative to the noise level, is at least 10. This ratio indicates that ice of least 1 km depth could be been profiled and still keep the signal well above noise. This depth is sufficient to cover most temperate glaciers.

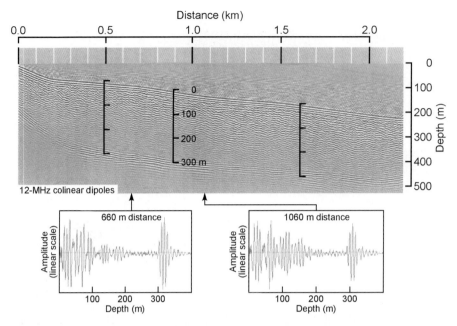

Figure 12.25 A 12-MHz radar profile along the axial transect seen in Figure 12.24. The wavelength in ice is about 14 m, which precludes seeing much detail along the bottom. The bifurcated bottom reflection starting at about 1.6 km might be events from two aspects of the bottom surface, such as one directly beneath, and one to the side of the transect.

12.3.3. Temperate firn: Bagley Ice Field, Alaska

Significant wetting of a temperate glacier occurs in the snow and firn. Maximum water contents are about 9% by volume; above amount drainage occurs. In the Chugach Mountains, soaked firn will exist by late spring below about 2000 m elevation, and to over 4000 m by late summer. The water drains down to the solid ice where it can pond to form a water table or continue to drain into conduits or, mainly, crevasses. Firn depths in temperate glaciers may only be 10–30 m thick. The drainage and refreezing process makes for complicated structures of ice pipes and lenses, which, when augmented by crevasses, suggests an almost hopeless medium in which to identify stratification with GPR. Nevertheless, a GPR profile, when subject to high rates of trace stacking over long distances, can pick out stratification, as shown here for the Bagley Ice Field.

A 50-km airborne profile along the transect shown in Figure 12.26 was recorded on June 23, 1994, when melting had long been underway (Arcone, 2002b). At the time, the glacier was surging, moving about 3 m day^{-1}, and heavily crevassed. Figure 12.27 shows a 15-km segment recorded at 135-MHz dominant frequency using an antenna slung from a helicopter. The segment shown is over the lower reaches of the east arm. The profile is stacked over 50-fold from the raw data, which eliminated the crevasse diffractions. Although the layering in the radar profile is not clear, most of the prominent layers appear to converge at the surface about 4 km down ice from

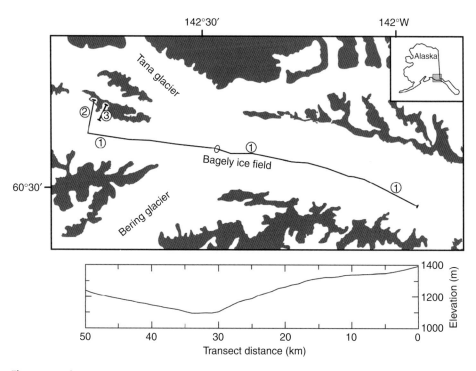

Figure 12.26 Location of radar flight line transects along the Bagley Ice Field, Alaska. The profile discussed was recorded from 0 to 15 km along transect 1. Profiles recorded along all transects 1–3 are discussed by Arcone (2002).

the "snowline." This convergence point is approximately along the firn line, which rapidly moved downice from the snowline because of the surge.

Within the unstacked data, selected crevasse diffractions with apices at the top of progressively deeper layers were modeled and interpreted for the average velocity of the firn above. By working through the layers a velocity structure was then constructed. The results show that the permittivities of the layers were significantly higher than that of temperate ice ($\varepsilon = 3.18$), and the values of several layers were interpreted for water content (Arcone, 2002b). The average firn refractive index ($n = \sqrt{\varepsilon}$) was $n = 2.5$ ($\varepsilon = 6.3$), which determined that the stratification was about 15 m thick at the east end of the profile. The loss of stratification below this depth suggests a transition to ice.

12.3.4. Temperate hydrology: Black Rapids Glacier

Temperate and some arctic glaciers are known to contain hydraulic drainage systems. Some may consist of single tunnels, whereas others may be extensive and complex. Hydraulic systems usually start at the surface in crevasses and moulins, and then branch while making their way to the bottom. Conduits are localized features and so must produce characteristic hyperbolic diffraction responses. Although

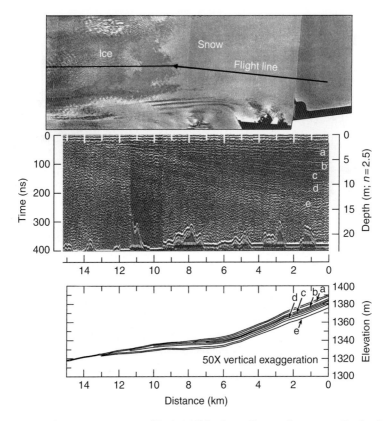

Figure 12.27 Airborne 135-MHz profile (middle) of wet firn on the eastern Bagley Ice Field, recorded June 23, 1994. The aerial photo (top) was taken September 7, 1994. The glacier was surging, and so the firn line, at 13 km along the transect, is 4 km below where the profile crossed the snowline. The interpretation (bottom) gives the perspective of the firn regime. The lettered horizons may represent summer surfaces.

conduits may seem to be an elusive target, Moorman and Michel (2000) used GPR to follow a single one along its englacial path. Obtaining a GPR image of a complex network would seem to be a matter of luck, but such a case seems evident from an investigation carried out during March 1996 on the Black Rapids Glacier.

Airborne GPR profiles of the upper reaches of the main trunk of the Black Rapids Glacier, Alaska Range, were recorded using conventional, 100-MHz GPR, and an experimental, 1.1–1.7 GHz, FMCW system (Arcone and Yankielun, 2002). The 100-MHz recordings validate the FMCW results. The cold, flat, and smooth surface conditions allowed the higher frequency FMCW signals to penetrate the surface. Figure 12.28 shows aerial photographs of the area, which contains an extensive network of potholes, some of which are interconnected from observations of water flowing into and out of them.

Figure 12.29 shows profiles recorded downglacier from the potholes. The two radar systems were used on separate runs, but both followed the same path by GPS

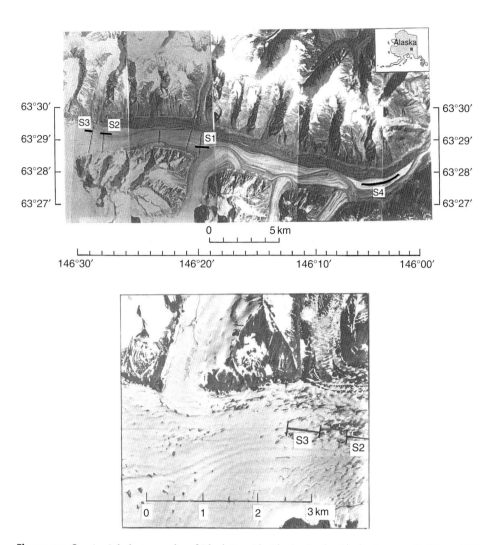

Figure 12.28 Aerial photographs of Black Rapids Glacier in the Alaska Range. S1, S2, and S3 are short axial transects and S4 is a longer one. The detail of the bottom image shows a field of potholes that are known to drain water in the summer. Arcone and Yankielun (2002) discuss profiles along all four transects.

positioning and landmarks on the ice. The profiles reveal images interpreted to be drainage networks within the near-surface (Arcone and Yankielun; 2002). Although the FMCW profile is far clearer, the 100-MHz profile reveals all the same features, and could have been improved if not for the clutter that resulted from the mounting of the antenna directly below the helicopter fuselage. The arborescent nature of the features is impressive and difficult to explain theoretically. The distance of the profile downglacier from the pothole field suggests that this network has lasted, and probably evolved, over many years.

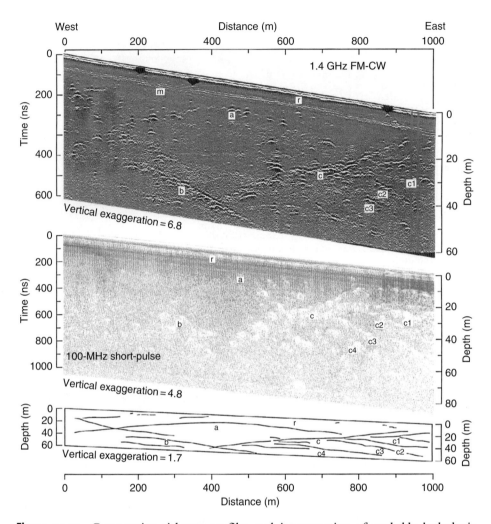

Figure 12.29 Comparative airborne profiles and interpretation of probable hydrologic features within 60 m of the Black Rapids Glacier surface. The profiles were recorded along the transect labeled S1 in Figure 12.28. The upper profile was recorded with an FM-CW system and the lower with a conventional 100-MHz GPR. The interpretation gives the perspective of the feature geometry. Corresponding branches of the structure between profiles are labeled.

12.4. SUMMARY

GPR is capable of sounding the depths and of profiling the stratigraphy of wet or dry firn, ice shelves, most small valley glaciers, and deep polar ice. Most significant is its ability to respond to the weak stratification of physical or chemical contrasts. The causes of these contrasts, such as density, ice fabric, conductivity, or even non-conductive changes in the imaginary component of permittivity, may not always be obvious or even measurable in a snow pit, trench, or ice core. In addition these are only point observations over a horizontal area, whereas a GPR horizon is an integration of many returns from many points. The ever-increasing GPR and coring activity in Antarctica by many countries may shed more light on this topic.

A significant GPR research topic would be to test its ability to sound the depths and profile the strata of temperate glaciers of more than 500 m depth. Access, weather, and crevassing over deep glacial basins may preclude obtaining such profile depths, for which even quality, airborne RES data are not available. Low-frequency 3–5 MHz radars are well suited for this purpose, but if penetration of hundreds of meters, as discussed here, is possible at 12–30 MHz, then soundings near 1-km depth should also be possible. The advent of front-end digitization of the received RF signal using faster, 16-bit analog-to-digital chip technology should make deep profiling at these frequencies possible. Such higher frequencies should also make profiling of englacial strata possible, but englacial stratification in any valley glacier may be precluded by either melting or wind mixing of chemical impurities upon snow deposition.

REFERENCES

Alley, R.B. and Bindschadler, R. A. (eds), 2001, The west Antarctic ice sheet. Behavior and Environment, Washington, DC, American Geophysical Union, Antarctic Research Series, Vol. 77, 296 p.

Arcone, S.A., 1995, Numerical studies of the radiation patterns of resistively-loaded dipoles. Journal of Applied Geophysics, Vol. 33, pp. 39–52.

Arcone, S.A., 1996, High resolution of glacial ice stratigraphy: A ground-penetrating radar study of Pegasus Runway, McMurdo Station, Antarctica. Geophysics, Vol. 61, pp. 1653–1663.

Arcone, S.A., 2002a, Stratigraphic profiling of Antarctic firn with 400-MHz GPR at 1500 ns. Proceedings, International Conference on Ground-Penetrating Radar, 9th, Santa Barbara, California, pp. 433–437.

Arcone, S.A., 2002b, Airborne-radar stratigraphy and electrical structure of temperate firn: Bagley Ice Field, Alaska. Journal of Glaciology, Vol. 48, pp. 317–334.

Arcone, S.A. and Delaney, A. J., 2000, GPR images of hidden crevasses in Antarctica. Proceedings, International Conference on Ground-Penetrating Radar, 8th, Gold Coast, Australia, pp. 760–765.

Arcone, S.A. and Yankielun, N.E., 2000, 1.4-GHz radar penetration, and evidence of drainage structures in temperate ice: Black Rapids Glacier, Alaska, U.S.A. Journal of Glaciology, Vol. 46, pp. 477–489.

Arcone, S.A. and Laatsch, J. G., 2004, Reversing the roles of high and low frequency to profile the dynamics of an ice shelf. Proceedings, International Conference on Ground-Penetrating Radar, 10th, Delft, Netherlands, pp. 781–784.

Arcone, S.A., Lawson, D.E., and Delaney, A.J., 1995, Short-pulse radar wavelet recovery and resolution of dielectric contrasts within englacial and basal ice of Matanuska Glacier, Alaska. Journal of Glaciology, Vol. 41, pp. 68–86.

Arcone, S.A., Lawson, D.E., Delaney, A.J. and Moran, M.L., 2000, 12–100 MHz depth and stratigraphic profiles of temperate glaciers. Proceedings, International Conference on Ground-Penetrating Radar, 8th, Gold Coast, Australia, pp. 377–382.

Arcone, S.A., Spikes, V.B., Hamilton, G.S. and Mayewski, P.A., 2004, Stratigraphic continuity in 400-MHz radar profiles in West Antarctica. Annals of Glaciology, Vol. 39, pp. 195–200.

Arcone, S.A., Spikes, V.B. and Hamilton, G.S., 2005a, Phase structure of radar stratigraphic horizons within Antarctic firn. Annals of Glaciology, Vol. 41, pp. 10–16.

Arcone, S.A., Spikes, V.B. and Hamilton, G.S., 2005b, Stratigraphic variation in polar firn caused by differential accumulation and ice flow: Interpretation of a 400-MHz short-pulse radar profile from West Antarctica. Journal of Glaciology, Vol. 51, pp. 407–422.

Barwick, S., Besson, D., Gorham, P. and Satzberg, D., 2005, South polar in situ radio-frequency ice attenuation. Journal of Glaciology, Vol. 51, pp. 231–238.

Bjornsson, H., Gjessing, Y., Hamran, S.-E., Hagen, J.O., Liestol, O., Palsson, F. and Erlingsson, B., 1996, The thermal regime of sub-polar glaciers mapped by multi-frequency radio-echo sounding. Journal of Glaciology, Vol. 42, pp. 23–32.

Black, H.P. and Budd, W., 1964, Accumulation in the region of Wilkes, Wilkes Land, Antarctica. Journal of Glaciology, Vol. 5, pp. 3–15.

Blindow, N., 1994, The central part of the Filchner-Ronne Ice Shelf, Antarctica: Internal structures revealed by 40 MHz monopulse RES. Annals of Glaciology, Vol. 20, pp. 365–371.

Bogorodsky, V.V., Bentley, C.R. and Gudmandsen, P.E., 1985, Radioglaciology, Dordrecht, D. Reidel Publishing Co., 254 p.

Clarke, T.S. and Bentley, C.R., 1994. High-resolution radar on Ice Stream B2, Antarctica: Measurements of electromagnetic wave speed in firn and strain history from buried crevasses. Annals of Glaciology, Vol. 20, pp. 153–159.

Daniels, D. (ed.), 2004, Ground Penetrating Radar, 2nd edition, The Institution of Electrical Engineers, London, UK, 726 p.

Fahenstock, M., Abdalati, W., Luo, S. and Gogineni, S., 2001, Internal layer tracking and age-depth-accumulation relationships for the northern Greenland ice sheet. Journal of Geophysical Research, Vol. 106(D24), pp. 33789–33797.

Fujita, S., Maeno, H., Uratsuke, S., Furukawa, T., Mae, S., Fujii, Y. and Watanabe, O., 1999, Nature of radio echo layering in the Antarctic ice sheet detected by a two-frequency experiment. Journal of Geophysical Research, Vol. 104(B6), pp. 13013–13024.

Glover, J.M. and Rees, H.V., 1992, Radar investigations of firn structures and crevasses, in Pilon, J. (ed.), Ground Penetrating Radar, Ottawa, CA, May 1988, Canada Communication Group–Publishing, Geological Survey of Canada, Paper 90-4, pp. 75–84.

Gogineni, S., Chuah, T., Allen, C., Jezek, K. and Moore, R.K., 1998, An improved coherent radar depth sounder. Journal of Glaciology, Vol. 44, pp. 659–669.

Hamran, S.-E., Gjessing, D.T., Hjelmstad, J. and Aarholt, E., 1995, Ground penetration synthetic pulse radar; dynamic range and modes of operation. Journal of Applied Geophysics, Vol. 33, pp. 7–14.

Hempel, L. and Thyssen, F., 1993, Deep radio echo soundings in the vicinity of GRIP and GISP2 Drill sites, Greenland. Polarforschung, Vol. 62, pp. 11–16.

Hempel, L., Thyssen, F., Gundestrup, N., Clausen, H.B. and Miller, H., 2000, A comparison of radio-echo sounding data and electrical conductivity of the GRIP ice core. Journal of Glaciology, Vol. 46, pp. 369–374.

Jacobel, R.W. and Anderson, S.K., 1987, Interpretation of radio-echo returns from internal water bodies in Variegated Glacier, Alaska, U.S.A. Journal of Glaciology, Vol. 24, pp. 321–330.

Jacobel, R.W., Gades, A.M., Gottschling, D.L., Hodge, S.M. and Wright, D.L., 1993, Interpretation of radar-detected internal layer folding in West Antarctic ice streams. Journal of Glaciology, Vol. 39, pp. 528–537.

Jezek, K.C., Bentley, C.R. and Clough, J.W., 1979, Electromagnetic sounding of bottom crevasses on the Ross Ice Shelf, Antarctica: Journal of Glaciology, Vol. 24, pp. 321–330.

Kanagaratnam, P., Gogineni, S.P., Gundestrup, N. and Larsen, L., 2001, High-resolution radar mapping of internal layers at NGRIP. Journal of Geophysical Research, Vol. 106(D24), pp. 33799–33811.

Kanagaratnam, P., Gogineni, S.P., Ramasami, V. and Braaten, D., 2004, A Wideband radar for high-resolution mapping of near-surface internal layers in glacial ice. IEEE Transactions on Geoscience and Remote Sensing, Vol. 42, pp. 483–490.

Kovacs, A. and Gow, A.J., 1975, Brine infiltration in the McMurdo Ice Shelf, McMurdo Sound, Antartica. Journal of Geophysical Research, Vol. 80, pp. 1957–1961.

Kohler, J., Moore, J., Kennett, M., Engeset, R. and Elvehoy, H., 1997, Using ground-penetrating radar to image previous years? summer surfaces for mass balance measurements. Annals of Glaciology, Vol. 24, pp. 255–260.

Lawson, D.E., 1993, Glaciohydrologic and glaciohydraulic effects on runoff and sediment yield in glacierized basins: CRREL Monograph 93-2, U.S. Army Cold Regions Research and Engineering Laboratory, Hanover, NH, 108 p.

Lawson, D.E., Strasser, J.C., Evenson, E.B., Alley, R.B., Larson, G.J. and Arcone, S.A., 1998, Glaciohydraulic supercooling: A freeze-on mechanism to create stratified, debris-rich basal ice: I. Field evidence. Journal of Glaciology, Vol. 44, pp. 547–562.

LeGrand, M. and Mayewski, P.A., 1997, Glaciochemistry of polar ice cores: A review. Reviews of Geophysics, Vol. 35, pp. 219–244.

Maijala, P., Moore, J.C., Hjelt, S.-E., Palli, A. and Sinisalo, A., 1998, GPR investigations of glaciers and sea ice in the Scandinavian arctic. Proceedings International Conference on Ground-Penetrating Radar, 7th, Lawrence, Kansas, pp. 143–148.

Mayewski, P.A., 2003, Antarctic oversnow traverse-based Southern Hemisphere climate reconstruction: EOS. Transactions of the American Geophysical Union, Vol. 84, No. 22, 3 June.

Moorman, B.J. and Michel, F.A., 2000, Glacial hydrological system characterization using ground-penetrating radar. Hydrological Processes, Vol. 14, pp. 2645–2667.

Moran, M.L., Greenfield, R.J., Arcone, S.A. and Delaney, A.J., 2000, Delineation of a complexly dipping temperate glacier bed using short-pulse radar arrays. Journal of Glaciology, Vol. 46, pp. 274–286.

Morse, D.L. and Waddington, E.D., 1994, Recent survey of brine infiltration in McMurdo Ice shelf, Antarctica. Annals of Glaciology, Vol. 20, pp. 215–218.

Murray, T., Gooch, D.L. and Stuart, G.W., 1997, Structures within the surge front at Bakaninbreen, Svalbard, using ground-penetrating radar. Annals of Glaciology, Vol. 24, pp. 122–129.

Murray, T., Stuart, G.W., Fry, M., Gamble, N.H. and Crabtree, M.D., 2000, Englacial water distribution in a temperate glacier from surface and borehole radar velocity analysis. Journal of Glaciology, Vol. 46, pp. 389–398.

Neal, C.S., 1979, The dynamics of the Ross Ice Shelf revealed by radio-echo sounding. Journal of Glaciology, Vol. 24, pp. 295–307.

Nereson, N.A., Raymond, C.F., Jacobel, R.W. and Waddington, E.D., 2000, The accumulation pattern across Siple Dome, West Antarctica, inferred from radar-detected internal layers. Journal of Glaciology, Vol. 46, pp. 75–87.

Nicollin, F. and Kofman, W., 1994, Ground penetrating radar sounding of a temperate glacier; modeling of a multilayered medium. Geophysical Prospecting, Vol. 42, pp. 715–734.

Nolan, M., Motyka, R.J., Echelmeyer, K. and Trabant, D.C., 1995, Ice-thickness measurements of Taku Glacier, Alaska, U.S.A., and their relevance to its recent behavior. Journal of Glaciology, Vol. 41, pp. 541–553.

Odegard, R.S., Hagen, J.O. and Hamren, S.-E., 1997, Comparison of radio-echo sounding (30–1000 MHz) and high resolution borehole-temperature measurements at finsterwalderbreen, southern Spitsbergen, Svalbard. Annals of Glaciology, Vol. 24, pp. 262–267.

Paden, J., Allen, C., Gogineni, S., Dahl-Jensen, D., Larsen, L. and Jezek, K., 2005, Wideband measurements of ice sheet attenuation and basal scattering. IEEE Transactions on Geoscience and Remote Sensing Letters, Vol. 2, pp. 164–168.

Richardson, C., Aarholt, E., Hamran, S.-E., Holmlund, P. and Isaksson, E., 1997, Spatial distribution of snow in western Dronning Maud Land, East Antarctica, mapped by a ground-based snow radar. Journal of Geophysical Research, Vol. 102(B9), pp. 20343–20353.

Shabtaie, S. and Bentley, C.R., 1995, Electrical resistivity sounding of the East Antarctic Ice Sheet. Journal of Geophysical Research, Vol. 100(B2), pp. 1933–1954.

Vaughn, D.G., Corr, H.F.J., Doake, C.S.M. and Waddington, E.D., 1999, Distortion of isochronous layers in ice revealed by ground-penetrating radar. Nature, Vol. 398, 25 March.

Watts, R.D. and England, A.W., 1976, Radio-echo sounding of temperate glaciers: Ice properties and sounder design criteria. Journal of Glaciology, Vol. 17, pp. 39–48.

Welch, B.C. and Jacobel, R.W., 2003, Analysis of deep-penetrating radar surveys of West Antarctica, US-ITASE 2001. Geophysical Research Letters, Vol. 30, p. 1444, doi: 10.1029/2003GL017210.

Welch, B.C., Pfeffer, W.T., Harper, J.T. and Humphrey, N.F., 1998, Mapping subglacial surfaces of temperate valley glaciers by two-pass migration of a radio-echo sounding survey. Journal of Glaciology, Vol. 44, pp. 164–170.

Yankielun, N.E., Arcone, S.A. and Crane, R.K., 1992, Thickness profiling of freshwater ice using a millimeter-wave FM-CW radar. IEEE Transactions on Geoscience and Remote Sensing, Vol. 80, pp. 1094–1100.

PART IV

ENGINEERING AND SOCIETAL APPLICATIONS

CHAPTER 13

NDT Transportation

Timo Saarenketo

Contents

13.1.	Introduction	396
13.2.	GPR Hardware and Accessories	397
	13.2.1. General	397
	13.2.2. Air-coupled systems	398
	13.2.3. Ground-coupled systems	398
	13.2.4. Antenna configurations	399
	13.2.5. Antenna and GPR system testing	399
	13.2.6. Accessory equipment	400
13.3.	Data Collection	401
	13.3.1. General	401
	13.3.2. Data collection setups and files	403
	13.3.3. Positioning	404
	13.3.4. Reference sampling	405
13.4.	Data Processing and Interpretation	405
	13.4.1. General	405
	13.4.2. GPR data preprocessing	406
	13.4.3. Air-coupled antenna data processing	407
	13.4.4. Ground-coupled data processing	408
	13.4.5. Determining dielectric values or signal velocities	410
	13.4.6. Interpretation – automated vs. user controlled systems	411
	13.4.7. Interpretation of structures and other objects	411
13.5.	Integrated GPR Data Analysis with Other Road Survey Data	413
	13.5.1. General	413
	13.5.2. GPR and FWD	413
	13.5.3. Profilometer data	414
	13.5.4. GPS, digital video and photos	415
	13.5.5. Other data	416
13.6.	GPR Applications on Roads and Streets	416
	13.6.1. General	416
	13.6.2. Subgrade surveys, site investigations	416
	13.6.3. Unbound pavement structures	419
	13.6.4. Bound pavement structures and wearing courses	420
	13.6.5. GPR in QC/QA	423
	13.6.6. Special applications	425
13.7.	Bridges	425
	13.7.1. General	425
	13.7.2. Bridge deck surveys	426
	13.7.3. Other bridge applications	428

13.8. Railways 429
    13.8.1. General 429
    13.8.2. Data collection from railway structures 430
    13.8.3. Ballast surveys 431
    13.8.4. Subgrade surveys, site investigations 432
13.9. Airfields 433
13.10. Summary and Recommendations 435
References 436

13.1. INTRODUCTION

The history of ground penetrating radar (GPR) tests in traffic infrastructure surveys dates back to the early- and mid-1970s, when according to Morey (1998) the Federal Highway Administration (FHWA) in the USA tested the feasibility of GPR in tunnel applications and later on bridge decks (Saarenketo, 2006). The first vehicle-mounted GPR system for highways was developed under an FHWA contract in 1985 (Morey, 1998). In the early 1980s, GPR surveys were also started in Canada (see Manning and Holt, 1983; Carter et al., 1992). The other active area in the late 1970s and early 1980s was Scandinavia, where the first GPR tests with ground-coupled antennas were performed in Sweden (Ulriksen, 1982; Johansson, 1987) and Denmark (Berg, 1984), but the method did not receive general acceptance at that time. However, after the first tests were conducted in Finland in 1986 (Saarenketo, 1992), the method rapidly became a routine survey tool in various road design and rehabilitation projects in Finland (Saarenketo, 1992; Saarenketo and Maijala, 1994; Saarenketo and Scullion, 1994; Saarenketo and Scullion, 2000) and later as a pavement design and quality control tool (Saarenketo and Roimela, 1998; Scullion and Saarenketo, 1998; Saarenketo, 1999; Pälli et al., 2005).

In the late 1980s and early 1990s, most infrastructure applications in North America focused on pavement thickness measurements (Maser, 1994), detecting voids under concrete slabs (Scullion et al., 1994) and detecting deteriorated areas in bridge decks (Alongi et al., 1992). These surveys were mainly conducted with high-frequency (1.0 GHz) air-launched antennas (see Scullion et al., 1992). In the mid- and late-1990s the most common GPR applications by highway agencies were surveys to measure pavement layer thickness, detect voids and bridge delamination; followed by measuring depth to steel dowels and depth to bedrock, detection of buried objects, asphalt stripping and scour around bridge support. Of the various applications, GPR seemed to be the most successful for pavement layer thickness measurements, whereas agencies report less satisfactory results with void detection and questionable results locating areas of asphalt stripping (Morey, 1998).

According to Hobbs et al. (1993), the first civil engineering tests with GPR in the UK were done in the UK in 1984. Since then the published GPR research has focused especially on concrete structures (see Millard et al., 1993), pavement testing (Ballard, 1992, 1993; Daniels, 1996) and, recently, railway surveys (see Clark et al.,

2003a). In France, the main focus was on pavement testing (see Daniels, 1996). In the Netherlands, the main application on roads has been layer thickness measurements (Hopman and Beuving, 2002). In other parts of the world GPR techniques have been used for monitoring roads in more than 20 countries, and according to the knowledge of the author, GPR surveys on roads are quite widely used in Australia, Canada, China, Estonia, Germany, Italy, Lithuania, New Zealand, Spain, Sweden, Switzerland and the UK.

Currently among the geophysical engineering tools that provide information about the physical properties of a site which in turn can be related to highway problems GPR has the greatest number of applications (Anderson and Ismael, 2002). In traffic infrastructure surveys the major advantages of GPR testing are continuous profile, speed and accuracy. According to Hall et al. (2002), it continues to be the only technology that can provide meaningful subsurface information at close to highway speed. Its disadvantages include the complexity of the GPR data, and as such good software products are needed to make the GPR signals meaningful to engineers (Saarenketo and Scullion, 2000; Hall et al., 2002).

13.2. GPR Hardware and Accessories

13.2.1. General

The GPR systems used in road surveys are mainly impulse radars, but recently stepped frequency radar systems have also been tested in road surveys (Dérobert et al., 2001, 2002a; Eide, 2002). The GPR hardware, mounted on a survey van and used in traffic infrastructure surveys normally have the following components: (1) ground-coupled and/or air-coupled antennas with transmitter/receiver electronics; (2) cables; (3) GPR control unit; (4) pulse encoder and other positioning units; and (5) accessory equipment. Normally a GPR road survey unit has an additional control unit, normally a PC, to facilitate the combined use of GPR and accessory equipment or make log files that allow subsequent linking of the data sets to one another (Figure 13.1).

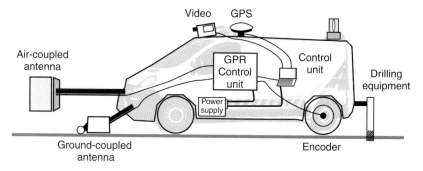

Figure 13.1 A GPR system for road surveys. Antennas are normally mounted in front of the car, which allows the driver to control and maneuver the antennas and add markers precisely when the antenna passes reference points.

13.2.2. Air-coupled systems

The air-coupled GPR systems are increasingly being used for evaluation of the upper part of the pavement structure. They produce relatively clean signals and can operate at close to highway speed. Furthermore, with defect-free pavements the signals can be processed to compute both layer thickness and layer dielectrics (Saarenketo and Scullion, 2000). Air-coupled antenna systems are pulse radar systems, and they generally operate in the range from 500 MHz to 2.5 GHz, the most common central frequency being 1.0 GHz. Their depth penetration is typically 0.5–0.9 m. During data acquisition these antennas are suspended 0.3–0.5 m above the pavement surface. Most air-coupled antenna types are transverse electromagnetic (TEM) horn antennas but hemispherical butterfly dipole (HBD) types have also been used in road surveys.

The greatest advantage of air-coupled systems is, because antenna coupling does not change with the changes in pavement properties, their repeatability, which allows them to be used for measuring changes in material properties for instance in asphalt quality control surveys (Saarenketo, 1998). Another advantage is, because they are mounted above the pavement, data collection can be done at full speed (up to $100 \, km \, hr^{-1}$) without interfering with traffic. That is why air-coupled systems are recommended to be used in network level PMS surveys in Germany (Golgowski, 2003). Currently horn antenna-type air-coupled systems are manufactured by GSSI, Penetradar; Pulse Radar and Wavebounce, all from the USA, and butterfly dipole systems by Radar Team Sweden Ab. Other air-coupled systems reported by Hopman and Beuving (2002) to be in use in pavement surveys in the Netherlands are Euradar and IRIS.

The Texas Transportation Institute has conducted a large amount of research and development work both in improving air-coupled system performance and in specifications as well as developing new pavement-testing applications (see Lau et al., 1992; Scullion et al., 1992; Scullion et al., 1994; Scullion et al., 1997; Scullion, 2001).

13.2.3. Ground-coupled systems

Ground-coupled antennas operate in a wide range of central frequencies from 80 to 1500 MHz, and the signal penetration in traffic infrastructure surveys can be up to 20–30 m. During data acquisition these antennas maintain contact with pavement or they are suspended just above it. If they are not in contact, the distance to structure surface must be kept constant because the coupling changes as a function of distance. The clear advantage of ground-coupled systems is the better signal penetration compared with that of air-coupled systems, although surface coupling and antenna ringing present problems, which make it difficult to obtain any quantitative information from the near surface without signal processing. Another advantage is better vertical resolution compared with air-coupled antenna systems, which allows these antennas to be used, for example, to detect pavement cracks, cables and reinforcement bars in concrete structures. Data collection speed with ground-coupled systems is normally $5–30 \, km \, hr^{-1}$.

The leading commercial manufacturers of ground-coupled antennas used in road, airport and railway surveys are GSSI (USA), IDS (Italy), MALA (Sweden), Penetradar (USA), Sensors and Software (Canada) and UTSI Electronics (UK).

13.2.4. Antenna configurations

Most GPR antennas used in traffic infrastructure surveys are bistatic even though the antenna elements are mainly installed in the same antenna box with transmitter and receiver electronics. Bistatic antennas in different boxes have the advantage that they can be used for determining the dielectric properties of the pavement structure using, for example wide angle reflection and refraction (WARR) or common mid point (CMP)-sounding techniques, but these sounding techniques can also be used with multichannel GPR systems.

Due to the rapid development of data processors and data storage capabilities, multichannel systems are more and more popular in road surveys. There are several advantages, when data are collected using several antennas simultaneously: (1) high-frequency antennas with good resolution near the pavement surface and lower frequency antennas with greater signal penetration can be used at the same time, (2) multiple channels allow the use of antenna array techniques to determine signal velocities (Davis et al., 1994; Mesher et al., 1995; Emilsson et al., 2002) and (3) multichannel systems allow data collection using many antennas with same frequency to collect several survey lines simultaneously which facilitates the preparation of a 3D model of the surveyed structures (Davidson and Chase, 1998; Manacorda et al., 2002), or the configuration can be a mixture of any of these three. Figure 13.2 presents different GPR antenna systems used in road surveys.

13.2.5. Antenna and GPR system testing

One problem with GPR hardware systems is that all the systems are unique, and especially in the early 1990s, there were major differences when different systems using the same model of antenna were compared on the same test track. The other problem was that antennas' performance was changing over time, and GPR users had difficulty testing if their older GPR systems were still functioning accurately in the field. The accuracy of GPR is especially important in asphalt quality control surveys because large fines can be imposed on asphalt contractors based on the GPR survey results.

To compare the performance of different GPR systems, the Texas Department of Transportation (TxDOT) requested that the Texas Transportation Institute develop a series of performance specifications for 1.0 GHz air-coupled systems (Scullion et al., 1996). The proposed tests were (1) noise-to-signal ratio (N/S ratio), (2) signal stability (amplitude and time jitters), (3) travel-time linearity, (4) long-term stability (time widow shifting and amplitude stability) and (5) penetration depth. All of these tests, except the penetration depth test, have become common methods for testing air-coupled antenna systems; these tests can also be done with ground-coupled systems.

Figure 13.2 Different GPR systems and antenna configurations used in road surveys. Photo (a) presents a multichannel stepped frequency radar by 3D Radar from Norway; photo (b) presents a 100 MHz GSSI ground-coupled antenna and a 1.0 GHz Pulse Radar antenna used by the Texas Transportation Institute (TTI); photo (c) presents a Canadian Road Radar system that has a horn antenna system and a multichannel ground-coupled antenna array (see Davis et al., 1994), and photo (d) presents a multichannel air-coupled horn antenna system by Penetradar.

13.2.6. Accessory equipment

There are many different accessories that can be used with GPR systems in traffic infrastructure surveys, but the two most popular systems, used on most survey vehicles, are digital video systems and GPS. A third recommended accessory is a drilling system for taking reference samples.

Digital video allows the interpreter to see the antenna's surroundings during data collection, which further helps in comprehension of the GPR signal which in turn leads to more accurate interpretations of the structure or individual reflectors such as a culvert. Video is especially useful in pavement rehabilitation projects or forensic surveys where it is important to correctly diagnose the reasons for a defect. Infrared thermography cameras have also been used together with the collection of GPR data especially on bridge decks and concrete runways (Manning and Holt, 1986; Maser and Roddis, 1990; Weil, 1992). Recently, infrared thermal cameras and temperature sensors have been used together with GPR in asphalt quality control (Sebesta and Scullion, 2002b) and on railways (Clark et al., 2004).

Another, almost compulsory, system in traffic infrastructure surveys is a global positioning system (GPS) because, in most cases, the survey results need to be projected onto a survey line. This line is mainly a road register address that is calculated to the road centre because when data are collected from the outer lane or even the outer wheelpath in two-lane roads there are distance shifts on curves. Good-quality GPS data permit distance data corrections and comparison of data collected from different lanes. Also, a GPS system with accurate z-coordinates helps interpreters and pavement engineers to better understand the GPR data. A GPS system is also very useful when data are collected from wide areas with no specific visual position referencing such as airport runways and taxiways. Normally GPS coordinates are linked with GPR scan numbers, and data-processing software is used to make distance corrections.

There are also many types of road survey vans that have integrated different measurement devices into the same vehicle. Profilometers and pavement distress-mapping systems especially have been instrumented together with GPR into the same van. Some attempts have also been made to integrate GPR system into a falling weight deflectometer (FWD) vehicle.

13.3. Data Collection

13.3.1. General

GPR survey design is a process where close co-operation between the customer and the GPR survey contractor is strongly recommended. If the customer manages the project through a competition between different GPR contractors, then detailed project descriptions and GPR specification documents are necessary in the tender documents (see Golgowski, 2003).

The key issue in the project description is a detailed outline of the nature of the problem with which the survey is concerned. The textbook *Ground Penetrating Radar* (1992) provides a checklist of points that customers and GPR consultants should resolve before the survey contract is signed.

In planning a GPR survey, the number of survey lines should be defined. In routine road surveys, one longitudinal survey line is made and most commonly, in two-lane roads, the right wheelpath of the right lane (increasing road data bank distance) is used. However, recent results suggest that in asphalt quality control surveys utilizing a horn antenna, data should be collected from between the wheelpaths because of the effects of compaction on the wheelpaths caused by heavy traffic (Saarenketo, 2006). If a two antenna system with one air-coupled and one ground-coupled antenna is used, they can be placed in a line or beside each other in which case the ground-coupled antenna should be placed to measure between the wheelpaths. In railway surveys, if only one line is to be measured, it should be done between the rails, but in many surveys data are also collected from both sides of the rails. In bridge deck surveys the recommended interval between survey lines is 0.5 m.

Cross sections always provide very useful information concerning the road structures and help with the understanding of road failure mechanisms but on highly trafficked roads data collection is especially difficult and dangerous and, as such, special safety arrangements are a requirement. With new 3D GPR techniques, analysing road structures through cross sections will become much easier.

Collection of GPR data does have some weather restrictions. Collecting data during the rainfall or when the pavement is wet is not recommended. When collecting the data with air-coupled antenna systems for quality control purposes the pavement must be very dry with no visual moist spots, and the pavement temperature should be above 0°C if dielectric information is to be used in the analysis. On the other hand, GPR data collection with ground-coupled antennas during the winter can provide even better results especially on gravel roads, because the frozen wearing course with dust-binding chlorides, which increases electrical conductivity, will not cause so much increased attenuation (Figure 13.3). On high traffic volume roads, airports and busy railways data are often collected during the night. This does not cause any problems other than video data collection cannot be carried out at the same time.

To collect good-quality data, the collection speed should be kept as slow as possible without interfering too much with traffic. Higher collection speed reduces the accuracy of the results. Hopman and Beuwing (2002) have compared GPR data collected by Dutch GPR contractors at different speeds with the drill core data and according to the results mean error at low speed was 5% while at a speed of 80 km hr^{-1} the error was 9%.

The influence of surface roughness has also been discussed in some papers. Davis et al. (1994) suggest that roughness decreases the amplitude of the reflected signal from the pavement. Spagnolini (1997) discusses the scattering of the electromagnetic (EM) waves on rough surfaces but concludes that the phenomenon can be taken into account during the data processing.

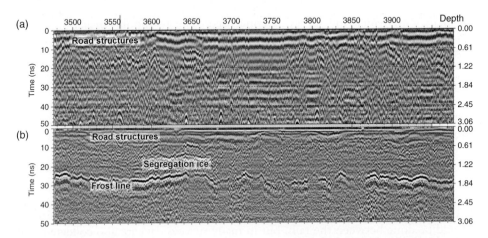

Figure 13.3 A gravel road section from the Vaasa area in Finland measured using a 400-MHz ground-coupled antenna in summer (a) and in winter during the time of maximum frost depth (b). The GPR data collected during the winter shows 0.5 m thick road structures while it is quite impossible to define them in the summer data. Winter data also present, very clearly, the frost line as well as the presence of segregation ice (ice lenses) that cause differential frost heave and spring thaw-weakening problems in the road.

13.3.2. Data collection setups and files

Data collection setups can be site and problem specific, for instance the number of samples from bridges is much higher than from roads. However, the following setups have proven to provide good quality GPR data.

When measuring longitudinal sections on roads, railways and airports a good sampling density is 10 scan m^{-1} for both air-coupled and ground-coupled systems. This sampling density provides information about cracks and crack propagation in pavement (Scullion and Saarenketo, 1995) and segregation and enables detection of point-like objects such as cables and pipes. When measuring cross sections on roads or conducting bridge deck surveys the recommended sampling density is 40 scan m^{-1}.

The gain setting on air-coupled systems should be one-point gain (flat gain) because amplitude parameters are used in GPR analysis. The metal plate reflection sets a limit for the maximum gain with air-coupled antennas. With ground-coupled systems, gain with several gain points can be used but the gain curve should be smooth. The recommended interval between gain points is 20 ns. The most common mistake during data collection on roads is that too much gain has been used and there is clipping of the GPR signal. Most GPR data collection software packages have an 'autogain', but this feature should not be used. The optimal gain settings for each pavement type or bridge deck can be determined through testing; once defined these settings should be stored in the memory of the control unit. During data collection it may appear that the gain is not high enough but when the data density is 16 bit these problems can be handled with post-processing software packages. New GPR control units have high data-storage capacity and thus 8-bit sampling is no longer recommended. The recommended sample/scan density is 512 in most traffic infrastructure surveys.

During the data collection it is recommended that certain filters, to remove noise and ringing from the data, be used. However, this filtering should be very light so that actual structural or point-like object reflections are not removed. It should also be kept in mind that filtering can be done afterwards and many GPR systems, such as MALA, record only raw data during the data collection. However, if the filtering is done the following filter setting have proven to work well:

- Ground-coupled antennas: IIR filters with high-pass filter one-fifth of the central frequency and low-pass filter ×5 central frequency (i.e. with 400 MHz antenna HP 80 MHz and LP 2500 MHz);
- Air-coupled antennas: FIR filters with high-pass filter one-half of the central frequency and low-pass filters ×3 central frequency (i.e. with 1.0 GHz antenna HP 500 MHz and LP 3.0 GHz).

Each GPR manufacturer has their own recommendations regarding filter settings.

The measurement time depends on the target under examination, but normally a 20-ns time range has been used with high-frequency pavement radar systems, where at least a 12-ns time window is collected from the pavement structure. With 400–600 MHz ground-coupled systems normally time window of 60 or 80 ns is used. With lower frequency antennas the time window should be defined based on

Figure 13.4 GPR antenna calibration tests before or after data collection: photo (a) presents the lifting test for a ground-coupled antenna, photo (b) the metal plate test for a horn antenna and photo (c) the antenna-bouncing test for a horn antenna.

the target depth of the survey so that the window should be about one-third longer than the maximum calculated depth.

When defining the position of a time window it is important that with air-coupled systems direct pulse from transmitter to receiver is collected because this pulse is used as a reference pulse in most post-processing software packages. When selecting a time window with ground-coupled systems it has to be ensured that the surface reflection is within the window. This can be checked with an antenna-lifting test (Figure 13.4) that should be done after each survey session. Because of coupling, it is difficult to define the correct surface reflection in the data-processing phase and lifting test provides a solution for this problem.

When collecting air-coupled data a static metal plate reflection should always be collected (Figure 13.4). In this survey a metal plate (about 100×100 cm) is placed under the antenna and 100–200 scans are collected. It is highly recommended that this be done before and after the survey, but at the very least after the survey. Depending on the antenna quality, the so-called air pulse should also be collected. To do this, the antenna is pointed upwards or sideways and a GPR signal without any reflections in the time window is collected. A third data file that is required to be collected is a height-calibration file, which is used to correct the reflection amplitude as a function of the antenna's height above the ground. Placing a metal plate under the antenna at different depths is one method of preparing a height-calibration file. Calibration can also be done using the so-called bouncing test whereby the road survey vehicle is bounced while the metal plate reflection is collected (Figure 13.4).

Before starting the data collection with most antennas, especially with air-coupled antennas, it is important to allow an antenna to warm up at least 15–20 min to avoid amplitude and time drift of the GPR signal during data collection.

13.3.3. Positioning

Accurate positioning of the GPR data is the most important thing in the data collection process. Data with incorrect referencing are worthless to the customer and damaging for the GPR industry. A great number of GPR tests on roads have been classified as failures when the GPR data was compared with the ground truth data; many of these failures have originated from false positioning – either from GPR or from reference sampling.

Positioning can be done (1) using encoders that control the sampling interval, (2) adding markers to the GPR data at known reference points and (3) using GPS techniques. Digital video linked to GPR scan numbers also helps to ensure correct positioning. The best way is to use all these methods in combination. Encoders should be installed on vehicle wheels or distance measurement instruments. The so-called fifth wheel systems have proven to be unreliable because of bouncing or movement in sharp curves. Markers should be inserted in the data at known reference points, such as culverts, bridge joints and access road intersections. Starting and ending place of the survey should be always marked on the road with road paint. On a short special survey section, such as bridge deck or forensic surveys, it is recommended that reference points be painted on the pavement at 20–100 m intervals and markers be inserted at these points. In routine road surveys, it is recommended that surveys be started and ended at known road registry-referencing points.

Especially in bridge deck surveys accurate positioning in the transverse direction is also extremely critical and to do this different kinds of aiming systems, such as painted marks, lasers or video cameras have been used successfully.

Over the last few years, there has been a fast development of GPS technology that has significantly increased the accuracy of the positioning. With a standard differential GPS system, it is easy to gain ± 1 m accuracy, while with the latest virtual reference station (VRS) that uses both GPS and Glonass satellites the accuracy of the GPR survey can be improved to ± 0.05 m.

13.3.4. Reference sampling

In many projects, collecting reference samples at the same time as the GPR data are recommended, and if this is not possible it is recommended that the GPR crew mark the points for reference sampling on the road. The problem with this technique is, however, that preliminary data processing has to be carried out in the field.

In network level surveys and rehabilitation projects the points for reference samples should be selected from sections that present typical structures in the survey project. These points should be selected from sections where there are no changes in structure thickness at least ± 10 m around the point. This ensures that there will be no errors in backcalculating dielectric values for each layer. Other points for reference sampling are anomalous areas indicating possible structural problems where samples can be taken to verify the problem and its severity.

13.4. DATA PROCESSING AND INTERPRETATION

13.4.1. General

GPR data-preprocessing, interpretation and visualization software for roads are used for detecting layer interfaces and individual objects from the GPR data and transforming the GPR data time scale into depth scale.

Despite the fact that computer processors are becoming more efficient and GPR software packages are becoming more user friendly, the processing and interpretation of the GPR data from roads, railways and airports is still the most time-consuming phase and an interpreter's skills play a key role in the success of a GPR project. GPR data processing can be divided into four phases: (1) preprocessing, (2) data processing, (3) interpretation and visualization and (4) reporting. During recent years, many software packages have developed new features allowing integration and viewing of other road survey data, video and also facilitate making road analysis (Roimela et al., 2000) and integrated rehabilitation design (Saarenketo, 2001). Many of the previously mentioned GPR manufacturers are also producing software packages or modules for road and bridge surveys, but there are also special commercial packages that have specifically been developed for use in road surveys, such as "Road Doctor Pro" or "Road Doctor GPR Pro A" by Roadscanners (Finland) or "Reflex" by Carl Sandmayer (Germany). In addition, many road research laboratories have developed their own software packages.

13.4.2. GPR data preprocessing

In the preprocessing phase GPR data are edited in such a way that the raw data itself will not be changed. In this phase the GPR data-editing features normally used are file reversal (allows comparison of data collected from right and left lane in two-lane roads or bridges), cutting and combining, and scaling and linking data to GPS coordinates or to different road address data bases. In the distance-scaling phase it is important to check that markers match with known landmarks – this can be done using video and maps, for example. Most of the new GPR software packages for road data analysis now have project options that allow the interpreter to link several GPR survey lines to the same project and thus control surveys. Useful project-handling features often include processing templates to be used with every data file. In the preprocessing phase other road survey data and ground truth data, such as drill core data, should also be linked to the project.

Distance scaling is, at its simplest, a phase in which the starting and ending points and the distance between these points is calculated and the software scales all the scans between these two points using a constant scale parameter (scan m^{-1}). The problem with this scaling is distance errors in curves, which can be solved by using the previously mentioned markers technique and/or using GPS data linked to scan numbers. Major changes in z-coordinates (height) can also cause problems in scaling. GPR data-preprocessing software can the calculate real x, y, z coordinates and distance from starting or reference point to each scan.

In GPR data preprocessing it should be ensured that GPR signal polarity and colour scales follow certain rules agreed upon by GPR consultants, performing road surveys, in the early 1990s. In single scan mode a metal plate the maximum peak (see Figure 13.5) should always point to the right and it should be called as "positive reflection" (even though it is not necessarily positive) and in line scan format in a grey scale it should be white and in blue/red colour scale format red. The two smaller peaks around the "positive" peaks are called negative peaks and in a grey scale they should be black and in a colour scale they should be blue. The same rule

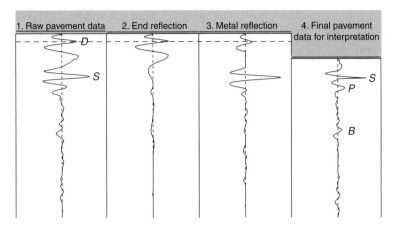

Figure 13.5 Different air-coupled antenna signals during the antenna processing. Signal 1 presents raw GPR data collected from the road, D is a direct signal from transmitter to receiver and S presents a surface reflection. Signal 2 presents an end reflection, where the antenna is pointed upwards or down from a dock, for instance, with no reflectors below. Signal 3 presents a metal reflection signal where the end reflection has been subtracted after the direct wave (see the straight signal above the surface reflection). Signal 4 presents processed GPR data ready for interpretation, where S represents the pavement surface, P presents a reflection from pavement bottom and B presents a reflector from the bottom of the base course.

also corresponds for ground-coupled systems so that positive (white) reflections are registered when the lower layer has higher dielectric value and negative (black) reflections appear when the lower layer has lower dielectric value than the upper layer. If the polarity is reversed during data collection, it can be changed by multiplying the GPR signals by a factor of -1. These basic rules allow engineers to read and compare the data collected from the traffic infrastructure consistently and help with observation of changes in moisture content in different layers or if the voids are filled with air or water.

13.4.3. Air-coupled antenna data processing

The different GPR signals needed in each phase of air-coupled data processing are presented in Figure 13.5 and signal processing is described in detail in different software manuals. In the first phase a template subtraction is made, that is air pulse is removed from the metal plate file and from the raw data. It is only necessary to do this if the direct wave has clutter below the pavement surface level (S). If the air-coupled antenna signal quality is good and the pulse shape is straight above the pavement surface reflection this process is not needed and the processing will be done using only the metal plate reflection. After the metal plate reflection is defined, the software flattens the changes in the surface reflection caused by bouncing of the survey vehicle and calculates the dielectric value of the road surface. Many software packages also improve the surface resolution in this phase by using different algorithms.

Data filtering should also be done in this processing phase. Different filter settings can be tested to see if they improve the data quality. After the metal plate calibration and template subtraction other filters can be used. A special "background removal" filter described by Maijala et al. (1994) has also proven to improve the data quality. However, the background signal, which is to be removed from the data, has to be calculated from a road section where structure thickness changes, otherwise structural data can also be removed. Background removal cannot be used when the dielectric values of lower layers are calculated using the reflection technique.

13.4.4. Ground-coupled data processing

The basic processing of ground-coupled data in traffic infrastructure surveys normally has only two steps: (1) defining the pavement surface level and (2) background removal filtering. The pavement surface level can be defined by using the lifting test data. In process the background is removed from the lifting file leaving only the position of the surface reflection (Figure 13.6). There are also other techniques to detect this level, also known as the zero level, and Yelf (2004)

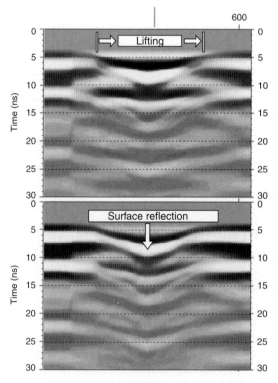

Figure 13.6 Defining the pavement surface reflection position using the lifting test. These data were produced using a 200-MHz ground-coupled antenna lifted from the pavement surface. In the case where there is an air gap between the antenna and pavement, before starting to lift the difference would have been greater.

provides a good description and test results of different ways to detect the true time zero level. Surface coupling also causes problems in the detection of structural layers near the road surface. A background removal filter is a good solution to this problem (Figure 13.7). Another processing technique that can be useful in road survey data analysis is viewing the data in the frequency domain by using a Fourier transformation, which provides information regarding changes in electrical properties in structural layers and subgrade soils. Migration has been used occasionally in bridge deck surveys to filter the hyperbolas caused by the reinforcements. Migration is also needed when making time slices of the 3D GPR data.

Dielectric dispersion is significant and has to be taken into account when GPR surveys are carried out in wet soil conditions (see Goodman et al., 1994; Saarenketo, 1998). In wet soils this usually results in the higher frequency components of a pulse attenuating and propagating faster than the lower frequency components resulting in pulse broadening (Olhoeft and Capron, 1994). This result causes problems in deconvolution and migration of the data because the pulse is not in same shape and phase everywhere.

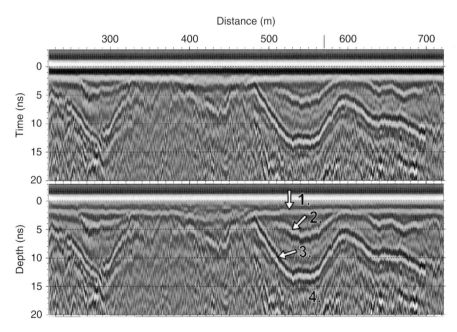

Figure 13.7 An example of the effect of background removal in improving the 400 MHz ground-coupled data quality. The top profile is raw data and below is the same data after background subtraction. The pavement bottom reflection (1.) can easily be defined from the filtered data, whereas this cannot be done in the raw data. The figure also illustrates how the polarity of the reflections provides information on the dielectric properties (and moisture) of the road structures. Positive (white in the middle) reflectors 1 and 2 reveal that the dielectric value of the base course, below the pavement, and sub-base, below the base course, is higher than the dielectric value of the layers above. However, reflection 3 has negative polarity (black in the middle) and this shows that layer 4, which is sandy gravel, has a lower dielectric value and lower moisture content than the sub-base layer.

13.4.5. Determining dielectric values or signal velocities

An accurate estimation of layer dielectric values or signal velocities is a key issue in successful traffic infrastructure GPR data processing. An interpreter, analyzing traffic infrastructure data, needs information concerning the dielectric properties of structures and subgrade soils in order to (1) calculate the correct layer thickness of structural layers and subgrade soil layers, (2) calculate the moisture content, (3) calculate the asphalt air voids content, (4) estimate the moisture susceptibility and sensitivity which is directly related to permanent deformation of unbound materials, (5) estimate the frost susceptibility of subgrade soils, (6) estimate the compressibility of subgrade soils and (7) estimate the homogeneity and fatigue of bound layers. In many surveys, especially in quality control/quality assurance (QC/QA) projects, there are major economic factors attached to the surveys results and as such there is a requirement for high-quality data.

The traditional method for determining the dielectric value of pavement is backcalculating the value using reference drill cores. This method is still the most common especially when using ground-coupled systems. The other very popular method is surface reflection method (Maser and Scullion, 1991), which can be used with air-coupled antenna systems. In this method reflection amplitude from the pavement surface is compared with the metal plate reflection representing a total reflector. By calculating the amplitudes, it is possible to calculate both layer dielectrics. Equation (13.1) presents the algorithm for the surface dielectric value calculation and Equation (13.2) for second layer (base course) surface dielectric value.

$$\varepsilon_a = \left(\frac{1 + A_1/A_m}{1 - A_1/A_m}\right)^2 \tag{13.1}$$

where

ε_a = the dielectric value of the asphalt surfacing layer
A_1 = the amplitude of the reflection from the surface
A_m = the amplitude of the reflection from a large metal plate (100% reflection case)

$$\sqrt{\varepsilon_b} = \sqrt{\varepsilon_a} \frac{\left[1 - \left(\frac{A_1}{A_m}\right)^2 + \left(\frac{A_2}{A_m}\right)\right]}{\left[1 - \left(\frac{A_1}{A_m}\right) - \left(\frac{A_2}{A_m}\right)\right]} \tag{13.2}$$

where

ε_b = the dielectric of the layer 2 (base layer)
A_2 = the amplitude of reflection from the top of layer 2

These equations have proven to work well for estimating dielectric values for the first layer in homogenous asphalt pavement and most concrete pavements. Equation (13.2) assumes that no attenuation of the GPR signal occurs in the surface layer. This assumption appears to be reasonable for asphalt pavements and also provides reasonable dielectric values for the base layer, if the asphalt is thicker than 60 mm and there are no thin layers with different dielectric properties on the bottom of the asphalt on top of the base. However, the computations are less reliable for certain concrete pavements. To improve this method in the future, signal attenuation has to be incorporated into the calculation process.

For new or defect-free pavements, the surface and base dielectrics together with the surface thickness can be calculated easily. The factor, which primarily impacts the surface dielectric, is the density of the asphalt layer and the factor impacting the base dielectric is the volumetric moisture content of the base. The GPR reflections can also be used to judge the homogeneity of the pavement layers.

Another method used in estimating the dielectric values of road structures is the common middle point (CMP) method (Al-Qadi et al., 2002; Maser, 2002a). According to Fauchard et al. (2003) this method provides sufficient accuracy of dielectric values for first two or three layers in road structures. The CMP theory is described in another chapter of this textbook. Dielectric values have been also measured using different antenna array techniques (Davis et al., 1994; Emilsson et al., 2002).

13.4.6. Interpretation – automated vs. user controlled systems

Many efforts, including the use of neural networks, have been made to develop automatic interpretation software for roads and bridges. However, the results of these development projects have not been encouraging and have also resulted in confusion amongst highway engineers. The reason that automatic interpretation software packages will, most likely, never work on older roads, railways or airports is that these structures are typically historical structures with discontinuities in longitudinal, vertical and transverse directions (see Saarenketo and Scullion, 2000; Hugenschmidt, 2003). Manually controlled semiautomatic interpretation software used by well-trained and experienced interpretation staff and utilizing different kinds of reference survey results has proven to be the only reliable solution in traffic transport infrastructure surveys (Mesher et al., 1996; Roberts and Petroy, 1996; Saarenketo and Scullion, 2000). The only cases where automatic interpretation seems to calculate the correct thickness and dielectric values are surveys on new and defect-free pavements (Saarenketo and Scullion, 2002).

13.4.7. Interpretation of structures and other objects

The interpreter's knowledge of the road, bridge, railways and airport runway or taxiway structures and their damage mechanisms play a key role in the interpretation process. The GPR data are filled with information from reflectors, individual objects, amplitude anomalies (moisture, attenuation, etc.) and the interpreter has to select and report on those that are essential to the final report. As such the interpreter should always, at least, interpret the key structures and, most importantly, the individual objects found in the data. Based on the goals of the project as determined in the project design phase, it should be determined what other structures are to be reported.

In road and airport runway and taxiway surveys the key structures that should always be interpreted are (1) bottom of the bound layers (pavement and bound base), (2) bottom of unbound base, (3) bottom of the entire pavement structure and (4) bottom of the embankment, if the road is on an embankment. The layer thickness data are needed to backcalculate layer moduli values from FWD data (Briggs et al., 1991), and also for dimensioning a new pavement structure. In addition in routine GPR surveys an estimate of subgrade soil quality should always be made. If bedrock is close to surface and it can be identified, it should also be interpreted. Additional information that should be reported is the location of culverts and bridges and damaged road sections and reasons for the damage. Cables and pipelines should be reported if this has been written in the contract. With regard to road structures usually the thickness of the sub-base and filter course and the location of old road structures should also be reported.

In bridge deck surveys a standard interpretation includes identification of the following interfaces: (1) pavement bottom, (2) bottom of the protective concrete (if it exists) and (3) the level of the top reinforcement of the bridge slab. In railway surveys the key structures are normally (1) ballast, (2) sub-ballast, (3) thickness of the whole structure and (4) embankment, if it exists.

The exact location in the GPR signal where a layer interface is selected and followed varies. Most GPR consultants and software use the maximum amplitude of each reflection. This technique is the most accurate in measuring different layer thickness, but the interpreter must always be aware of changes in polarity (see Figure 13.7). Another technique, still in use, is to define the layer interface to the position where the signal passes the zero amplitude level after the first reflection peak of that layer. This technique is independent of the changes in signal polarity between each layer but, on the other hand, easily gives false thickness information if there are thin layer reflectors near the key interfaces causing overlapping reflections.

Interpretation of a layer interface of new structures is generally a straightforward process as they are quite easy to identify. Problems however can occur when dealing with older fatigued structures. Frost action and heavy traffic especially can cause mixing in the layers such that the layer interfaces are very hard to define precisely. However, in rehabilitation design surveys it is very important that these layers are interpreted and usually it is important to report that there might be an important layer or an object, such as bedrock or old frost insulation board, that the design engineer should be aware of. Finnish GPR users have agreed on a technique to distinguish the degree of certainty regarding an interpreted layer (Ground Penetrating Radar, 1992). With this system GPR consultants can illustrate the reliability of an interpretation. This classification has three classes (Table 13.1).

Table 13.1 Layer interface identification class (Individual objects about which the interpreter is uncertain are marked with a question mark (?) after the object description.)

Symbol	Degree of distinction
————	Obvious interface
– – – – – –	Distinguishable interface (lines longer than spaces)
- - - - -	Possible interface (spaces longer than lines)

The easiest option with difficult GPR data cases, such as fuzzy layer interfaces, is to leave them uninterpreted. However, the professional skill of an interpreter is reflected in how well one can identify these layers, and in Finland, according to Finnish GPR specifications, it is required that all of the layers be interpreted (Saarenketo, 2006).

During the interpretation the layer thickness should be always verified with drilling and sampling data and crosschecked with the GPR data to see if the interpreted layers match with the ground truth thickness information.

13.5. INTEGRATED GPR DATA ANALYSIS WITH OTHER ROAD SURVEY DATA

13.5.1. General

As with many other geophysical applications GPR survey results become much more easier to adopt and results are more reliable if there is other supporting data available when analyzing the data. Roads can be surveyed using several other non-destructive techniques, such as FWDs, and profilometers and combining GPR with other non-destructive road survey techniques provides a powerful tool for diagnosing current pavement problems and selecting the optimum repair technique (Saarenketo, 1999; Roimela et al., 2000; Saarenketo, 2001; Johansson et al., 2005). Integrated analysis has also been used in estimating the need for load restrictions on low-volume roads (see Mohajeri, 2002; Saarenketo and Aho, 2005).

13.5.2. GPR and FWD

The integrated analysis of GPR and FWD data, used in support of one another, offers many advantages for pavement evaluation. Especially FWD data backcalculations require the kind of accurate pavement structure thickness information that GPR can provide (Briggs et al., 1991; Lenngren et al., 2000; Al-Qadi et al., 2003a; Noureldin et al., 2003). Saeed and Hall (2002) compared several testing methods for characterizing the in situ properties of pavement materials. The best combination was FWD and GPR. Noureldin et al. (2003) also recommend that FWD and GPR be used as part of the Indiana Department of Transportation's pavement management system. Using moduli values backcalculated from the FWD data and GPR thickness data it is also possible to check the quality of the GPR interpretation and, based on the GPR profile, ignore those FWD data points that were collected from points that do not represent the structure well. In pavement condition evaluation the FWD data helps to verify disintegration in the pavement layers and/or check if the problems are related to the base course or sub-base (Saarenketo et al., 2000). GPR data can be used to locate water susceptible base course sections where the FWD data, collected during dry summer months, would not indicate any problems. The FWD data provide valuable information for GPR analysis about the subgrade soil type. In addition, the shape of the deflection bowl in combination with the GPR data indicates immediately if bedrock is present and close to the surface or if the road has been constructed over peat. Using FWD backcalculation software or by using other subgrade moduli calculation methods it is also possible to estimate with GPR data where the road has been constructed on peat, silt or clay or sand and gravel (Saarenketo, 2001)

13.5.3. Profilometer data

Road profilometers are used to measure parameters, which describe longitudinal and transverse evenness and the cross fall of a road surface. The most common parameter describing the longitudinal evenness is called International Roughness Index (IRI) and transverse evenness is described with different rut depth parameters. Analyzing GPR data with profilometer data helps interpretation personnel to locate problem areas and identify the reason for the problems, for instance whether problems are due to differential frost heave or settlements or caused by permanent deformation because of moisture susceptible road materials.

If profilometer data are not available and information on the unevenness of the road is needed a good indicator for sections with unevenness problems is the air-coupled antenna height file, which can be calculated from the raw data. The data from this file present the height of the air-coupled antenna from the pavement surface and changes in this height demonstrate areas where the survey vehicle has been driving over uneven spots on pavement (Figure 13.8).

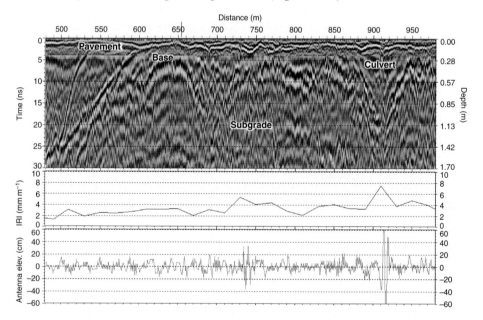

Figure 13.8 Integration of GPR data, profilometer data and air-coupled antenna height data, Road 752, Sweden. The first 4 ns in the GPR profile is 1.0 GHz horn antenna data and 4–30 ns is 400 MHz ground-coupled data. The profile in the center is International Roughness Index (IRI) presented at 10 m intervals. The data in the bottom present the distance from the horn antenna to the pavement surface. The biggest differential frost heave sections can be related to section 700–770 m where frost fatigue has also mixed the layer interfaces in the GPR data and to 900–920 m, where the uneven frost bump can be related to a culvert.

In the future, profilometers or road surface scanners will be used with multi-channel GPR systems to prepare 3D models of roads, railways and airports.

13.5.4. GPS, digital video and photos

GPS data are mainly used to position the measurements correctly but it also helps with interpretation. If the correct GPS z-coordinates are available and the software has a topography correction feature, this helps an interpreter to check that the interpretation is logical (Figure 13.9). If the GPR data are linked with GPS coordinates the GPR survey line can be linked to geological maps, like soil maps, and this helps with verification of the subgrade soil quality.

As previously mentioned, having video with the GPR data is a "must" for most organizations performing GPR surveys on roads, railways and airports. Because, without digital video, it is very difficult to know whether the structures are on an embankment or a road cut. Video also helps to identify problems related to drainage and the presence of bedrock. Video shows the pavement damage and helps the interpreter to make a correct diagnosis of the problems. Video can be viewed from a separate video player, but the best option is when digital video is linked with the GPR data.

Digital photos are also useful during interpretation. Detailed photos of the drill cores have been especially useful during the interpretation.

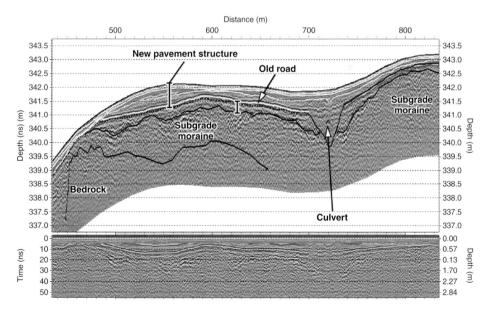

Figure 13.9 About 400 MHz ground-coupled and 1.0 GHz air-coupled data (first 5 ns) from HWY 21 in Finland presented as uninterpreted data (*bottom*) and height-corrected and interpreted data (*top*). The height-corrected profile demonstrates well how the vertical geometry of the road has been improved during the last strengthening project.

13.5.5. Other data

Road, railway and airport authorities have many kinds of databases, and these data are often very useful in GPR data analysis. An interpretation can be greatly improved if an interpreter has access to pavement distress and condition databases, paving history and thickness databases

In special projects it is worthwhile collecting other geophysical data to support the GPR data interpretation (see Wilson and Garman, 2002). Electrical resistivity sounding or profiling data support the GPR analysis if the GPR signal has penetration problems because of high conductivity of subgrade soils. Resistivity profiles help with identification of soil types and their moisture contents. Soil type and moisture content can also be verified through measurement of soil dielectric value and electrical conductivity data collected using a dielectric probe, such as a Percometer (Saarenketo, 1995, 2001). EM measurements have aided in the identification of cables and other objects under the pavement.

13.6. GPR Applications on Roads and Streets

13.6.1. General

GPR applications on roads and streets can be divided roughly into four main categories: (1) surveys carried out in design of a new road, (2) surveys needed for the rehabilitation design of an existing road, (3) quality control or quality assurance surveys on a road project and (4) surveys for pavement management systems and similar purposes (see Saarenketo, 1992; Fernando et al., 1994; Hugenschmidt et al., 1996; Scullion and Saarenketo, 1998; Saarenketo and Scullion, 2000; Saarenketo, 2001; Scullion, 2001; Maser, 2002a,b; Ahmed et al., 2003; Al-Qadi et al., 2003c; Johansson et al., 2005). GPR can provide different types of information regarding the bound and unbound pavement structures, subgrade soils, moisture contents and other features of interest to these projects. A general description of each application is given in the following sections.

13.6.2. Subgrade surveys, site investigations

Subgrade surveys and site investigations with GPR have been classified into the following three categories (Saarenketo and Scullion, 1994): (1) new road alignment and site investigations, (2) strengthening and widening of an existing road, and (3) using the existing road as an information source for the design of a new roadway alongside the existing road. In each case the basic problem is similar, but the way in which the GPR techniques are applied varies.

13.6.2.1. Subgrade quality and presence of bedrock
When evaluating the subgrade soil type in road projects it is usually quite easy to identify coarse grained gravel, sand and glacial till soils from the GPR data. GPR also works well for identifying most organic peat soils (Ulriksen, 1982; Doolittle

and Rebertus, 1988; Ground Penetrating Radar, 1992; Saarenketo et al., 1992). GPR signals have relatively good signal penetration in most silty soils, but problems arise when surveys are carried out on clay soils. In Scandinavia, GPR signal penetration in clay soil areas is normally about 2 m, which is adequate for cable and pipeline surveys but not for highway design purposes. In the USA, penetration depth depends on the mineralogy and clay content of the soils. According to Doolittle and Rebertus (1988) a penetration depth of 5 m has been achieved in areas of Site Oxidic soils, while radar signals penetrate only 0.15 m in Vaiden type Montmorillonitic soils.

In many cases the soil type can be determined from the GPR data, because each soil has its own specific geological structure, dielectric, and electrical conductivity properties (see Saarenketo, 1998; Benedetto and Benedetto, 2002); these properties produce a special "finger print texture" in a GPR profile. Soil type evaluations always require some ground truth data to confirm the GPR interpretation. Excellent supporting information can also be obtained from the FWD survey data.

In road surveys GPR information concerning the depth of overburden and location of bedrock is used in the design of grade lines, or in design against uneven frost heave and other specific technical problems (Saarenketo and Scullion, 2000; Fish, 2002). If the bedrock is close to the surface, the GPR interpretation can be confirmed with the FWD deflection bowl shape and FWD backcalculation algorithms (Scullion and Saarenketo, 1999; Saarenketo and Scullion, 2000). If GPR surveys are performed in the wintertime, the areas of bedrock closer to the surface than the frost level are easy to identify because there are no frost line reflections in the bedrock in the radar profile. GPR also allows observation of bedrock stratification and major fracture zones when evaluating the stability of *highway cutting walls* (Saarenketo, 1992; Fish, 2002). Similar information can also be obtained in highway tunnel surveys where both ground-coupled and drill-hole antennas have been used (Westerdahl et al., 1992).

When calculating the thickness of subgrade soil layers, it should be kept in mind that dielectric properties of soil correlate highly with water content and type of water in the soils. In soil surveys dielectric dispersion is quite significant, and different dielectric values will be obtained if the measurements are performed with 400 MHz and 1.5 GHz antennas (see Saarenketo, 1998; Benedetto and Benedetto, 2002).

13.6.2.2. Soil moisture and frost susceptibility

Moisture content has a great effect on the strength and deformation properties of the road structure and subgrade soils. Information about the subgrade soil moisture content is needed when estimating the stability (Ékes and Friele, 2004) and compressibility of subgrade soils (Jung et al., 2004), when designing highways in areas with expansive clays and when evaluating their frost susceptibility. In soils with low moisture content suction, which generates tension in the pore water between soil particles, can increase the stiffness of soils and unbound aggregates and lead to high modulus values, but when moisture content increases suction decreases (Fredlund and Rahardjo, 1993). At high moisture content positive pore water pressure can decrease the material resistance to permanent deformation.

In the case of widening and/or strengthening an existing highway, or when constructing new lanes, the best information regarding changes in the compressibility of the subgrade soil can be obtained directly by surveying the existing road with GPR. This technique is especially useful when estimating the extent of settlements in a new road and when designing preloading embankments over clay, silt or peat subgrade (Saarenketo et al., 1992; Saarenketo and Scullion, 2000).

The GPR technique has been used to locate road sections with excess moisture in the subgrade and help pavement engineers to design proper drainage (Wimsatt et al., 1998).

A GPR profile can present areas, where frost action has caused damage to the road. The damage, or early phase frost fatigue, can be observed as permanent deformation of road structures taking place during the spring frost thawing period or in the form of an uneven frost heave related to subgrade soil and moisture transition areas, or to the presence of bedrock or boulders. Structural elements in the road body, such as culverts, have also caused surface damage if the transition wedges have not been properly constructed (see Figure 13.8) (Saarenketo and Scullion, 2000).

Frost susceptibility is also closely related to the moisture content and drainage characteristics of the subgrade, which can be estimated with GPR and other dielectric measurement devices. Saarenketo (1995) has proposed a frost susceptibility and compressibility classification for the subgrade soils in Finland, which is based on in situ measurements of dielectric value and electrical conductivity. The evaluation of potential frost action areas and presence of segregation ice using GPR can be performed in winter when the frost has penetrated the subgrade soil and when the dielectric value of frozen soils is closer to the relative dielectric value of frozen water (3.6–4.0). When analyzing the GPR data the interpreter has to pay attention to the following phenomena: (1) the appearance of the frozen/non-frozen soil interface reflection, (2) the depth of the frost table and clarity of the reflection, and (3) the effect of the frost action upon the road structures (see Saarenketo and Scullion, 1994; Saarenketo, 1995; Saarenketo, 1999). If the frost level cannot be identified from the radar data, then the subgrade soils have a low dielectric value and thus are non-frost susceptible. If the reflection from the frost level is very clear in the GPR data, then the frost has penetrated the subgrade without forming segregation ice lenses that cause frost heave. High and uneven frost heave can be found in areas where the frost level comes close to the surface (see Figure 13.3) (Saarenketo and Scullion, 2000).

Saarenketo and Scullion (2000) have proposed a quality assessment of the strength and deformation and frost-susceptibility properties of soils and unbound road materials based on their dielectric properties, which is presented in Table 13.2 (see also Hänninen and Sutinen, 1994; Hänninen, 1997).

13.6.2.3. Other subgrade applications

GPR has great potential for aggregate prospecting (Saarenketo and Maijala, 1994; Cardimona and Newton, 2002). GPR techniques have also been used to locate sinkholes (Beck and Ronen, 1994; Saarenketo and Scullion, 1994; Casas et al., 1996; Geraads and Omnes, 2002; Wilson and Garman, 2002) and washouts under the road (Adams et al., 1998; Lewis et al., 2002). After locating the areas of voids,

Table 13.2 Quality assessment of mineral soils and unbound road materials according to their dielectric properties

Dielectric value	Interpretation
4–9	Dry and non-frost susceptible soils and road material, in most cases good bearing capacity (excluding some sands)
9–16	Moist and slightly frost susceptible soil, reduced but in most cases adequate bearing capacity
16–28	Highly frost susceptible and water susceptible soil, low bearing capacity, under repeated dynamic load positive water pressure causes permanent deformation
28–	Plastic and unstable soil

GPR has also been used to monitor the injection of grout into the voids (Ballard, 1992). Another GPR application related to subgrade surveys is locating underground utilities under an existing road (Bae et al., 1996). Numerous authors have also had success with detecting buried tanks close to the edge of existing pavement structures prior to widening of the highway.

13.6.3. Unbound pavement structures

Unbound pavement structures are situated between the subgrade soil and top bound layers. Unbound pavement structure is normally made of crushed gravel, crushed hard rock, ballast or macadam and non-frost susceptible and non-water susceptible natural soils, such as gravel and sand. The key road structures in which these materials are used are presented in Section 13.4.7. The accuracy of base layer thickness measurements using GPR have been reported to be close to 8–12% (Al-Qadi et al., 2002; Maser, 2002b).

The base course layer, made of unbound aggregates, which supports asphalt or concrete pavement is one of the most critical layers in a road structure. A series of research projects has been carried out in Finland and Texas where the relationship of the dielectric properties of Finnish and Texas base materials to their water content and strength and deformation properties was studied using a dynamic cone penetration (DCP) test, dynamic triaxial tests and a permanent deformation test (see Saarenketo and Scullion, 1995, 1996; Saarenketo et al., 1998; Kolisoja et al., 2002). A special tube suction laboratory test, based on the dielectric measurements of the material at its moisture equilibrium level, was also developed in order to test the moisture susceptibility of road materials (Saarenketo and Scullion, 1995; Scullion and Saarenketo, 1997; Syed et al., 2000; Kolisoja et al., 2002).

The results from the surveys have shown that the dielectric value, which is a measure of how well the water molecules are arranged around and between the aggregate mineral surfaces and how much free water or "loosely bound water" exists in materials, is a much better indicator of the strength and deformation properties of road aggregates than the moisture content. Each aggregate type has a unique relationship between material dielectric and moisture content. Furthermore, high dielectric values (9–16, >16), calculated from the GPR data, for example, are good indicators for potential problems in the layer (Saarenketo et al., 1998; Saarenketo and Scullion, 2000; Scullion, 2001; Scullion and Saarenketo, 2002).

13.6.4. Bound pavement structures and wearing courses

13.6.4.1. Bound structures

The top part of the pavement structure of roads and airfields is mainly made of the so-called bound materials. These bound layers can be bound base course, pavement and wearing course – or only one of them. The most common binding agents in road pavements are bitumen (asphalt) and cement (concrete), but other agents have also been used marginally. Gravel road-wearing course is defined as an "unbound" layer or "water bound" layer but related GPR applications are also described in this chapter.

13.6.4.2. Bituminous pavement thickness and moisture

The GPR pavement thickness data have been collected for (1) network level surveys where thickness is required for pavement management system (PMS) data bases, (2) to supplement FWD data in calculation of layer moduli, (3) pavement design purposes, e.g. divide road into homogenous sub-sections, or to check if the pavement is thick enough for recycling milling and (4) for quality control purposes (see Briggs et al., 1991; Fernando et al., 1994; Saarenketo, 1997; Saarenketo and Roimela, 1998; Scullion and Saarenketo, 1998; Berthelot et al., 2001; Maser, 2002b; Saeed and Hall, 2002; Sebesta and Scullion, 2002a,b; Hugenschmidt, 2003; Noureldin et al., 2003).

Using the techniques for new pavements described earlier the accuracy of GPR thickness predictions has been around 3–5%, without taking a validation core. The highest accuracy has been reported by Maser et al. (2003), who tested both the horn antenna technique and the ground-coupled antenna with CMP technique on quality assurance on new asphalt pavements and reported an accuracy of 2.5 mm. The problem with older pavement, when using the surface reflection technique, is that the surface dielectric value of the pavement is estimated from the asphalt surface and this can sometimes lead to overestimation of asphalt thickness (see Al-Qadi et al., 2002). As such validation cores are recommended for older pavements (see Maser and Scullion, 1991; Morey, 1998; Maser, 2002b; Saarenketo and Scullion, 2000) with the accuracy then varying between 5 and 10%. The thinnest pavement layers that can be detected with the newest 2.0 GHz GPR antennas and good processing software are less than 20 mm. Dérobert et al. (2002a) have been developing a stepped frequency radar with an ultra wide band Vivaldi antenna to detect the thickness of very thin pavements.

GPR can also be used to monitor changes in the moisture content of asphalt pavement (Liu, 2003). In these surveys the changes in signal velocities are so small (3–5%) that it corresponds to only 2–3 samples in one scan (Liu, 2003).

13.6.4.3. Defect in bituminous pavements

In bituminous pavements GPR techniques have been used to detect different kinds of defects, such as segregation, stripping, crack detection and moisture barriers

(Saarenketo, 1995; Saarenketo and Scullion, 2000; Sebesta and Scullion, 2002a, 2002b). The following provides a brief description of these problems:

Segregation manifests itself as localized periodic small areas of low-density material in the compacted surfacing layer (Saarenketo and Scullion, 2000). Upon close inspection of these small, localized areas an excess of coarse aggregates will be found. The causes are often traced to improper handling or construction techniques. If an asphalt surface is uniformly compacted, the surface dielectric should be constant; however, if an area of low permeability has excessive air voids this will be observable in the surface dielectric plot as a decrease in measured dielectric value (Sebesta and Scullion, 2002a, 2002b) as seen in Figure 13.10.

The most common asphalt pavement damage inside the pavement is stripping, which is a moisture-related mechanism where the bond between asphalt and aggregate is broken leaving an unstable low-density layer in the asphalt. Stripped

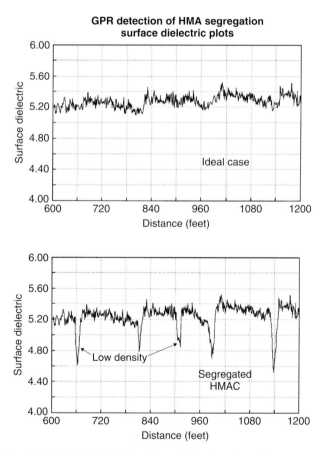

Figure 13.10 Hot Mix Asphalt (HMA) surface dielectric values from new pavements from US 79, Texas, USA. The upper profile presents an ideal case with homogenous asphalt density. The lower profile presents segregation problems and a low-density asphalt surface (Sebesta and Scullion, 2003).

layers should always be detected and removed before placing a new overlay. The GPR technique has been widely used to detect stripping with varying success (see Rmeili and Scullion, 1997; Morey, 1998; Saarenketo and Scullion, 2000; Cardimona et al., 2003). Stripping can be seen, in most cases, as an additional positive or negative (reversed reflection polarity) reflection in the pavement. However, similar reflections can also be received from an internal asphalt layer with different electrical properties, which is why reference data, such as drill cores and FWD data, should always be used to confirm the interpretation (Saarenketo et al., 2000).

Another major cause of surface distress is moisture becoming trapped within the surface layer (moisture barriers). This happens when impermeable fabrics or chip seals are placed between asphalt layers or when the existing surface is milled and replaced with a less-dense layer. GPR signals are highly sensitive to variations in both moisture and density. GPR reflections from existing highways can be complex particularly if the old Hot Mix Asphalt (HMA) layer contains numerous thin layers constructed with different aggregates compacted to different densities. Moisture barriers within the layer and collecting data shortly after significant rain can also complicate the analysis (Saarenketo and Scullion, 2000). A special moisture barrier also causing problems for road performance are old bituminous layers left inside the old unbound base course. If they are closer than 40 cm from the current pavement surface, the interpreter should identify it in the GPR data.

Other asphalt defects include cracking, thermal cracking and debonding, which takes place when the bonding between separate asphalt layers comes loose. Of these defects, GPR has proven to be useful in detecting transverse cracking (Saarenketo and Scullion, 1994). In the UK, Forest and Utsi (2004) have introduced a Non Destructive Crack Depth Detector, based on slow-speed GPR, to detect the depth of top-down cracks in asphalt pavements.

13.6.4.4. Concrete pavements

The major defects in concrete pavement have been reported as voids beneath the joints, cracking and delamination of the concrete pavement. The formation of voids beneath concrete pavements is a serious problem, which is particularly noted on jointed concrete pavements built with stabilized bases (Saarenketo and Scullion, 1994). Over time bases erode and supporting material is frequently pumped out under the action of truckloads.

GPR was first reported to be in use on concrete pavements to locate voids under concrete pavement (Kovacs and Morey, 1983; Clemena et al., 1987); however, the problem has been in detecting small voids with several failures (see Morey, 1998; Al-Qadi et al., 2002). Glover (1992) reports that the smallest voids detectable with ground-coupled antennas are with a diameter of 0.25 m. Air-coupled GPR systems can detect voids larger than 15 mm (Saarenketo and Scullion, 1994). The problem with void detection is also knowing if the voids are air voids, water filled or partly water filled, in each case the GPR reflection pattern looks different (Saarenketo and Scullion, 2000).

Another GPR application on concrete pavements has been detecting and locating dowels and anchors around the concrete slab joints (Maierhofer and

Kind, 2002). Huston et al. (2000) have tested stepped frequency radar using 0.5–6.0 GHz air-coupled waves to detect delamination in concrete roads.

Concrete thickness measurements have caused problems sometimes, because detecting the reflection from the concrete–base course interface can be difficult (Cardimonda et al., 2003). The reason for this difficulty has been attributed to the similar properties of concrete pavement and base course. Higher signal attenuation in the concrete can also be another reason for this problem. Other reasons can be also using wrong antenna type and improper data processing. Clark and Crabb (2003) have been able to measure changes in sub-base moisture content under the concrete pavement using 450 and 900 MHz ground-coupled antennae. Often, the dielectric value of concrete (normally around 9) is higher than the base course (normally 6–7), and thus negative reflections can be obtained from the concrete–base interface (Saarenketo et al., 2000).

In many cases, concrete pavements have asphalt overlay, and in rehabilitation projects, GPR has been used to measure the overlay thickness and indicate areas with a likelihood of deterioration in the underlying concrete (Maser, 2002b).

13.6.4.5. Gravel road-wearing course

Gravel surfacing is still widely used on low-volume public roads around the world. The maintenance actions on these roads focus mainly on wearing course. The wearing course should have a proper thickness, and the material should have special suction properties to prevent dusting but at the same time not become plastic under wheel loads in wet conditions. Saarenketo and Vesa (2000) present a technique for classifying gravel road-wearing courses based on surface dielectric value and wearing course thickness. Because wearing course thickness also has great variation in the transverse direction of the road, the accuracy of GPR measurements in wearing course surveys is 25 mm. The optimum range of dielectric value for a 100-mm thick gravel road-wearing course is 12–16 (Saarenketo and Vesa, 2000). Tests performed on gravel roads in Finland in 2004–2005 have shown that the >2 GHz horn antennas work better when measuring wearing course thickness compared with the "traditional" 1.0 GHz antenna (Saarenketo, 2006).

13.6.5. GPR in QC/QA

Since the late 1990s GPR has gained increasing popularity in quality control surveys of new road structures. The traditional application of GPR in quality control surveys has only been road structure thickness verification (Al-Qadi et al., 2003b). New GPR quality control applications include measuring the air voids content of asphalt and detecting segregation in asphalt. The greatest advantages of GPR methods are that they are not destructive in comparison with the traditional drill core methods, costs are low and GPR surveys can be performed from a moving vehicle reducing safety hazards for highway personnel. The GPR method also presents the possibility of continuous linear data collection and thus 100% coverage of a new road structure under inspection can be acquired. Drill core methods only provide point-specific information, and thus they cannot reliably be used to find defective areas in new pavements (Scullion and Saarenketo, 2002).

Asphalt air voids content, i.e. the amount of air incorporated into the material or its function asphalt density, is one of the most important factors affecting the life span and deformation properties of pavements. Measuring voids content using its dielectric value relies on the fact that the dielectric value of the asphalt pavement is a function of volumetric proportions of the dielectric values of its components. Compaction of the asphalt reduces the proportion of low-dielectric value air in the asphalt mixture and increases the volumetric proportions of bitumen and rock and thus results in higher dielectric values of asphalt (Saarenketo, 1997; Saarenketo and Roimela, 1998; Scullion and Saarenketo, 2002). Figure 13.11 presents the relationship between dielectric value of asphalt and air voids content measured in Texas (Sebesta and Scullion, 2002b).

The GPR measurements in the field are performed using a 1.0-GHz horn antenna, and at that frequency the thickness range of measured density is normally 0–30 mm. Higher frequency antennas can be used to measure the density of thinner overlays, but on the other hand, they are also more sensitive to variations in asphalt surface texture. Dielectric values of asphalt surfacing are calculated by using the surface reflection techniques described earlier. Following the GPR field evaluation one or two calibration cores are taken and these cores are returned to the laboratory for traditional void content determination. For each type of aggregate and mix design, similarly shaped relationships have been developed (Saarenketo and Roimela, 1998). The calibration cores are used to establish the link for each specific project. In 1999 GPR was accepted for use as a quality control tool among other pavement density measurement techniques on all new surfacing projects in Finland, and in 2004 it was the only method allowed on high traffic volume roads because it was not intrusive to traffic. Figure 13.12 presents a contour map of asphalt air voids content measured in US 290 in Texas, USA.

Asphalt quality control must always be conducted in dry conditions and not when the pavement is wet or frozen (see Saarenketo and Roimela, 1998; Liu and Guo, 2002).

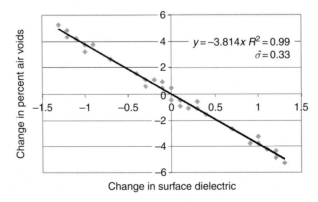

Figure 13.11 Correlation between changes in the surface dielectric value (x-axis) and changes in the percentage of air voids (y-axis) on US 79, Texas, USA (from Sebesta and Scullion, 2003).

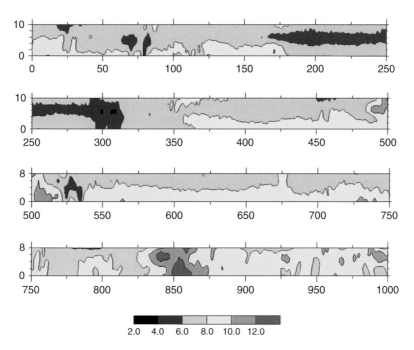

Figure 13.12 A contour map presenting air voids content (%) measured using a GPR horn antenna in a test section on US 290, Texas, USA. Distance is presented in feet (from Sebesta and Scullion, 2002).

In addition to quality control or quality assurance surveys of new asphalt pavement, GPR has, during the last few years, been applied increasingly in quality control surveys of other road structures (Pälli et al., 2005).

13.6.6. Special applications

GPR has been used for tunnel inspection in both roadway and railway tunnels. The focus of these surveys has been to detect fracture zones (Davis and Annan, 1992) and measuring concrete wall thickness, locating rebar or detecting voids between the concrete and the bedrock, detecting water leakage and other defects (Uomoto and Misra, 1993; Daniels, 1996; Hugenschmidt, 2003) as well as testing grouting behind the lining of the shield tunnels (Xie et al., 2004).

13.7. BRIDGES

13.7.1. General

GPR was first used in bridge deck surveys in the USA and Canada in the early 1980s (Cantor and Kneeter, 1982; Clemena, 1983; Manning and Holt, 1983). Quite soon the research was focused on developing automated or semiautomated GPR data analysis (Maijala et al., 1994; Mesher et al., 1996). In the late 1990s, research and development work was done especially in developing multichannel GPR system for mapping bridge deck deterioration (Azevedo et al., 1996) but the methods did not become popular. During recent years, the focus has been on collecting reflection amplitude data from bridge decks and preparing maps that present damaged areas in the bridge structures (Romero and Roberts, 2004), testing multichannel high-frequency GPR systems (Washer, 2003) and testing high-frequency (0.5–6 GHz) GPR systems to detect subsurface delamination (Huston et al., 2002).

GPR applications related to bridge surveys can be divided roughly into (1) bridge foundation-related problems such as site investigations and detecting scours around bridge piers (Haeni et al., 1992; Forde et al., 1999; Fish, 2002), (2) bridge decks and bridge beams surveys (Hugenschmidt, 2004; Romero and Roberts, 2004) and (3) other surveys such as surveys on approaching slabs and bridge abutments (Lewis et al., 2002; Hugenschmidt, 2003).

A majority of the reported GPR bridge surveys have been done on concrete bridges, but there has also been some testing done on masonry bridges (see Clark et al., 2003b) and wood bridges (Muller, 2003).

13.7.2. Bridge deck surveys

A bridge deck can be examined in many ways, and Hugenschmidt (2004) lists the following issues that can be addressed using GPR: pavement thickness, thickness of single pavement layer, pavement damage, concrete cover of top layer of reinforcement, spacing between re-bars (reinforcement bars), position of tendons or tendon ducts, concrete damage, concrete and pavement properties.

Parry and Davis (1992) have compiled a test parameter matrix for bridge decks, where they prioritized the different types of survey parameters that can be identified by measuring dielectric contrast, signal velocity, reflection coefficient and attenuation estimate. Highest priority, class 1 parameters were asphalt/pavement thickness, rebar covering, debonding, delamination and scaling; class 2 parameters were chloride content, moisture content (free moisture), moisture content (bound in concrete); class 3 parameters: voids; and class 5 (lowest) parameters were cracking (surface) and cracking (subsurface).

In general, the primary cause of deterioration in bridge decks is corrosion of the steel reinforcements which induces concrete cracking and frequently results in delamination. This corrosion is caused mainly by deicing salts and when the reinforcements corrode they expand and a horizontal crack is formed in the reinforcement level. Another primary cause of deterioration in cold climate areas is freeze–thaw cycles in chloride-contaminated concrete (Saarenketo and Söderqvist, 1993). This deterioration, also called scaling, normally starts at the concrete surface and progresses downwards. According to Manning and Holt (1986) deterioration can be rapid if the cover layer over the top reinforcements is too thin. A third damage type found on bridges is debonding which takes place when asphalt or concrete overlay debonds from the concrete bridge deck.

Both high-frequency ground-coupled and air-coupled antennas can be used in bridge deck surveys. Ground-coupled systems can provide very detailed information about the bridge deck's structures and reinforcement bars. The major problems are the slow speed of data collection and that bridge lanes have to be closed during the data collection. As such, the use of air-coupled antenna systems, which can perform data collection without causing major traffic problems, is highly recommended especially on high traffic volume roads.

Bridge deck deterioration mapping using the reflection amplitude from layer interfaces and especially from top rebar reflection amplitude as measure of deterioration has become a very popular method in deck condition assessment. Romero and Roberts (2002, 2004) have introduced a special dual polarization horn antenna setup (Figure 13.13) and processing technique that can be used to eliminate the effect of longitudinal reinforcement bars, which has been the problem when collecting data with air-coupled antennas. Figure 13.14 presents bridge deterioration maps made using ground-coupled antennas, dual polarization horn antenna setup and ground truth testing.

Shin and Grivas (2003) reported a statistical method for evaluating the accuracy of GPR results on bridge decks. The results showed that rebar reflection data detect defects at 75% true detection rate with a 15% false detection rate. Surface dielectric value failed to discriminate defects from the decks. There is still a fair amount of controversy over publications and research reports concerning the reliability of GPR especially in detecting delamination in bridge decks (see Morey, 1998). Some research has attributed this difficulty to GPR resolution problems (Rhazi et al., 2003; Washer, 2003). Because of the nature of the problems in bridge deck surveys, GPR alone cannot provide reliable enough information on the damage in the deck but it is an excellent tool for the initial mapping and specifying of locations where other non-destructive evaluation methods and limited ground truth (sampling)

Figure 13.13 Dual polarization configuration of horn antennas for high-speed, high-resolution bridge deck surveys (photo Francisco Romero).

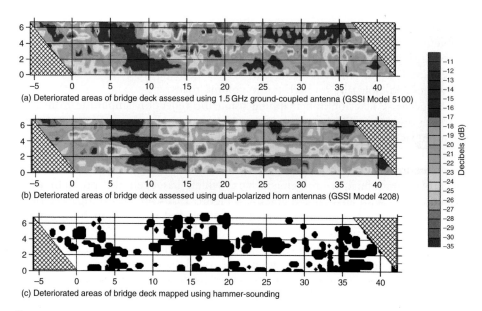

Figure 13.14 Comparison of deterioration maps obtained from (a) 1.5 GHz ground-coupled data and (b) 1.0 GHz dual polarization horn antenna data with (c) ground truth data obtained from hammer sounding. Data were obtained from IS 93 North in New Hampshire, USA (from Romero and Roberts, 2002).

testing can be used to verify the problems. GPR is sensitive to electrical and chemical changes that are present in early phase corrosion and concrete disintegration. Often, anomalies detected by GPR cannot be verified visually from drill cores but only by using thin slides or chemical analysis (Figure 13.15).

GPR can also be used as an evaluation tool for quality assurance regarding placement, density and pattern of steel reinforcement. This can be done quite precisely and Hugenschmidt (2003) reported a mean error of 3 and 17 mm for concrete cover.

Another promising application of GPR on bridges is detecting voids in post-tensioned concrete beams. Giannopoulos et al. (2002) suggest that the optimum orientation of GPR antennas is perpendicular to the long axis of the ducts containing the post-tensioning tendons. Dérobert et al. (2002b) have tested several NDT methods in testing post-tensioned bridge beams and suggest that the best combination of current techniques is GPR before gammagraphy.

13.7.3. Other bridge applications

In highway bridge site investigations, GPR has been used to monitor the bottom topography of rivers and lakes to map the quality of underwater sediments (Ground Penetrating Radar, 1992). Another application that has been tested in various projects has been detecting scours around bridge piers or abutments caused by the water flow erosion of a riverbed (Haeni et al., 1992; Forde et al., 1999; Fish,

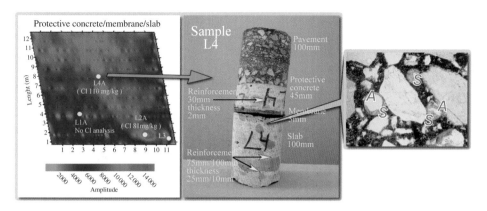

Figure 13.15 Case from Långsvedjan bridge in Sweden presents an example of how GPR can be used to predict problems, in the early phase, in bridge decks. The bridge structure has asphalt on the top, protective concrete in the middle and then a slab. The contour maps on the left present reflection amplitude from the protective concrete/slab interface, where a 5-mm membrane is also located. The photo in the middle is of a drill core with no visual defects in any layer. The photo on the right is taken from a thin slice taken from the slab below the membrane. This thin slice shows the first indicators of deterioration where secondary minerals (S) have started to fill air pores (A) which appear as yellowish material in the photo. Chloride content of the concrete in this place was also high, 110 mg kg^{-1} (figure modified from Report "Tillståndsvärdering av brodeck med georadar på E4 I Härnösand" by Mika Silvast and Svante Johansson, Roadscanners Sweden AB).

2002; Webb et al., 2002). These scours have been tested using low-frequency antennas mounted in plastic or rubber boats or antennas have been moved over the river hanging from a cable. In Finland, these surveys have mainly been done in the wintertime when the river is frozen. The best results are obtained if the conductivity of the river water is less than 1000 µs cm^{-1} (Forde et al., 1999).

13.8. RAILWAYS

13.8.1. General

With regard to traffic infrastructure, GPR applications in railway surveys have had the fastest growth in recent years. In Finland, GPR was tested on railways in the mid-1980s but the results were not encouraging mainly because of problems with data collection and processing. In Germany, the GPR was first used to measure ballast thickness and to locate mudholes and ballast pockets and define the subgrade soil boundaries (Göbel et al., 1994).

GPR started to become more widely accepted among railway engineers in the mid-1990s. The first reports of successful track inspection tests were reported by Hugenschmidt in Switzerland (1998, 2000) and by Galagher et al. in the UK (1998). Since 1998, a number of publications have been made regarding different GPR applications on railways. The system has been used in North America (Olhoeft and Selig, 2002; Sussman et al., 2002) and in Europe, more specifically, in the UK (Brightwell and Thomas, 2003; Clark et al., 2003), Germany (Manacorda et al., 2002), Austria, Switzerland, France, the Netherlands, Slovenia (see Staccone and de Haan, 2003), Sweden (Smekal et al., 2003) and Finland (Saarenketo et al., 2003; Silvast et al., 2006).

GPR applications on railways can be classified as (1) ballast surveys (Clark et al., 2001; Sussman et al., 2002; Clark et al., 2003c), (2) geotechnical investigations (Saarenketo et al., 2003; Sussman et al., 2003; Carpenter et al., 2004), and (3) structural quality assurance of new non-ballasted railway trackbeds (Maierhofer and Kind, 2002; Gardei et al., 2003). GPR has also been used in railway bridge and tunnel surveys.

13.8.2. Data collection from railway structures

Compared with pavement surveys, there are many more complications with obtaining a good-quality GPR signal from railways. GPR operators have to struggle with interference from the rails and sleepers, especially concrete sleepers. Electrical wires have also created interference in the GPR data especially when unshielded antennas are used. Researchers have solved these problems by using different kinds of antenna configurations (Figure 13.16). Another problem has

Figure 13.16 Different GPR antennas used in railway surveys. Photo (a) presents a Swedish Mala ground-coupled antenna, photo (b) presents a multichannel GSSI air-coupled system (from Olhoeft et al., 2004), photo (c) presents a GSSI 400-MHz ground-coupled antenna and photo (d) presents Radar Team Sweden 350 MHz 1.2 GHz Sub-Echo HBD antennas.

been the data collection speed. Railway surveys require high sampling density (scans m^{-1}), and thus the data collection speed, with many systems, has been slow and has caused problems with scheduling traffic. However, new GPR systems with fast processors allow high-speed data collection with no problems with scheduling surveys (Clark et al., 2004).

The optimum antenna for the railway surveys varies. In North America many successful tests have been carried out using 1.0 GHz horn antennas (Olhoeft and Selig, 2002), while in the UK (Clark et al., 2003c) and Finland (Saarenketo et al., 2003) the best quality data in railway structure thickness measurements were collected using low-frequency 400–500 mHz antennas. There have been problems using high-frequency air-coupled antennas to measure ballast thickness but the antennas were good for detecting frost insulation boards (Saarenketo et al., 2003). Higher frequency antenna pulses with shorter wavelength are scattered in ballast with stones larger than 50 mm in diameter (see Clark et al., 2001). In railway surveys, antennas have to be kept relatively high above the surface level and as such antenna central frequencies are also slightly higher compared with when they are in contact with the surface (Clark et al., 2003).

In railway surveys, the best results have been achieved when GPR results are combined with the results from other NDT methods. In the UK, good results have been obtained from the integrated analysis of GPR and Infrared thermography (Clark et al., 2003c, 2004). Smekal et al. (2003) have tested GPR and a Track Loading Vehicle on the Swedish Western Main Line with a focus on sections with excessive settlements, slides and environmental vibration problems. In these surveys they used a 100-MHz ground-coupled antenna in monostatic and array mode and 500 MHz antennas in bistatic mode. However, they did not find any correlation between vertical track stiffness and the average amplitude of GPR signal. Later the integrated analysis of the GPR data and track stiffness data has given more promising results (Berggren et al., 2006). Grainger and Armitage (2002, referred by Clark et al., 2003c) also report the combined use of GPR, an automated ballast sampler and an FWD in railway trackbed evaluation.

13.8.3. Ballast surveys

Railway ballast thickness measurements and quality evaluations are the main applications of GPR on railways. Ballast is made of crushed hard rock or, sometimes, crushed gravel material, where smaller mineral particles have been sieved away. The ballast of a railway line must perform many different functions some of which are reduce stresses applied to weaker interfaces, resist vertical, lateral and longitudinal forces applied to sleepers to maintain track position; and to provide drainage for water from the track structure (Clark et al., 2001). Ballast quality surveys focus on locating sections of clean and spent ballast. Spent ballast normally has a higher amount of fine particles than is allowed and can no longer fulfil the requirements for which it is being used (Clark et al., 2001). Fines are mainly formed through the mechanical wear, imposed by vibrations and loads from passing trains, and chemical wear caused by pollution and the effects of weather erosion on the

Table 13.3 Published dielectric values of ballast

Material	Dielectric value (Clark, 2001)	Dielectric value of granite ballasts (Sussmann et al., 2002)
Dry clean ballast	3	3.6
Moist clean ballast	3.5	4
Dry spent ballast	4.3	3.7
Moist spent ballast	7.8	5.1
Wet spent ballast	38.5	7.2

Source: Prepared by Clark et al. (2003a).

larger particles (Clark et al., 2001). According to Nurmikolu (2005) mechanical wear is the most important factor increasing the fines content in ballast in Finland but organic material from external sources also has an important role in increasing water adsorption properties of ballast. However, in some countries coal dust from cargo trains has been the main source of fines in ballast. GPR can also detect, very reliably, if subgrade soil material has penetrated or mixed with the ballast (Hugenschmidt, 2000; Brightwell and Thomas, 2003).

Dielectric value is a good indicator of ballast quality. The dielectric properties of ballast materials have been surveyed by Clark et al. (2001) and Sussman et al. (2002). The main parameters affecting the dielectric properties are moisture content and the level of fouling (Sussman et al., 2002). Clark et al. (2003a) have presented a table (Table 13.3) where the dielectric values of good- and poor-quality ballast materials have been compared in dry, moist and wet conditions.

The above table also shows that in ballast thickness and quality surveys it is important to know the dielectric value of the material. Since the surface reflection technique cannot be used the options for determining the dielectric value are the CMP or WARR method. In Finland, if the goal is only to determine ballast thickness, then the potential error caused by changes in dielectric values are eliminated by measuring the ballast thickness when it is frozen (Saarenketo et al., 2003).

13.8.4. Subgrade surveys, site investigations

GPR can be used to detect problems related to embankment instability with a risk for track settlement (Sussman et al., 2003). Olhoeft and Selig (2002) have collected data from the railway substructure using 1.0 GHz horn antennas. Data were collected from the centreline and both sides of the rails and then compared. If the data are similar all is well but when they differ dramatically, that should cause concern for track stability. In construction of new railways, Carpenter et al. (2004) have reported the use of a specialized geosynthetic, detectable with radar, in railway structures, which enables the use of GPR to monitor track stability in high-risk areas with karstic landscapes or old mineworkings for instance.

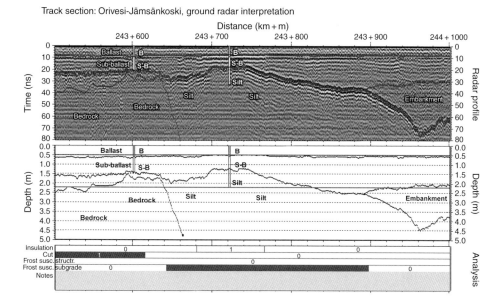

Figure 13.17 A GPR profile of a railway section from Jämsä to Orivesi in Finland. The profile on the top presents interpreted 400 MHz ground-coupled GPR data with ground truth information. The profile in the middle presents thickness interpretation together with ground truth data. The top horizontal line presents the minimum ballast thickness level and the lower line the minimum total thickness of the whole structure against frost heave, if frost insulation is not used. The colour bars in the bottom frame present the location of frost insulation, the location of sections where the railway is in a cut, the location of frost susceptible ballast and sub-ballast structures and the location of frost-susceptible subgrade.

In addition to the risk of settlement, differential frost heave also causes problems in cold climate areas. Saarenketo et al. (2003) have tested different antenna types and frequencies for obtaining information from railway ballast and substructures up to maximum frost penetration depth of 3–4 m (Figure 13.17). Tests showed that GPR can be used to detect frost-susceptible areas and structures (see also Saarenketo and Scullion, 2000).

13.9. AIRFIELDS

There are not many publications available about the application of GPR on airfields even though the system has been widely used in different airfield-related surveys. Beck and Ronen (1994) have tested 1.0 GHz ground-coupled antennas for runway substructure stratigraphic mapping and lateral variation estimations used in the planning of the reconstructed runway. Malvar and Cline (2002) have tested GPR together with other NDT methods in detecting voids under airfield pavements. In these projects GPR was useful in 25–75% of the cases, but on the other hand, the advantage of the GPR resides in the ability to process a large amount of data quickly and its ability to pinpoint drainpipe locations (Malvar and Cline,

2002). Szynkiewicz and Grabowski (2004) have used GPR and heavy weight deflectometer (HWD) on airfields in Poland to map different pavement structures and their condition, and to locate damages and their causes.

According to Weil (1992, 1998) the two prime NDT techniques in airfields are GPR and/or infrared thermography. He presents a list of applications in which these techniques can be used individually or to complement each other (Weil, 1998): (1) locating voids within, and below, concrete runways and taxiways,

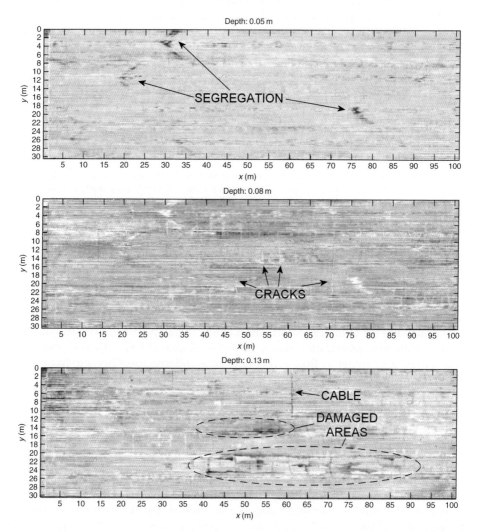

Figure 13.18 A 3D radar time slice from three asphalt layers of an airport runway in Finland. The first time slice calculated from a depth of 40–50 mm from the new asphalt overlay presents segregation in three places. Transverse cracks can be seen in the second time slice from 70 to 80 mm. The third time slice from 120 to 130 mm presents areas of cracking with moisture problems highlighted with the circles. A cable installed in the asphalt can also be seen in the lowest time slice.

(2) locating moisture trails within and below concrete runways and taxiways, (3) locating post-tensioning cables, in concrete garages, bridges and buildings, (4) locating voids and delaminations caused by corrosion or poor mixing of concrete roofs, (5) locating cables, conduits and reinforcing steel in concrete floors to assist in equipment placement, (6) locating underground storage tanks below concrete and asphalt pavements, and (7) locating sewer, water, gasoline, jet fuel, natural gas, glycol, steam and chemical buried pipelines and leaks in those lines.

GPR has also been used in airport QC/QA surveys (Scullion and Saarenketo, 2002). Because the depth of interest under airfield pavements can vary up to 3 m, use of two antenna systems, where one is high and the other a low-frequency antenna, is recommended (see Malvar and Cline, 2002).

A very good new tool for airfield surveys is the multichannel 3D GPR technique which enables quick and economical coverage of the wide areas of runways and taxiways. Figure 13.18 presents an example of a runway pavement condition test survey done in Finland, where time slices from three asphalt layers show different types of defects. The time slice from the depth of 130 mm shows a problem zone in the third asphalt layer which can, in the long term, cause reflection cracking problems even in the pavement surface. The lower maps also show the location of a cable in the pavement (Saarenketo, 2006).

13.10. SUMMARY AND RECOMMENDATIONS

The history of GPR in traffic infrastructure surveys is very short, but over this period the technique has grown from an uncommon geophysical technique to a more and more routine survey tool when detailed information about roads, streets, railways, runways or bridge structures or subgrade soils and their properties is needed. In traffic infrastructure surveys the major advantages of GPR techniques are continuous profile, speed and accuracy. Its disadvantages include the complexity of the GPR techniques, GPR data and, as such, more education and good software products are needed in order to make the GPR signals meaningful to engineers. Specifications and standards for the GPR equipment, data collection, processing and interpretation as well as output formats are also needed to ensure the high quality of the surveys results.

The future of GPR on roads looks even more promising. The technology so far has gone through two generations in the development of GPR systems; first generation was analogue GPR systems and the second generation started when the first digital systems came to the market in the mid-1980s. Currently the third-generation 3D GPR systems are entering the market with multiple antennas, faster processors and larger data storage capabilities. This opens a whole new range of applications in which structures and other objects, and their properties can be analysed in a 3D format. However, GPR techniques alone cannot make traffic infrastructure management more economical if the management process does not find ways to make good use of the results. This is why GPR surveys need to be more closely integrated to entire traffic infrastructure rehabilitation process, starting from data collection during the monitoring phase and ending with the downloading of results to automated road construction or rehabilitation machinery.

REFERENCES

Adams, G., Anderson, N., Baker, J., Shoemaker, M. and Hatheway, A., 1998, Ground-penetrating radar study of subsidence along U.S. interstate 44, Springfield, Missouri. Proceedings of the Seventh International Conference on Ground Penetrating Radar, May 27–30, 1998, Lawrence, Kansas, USA, pp. 619–623.

Ahmed, Z., Helali, K., Jumikis, A.A. and Khan, R.A., 2003, Enhancing the Pavement Management Systems Database through incorporation of GPR and Core Data. Transportation Research Board Meeting Records, pp. 20p.

Alongi, T., Clemena, G.G. and Cady, P.D., 1992, Condition Evaluation of Concrete Bridges relative to Reinforcement Corrosion, Volume 3, SHRP Report SHRP-S/FR-92-105, Washington, DC.

Al-Qadi, I.L., Clark, T.M., Lee, D.T., Lahouar, S. and Loulizi, A., 2003a, Combining Traditional and Non-Traditional NDT Techniques to Evaluate Virginia's Interstate 81. Transportation Research Board Paper 03-4214, pp. 17p.

Al-Qadi, I.L., Lahouar, S. and Loulizi, A., 2002, Ground penetrating radar evaluation for flexible pavement thickness estimation, Al-Qadi, I.L. and Clark, T.M. (eds). Proceedings of the Pavement Evaluation 2002 Conference, October 21–25, 2002, Roanoke, VA, USA, pp. 15p.

Al-Qadi, I.L., Lahouar, S. and Loulizi, A., 2003c, GPR: From the state-of-the-art to the state-of-the-practice. Proceedings of International Symposium on Non-Destructive Testing in Civil Engineering (NDT-CE), September 16–19, 2003, Berlin, Germany.

Al-Qadi, I., Lahouar, S. and Loulizi, A., 2003b, Successful Application of GPR for Quality Assurance/Quality Control of New Pavements. Transportation Research Board Paper 03-3512, pp. 21p.

Anderson, N. and Ismael, A. ,2002, A protocol for selecting appropriate geophysical surveying tools based on engineering objectives and site characterization. Proceedings of the 2nd Annual Conference on the Application of Geophysical and NDT Methodologies and NDT Methodologies to Transportation Facilities and Infrastructure, April 15–19, 2002, California, USA, FHWA-WCR-02-001. pp. 16p.

Azevedo, S.G., Mast, J.E., Nelson, S.D., Rosenbury, E.T., Jones, H.E., McEwan, T.E., Mullenhof, D.J., Hugenberger, R.E., Stever, R.D., Warhus, J.P. and Wieting, M.G., 1996, HERMES, A high-speed radar imaging system for inspection of bridge decks. Nondestructive evaluation techniques for aging infrastructure and manufacturing. SPIE Vol. 2946, pp. 195–204.

Bae, S.H., Kim, H.S. and Yoon, W.S., 1996, Cases studies on the application of ground penetrating technology in detection of underground utilities and structure safety diagnosis. Proceedings of the Sixth International Conference on Ground Penetrating Radar, September 30–October 3, 1996, Sendai, Japan, pp. 467–472.

Ballard, G., 1992, Under the Skin. World Highways/Routes du Monde. Jan/Feb 1992, pp. 37–39.

Ballard, G., 1993, Non destructive assessment of pavement design & new build quality, Bungey, J. (ed.). Proceedings of Conference on Non-Destructive Testing in Civil Engineering, April 14–16, 1993, Liverpool, U.K., . pp. 391–404.

Beck, A. and Ronen, A., 1994, The application of ground penetrating radar in Israel: Selected case histories. Proceeding of the Fifth International Conference on Ground Penetrating Radar, June 12–16, 1994, Kitchener, Ontario, Canada, Vol. 13 of 3, pp. 1101–1106.

Benedetto, A. and Benedetto, F., 2002, GPR experimental evaluation of subgrade soil characteristics for rehabilitation of roads. Proceedings of the Ninth International Conference on Ground Penetrating Radar, April 29–May 2, 2002, Santa Barbara, CA, S. Koppenjan and H. Lee (eds). Proceedings of SPIE, Vol. 4758, pp. 708–714.

Berg, F., 1984, Ikke destruktiv måling af lagstykkelser I vejbefastelser. Vejdirektoratet, Statens vejlaboratorium, Notat 157, Denmark.

Berggren, E.G., Smekal, A. and Silvast, M., 2006, Monitoring and substructure condition assessment of existing railway lines for upgrading to higher axle loads and speeds. Proceeding of 7th World Congress on Railway Research, June 4–8, 2006, Montréal, Canada.

Berthelot, C., Scullion, T., Gerbrandt, R. and Safronetz, L., 2001, Application of ground penetrating radar for cold in-place recycled road systems. Journal of Transportation Engineering. American Society of Civil Engineers, Vol. 127, No. 4.

Briggs, R.C., Scullion, T. and Maser, K., 1991, Asphalt thickness variation an effect on backcalculation. 2nd International Conference on Backcalculation, Nashville, USA.

Brightwell, S. and Thomas, M., 2003, Geophysical assessment of railway assets – A practical view. Proceedings of 6th International Conference on Railway Engineering, April 30–May 1, 2003, London, UK, pp. 8p.

Cantor, T.R. and Kneeter, C.P., 1982, Radars as Applied to Evaluation of Bridge Decks. Transportation Research Records. Transportation Research Record, Vol. 853, pp. 37–42.

Cardimona, S. and Newton, T., 2002, Evaluation of GPR as a tool for determination of granular material deposit volumes. Proceedings of the 2nd Annual Conference on the Application of Geophysical and NDT Methodologies and NDT Methodologies to Transportation Facilities and Infrastructure, April 15–19, 2002, California, USA, FHWA-WCR-02-001. pp. 10p.

Cardimona, S., Willeford, B., Webb, D., Hickman, S., Wenzlick, J. and Anderson, N., 2003, Automated pavement analysis in Missouri using ground penetrating radar. Proceedings of the 2nd Annual Conference on the Application of Geophysical and NDT Methodologies and NDT Methodologies to Transportation Facilities and Infrastructure, April 15–19, 2003, California, USA, FHWA-WCR-02-001. pp. 10p.

Carpenter, D., Jackson, J. and Jay, A., 2004, Enhancement of the GPR method of railway trackbed investigation by installation of radar detectable geosynthetics. NDT and E International, Vol. 37, No. 2, pp. 95–103.

Carter, C.R., Chung, T., Masliwec, T. and Manning, D.G., 1992, Analysis of radar reflections from Asphalt covered Bridge Deck structures, in Pilon, J. (ed.), Ground Penetrating Radar, Geological Survey of Canada, Paper 90–94. pp. 33–40.

Casas, A., Lazaro, R. and Vilas, M., 1996, Detecting karstic cavities with ground penetrating radar at different geological environment in Spain. Proceedings of the Sixth International Conference on Ground Penetrating Radar, September 30–October 3, 1996, Sendai, Japan, pp. 455–460.

Clark, M.R. and Crabb, G., 2003, Detection of water movement below a concrete road using ground penetrating radar. Proceedings of 10th International Conference on Structural Faults and Repair 2003, Engineering Tech Press, London.

Clark, M.R., Gillespie, R., Kemp, T., McCann, D.M. and Forde, M.C., 2001, Electromagnetic Properties of Railway Ballast. NDT&E International, Vol. 34, pp. 305–311.

Clark, M., Gordon, M. and Forde, M.C., 2004, Issues over High-Speed Non-Invasive Monitoring of Railway Trackbed. NDT&E International, Vol. 37, pp. 131–139.

Clark, M.R., Gordon, M.O., Giannopoulos, A. and Forde, M.C., 2003a, Experimental and computer modelling of GPR to characterize trackbed ballast. Proceedings of World Congress on Railway Research, Scotland. pp. 16p.

Clark, M.R., McCann, D.M. and Forde, M.C., 2003c, GPR as a tool for the characterization of ballast. Proceedings of 6th International Conference on Railway Engineering, April 30–May 1, 2003, London, UK, pp. 9p.

Clark, M.R., Parsons, P., Hull, T. and Forde, M.C., 2003b, Case study of radar laboratory work on the Masonry arch bridge. Proceedings of 10th International Conference on Structural Faults and Repair, Engineering Technics press, London, pp. 12p.

Clemena, G., 1983, Nondestructive Inspection of Overlaid Bridge Decks with Ground Penetrating Radar. Transportation Research Record, Vol. 899, pp. 31–37.

Clemena, G., Sprinkel, M. and Long, R.J., Jr., 1987, Use of Ground Penetrating Radar for Detecting Voids Under a Jointed Concrete Pavement. Transportation Research Record, Vol. 1109, pp. 1–10.

Daniels, D.J., 1996, Surface Penetrating Radar. Institution of Electrical Engineers, U.K., pp. 300p.

Davidson, N.C. and Chase, S.B., 1998, Radar systems for tomographic bridge deck inspection. Proceeding of the Seventh International Conference on Ground Penetrating Radar, May 27–30, 1998, Lawrence Kansas, Vol. 2, pp. 747–751.

Davis, J.L. and Annan, A.P., 1992, Applications of ground penetrating radar to mining, ground water and geotechnical projects: Selected case histories, in Pilon, J. (ed.), Ground Penetrating Radar, Geological Survey of Canada, Paper 90–94. pp. 49–55.

Davis, J.L., Rossiter, J.R., Mesher, D.E. and Dawley, C.B., 1994, Quantitative measurement of pavement structures using radar. Proceedings of 5th International Conference on Ground Penetrating Radar, Kitchener, Ontario, Canada.

Dérobert, X., Abraham, O. and Aubagnac, C., 2002b, Review of NDT methods on a weak post-tensioned beam before autopsy. Proceedings of the Ninth International Conference on Ground Penetrating Radar, April 29–May 2, 2002, Santa Barbara, California, Koppenjan, S. and Lee, H. (eds). Proceedings of SPIE, Vol. 4758, pp. 365–370.

Dérobert, X., Fauchard, C., Cote, Ph., Le Brusc, E., Guillanton, E., Dauvignac, J.Y. and Pichot, Ch., 2001, Stepped-frequency radar applied on thin road layers. Journal of Applied Geophysics, Vol. 47, Nos. 3–4, pp. 317–325.

Dérobert, X., Simonin, J.M. and Laguerre, L., 2002a, Step-frequency radar applied on thin pavement. Proceedings of the Ninth International Conference on Asphalt Pavements, August 17–22, 2002, Copenhagen, Denmark. pp. 16p.

Doolittle, J.A. and Rebertus, R.A., 1988, Ground-Penetrating Radar as Means of Quality Control for Soil Surveys. Transportation Research Board Record, Vol. 1192.

Eide, E., 2002, Ultra-wideband 3d-imaging ground penetrating radar using synthetic waveforms. Proceedings of 32nd European Microwave Conference, September 6–26, 2002, Milan, Italy, pp. 4p.

Ékes, C. and Friele, P., 2004, Ground penetrating radar and its use in forest road stability analysis, Slob, E., Yarovov, A. and Rhebergen, J. (eds). Proceedings of the Tenth International Conference on Ground Penetrating Radar, Vol. II, pp. 639–642.

Emilsson, J., Englund, P. and Friborg, J., 2002, A simple method for estimation of water content of roadbeds using multi-offset GPR. Proceedings of the Ninth International Conference on Ground Penetrating Radar, April 29–May 2, 2002, Santa Barbara, CA,. Koppenjan, S. and Lee. H. (eds). Proceedings of SPIE, Vol. 4758, pp. 422–426.

Fauchard, C., Xerobert, X., Cariou, J. and Cote, Ph., 2003, GPR performance for thickness calibration on road test sites. NDT and E International, Vol. 36, No. 2, pp. 67–75.

Fernando, E.G., Maser, K.R. and Dietrich, B., 1994, Implementation of ground penetrating radar of network level pavement evaluation in Florida. Proceedings of the Fifth International Conference on Ground Penetrating Radar, June 12–16, 1994, Kitchener, Ontario, Canada, Vol. 1 of 3, pp. 351–365.

Fish, M., 2002, Enhancing geotechnical information with ground penetrating radar. Proceedings of the 2nd Annual Conference on the Application of Geophysical and NDT Methodologies and NDT Methodologies to Transportation Facilities and Infrastructure, April 15–19, 2002, California, USA, FHWA-WCR-02-001, pp. 15p.

Forde, M.C., McCann, D.M., Clark, M.R., Broughton, K.J., Fenning, P.J. and Brown, A., 1999, Radar measurement of bridge scour. NDT&E International, Vol. 32, pp. 481–492.

Forest, R. and Utsi, V., 2004, Non destructive crack depth measurements with ground penetrating radar, Slob, E., Yarovov, A. and Rhebergen, J. (eds). Proceedings of the Tenth International Conference on Ground Penetrating Radar, Vol. II,. pp. 799–802.

Fredlund, D.G. and Rahardjo, H., 1993, Soil Mechanics for Unsaturated Soils, John Wiley & Sons, New York, p. 517.

Galagher, G.P., Leiper, Q., Williamson, R., Clark, M.R. and Forde, M.C., 1998, Application of ground penetrating radar in the railway industry. Proceedings of the 1st International Conference Railway Engineering, Technical Press, London, 1999, pp. 33–36.

Gardei, A., Mittag, K., Wiggenhauser, H., Ripke, B. and Jovanovic, M., 2003, Inspection of concrete-embedded tracks process development for the quality assurance of concrete-embedded tracks using non-destructive testing methods. Proceedings of International Symposium on Non-Destructive Testing in Civil Engineering (NDT-CE), September 16–19, 2003, Berlin, Germany.

Geraads, S. and Omnes, G., 2002, Imaging karstic structures with GPR along a motorway under construction. Proceedings of the Ninth International Conference on Ground Penetrating Radar,

April 29–May 2, 2002, Santa Barbara, CA, Koppenjan, S. and Lee, H. (eds). Proceedings of SPIE, Vol. 4758, pp. 493–495.

Giannopoulos, A., MacIntyre, P., Rodgers, S. and Forde, M.C., 2002, GPR detection of voids in post tensioned concrete bridge beams. Proceedings of the Ninth International Conference on Ground Penetrating Radar, April 29–May 2, 2002, Santa Barbara, CA, Koppenjan, S. and Lee, H. (eds). Proceedings of SPIE, Vol. 4758, pp. 376–381.

Glover, J.M., 1992, Void detection using standing wave analysis. In Ground Penetrating Radar, Pilon, J. (ed.), Geological Survey of Canada, Paper 90–94. pp. 55–73.

Göbel, C., Hellman, R. and Petzold, H., 1994, Georadar – Model and in-situ investigations for inspection of railway tracks. Proceedings of the Fifth International Conference on Ground Penetrating Radar, June 12–16, 1994, Kitchener, Ontario, Canada, Vol. 3 of 3, pp. 1101–1106.

Golgowski, G., 2003, Arbeitsanleitung fur den Einsatz des Georadars zur Gewinnung von Bestandsdaten des Fahrbahnaufbaues. Berichte der Bundesanstalt fr StraÔenwesen (bast), StraÔenbau, Heft S 31, pp. 22p.

Goodman, D., Nishimura, Y. and Tobita, K., 1994, GPRSIM forward modeling software and time slices in ground penetrating radar simulations. Proceedings of the Fifth International Conference on Ground Penetrating Radar, Kitchener, Ontario. Waterloo Centre of Groundwater Research, Waterloo, pp. 31–43.

Grainger, P.S. and Armitage, R.J., 2002, Trackbed investigatuion and interpretation in practise. Proceeding of the Railway Technology Conference, NEC, Birmingham, November 26–28, 2002.

Ground Penetrating Radar, 1992, Geophysical Survey Methods, The Finnish Geotechnical Society and The Finnish Building Centre Ltd., Finland, pp. 64p.

Haeni, F.P., Placzek, G. and Trent, R.E., 1992, Use of Ground Penetrating Radar to Investigate Refilled Scour Holes at Bridge Foundations. Geological Survey of Finland, Special Paper 16, pp. 285–292.

Hall, K.T., Correa, C.E., Carpenter, S.H. and Elliot, R.P., 2002, Guidelines for evaluation of highway pavements for rehabilitation, Al-Qadi, I.L. and Clark, T.M. (eds). Proceedings of the Pavement Evaluation 2002 Conference, October 21–25, 2002, Roanoke, VA, USA, pp. 23p.

Hänninen, P., 1997, Dielectric Coefficient Surveying for Overburden Classification. Geological Survey of Finland, Bulletin 396, Helsinki, pp. 72p.

Hänninen, P. and Sutinen, R., 1994, Dielectric prediction of landslide susceptibility: A model applied to recent sediment flow deposits. Proceedings of VII IAEG, Lisboa, Vol. 1, pp. 137–144.

Hobbs, C.P., Temple, J.A.G. and Hillier, M.J., 1993, Radar inspection of civil engineering structures, Bungey, J. (ed.). Proceedings of Conference on Non-Destructive Testing in Civil Engineering, April 14–16, 1993, Liverpool, U.K., pp. 79–96.

Hopman, V. and Beuving, E., 2002, Repeatability, reproducibility and accuracy of GPR measurements. Proceedings of the 6th International Conference on the Bearing capacity of Roads, Railways and Airfields, Lisbon, Portugal, June 24–26, 2002. pp. 637–645.

Hugenschmidt, J., 1998, Railway track inspection using GPR – Some examples from Switzerland. Proceeding of Seventh International Conference on Ground Penetrating Radar, May 27–30, 1998, Lawrence, KS, USA, Vol. 1 of 2, pp. 197–202.

Hugenschmidt, J., 2000, Railway track inspection using GPR. Journal of Applied Geophysics, Vol. 43, pp. 147–155.

Hugenschmidt, J., 2003, Non-destructive-testing of traffic-infrastructure using GPR. Proceedings of International Symposium on Non-Destructive Testing in Civil Engineering (NDT-CE), September 16–19, 2003, Berlin, Germany.

Hugenschmidt, J., 2004, Accuracy and reliability of radar results on bridge decks. Proceedings of the 10th International Conference on Ground Penetrating Radar, June 21–24, 2004, Delft, The Netherlands (in print).

Hugenschmidt, J., Partl, M.M. and de Witte, H., 1996, GPR inspection of a Mountain Motorway – A case study. Proceedings of the Sixth International Conference on Ground Penetrating Radar, September 30–October 3, 1996, Sendai, Japan, pp. 365–370.

Huston, D., Fuhr, P., Maser, K. and Weedon, W., 2002, Nondestructive Testing of Reinforced Concrete Bridges Using Radar Imaging Techniques, The New England Transportation Consortium Report, 94–92, pp. 182p.

Huston, D., Pelczarski, N., Esser, B. and Maser, K., 2000, Damage detection in roadways with ground penetrating radar. Proceeding of Eight International Conference on Ground Penetrating Radar, May 23–26, 2000, Gold Coast Australia, Noon, D.A., Stickley, G.F. and Longstaff, D., SPIE, Vol. 4084, pp. 91–94.

Johansson, H.G., 1987, Använding av georadar I olika vägverksprojekt. Vägverket, Serviceavdelning Väg och brokonstruktion, Sektionen för geoteknik, Publ. 1987:59, Sweden.

Johansson, S., Saarenketo, T. and Persson, L., 2005, Network and project level bearing capacity surveys and analysis using modern techniques. Proceedings of the Seventh International Conference on Bearing Capacity of Roads, Railways and Airfields, Trondheim, Norway (CD-rom), pp. 10p.

Jung, G., Jung, J. and Cho, S.-M., 2004, Evaluation of road settlements on soft ground from GPR investigations, Slob, E., Yarovov, A. and Rhebergen, J. (eds). Proceedings of the Tenth International Conference on Ground Penetrating Radar, Vol II, pp. 651–654.

Kolisoja, P., Saarenketo, T., Peltoniemi, H. and Vuorimies, N., 2002, Laboratory Testing of Suction and deformation Properties of Base Course Aggregates. Transportation Research Board Record, Vol. 1787, Paper No. 02-3432. pp. 83–89.

Kovacs, A. and Morey, R.M., 1983, Detection of Cavities under Concrete Pavement. US Army Corps of Engineers, Cold Region Research and Engineering Laboratory, Hanover, New Hampshire, CRREL Report No 83-18.

Lau, C.-L., Scullion, T. and Chan, P., 1992, Using Ground Penetrating Radar Technology for Pavement Evaluations in Texas, USA. Geological Survey of Finland, Special Paper 16, pp. 277–283.

Lenngren, C.A., Bergström, J. and Ersson, B., 2000, Using ground penetrating radar for assessing highway pavement thickness. Proceedings of SPIE, Vol. 4129, pp. 474–483.

Lewis, J.S., Owen, W.P. and Narwold, C., 2002, GPR as a tool for detecting problems in highway-related construction and maintenance. Proceedings of the 2nd Annual Conference on the Application of Geophysical and NDT Methodologies and NDT Methodologies to Transportation Facilities and Infrastructure, April 15–19, 2002, California, USA, FHWA-WCR-02-001, pp. 11p.

Liu, L., 2003, Ground Penetrating Radar: What Can It Tell about the moisture Content of the hot Mix Asphalt Pavement? Transportation Research Board Paper, pp. 9p.

Liu, L. and Guo, T., 2002, Dielectric property of asphalt pavement specimens in dry, water-saturated and frozen conditions. Proceedings of the Ninth International Conference on Ground Penetrating Radar, April 29–May 2, 2002, Santa Barbara, CA, Koppenjan, S. and Lee, H. (eds), Proceedings of SPIE, Vol. 4758, pp. 410–415.

Maierhofer, K. and Kind, Th., 2002, Application of impulse radar for non-destructive investigation of concrete structures. Proceedings of the Ninth International Conference on Ground Penetrating Radar, April 29–May 2, 2002, Santa Barbara, CA, Koppenjan, S. and Lee, H. (eds). Proceedings of SPIE Vol. 4758, pp. 382–387.

Maijala, P., Saarenketo, T. and Valtanen, P., 1994, Correlation of some parameters in GPR measurement data with quality properties of pavements and concrete bridge decks. Proceedings of the Fifth International Conference on Ground Penetrating Radar, June 12–16, 1994, Kitchener, Ontario, Canada, Vol. 2 of 3, pp. 393–406.

Malvar, L.J. and Cline, G.D., 2002, Void detection under airfield pavements, Al-Qadi, I.L. and Clark, T.M. (eds). Proceedings of the Pavement Evaluation 2002 Conference, October 21–25, 2002, Roanoke, VA, USA, pp. 15p.

Manacorda, G., Morandi, D., Sparri, A. and Staccone, G., 2002, A customized GPR system for railroad tracks verification. Proceedings of the Ninth International Conference on Ground Penetrating Radar, April 29–May 2, 2002, Santa Barbara, CA, Koppenjan, S. and Lee, H. (eds). Proceedings of SPIE, Vol. 4758, pp. 719–723.

Manning, D.G. and Holt, F.B., 1983, Detecting Deterioration in Asphalt Covered Bridge Decks. Transportation Research Record, Vol. 899, pp. 10–19.

Manning, D.G. and Holt, F.B., 1986, The Development of Deck Assessment by Radar and Thermography. Transportation Research Record, Vol. 1083, pp. 13–20.

Maser, K., 2002a, Non-destrcutive measurement of layer thickness of newly constructed asphalt pavement, Al-Qadi, I.L. and Clark, T.M. (eds). Proceedings of the Pavement Evaluation 2002 Conference, October 21–25, 2002, Roanoke, VA, USA, pp. 24p.

Maser, K., 2002b, Use of Ground Penetrating Radar Data for Rehabilitation of Composite Pavements on High-Volume Roads. Transportation Research Record, Vol. 1808, pp. 122–126.

Maser, K.R., 1994, Highway speed radar for pavement thickness evaluation. Proceedings of the Fifth International Conference on Ground Penetrating Radar, June 12–16, 1994, Kitchener, Ontario, Canada, Vol. 2 of 3. pp. 423–432.

Maser, K.R. and Roddis, W.M.K., 1990, Principles of thermography and radar for bridge deck assessment. Proceedings of the American Society of Civil Engineers. Journal of Transportation Engineering, Vol. 116, No. 5, pp. 583–601.

Maser, K.R. and Scullion, T., 1991, Automated Detection of Pavement Layer Thicknesses and Subsurface Moisture Using Ground Penetrating Radar. TRB paper.

Maser, K.R., Holland, T.J., Roberts, R. and Popovics, J., 2003, Technology for quality assurance for new pavement thickness. Proceedings of International Symposium on Non-Destructive Testing in Civil Engineering (NDT-CE), September 16–19, 2003, Berlin, Germany.

Mesher, D.E., Dawley, C.B. and Pulles, P.C., 1996, A comprehensive radar hardware, interpretation software and survey methodology paradigm for bridges deck assessment. Proceedings of the Sixth International Conference on Ground Penetrating Radar, September 30–October 3, 1996, Sendai, Japan, pp. 353–358.

Mesher, D.E., Dawley, C.B., Davies, J.L. and Rossiter, J.R., 1995, Evaluation of New Ground Penetrating Radar Technology to Quantify Pavement Structures. Transportation Research Record, Vol. 1505, pp. 17–26.

Millard, S.G., Bungey, J.H. and Shaw, M.R., 1993, The assessment of concrete quality using pulsed radar. Proceeding of the Non-Destructive Testing in Civil Engineering, April 14–16, 1993, The University of Liverpool, Vol. 1 of 2, pp. 161–186.

Mohajeri, J.H., 2002, Loading zoning study, Elkhart County, Indiana, Al-Qadi, I.L. and Clark, T.M. (eds). Proceedings of the Pavement Evaluation 2002 Conference, October 21–25, 2002, Roanoke, VA, USA, pp. 11p.

Morey, R., 1998, Ground Penetrating radar for Evaluating Subsurface Conditions for Transportation Facilities. Synthesis of Highway Practice 255, National Cooperative Highway Research Program, Transportation Research Board. National Academy Press.

Muller, W., 2003, Timber girder inspection using ground penetrating radar. Proceedings of International Symposium on Non-Destructive Testing in Civil Engineering (NDT-CE), September 16–19, 2003, Berlin, Germany.

Noureldin, A.S., Zhu, K., Li, S. and Harris, D., 2003, Network Pavement Evaluation Using Falling Weight Deflectometer and Ground Penetrating Radar. Transportation Research Board Paper, pp. 25p.

Nurmikolu, A., 2005, Degradation and Frost Susceptibility of Crushed Rock Aggregates Used in Structural Layers of Railway Track. PhD Thesis, Tampere University of Technology, Publication 567, Pp. 235p.

Olhoeft, G.R. and Capron, D.E., 1994, Petrophysical causes of electromagnetic dispersion. Proceedings of the Fifth International Conference on Ground Penetrating Radar, Kitchener, Ontario, Waterloo Centre of Groundwater Research, Waterloo, Vol. 1 of 3, pp. 145–152.

Olhoeft, G.R. and Selig, E.T., 2002, Ground penetrating radar evaluation of railway track substructure conditions. Proceedings of the Ninth International Conference on Ground Penetrating Radar, April 29–May 2, 2002, Santa Barbara, CA, Koppenjan, S. and Lee, H. (eds). Proceedings of SPIE, Vol. 4758, pp. 48–53.

Olhoeft, G.R., Smith III, S., Hyslip, J.P. and Selig Jr., E.T., 2004, GPR in railroad investigations. 10th International Conference on Ground Penetrating Radar, June 21–24, 2004, Delft, The Netherlands, in press.

Pälli, A., Aho, S. and Pesonen, E., 2005, Ground penetrating radar as a quality assurance method for paved and gravel roads in Finland, Lambot, S. and Gorriti, A.G. (eds). Proceedings of the 3rd International Workshop on Advanced Ground Penetrating Radar, May 2–3, 2005, Delft, The Netherlands.

Parry, N. and Davis, J.L., 1992, GPR Systems for Roads and Bridges. Geological Survey of Finland, Special Paper 16, pp. 247–257.

Rhazi, J., Dous, O., Ballivy, G., Laurens, S. and Balayssac, J.-P., 2003, Non destructive health evaluation of concrete bridge decks by GPR and half cell potential techniques. Proceedings of

International Symposium on Non-Destructive Testing in Civil Engineering (NDT-CE), September 16–19, 2003, Berlin, Germany.

Rmeili, E. and Scullion, T., 1997, Detecting Stripping in Asphalt Concrete Layers Using Ground Penetrating Radar. Transportation Research Board Paper No 97-0508, pp. 25p.

Roberts, R. and Petroy, D., 1996, Semi-automatic processing of GPR data collected over pavement. Proceedings of the Sixth International Conference on Ground Penetrating Radar, September 30–October 3, 1997, Sendai, Japan, pp. 347–352.

Roimela, P., Salmenkaita, S., Maijala, P. and Saarenketo, T., 2000, Road analysis – A tool for cost effective rehabilitation measures for Finnish Roads. GPR'2000. Proceeding of the Eighth International Conference on Ground Penetrating Radar, 23–36 May, 2000, Gold Coast, Australia, pp. 107–112.

Romero, F.A. and Roberts, R.L., 2002, Mapping concrete deterioration: High speed ground penetrating radar surveys on bridge decks using new analysis method based on dual polarization deployment of horn antennae. Proceedings of International Bridge Conference, June 10–12, 2002, Pittsburgh, PA, USA, Paper IBC-02-12.

Romero, F.A. and Roberts, R.L., 2004, Data collection and analysis challenges – GPR bridge deck deterioration assessment of two unique bridge deck systems. Proceedings of SAGEEP 2004.

Saarenketo, T., 1992, Ground Penetrating Radar Applications in Road Design and Construction in Finnish Lapland. Geological Survey of Finland, Special Paper 15, pp. 161–167.

Saarenketo, T., 1995, The use of dielectric and electrical conductivity measurement and ground penetrating radar for frost susceptibility evaluations of subgrade soils. Proceedings of the Application of Geophysics to Engineering and Environmental Problems, Comp. by R.S. Bell, Orlando, FL, pp. 73–85.

Saarenketo, T., 1997, Using Ground Penetrating Radar and Dielectric Probe Measurements in Pavement Density Quality Control. Transportation Research Record, Vol. 1997, pp. 34–41.

Saarenketo, T., 1998, Electrical properties of water in clay and silty soils. Journal of Applied Geophysics, Vol. 40, pp. 73–88.

Saarenketo, T., 1999, Road analysis, an advanced integrated survey method for road condition evaluation. Proceedings of the COST Workshop on "Modelling and Advanced Testing for Unbound and Granular materials", January 21–22, 1999, Lisboa, Portugal.

Saarenketo, T., 2001, GPR based road analysis – A cost effective tool for road rehabilitation – Case history from Highway 21, Finland. Proceedings of 20th ARRB Conference, Melbourne, Australia, March 19–21, 2001, pp. 19p.

Saarenketo, T., 2006, Electrical Properties of Road Materials and Subgrade Soils and the Use of Ground Penetrating Radar in Traffic Infrastructure Surveys. PhD Thesis, Faculty of Science, Department of Geosciences, University of Oulu. Acta Universitatis Ouluensis, A 471, pp. 121p.

Saarenketo, T. and Aho, S., 2005, Monitoring and classifying spring thaw weakening on low volume roads in northern periphery. Proceeding of the Seventh International Conference on Bearing Capacity of Roads, Railways and Airfields, Trondheim, Norway (CD-rom), pp. 11p.

Saarenketo, T. and Maijala, P., 1994, Applications of Geophysical Methods to Sand, Gravel and Hard Rock Aggregate Prospecting in Northern Finland. In Aggregates – Raw Materials' Giant, Report on the 2nd International Aggregates Symposium, Erlangen, Germany October 22–27, 1994, G.W. Luttig (ed.), pp. 109–123.

Saarenketo, T. and Roimela, P., 1998, Ground penetrating radar technique in asphalt pavement density quality control. Proceedings of the Seventh International Conference on Ground Penetrating Radar, May 27–30, 1998, Lawrence Kansas, Vol. 2, pp. 461–466.

Saarenketo, T. and Scullion, T., 1995, Using Electrical Properties to Classify the Strength Properties of Base Course Aggregates, Research Report 1341-2, Texas Transportation Institute, College Station, Texas.

Saarenketo, T. and Scullion, T., 1994, Ground Penetrating Radar Applications on Roads and Highways, Research Report 1923-2F, Texas Transportation Institute, College Station, Texas, pp. 36p.

Saarenketo, T. and Scullion, T., 1996, Laboratory and GPR tests to evaluate electrical and mechanical properties of Texas and Finnish base course aggregates. Proceedings of the Sixth International Conference on Ground Penetrating Radar, September 30–October 3, 1996, Sendai, Japan, pp. 477–482.

Saarenketo, T. and Scullion, T., 2000, Road evaluation with ground penetrating radar. Journal of Applied Geophysics, Vol. 43, pp. 119–138.

Saarenketo, T., Scullion, T. and Kolisoja, P., 1998, Moisture susceptibility and electrical properties of base course aggregates. Proceeding of BCRA'98, July 6–8, 1998, Trondheim, Norway, Vol. 3, pp. 1401–1410.

Saarenketo, T. and Söderqvist, M.-K., 1993, GPR applications for bridge deck evaluations in Finland. Proceedings of the Non-Destructive Testing in Civil Engineering, April 14–16, 1993, The University of Liverpool, Vol. 1 of 2, pp. 211–226.

Saarenketo, T., Hietala, K. and Salmi, T., 1992, GPR Applications in Geotechnical Investigations of Peat for Road Survey Purposes. Geological Survey of Finland, Special Paper 16, pp. 293–305.

Saarenketo, T., Silvast, M. and Noukka, J., 2003, Using GPR on railways to identify frost susceptible areas. In Proceedings of 6th International Conference on Railway Engineering, April 30–May 1, 2003, London, UK, pp. 11p.

Saarenketo, T., van Deusen, D. and Maijala, P., 2000, Minnesota GPR project 1998 – Testing ground penetrating radar technology on Minnesota roads and highways. In Eighth International Conference on Ground Penetrating Radar, Noon, D.A., Stickley, G.F. and Longstaff, D. (eds), SPIEE, Vol. 4084, pp. 396–401.

Saarenketo, T. and Vesa, H., 2000, The use of GPR technique in surveying gravel road wearing course. Eighth International Conference on Ground Penetrating Radar, Noon, D.A., Stickley, G.F. and Longstaff, D. (eds), SPIEE, Vol. 4084, pp. 182–187.

Saeed, A. and Hall, J.W. Jr., 2002, Comparison of non-destructive testing devices to determine in situ properties of asphalt concrete pavement layers, Al-Qadi, I.L. and Clark, T.M. (eds). Proceedings of the Pavement Evaluation 2002 Conference, October 21–25, 2002, Roanoke, VA, USA, pp. 14p.

Scullion, T., 2001, Development and implementation of the Texas ground penetrating radar system. Proceedings of 20th ARRB Conference, Melbourne, Australia, March 19–21, 2001, pp. 19p.

Scullion, T. and Saarenketo, T., 1995, Ground penetrating radar technique in monitoring defects in roads and highways. Proceedings of the Symposium on the Application of Geophysics to Engineering and Environmental Problems, SAGEEP, Comp. Ronald S. Bell, April 23–26, 1995, Orlando, FL, pp. 63–72.

Scullion, T. and Saarenketo, T., 1997, Using Suction and Dielectric Measurements as Performance Indicators for Aggregate Base Materials. Transportation Research Board Record, Vol. 1577, pp. 37–44.

Scullion, T. and Saarenketo, T., 1998, Applications of ground penetrating radar technology for network and project level pavement management survey systems. Proceedings of the 4th International Conference on Managing Pavements, Durban, South Africa, May 17–21, 1998.

Scullion, T. and Saarenketo, T., 1999, Integrating ground penetrating radar and falling weight deflectometer technologies in pavement evaluation. Proceedings of the Third ASTM Symposium of Nondestructive Testing of Pavements and Backcalculation of Moduli, June 30–July 1, 1999, Seattle, Washington.

Scullion, T. and Saarenketo, T., 2002, Implementation of ground penetrating radar technology in asphalt pavement testing. Proceedings of the Ninth International Conference on Asphalt Pavements, August 17–22, 2002, Copenhagen, Denmark, pp. 16p.

Scullion, T., Lau, C.-L. and Saarenketo, T., 1996, Performance specifications of ground penetrating radar. In GPR 1996: 6th International Conference on Ground Penetrating Radar, September 30–October 3, 1996, Sendai, Japan. Proceedings, Sendai: Tohoku University. pp. 341–346.

Scullion, T., Lau, C.L. and Chen, Y., 1992, Implementation of the Texas Ground Penetrating Radar System, Report No. FHWA/TX-92/1233-1, Texas Department of Transportation.

Scullion, T., Lau, C.L. and Chen, Y., 1994, Pavement evaluations using ground penetrating radar. Proceedings of the Fifth International Conference on Ground Penetrating Radar, June 12–16, 1994, Kitchener, ON, Canada, Vol. 1 of 3, pp. 449–463.

Scullion, T., Servos, S., Ragsdale, J. and Saarenketo, T., 1997, Applications of Ground-Coupled GPR to Pavement Evaluation, TTI Report 2947-S.

Sebesta, S. and Scullion, T., 2002a, Using Infrared Imaging and Ground-Penetrating Radar to Detect Segregation in Hot-Mix Asphalt Overlays, Research Report 4126-1, Texas Department of Transportation, Texas A&M University, College Station, TX.

Sebesta, S. and Scullion, T., 2002b, Application of Infrared Imaging and Ground Penetrating Radar for Detecting Segregation in Hot-Mix Asphalt Overlays. Transportation Research Board Paper, pp. 15p.

Shin, H. and Grivas, D., 2003, How Accurate is Ground Penetrating Radar (GPR) for Bridge Condition Assessment? Transportation Research Board Paper, pp. 22p.

Silvast, M., Levomäki, M., Nurmikolu, A. and Noukka, J., 2006, NDT techniques in railway structure analysis. Proceedings of 7th World Congress on Railway Research, June 4–8, 2006, Montréal, Canada.

Smekal, A., Berggren, A. and Hrubec, K., 2003, Track-substructure investigations using ground penetrating radar and track loading vehicle. Proceedings of 6th International Conference on Railway Engineering, April 30–May 1, 2003, London, UK, pp. 15p.

Spagnolini, U., 1997, Permittivity measurements of multilayered media with monostatic pulse radar. IEEE Transactions on Geoscience and remote Sensing, Vol. 35, No. 2, pp. 454–463.

Staccone, G. and de Haan, E., 2003, Safe rail track bed investigation by high speed radar. Proceedings of 6th International Conference on Railway Engineering, April 30–May 1, 2003, London, UK, pp. 4p.

Sussman, T.R., O'Hara, K.R. and Selig, E.T., 2002, Development of material properties for railway application of ground penetrating radar. Proceedings of the Ninth International Conference on Ground Penetrating Radar, April 29–May 2, 2002, Santa Barbara, California, Koppenjan, S. and Lee, H. (eds). Proceedings of SPIE Vol. 4758, pp. 42–47.

Sussman, T.R., Selig, E.T. and Hyslip, J.P., 2003, Railways track condition indicators from ground penetrating radar. NDT and E International, Vol. 36, No. 3, pp. 157–167.

Syed, I., Scullion, T. and Randolph, R.B., 2000, Use of Tube Suction Test for Evaluation of Aggregate Base Materials in Frost/Moisture Susceptible Environment. Transportation Research Board Paper 00.-1147, pp. 27p.

Szynkiewicz, A. and Grabowski, P., 2004, GPR monitoring of pavement on airfields, Slob, E., Yarovoy, A. and Rhebergen, J. (eds). Proceedings of the Tenth International Conference on Ground Penetrating Radar, Vol. II, pp. 803–805.

Ulriksen, C.P.F., 1982, Application of Impulse Radar to Civil Engineering. Doctoral Thesis, Lund University of Technology, Department of Engineering Geology, pp. 179p.

Uomoto, T. and Misra, S., 1993, Non destructive testing of RC structures, recent developments in Japan, Bungey, J. (ed.). Proceedings of Conference on Non-Destructive Testing in Civil Engineering, April 14–16, 1993, Liverpool, UK, pp. 65–77.

Washer, G., 2003, Nondestructive evaluation for highway bridges in the United States. Proceedings of International Symposium on Non-Destructive Testing in Civil Engineering (NDT-CE), September 16–19, 2003, Berlin, Germany.

Webb, D.J., Anderson, N.L., Newton, T., Cardimona, S. and Ismail, A., 2002, Ground penetrating radar (GPR): A tool for monitoring bridge scour. Proceedings of the 2nd Annual Conference on the Application of Geophysical and NDT Methodologies and NDT Methodologies to Transportation Facilities and Infrastructure, April 15–19, 2002, California, USA, FHWA-WCR-02-001, pp. 24p.

Weil, G.J., 1992, Non-destructive testing of bridge, highway and airport pavements. Proceedings of Fourth International Conference on Ground Penetrating Radar, June 8–13, 1992, Rovaniemi, Finland, Geological Survey of Finland, Special Paper 16, pp 259–266.

Weil, G.J., 1998, Nondestructive testing of airport concrete structures. Proceeding of the Seventh International Conference on Ground Penetrating Radar, May 27–30, 1998, Lawrence, KS, USA, pp. 443–448.

Westerdahl, H.R., Austvik, R. and Kong, F.-N., 1992, Geo-Radar in Tunnelling – The Tunnel Radar. Geological Survey of Finland, Special Paper 16, pp. 41–45.

Wilson, L.W. and Garman, K.M., 2002, Identification and delineation of sinkhole collapse hazards in Florida using ground penetrating radar and electrical resistivity imaging. Proceedings of the 2nd Annual Conference on the Application of Geophysical and NDT Methodologies and NDT Methodologies to Transportation Facilities and Infrastructure, April 15–19, 2002, California, USA, FHWA-WCR-02-001, pp. 13p.

Wimsatt, A., Scullion, T., Ragsdale, J. and Servos, S., 1998, The Use of Ground Penetrating Radar in Pavement Rehabilitation Strategy Selection and Pavement Condition Assessment, TRB-paper 980129, pp. 21p.

Xie, X., Liu, Y., Huang, H. and Shao, J., 2004, Research of GPR non-damaged test of grouting behind lining of shield, Slob, E., Yarovoy, A. and Rhebergen, J. (eds). Proceedings of the Tenth International Conference on Ground Penetrating Radar, Vol. I, pp. 419–422.

Yelf, R., 2004, Where is true time zero? Slob, E., Yarovoy, A. and Rhebergen, J. (eds). Proceedings of the Tenth International Conference on Ground Penetrating Radar, Vol. I, pp. 279–282.

CHAPTER 14

LANDMINE AND UNEXPLODED ORDNANCE DETECTION AND CLASSIFICATION WITH GROUND PENETRATING RADAR

Alexander Yarovoy

Contents

14.1. Introduction	445
14.2. Electromagnetic Analysis	446
14.3. System Design	455
14.4. GPR Data Processing for Landmine/UXO Detection and Classification	462
14.5. Fusion with Other Sensors	469
14.6. Overall Performance of GPR as an UXO/Landmine Sensor	472
14.7. Conclusion	473
References	474

14.1. INTRODUCTION

Contamination by landmines and all other types of unexploded ammunition (usually called unexploded ordnance or UXO) is a worldwide problem with enormous humanitarian impact. At least 67 countries are being affected by landmines, but reliable estimates of the area affected worldwide are not readily available. Estimates of the number of mines laid vary widely, from 50 to 150 million. In the mid-1990s, humanitarian demining became a hot topic on the political agenda. According to the 1997 Convention on the Prohibition of the Use, Stockpiling, Production and Transfer or Anti-Personnel Mines and on Their Destruction (often referred to as the Ottawa Treaty; http://www.un.org/Depts/mine/UNDocs/ban_trty.htm), which entered into force in 1999, all stockpiles of mines should be destroyed within 4 years and all minefields lifted in 10 years.

Despite the political willingness of the world community to solve the landmine problem in the short term, the situation in situ did not change fast because of limited performance of the detection means available before 2006 for operational deminers: the prodding sticks, dogs, and electromagnetic induction metal detectors (Mine Action Equipment: Study of Global Operational Needs, 2002). Being the most sophisticated demining tool until recently, the metal detector suffers from problems such as insufficient detection depth and a high false-alarm rate (FAR; false detection of subsurface inhomogeneities such as roots, rocks, and water pockets) for antipersonnel (AP) mines with low metal content. An example of metal-detector performance is given by the statistics of humanitarian demining in Cambodia between 1992 and 1998: only 0.3% of the 200 million items excavated by deminers were AP mines or UXO (McDonald et al., 2003, p. xvi). Without the use of reliable high-tech tools for humanitarian demining, it remains a dangerous, slow, and costly process. For military demining, some tools to trigger mines and cause their explosion (e.g., mechanical flails or rollers) do exist (Blagden, 1996); however, such mechanical demining does not meet the high safety standards for a cleared area established by the UN (an area is declared as safe if 99.6% of all mines have been cleared).

In principle, clearance of subsurface UXO is fraught with the same problems as humanitarian demining. The conventional approach of using only electromagnetic-induction metal detectors (Won et al., 1997) and passive magnetometers often results in a large number of false alarms due to the presence of shrapnel and other metal debris. Although recent technological advances in autonomous surveying (McDonald et al., 1997; Kerr, 1999) have greatly increased the detection sensitivity as well as the speed, the high FAR is still a challenging problem. This is also indicated by the final report on field tests of three Advanced Technology Demonstrators for UXO detection conducted in the USA between 1994 and 1997, which states that "...demonstrators lack a capability to distinguish ordnance and the implanted nonordnance" (UXO Technology Demonstration Program at Jefferson Proving Ground, 1997).

In the last two decades, a lot of attention has been paid to the application of Ground Penetrating Radar (GPR) as a landmine/UXO sensor. It was found that in many field conditions, a GPR sensor can detect all antitank and AP mines as well as UXO. Furthermore, a GPR sensor can support classification of detected objects and drastically reduce the FAR during demining operations. Finally, standoff GPR systems might play some role in the detection of landmine fields and in this way contribute to the reduction of suspected mine areas.

In this chapter, we shall mainly focus on such applications of GPR as a landmine sensor and briefly address issues related to the use of GPR for UXO detection. Major physical phenomena related to reflection of EM pulses from mines and UXO are described in Section 14.2. Hardware and software issues specific for this application are addressed in Sections 14.4 and Signal Processing, respectively. The next two sections deal with the performance of GPR combined with other electromagnetic sensors and the overall performance of GPR for landmine detection.

14.2. ELECTROMAGNETIC ANALYSIS

Electromagnetic analysis of landmine detection with GPR substantially differs from the analysis of a typical GPR scenario. Formally, two main differences between both scenarios are the following. First, in the landmine scenario, GPR antennas are elevated above the ground, while in a typical GPR scenario, they are placed right above the air–ground interface. Second, in the most of the situations in humanitarian demining, the maximum depth to be surveyed is defined as 20 cm; thus the typical target depth is about a few wavelengths at the central operational frequency or less, while in the typical GPR scenario, it is at least one order of magnitude larger.

Purposes of electromagnetic analysis in both scenarios are also different. The goal of a typical GPR scenario analysis is to calculate attenuation of EM waves in the ground. The latter is the most important factor as it determines the penetration depth and detectability of buried objects. The major task of a GPR sensor for landmine detection is however classification of detected targets rather than its detection. The classification can be done only by analyzing an object response. Furthermore, here the detectability of a mine is not determined by the attenuation in the soil (due to the relatively shallow target depth, ohmic losses usually do not play a decisive role) but limited mainly by surface and underground clutter (McDonald et al., 2003).

Landmines and UXO. We start electromagnetic analysis of the problem with the description of targets (landmines and UXO) as physical objects. The first AP mines have been used during the World War I, and after that the variety of AP drastically increased. At the end of the twentieth century, more than 350 types of AP mines were manufactured in more than 50 countries (Vines and Thompson, 1999). In general, AP mines are mostly cylindrical objects (Figure 14.1) with a diameter between 40 and 200 mm. The diameter of a typical modern AP mine varies between 55 and 100 mm, but roughly 50% of all laid mines have a diameter larger than 100 mm (MineFacts, 1995). The mines are laid in such a way that the cylinder's rotational axe is almost vertical. The height of an AP mine varies typically between 30 and 80 mm. Another typical mine shape is rectangular. Antitank mines typically also have a cylindrical geometry, but their dimensions are several times larger than those of AP mines. The operational burial depth of an AP mine is less than 200 mm, while that of AT mines is up to 100 cm.

From a construction point of view, mines consist of an explosive, a detonator, a casing, and (very often) a void. The casing may be made of plastic, wood, or metal. Small AP mines can contain only about 30 g of explosive. The detonator typically contains several metal parts (the only known exception is French bakelite landmines placed in Southern Lebanon in the late 1940s, which have no metal at all). According

Figure 14.1 Photos of antipersonnel (AP) mines: PMA3 (left), VS-50 (center), and PMN2 (right).

to GICHD (Mine Action Equipment: Study of Global Operational Needs, 2002), at present there are no known cases of mines laid with no metal content whatsoever. So-called minimum-metal mines, in which only the metal parts are related to the detonator, contain less than 5 g of metal (with a minimal weight of about 1 g) and are the most difficult to detect.

For an electromagnetic sensor such as GPR, the electromagnetic contrast between a mine and its environment is very important. The electromagnetic contrast depends mainly on the soil in which the mine has been laid. Except for magnetic soils, the electromagnetic contrast is the difference between the complex relative dielectric permittivity of a soil and those of a mine. The relative dielectric permittivity of a plastic or wooden (when dry) casing varies from 2 to 4. The relative dielectric permittivity of explosive depends on its type and typically has a value around 3. According to several studies (see, e.g., Redman et al., 2003), the largest reflectors in an AP mine are the detonator (for minimum-metal mines, the reflection from the detonator might be very small) and the void (filled with air or water, depending on the soil conditions).

Typical examples of UXO are shells, bombs, and projectiles. Despite enormous variety of sizes and shapes (see, e.g., a photo gallery at www.uxoinfo.com), the majority of UXOs are rotationally symmetric, strongly elongated objects with a metal casing. Typically its diameter is between 20 and 40 cm, while the length varies from 20 to 150 cm. Unexploded ordnance can be found at the depths from a few centimeters below the ground surface to a few meters. The orientation of UXO objects in the ground can differ (vertical orientation is relatively rare). Due to their metal casing and relatively large size, UXO objects are easier to detect than AP mines. However, similarly, as for landmines, the main task of the GPR sensor lies not in detecting an object but in classifying it as an UXO-like target.

Object responses. Despite the vast literature on radar responses of different targets, not many results are available for landmine and UXO responses. There are several reasons why available models on free-space radar cross sections (RCSs) cannot be applied directly to landmines/UXO. First, available results typically deal with metal targets, while the majority of AP mines and some AT mines are dielectric objects with some metal inclusions. Second, the influence of the environment on target response is quite complicated (especially that of the response of low-metal AP mines). Factors such as presence of air–ground interface, substantial difference between dielectric permittivity (and sometimes also magnetic permeability) of ground and air, losses in the ground, and dispersion of material parameters cause considerable changes in the target response of a buried object in comparison with the response of the same object but in free space (O'Neill, 2001). Finally, in a typical radar scenario, the target is situated in the far field of both the transmit and receive antennas. In typical landmine detection scenario (and often in UXO detection), the target is situated in a near field of the antenna system.

Object response can be analyzed either in the frequency domain or the time domain. In the former case, substantial work has been done on natural resonances of mines. The idea behind it is that the electromagnetic response of any object is determined by natural resonances at complex-valued frequencies, whose values are fully determined by the shape of the object (and by its internal structure if the object is transparent to electromagnetic waves) and invariant with target orientation relative to the probing wave (Baum, 1971; Baum et al., 1991). This approach has been shown to work reasonably well for UXO (Chen and Peters Jr, 1997) and metal antitank mines (Chan et al., 1979; Carin et al., 1999). The resonances of AP mines, in particular those without a metal case, are highly damped and very difficult to detect (Sullivan, et al., 1999). The relatively complex internal structure of AP mines makes detection of these resonances even more difficult. As has been shown in many studies (see, e.g., Kovalenko et al., 2003a), measured spectra of AP mine responses do not exhibit any signs of the resonances (Figure 14.2). The measured mines (types C and F) have the same cylindrical shape but different sizes.

The same as shown in Figure 14.2 AP mine responses in the time domain (frequently referred to as target impulse responses (IRs)) are shown in Figure 14.3. Theoretically, the IR of a small dielectric cylinder (which is widely used as a theoretical model for low-metal AP mines) looks like a time derivative of an incident EM pulse followed (with a small delay) by another time derivative with an opposite polarity (with respect to the first derivative) (Roth et al., 2002). Physically, such an IR corresponds to reflections of a probing wave from two flat surfaces (top and bottom) of the cylinder, which are parallel to each other and separated at a distance of the order of a few centimeters. The ratio between the magnitudes of both derivatives and the time delay between them are two parameters that distinguish responses of two different targets (Roth et al., 2003). Impulse responses for larger reflectors clearly exhibit a more complicated structure (Carin et al., 2002).

Polarimetry. Polarimetric features of the reflected wave in principle allow for classification of all detected objects as rotationally symmetric and asymmetric ones (Carin et al., 1999; Chen et al., 2001; Yarovoy et al., 2007b). As the majority of landmines are rotationally symmetric objects, this can be exploited to classify detected reflectors as mine-like targets (rotationally symmetric) and friendly targets

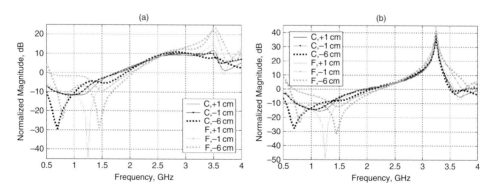

Figure 14.2 Frequency spectra of the AP mine reflections (from Kovalenko et al., 2003a). The measurements have been performed using full-polarimetric ground penetrating radar (GPR) (Yarovoy et al., 2002a).

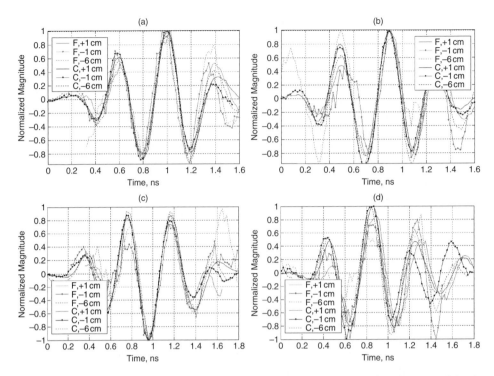

Figure 14.3 Experimentally determined reflections from two types of antipersonnel (AP) mines: surface-laid, flush and shallow buried in dry sand (from Kovalenko et al., 2003a).

(asymmetric ones). For a perfectly rotational symmetric target, the cross-polar components of this target signature are close to zero. As an example, the polarimetric signature of a circular metal plate is shown in Figure 14.4. However, despite their circular shape, some mines (e.g., PMN-2) show relatively strong cross-polar components $S_{xy}(t)$, $S_{yx}(t)$ (Roth et al., 2003) in the reflected signal (Figure 14.5). These cross-polar components can most likely be attributed to the presence of a horizontal detonator in PMN-2 mines.

Impact of environment. The impact of the antenna elevation above the ground, the soil dielectric permittivity, and losses on attenuation of reflection from a mine are described in Daniels and Martel (2001) and Redman et al. (2003). The latter shows how important the impact of the near field is on the landmine response.

For a fixed antenna elevation, the magnitude of mine reflections depends mainly on the electromagnetic contrast between the mine and its environment and on the size of the mine. Generally the burial depth of the mine does not influence the magnitude of the mine reflection considerably as typical propagation losses in a few centimeters of ground (even a lossy one) are not very large (below 10 dB). As an example, the mine reflection magnitudes from Figure 14.3 are shown in Figure 14.6. The reflections have been normalized to the magnitude of the direct

Landmine and UXO Detection and Classification with GPR 451

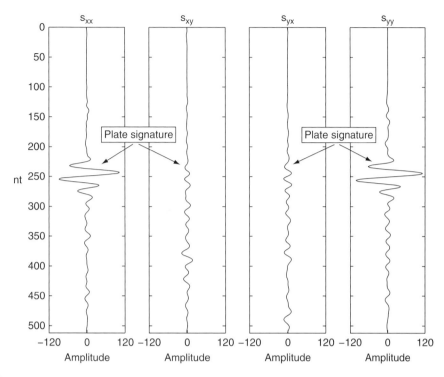

Figure 14.4 Polarimetric signature of the metal plate buried in dry sand at the depth of 6 cm (from Roth et al., 2003).

wave from the transmit antenna to the receive one. The main peak-to-peak values for both signals were taken as the magnitudes. The responses of mines buried at a depth of 6 cm were only a few decibels lower in magnitude than those of the same mines laid on the ground. The responses of flush-buried mines in these particular measurements were the lowest, because they were decreased by the strong influence of the ground reflection. However, the responses of all mines are well inside the dynamic range of any GPR sensor (which is about 60 dB or more). Reflections from mines laid in soils with higher losses are not necessarily smaller than those from mines buried in sand. This is because lossy soils usually have a higher dielectric permittivity (Daniels, 2004), and therefore the dielectric contrast between a mine and the soil (and thus the scattered-field magnitude) typically increases with the losses.

Soil properties as well as burial depth have been found to change the natural resonances of landmines and UXO (Vitebskiy and Carin, 1996; Carin et al., 1999). The natural resonances of UXO (especially for elongated targets) are simply shifted to lower frequencies according to averaged dielectric permittivity of the ground around UXO. However, for shallowly buried landmines, the situation is more complicated: one should take into account its burial depth, the orientation of the mine with respect to the ground, and the vertical distribution of dielectric

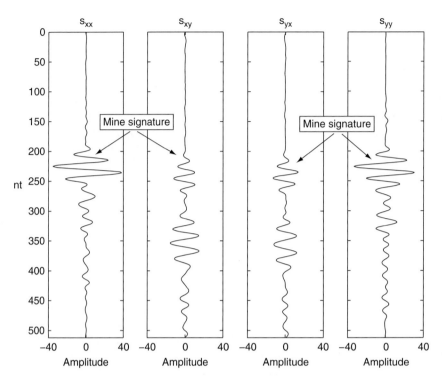

Figure 14.5 Polarimetric signature of a PMN-2 mine buried in dry sand at the depth of 1 cm (from Roth et al., 2003).

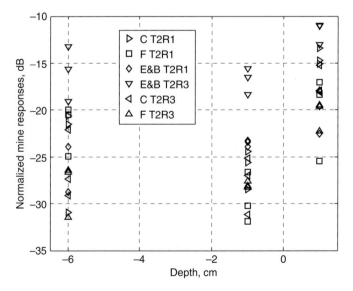

Figure 14.6 Magnitude of mine responses normalized to a magnitude of the direct wave from transmit to receive antenna (adapted from Kovalenko and Yarovoy, 2003c).

permittivity (for magnetic soils, the vertical distribution of magnetic permeability should be taken into account as well). Because of these dependencies, it is questionable whether natural resonances can be used as reliable features for classifying buried AP mines.

Clutter. Clutter is the major factor that limits the detectability of the landmines. Clutter is caused not only by random inhomogeneities in the soil and its topological variations but also by various deterministic reflectors, such as small metal fragments, shrapnel, spent bullet and cartridge cases, puddles of water, tufts of grass, animal burrows, cracks and fissures in the ground, rocks, and stones. On the basis of the arrival time of the clutter signal, the clutter can be classified as surface clutter (reflections from the air–ground interface) or subsurface clutter (reflections from all kinds of inhomogeneities in the ground).

For conventional GPR systems, ground reflection is well separated in time from reflections from targets of interest. In the case of landmine detection, however, the target response (particularly for flush-buried or shallowly buried AP mines) typically overlaps in time with the reflection from the air–ground interface. This results in masking of the mine response. As an example, Figure 14.7 shows typical signal-to-clutter ratios for AP mines (same as in Figures 14.2 and 14.3) buried in sand. Dry sand is a very homogeneous soil with a reasonably flat surface. Therefore both surface and subsurface clutter are reasonably low. Even in such a favorable for landmine detection scenario, the signal-to-clutter ratio for large AP mines (e.g., PMN-2) is only about 25 dB. This is not excessive, as most automatic target detection algorithms require at least 10 dB for a reliable detection. The signal-to-clutter ratio for small AP mines (e.g., NR-22 without metal ring) is about 5 dB.

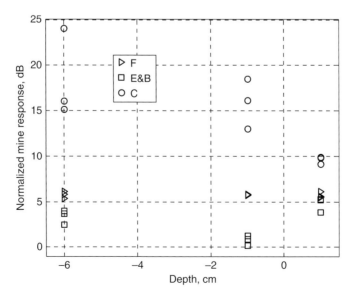

Figure 14.7 Mine responses normalized to the ground clutter (adapted from Kovalenko and Yarovoy, 2003c).

With the water content in sand, the spatial inhomogeneity of the dielectric permittivity increases, which drastically decreases the signal-to-clutter ratio. Other types of soils exhibit higher inhomogeneities of dielectric permittivity than those of sand at frequencies above 500 MHz. This explains why many GPR sensors for landmine detection that have been successfully tested in sand pits failed to detect landmines in field conditions. With frequency, the magnitude of the clutter caused by subsurface inhomogeneities and soil surface roughness increases faster than the landmine response, and at frequencies above 3 GHz, the clutter starts to dominate the AP mine responses.

Surface clutter depends on the surface roughness and dielectric permittivity of the soil. The more corrugated the ground surface (both in terms of larger root-mean-square (RMS) surface height and steeper surface slope) and the higher the dielectric permittivity of the surface, the stronger the clutter. The magnitude of clutter also depends on the polarization of incident and reflected waves. If an incident electromagnetic wave impinges on the ground at an incident angle different from normal, then the magnitude of backscattered surface clutter on vertical polarization is higher than that on horizontal polarization. With the increase in the incident angle, the discrepancy between clutter values at both polarizations increases (Dogaru et al., 2001). This angular and polarization dependence of surface clutter was the reason for the failure of the first forward-looking GPR sensors for landmine detection. It was expected that if vertically polarized electromagnetic wave impinges on the ground at the Brewster angle, then all electromagnetic energy will penetrate into the ground and the mine response will be large and easy to detect. However, in practice, it was found that the signal-to-clutter ratio is very low and targets are better visible on horizontal polarization. Differences in polarimetric properties of landmines and clutter can be used to discriminate between landmines and clutter (Cloude et al., 1996).

Although the impact of surface clutter on detectability of the landmines and UXO has been a subject of intensive research (see e.g., O'Neil et al., 1996; O'Neil, 2000; Dogaru et al., 2001), not much is known about properties of subsurface inhomogeneities and the subsurface clutter they cause.

The electromagnetic analysis shows that in principle a GPR sensor can detect buried landmines and UXO. In practice, detection is limited by the signal-to-clutter ratio. The clutter caused by subsurface inhomogeneities and soil surface roughness can be reduced by decreasing the size of the antenna footprint at the ground, by decreasing the operational frequencies (e.g., using frequencies below 3 GHz for AP mine detection), or by choosing an incident angle close to normal for the probing electromagnetic wave. Knowledge of the target IR (in particular, multiaspect one) can be used for detection with a low signal-to-clutter ratio. Even more important, target IR can also be used for classification of AP mines as well as AT mines and UXO. For identification of the latter, resonance frequencies can be used as well. In some cases (rotationally symmetric objects with a rotationally symmetric internal structure laid plainly in the ground), polarimetric features of the reflected field can be used to distinguish these targets from other reflectors and clutter.

14.3. SYSTEM DESIGN

There are three principal differences in GPR system design between conventional GPR sensors and GPR sensors for landmine detection (Chignell et al., 2000). First, the latter require a down-range resolution in the order of a few centimeters (in ground) to distinguish between reflection from a buried landmine and reflection from the air–ground interface. A down-range resolution of the same order is also required to distinguish between reflections from the top and from the bottom of a landmine, which is needed for target classification. Second, to avoid triggering of surface-laid or shallowly buried landmines, the antenna system of a landmine detection GPR sensor should be elevated above the ground. For a handheld system, the minimal elevation of the antenna system is in the order of a few centimeters, while for a vehicle-based sensor, such an elevation is typically several decimeters. Third, a new for a conventional GPR requirement to support classification of detected targets requires considerably higher stability and accuracy of the reflected-field measurements than that offered by conventional system. These three requirements lead to a qualitatively new system design of a GPR sensor for landmine detection. There is, however, a degree of similarity between the design of GPR sensors for landmine detection and those for road inspection: the first two above-mentioned requirements are in fact similar for both types of GPR sensors. Below we consider in detail the consequences of all three above demands for the landmine detection system design.

Operational bandwidth. We will start with the down-range resolution and transform it into the operating frequency band. Landmine detection requires a down-range resolution in the order of several centimeters in the ground (smaller than the typical size of an AP mine), which can be achieved using a bandwidth of the order of at least several gigahertz. The main question is where to allocate this frequency band. As shown in Cherniakov and Donskoi (1999), an ideal GPR should use all frequencies from DC till some upper frequency, which is determined by the down-range resolution required. Practically, however, very low frequencies are not accessible due to the limited size of an antenna system. So the lowest operational frequency varies from about 200 MHz (for video impulse systems) to about 2 GHz (for some stepped-frequency and frequency-modulated systems) (Witten, 1998; Daniels, 2006a). In contrast, at frequencies above 3 GHz, the propagation loss and clutter (caused by the air–ground interface roughness and inhomogeneity of the soil surface) considerably limits the applicability of GPR systems (Chen et al., 2000; Redman et al., 2003). So the most widely used operational bandwidth is from 200 MHz to 3 GHz, but there are some systems that use frequencies as low as 50 MHz and/or as high as 12 GHz.

Analysis of typical landmine detection scenarios (Yarovoy et al., 2002a) shows that there are two typical situations, which should be treated separately: mines lying on the ground (probably covered by vegetation) or flash buried in the ground and mines buried in the ground. To distinguish the mine from the environment in the first situation, a bandwidth of at least about 2 GHz is needed. Because the electromagnetic fields need not penetrate into the ground in this situation, it is not necessary to use low frequencies. The desirable operational frequency band for this scenario starts at around 1 GHz or above and goes up to at least several gigahertz. In the second situation, deep penetration into the ground and small propagation losses are essential, so low frequencies should be used. At the same time, the pulse length in soil is effectively compressed. Due to this phenomenon, without any loss of the down-range resolution, the bandwidth of the probing electromagnetic field can be reduced $\mathrm{Re}(\sqrt{\varepsilon})$ times, where ε is the complex-valued dielectric permittivity of the ground. However, soils with high dielectric permittivity usually have large losses (Daniels, 2004) and propagation in lossy medium causes increase in the pulse length. The lower the operational frequency, the smaller this effect. Thus the desirable operational frequency band for this scenario starts around 100 MHz and goes up to at least 1.5 GHz.

Like conventional GPRs, typical GPR sensors for landmine detection are either video impulse systems (which physically transmit a short pulse of electromagnetic energy) or stepped-frequency systems (which at any given moment of time physically transmit a monochromatic electromagnetic wave, but over some time period, the frequency of this wave varies within a large frequency band) (Daniels, 2004; Ch. 6). Owing to the requirement of high operational bandwidth, many GPR sensors for landmine detection are stepped-frequency systems [among them HSTAMIDS (Hatchard, 2003) and ALIS (Sato and Takahashi, 2007)]. This system approach allows realization of a relative bandwidth of typically not more than 10:1 (e.g., from 200 MHz to 2 GHz or from 1 to 10 GHz); however, such system as ALIS has a considerably large relative bandwidth of 40:1 (from 100 MHz to 4 GHz) (Sato and Takahashi, 2007).

The video impulse systems use frequency band from 500 MHz (some GPRs for AT mine detection) up to 7 GHz. The instantaneous relative bandwidth of such systems typically does not exceed 5:1 and the operational band chosen in such systems is a trade-off between desirable resolution and desirable penetration depth. The higher the centre frequency of the operational band, the smaller the penetration depth but the larger the absolute bandwidth and, consequently, the down-range resolution. In order to optimally combine high resolution with sufficient penetration, either very short pulses (e.g., the NIITEK radar has the instantaneous bandwidth from 50 MHz to 7 GHz) or several pulses with different pulse duration (Yarovoy et al., 2002a) should be used. In the latter case, a short-pulse generator provides high system down-range resolution for surface-laid and flash-buried targets, while a long-pulse generator provides deep penetration into the ground without degradation of the down-range resolution in the soil.

Two alternatives to the traditional GPR system design, namely noise radar and maximum-length binary sequence (MLBS) radar, have been also implemented for landmine detection. Introduced as a cheap technological alternative to stepped-frequency/frequency–modulated, continuous-wave systems, noise radar (Narayanan et al., 1995; Walton and Cai, 1998) has not demonstrated any operational advantages over traditional technology. Theoretically it should be able to measure simultaneously a full-polarimetric scattering matrix of a reflector; however, this promise has not been realized so far. At the same time, an MLBS radar (Ratcliffe et al., 2002; Sachs et al., 2003) developed within the DEMINE project has been proven to be a cheaper, lighter, and more compact alternative to the video impulse system and can be considered as an attractive option for a handheld system design. The main disadvantage of such a system is its relatively high level of autocorrelation function sidelobes for a reasonably long pulse sequence, which makes the operational down-range resolution of MLBS radar smaller than that of a video impulse system with the same bandwidth (Daniels, 1999).

Antenna system. To achieve the down- and cross-range resolution desired of the whole GPR sensor, it is not sufficient to realize a large generator bandwidth. The achievable resolution also depends on the GPR antenna system and the data processing method chosen (a high cross-range resolution can be achieved in GPR only via migration-like data processing). So it is commonly known that the antenna system is the most critical part of GPR hardware (Daniels, 2004). In landmine detection, the antenna system should be elevated above the ground, which makes for a principal difference in the GPR antenna design. Detaching antenna from the ground increases the antenna input impedance (up to its intrinsic value in free space), the antenna ringing, the transmit–receive antenna coupling within the antenna system, and finally the radiation into air. Moreover, it widens the choice of antenna types that can be used within the antenna system. Good candidates for standoff GPR sensor for landmine detection are TEM and rigid horns, Vivaldi antennas (and other tapered-slot antennas), the impulse-radiating antennas (IRA), and spiral antennas, which are not often used in conventional GPR systems. Also the problem of matching the antenna to the ground no longer exists. Such qualitatively different from conventional GPR situation requires new strategy in antenna system design. Below we illustrate such a strategy translating basic requirements for an antenna system into requirements for antennas themselves.

In order to achieve the necessary system performance, the transmit antenna should

- radiate a short, ultra-wideband pulse with small ringing (or for SFCW systems, allow synthesis of a short pulse with small ringing in post-processing). This demand is crucial as reflection of antenna ringing from the air–ground interface can easily mask the reflection of a main pulse from a shallowly buried mine. It is the antenna ringing that limits in practice the down-range resolution of the whole system.
- produce an optimal footprint on the ground surface and below it, that is, the size of the footprint should be sufficiently large to support data processing (the branches of hyperbolas in the raw data should be long enough for detection), but at the same time it should be as small as possible to reduce surface clutter and undesirable backscattering from surrounding objects.
- radiate pulse with the same waveform throughout the footprint (important when data focusing will be done in the data processing);
- be elevated above the ground surface to prevent triggering of mines.

The possibility to change polarization of the radiated field from one to orthogonal one is also desirable.

The primary purpose of a receive antenna is to measure the field scattered by the (sub)surface. So an ideal receive antenna should

- receive the ultra-wideband field scattered by a (sub)surface with minimal distortion;
- possess a relatively small effective aperture as the scattered field in a close proximity to a target is spatially nonhomogeneous and its averaging over a large antenna aperture (done by receive antenna) degrades the cross-range resolution;
- be elevated above the ground surface.

The possibility to simultaneously measure the backscattered field in two orthogonal polarizations is also desirable.

In addition to the above-formulated technical requirements for the separate transmit and receive antennas, the antenna system as a whole should possess high transmit–receive (Tx–Rx) antenna isolation. If this requirement is not satisfied, coupling in the transmit–receive antenna pair can obscure reflections from shallowly buried targets and can substantially limit the dynamic range of the whole GPR system.

Trying to satisfy all these demands, antenna designers have developed two approaches. In the first approach, the antenna system is physically small and placed very close to the ground. The transmit and the receive antennas are identical. Such antenna types as resistively loaded dipoles and bow ties (Daniels and Curtis, 2003; Ratcliffe et al., 2002), spiral antennas with an aperture of a few centimeters (Hatchard, 2003; Sato and Takahashi, 2007), and tapered-slot antennas are used. The resistive loading is responsible for low antenna ringing. The antenna shielding is widely used to decrease the transmit/receive antenna coupling. The small physical size of the antennas and their close proximity to the ground minimize the footprint and allow for measurements of the scattered field in a local point. This approach is typical for handheld systems (see, e.g., Figures 14.8 and 14.9) but is also used in

Figure 14.8 Photo of the MINETECT system with a video impulse (Courtesy: ERA Technology (Courtesy Dr. D. Daniels)). Two shielded dipole antennas are seen within the coil of a metal detector.

Figure 14.9 Ground survey with ALIS system (Courtesy: Prof. M. Sato). Two shielded spiral antennas are seen within the coil of a metal detector.

some vehicle-based systems (Chignell et al., 2000). The main advantage of this approach is that it offers accurate measurement of the scattered field in a near zone of the mine. The spatial spectrum of this field contains nonhomogeneous waves, which are of importance for target classification and to produce a high cross-range resolution in 3D images of the subsurface (Bloemenkamp and Slob, 2003).

In the majority of vehicle-based systems, however, another approach that is based on elevating the antenna system considerably above the ground is preferred (Witten, 1998). Such an approach allows antenna decoupling from the ground and focuses the radiated electromagnetic field within a relatively small area on the ground surface (typically, antennas are elevated so high that the ground is no longer situated in their near field and the electromagnetic energy propagates according to the antenna radiation patterns). The ground can be surveyed by mechanical or electronic scanning of the illuminated spot over the air–ground interface. A very clear description of such an approach is given in Chen et al. (2000). Elevation of the transmit antenna above the ground makes it possible not only to focus the radiated energy in a small spot on the surface but also to separate in time the antenna coupling and reflection from the ground. Furthermore, substantial antenna elevation allows realization of full-polarimetric antenna systems and does not put strict limitations on the physical size of the antenna aperture. The disadvantages of this approach are the large propagation losses of the electromagnetic waves (on their way from the transmitter to the ground and from the ground to the receive antenna) and the impossibility to measure nonhomogeneous waves from the spectrum of the electromagnetic field scattered by a target.

The elevated antennas are typically used within an antenna array. Use of an array allows speeding up the ground survey. Such an array is typically formed by a number of transmit–receive antenna pairs. Both antennas in such a pair are identical. TEM and rigid horns, V-dipoles, tapered-slot antennas, and spiral antennas are widely used as antenna elements. The mechanical scanning in the plane of the array is replaced by sequential operation of the constituting antenna pairs. Despite substantial increase in the survey speed in comparison with 2D mechanical scanning, this approach still limits the scanning speed up to a few kilometers per hour.

A hybrid approach that combines the advantages of both approaches mentioned above has been proposed in Yarovoy et al. (2001) and Sato (2003). The main idea of this hybrid approach is that the transmit antenna is elevated sufficiently high above the ground (exploiting all advantages of the second approach), while an array of receive antennas is placed as close as possible to the ground (thus exploiting all advantages of the first approach). For the transmit antenna, any ultra-wideband antenna with small ringing and narrow radiation patterns can be used, whilst for the receive antenna, the antenna should be small and transparent (for an electromagnetic field radiated from a transmit antenna) with small ringing, e.g., a dipole or a loop.

The hybrid antenna system described in Yarovoy et al. (2007a) consists of a single transmit and 13 receive antennas (Figure 14.10). The so-called dielectric wedge antenna (Yarovoy et al., 2002b) is used as transmit antenna and loop antennas are used as receive antennas. As the loops are transparent for the incident wave, they are placed below the transmit antenna in its H-plane. The antenna system is used with a multichannel GPR receiver, which allows for simultaneous measurements of the scattered by the subsurface field with all receive antennas and drastic (in more than 10 times) increase of the survey speed in comparison with a traditional array-based system.

Stability of the system. Finally, the last critical aspect of the GPR sensor for landmine detection is its time stability and accuracy of scattered-field measurements. Although most GPR applications need to detect only underground objects, a GPR for landmine detection should also be able to recognize detected objects to keep the FAR at an acceptably low level. Such object recognition requires precise quantitative measurement of the field scattered by underground objects. To be able to measure the electromagnetic field with a maximal error of about 1%, a GPR should

Figure 14.10 The hybrid antenna system (from Yarovoy et al., 2007a).

have a linearity better than 1% for both the amplitude and time axes of the sampling converter, and the receive channel should have a linear dynamic range above 40 dB. Nonlinear signal processing (like a time-varying gain) should not be used in the receiver chain. The measurement accuracy is also decreased by time jitter (short time instability) and discretization errors of an analog-to-digital converter (ADC). However, the errors introduced by these two factors are random with Gaussian distribution; their RMS value can be decreased to any desirable level by choosing a sufficient number of averaging at the cost of increased data acquisition time.

Polarimetry. Despite the vectorial nature of electromagnetic waves, they are used in the same way as acoustic (scalar) waves by the majority of GPR systems. Rarely are polarimetric properties of the reflected signal used in GPR sensors for landmine detection. The reason for this is the difficulty in measuring a full-polarimetric scattering matrix of a reflector from a moving platform (a handheld or vehicle-based one). The motion of the platform requires simultaneous or quasi-simultaneous transmission of two orthogonally polarized electromagnetic fields and measurements of the reflected field at two orthogonal polarizations. Fulfilling this requirement doubles system complexity and results in a very complicated antenna system.

However, promising theoretical results on polarimetric properties of landmines and the need to classify detected objects have pushed system designers to develop polarimetric systems. Therefore, fully polarimetric measurements have indeed been implemented in a number of forward-looking systems (Witten, 1998; Kositsky and Milanfar, 1999). Down-looking systems typically do not have sufficient physical space for a fully polarimetric antenna system [two exceptions are DEMAND radar (Sachs et al., 2003) and multiwaveform radar (Yarovoy et al., 2002b)]. For that reason some designers chose to use orthogonal polarizations for transmit and receive antennas (Chignell et al., 2000). In such systems, the surface reflection is considerably suppressed, improving the signal-to-clutter ratio.

The overview of system designs presented above is far from exhaustive. The challenges of the landmine detection problem have triggered an avalanche of R&D work in the area of GPR sensors in the last two decades. Many new ideas have been developed and implemented in GPR system design. As a result, a wider spectrum of GPR hardware has become available.

14.4. GPR Data Processing for Landmine/UXO Detection and Classification

Like the hardware of a GPR sensor for landmine detection, the signal processing of these sensors differs considerably from that of conventional GPR systems. In conventional systems, the signal processing chain typically consists of two steps: signal preprocessing (where raw data are enhanced by taking into account a particular hardware design and data acquisition strategy, e.g., compensation of all types of instabilities, data interpolation on a regular grid) and object detection. Interpretation of processed data is typically a task for a GPR operator. In the dedicated to landmine detection GPR, detection of weak signals in relatively strong clutter and classification of detected reflections are tasks too difficult and time consuming for an inexperienced operator. That is why automatic target recognition (ATR) algorithms are implemented to assist an operator (in handheld systems) or exclude a human operator from decision-making process at all (in vehicular-based systems). Typically the processing chain for landmine detection purposes consists of preprocessing (which is extended in comparison with conventional GPR case with a clutter suppression stage), initial detection (or primary detection or area selection), where all suspicious anomalies of the subsurface are selected, and discrimination, in which these anomalies are analyzed and classified as targets or friendly objects (Ho et al., 2004). The basis of ATR is a set of measurable features ("feature vectors"). Such a set should ideally uniquely characterize the target. Examples of such measurable features are the size and shape of the object, its dielectric permittivity, natural resonances, and particular patterns in the time–frequency plane. Theoretically, the reflector can be fully identified based on its shape, size, and spatial distribution of dielectric permittivity. All these features can be provided by solving the inverse problem (Berg and Abubakar, 2003). However, the huge total amount of necessary measurements of the reflected field and the very high computational costs involved prevent practical realization of this approach.

The requirements of the processing algorithms for the detection and discrimination parts of the processing chain are very different. The initial detection should be done in real time with a very fast algorithm, the probability of detection should be 100% in order not to miss any mine, and the FAR can be moderate. The discrimination step can be done in almost real time, a large set of data can be used to analyze each detected anomaly, and the processing should drastically reduce the overall FAR. Many algorithms originally developed for conventional GPRs are widely used for detection. However, due to differences in the scenario (elevation of antennas above the ground, low signal-to-clutter ratio, etc.), these algorithms should be modified. For the discrimination step, different ATR algorithms are used, and such algorithms are new to GPR processing. In the rest of this section we consider specificity of detection algorithms and briefly discuss ATR algorithms.

Initial detection. Detection of landmine and UXO responses in GPR data is complicated by the typically low signal-to-clutter ratio and the generally unknown response of targets. Elevation of antennas above the ground causes additional difficulties as the antenna coupling and the reflection from the air–ground interface are by far the strongest signals in a GPR profile and mask reflections from targets of interest.

If an object's response is known (from a model or a training dataset), it is possible to develop an algorithm searching for this known response in an A-scan. The most popular among such algorithms is the inverse-matched filter (Osumi and Ueno, 1984). The choice of the filter parameters is discussed in Osumi and Ueno (1984, 1985). When applied to a GPR profile, the inverse-matched filter efficiently compresses it. As an example of such compression, we present compression of the signal reflected from the air–ground interface in Figure 14.11. However, the sidelobe level of the compressed signal is relatively high; it masks reflections from flush and shallowly buried objects.

To overcome this particular problem, several other algorithms have recently been developed for object detection within a single A-scan (see, e.g., Roth et al., 2003; Kovalenko et al., 2003b). The searching algorithm of Kovalenko et al. is based on minimal discrepancy between the real signal and the template and does not require the use of deconvolution. Such search procedure can be applied not only as a stand-alone detector but also as a preprocessor for more complicated multistep detection algorithms. In such a case, the procedure considerably improves the detectability and reduces the FAR (Kovalenko et al., 2007a).

A principally different approach to target detection in a single A-scan is based on using different time–frequency methods. The localization of a specific frequency at a particular time is the basic principle of time–frequency analysis. Wavelet analysis

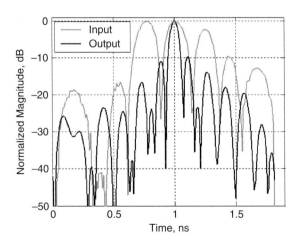

Figure 14.11 Application of the inverse-matched filter to the probing signal of the radar (from Kovalenko et al., 2003b).

(Le-Tien et al., 1997; Brooks, 2000) and Wigner transform analysis (Strifors et al., 1996) are the most often used tools.

Despite some successes achieved by target detection in a single GPR profile, the most common approach is target detection in a number of closely acquired GPR profiles (a B-scan or a C-scan). The simplest detection technique here is peak-energy detection, where all energy over each GPR profile is integrated over the profile and projected on the ground surface. Thresholding and binarization of this projection results in a 2D map of the horizontal positions of detected objects. This approach works well only for the detection of strong subsurface reflectors, as weak reflections are masked by the surface clutter.

A more efficient detection approach is based not on peak-energy detection but on detection of a certain spatial distribution of reflected energy. An object reflection in a 3D volume of GPR data has a shape similar to a hyperboloid. In the B-scans, it shows up as a hyperbola and in the time slices as an ellipse. Considerable work has been done on the detection of small objects with GPR using hyperbola detection (Milisavljević et al., 2001). This approach consists of five major steps: background removal, thresholding, edge detection, application of the randomized Hough transform (RHT) (Xu et al., 1990), and determining the burial depth of the reflector from the three hyperbola parameters. This approach works reasonably well for detection of relatively strong reflectors. Furthermore, homogeneity of the ground and positioning of GPR antenna system in direct contact with the ground are essential conditions for this method. Inhomogeneity of the ground results in another shape rather than a hyperbola for the reflector response.

Applied to landmine detection, such a detection approach has several limitations. First, elevation of GPR antennas above the ground changes the reflection shape. The original shape is described by a fourth-order curve, which can be approximated by a hyperbola near its top and at its far branches. Second, the parameters of this curve depend on the burial depth and horizontal displacement of a reflector from the GPR survey transect. This information is not known beforehand and can be determined only after object detection. To overcome these problems, the traditional hyperbola detection algorithm has been modified in Milisavljevic and Yarovoy (2002). The modification is based on simultaneous processing of the data acquired by two separate receive antennas and utilizes the fact that the targets of interest are flush or shallowly buried.

The above-mentioned problems due to antenna elevation can be also circumvented by choosing another cross section of C-scans, namely time slices, to search the target response. In time slices, the reflection shape shows up as an ellipse. An object detection algorithm, which is based on ellipse detection in time slices, is described in Yarovoy et al. (2003a). The algorithm is also based on the RHT. The detection algorithm searches for ellipses in a series of slices. When detected in a series of consecutive slices, an object response provides information about the horizontal position of the object (from the position of the ellipse centre), dielectric permittivity of the ground (from the increase in the ellipse size), and burial depth of the object (from a time delay and the calculated dielectric permittivity of the ground). The performance of the algorithm is demonstrated in Figure 14.12. It is important to mention that the ellipse detection algorithm is rigid against vertical

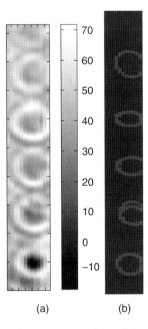

Figure 14.12 Time slice with the object responses (a) and detected ellipses (b).

variations of the soil dielectric permittivity and variation in the spatial positions of transmit and receive antennas (e.g., variation in the antenna elevation above the ground).

Furthermore, different statistical tests can be applied to B-scans (or C-scans) to detect variations from the background models including Kalman filter (Zoubir et al., 2002), HANOVA test (Xu et al., 2002), generalized likelihood ratio test (Gunatilaka and Baertlein, 2001; Ho and Gader, 2002), artificial neural networks (Plett et al., 1997), fuzzy logic (Gader et al., 2001b), hidden Markov models (Gader et al., 2001a), and abrupt change detection theory (Potin et al., 2006). The hidden Markov's model approach has been applied in Zhao et al. (2003) and Ho et al. (2004) for the detection of hyperboloid-like distributed features. A method of lowering the FAR for a handheld device, where hyperboloid traces of objects are disrupted by nonuniform sensor motion, is introduced in Gader et al. (2001a).

Similar approaches can be used for detecting targets in focused (migrated) GPR data or GPR images. The focusing algorithm places all reflectors in their correct physical positions and thus allows relatively simple separation of reflections coming from above surface reflectors, the air–ground interface and buried objects. This substantially simplifies the task of object detection. Furthermore, the focusing algorithm reconstructs the reflector shape (how accurately this is done depends on the focusing algorithm) (Slob, 2003). Below, we demonstrate advantages of GPR data focusing before target detection. However, implementing focusing algorithms in the processing chain is computationally expensive.

The simplest detection algorithm for focused data is very similar to the peak-energy detector described above. By concentrating reflected energy in positions where reflectors are physically situated, the focusing algorithm makes for a

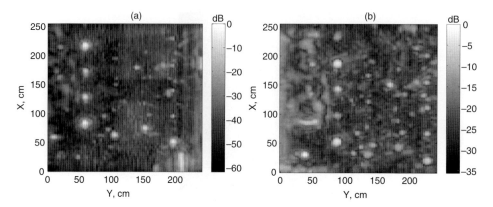

Figure 14.13 Energy projection of the SAR images for two orthogonal polarizations (adapted from Yarovoy et al., 2003b).

considerable change in the performance of the detection algorithm. A 2D map resulting from such an algorithm (Yarovoy et al., 2003b) is presented in Figure 14.13. It can be seen that the map represents the shapes of the reflectors relatively well. However, the surface-laid objects have considerably higher amplitude in comparison with identical but shallowly buried ones. To improve the visibility of a buried object, the projection algorithm (windowed projection) has been modified (Kovalenko et al., 2004). The windowed projection improves the detectability and reduces the FAR. Two-dimensional images resulting from projection can be further analyzed by means of well-developed image processing techniques for landmine image detection.

The disadvantage of any projection technique is that important information available in the depth dimension is lost. Several attempts to utilize this information have recently been made. Below, we shall demonstrate the high potential of 3D image analysis based on Ligthart et al. (2004). The first step of any 3D image analysis is volume selection, which can take the form of, for example, an area selection in consecutive depth slices. After thresholding and binarization, the 3D image looks like Figure 14.14. The image typically contains a lot of artifacts due to surface clutter, so the second step in image processing scheme is clutter removal. An example is removal of objects with large horizontal size and those with small height as surface clutter typically shows up in a few depth slices covering a depth range of a few millimeters, while the target reflection covers a range of about a few centimeters. The target reflection also has limited horizontal size (not larger than 15 cm), so objects with considerably larger horizontal size belong to the surface clutter. Removal of objects with a large horizontal size and those with small height resulted in a reduction of almost 70% of objects in the surveillance volume. The resulting image is shown in Figure 14.15.

Another approach based on the detection of a distinctive target signature, which appears in focused C-scans has been suggested in Cosgrove et al. (2004). If polarimetric information is available, it can be used to suppress clutter and improve object detection (Stiles et al., 1999; Sagues et al., 2001).

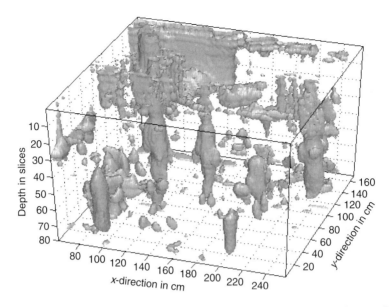

Figure 14.14 Binary 3D image volume after thresholding (adapted from Ligthart et al., 2004).

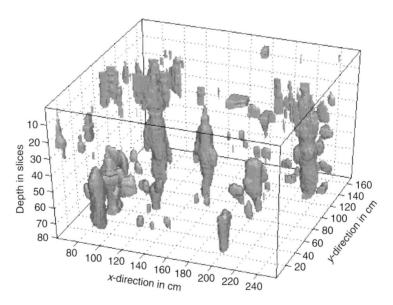

Figure 14.15 Binary 3D image volume after removing objects with large horizontal extend and small vertical size (adapted from Ligthart et al., 2004).

Object classification. As mentioned earlier, all processing steps that result just in object detection should be followed by a discrimination step. In this step, the friendly objects should be separated from targets. Such classification can be done based either on the analysis of the reflected signal or on the analysis of the reflector image. In both cases, classification requires feature extraction as a preliminary step. Correct selection of features is very important for obtaining good end result. Ideally the selected feature should be based on physical processes that distinguish landmines from other objects. Dielectric permittivity and regular shape (very often rotationally symmetric) can be given as examples of such distinctive features. The feature extraction step is followed by confidence assignment and decision-making steps. These steps are often done jointly by an algorithm called a classifier.

For reflection-based classification, either a template of the desired IR or some parameters of this response can be used. The simplest IR model (Roth et al., 2002) classifies objects by extracting three parameters of the model: the magnitude of reflection from the top of the object, the magnitude of reflection from the bottom of the object, and the separation in time between these two reflections, and fits the IR model to the measured reflection (Lijn et al., 2003). Features extracted from the early target response might be used for more advanced classification of targets (Savelyev et al., 2007).

Image-based classification typically requires considerably more features. These features can be grouped into several classes, e.g., shape-based features (dealing with object shape), statistical features (dealing with intensity distribution within the object), and template features (showing similarity between chosen templates and the measured image). Feature selection is discussed and some examples are given in Fukunaga (1990), Cremer (2003a), and Ligthart et al. (2004). Because classification time increases drastically with number of selected features, feature selection is an important issue. Selection of the best-performing features (selected as a single feature) typically does not result in optimal performance. Here it is more important to form a multidimensional feature space, in which a multidimensional volume-containing desired object is well separated from the rest of the reflectors.

Selecting a classifier is no trivial task either. The most often used classifiers are the supervised classifier, the Naive Bayes classifier, the Neyman–Pearson classifier, the learning vector quantization (LVQ) classifier, and statistical classifiers such as generalized-likelihood ratio test (Gunatilaka and Baertlein, 2001), discrete hidden Markov model (DHMM), (Zhao et al., 2003), ad hoc hierarchical decision schemes (Gader et al., 2004), and 2D the least mean squares (LMS) algorithm (Torrione et al., 2006]. The majority of advanced classifiers are based on neural networks, fuzzy logic, or statistical methods (Gader et al., 2001a; Collins et al., 2001). Performance of different classification algorithms applied to the sets of experimental data from NIITEK radar is described in Wilson et al. (2007). The considerable computational effort required for developing classification algorithms is typically paid off by good overall performance of the processing chain. Typically, a classification procedure outperforms any simple energy detection algorithm (energy can be viewed as a single feature with its value as a confidence level) (Gader et al., 2001b). As an example, Figure 14.16 shows the result of applying such a classification procedure to the image of Figure 14.14.

The final choice of processing algorithm (for both the detection and discrimination steps) depends largely on the type of system used. For a vehicle-based system, the application of an antenna array and the availability of accurate-positioning

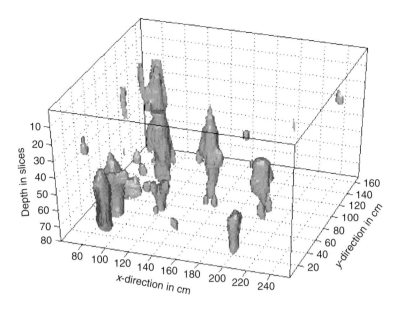

Figure 14.16 Binary 3D image volume (same as in Figure 14.13) after classification (adapted from Ligthart et al., 2004).

information together with the possibility of heavy signal processing implementation favor the selection of feature-based detection algorithms operating with 3D data volumes (C-scans). For the majority of handheld systems, the absence of accurate-positioning and sensor orientation information together with a varying antenna elevation above the ground favors a much simpler algorithm like the peak-energy detectors or hyperbola detectors. The final decision on object classification for handheld systems is typically left to the GPR operator.

14.5. FUSION WITH OTHER SENSORS

Numerous tests on different types of sensors for UXO/landmine detection have clearly demonstrated that at the moment, there is no single sensor that can do the work. Hence, a high-tech solution of the problem will have to be a combination of different sensors. The majority of multisensor systems developed so far combine a GPR sensor with a metal detector and a visual-light or infrared camera (McDonald et al., 2003). These three sensors form the "detection" kernel of the system. In addition, some systems include the so-called "confirmation" sensors (e.g., a thermal neutron detector), which are used only for final classification of suspicious detected objects. At the moment of writing of this Chapter, three dual-sensor, handheld systems were in production: AN/PSS-14 (formerly HSTAMIDS, which is a SFCW system with one transmit antenna and two spiral receive antennas; Hatchard, 2003), MINEHOUND (a video impulse radar with two shielded dipoles; Daniels et al., 2005), and ALIS (a SFCW system with two shielded spiral antennas; Sato and Takahashi, 2007). All these systems include a metal detector as a primary sensor, i.e., for mine detection, and a GPR sensor as a secondary sensor, i.e., for identification/classification/confirmation.

Mutual processing of the data coming from different sensors is called sensor fusion. It is the process in which information from different sensors is used for a unified declaration of objects as detected by these sensors. Commonly a distinction is made between three different levels of sensor fusion: data-level fusion, feature-level fusion, and decision-level fusion (Waltz and Llinas, 1990). In the majority of multisensor systems developed so far, data acquired by different sensors are fused at a decision level (McDonald et al., 2003). In the handheld systems, this fusion is actually done by an operator. Fusion at lower levels (feature fusion or data fusion) requires on the one hand considerably more computational power and on the other an in-depth understanding of the performance of each sensor and access to the preprocessed data of these sensors. Despite evident difficulties with its realization, feature-level fusion can provide much better results in terms of detectability and FAR.

The feature-level sensor fusion process starts with the selection of the regions of interest with their features as measured by the individual sensors and consists of three steps. The first step is initial detection and feature extraction. The second step is object association and feature reconciliation. In this step, the features from the selected regions from different sensors are combined to form an associated object. The third and final step is decision making. Object association and feature reconciliation is the most important part of feature-level sensor fusion. The remaining steps are general steps for detection and feature-based classification; they are not specific for feature-level fusion.

The simplest object association algorithm is as follows. For each object from one sensor, the object from the other sensor that is closest in distance is found. If this distance is within the maximum bound (smaller than a typical size of the target), then the two objects are associated. With this form of object association, there is always a maximum of two associated sensor objects. This one-to-one object association avoids the ambiguity that might arise with one-to-many object association. The disadvantage is that there is a chance that the wrong objects are associated. This chance can be minimized by including more information. One source of information may be the object depth (Schavemaker et al., 2001), but this requires a different object association approach. More advanced generic object association approach has been suggested in Kovalenko et al. (2007b).

In principle, feature-level fusion can handle all features from all sensors. For some sensors (e.g., polarimetric ones), a number of observations per sensor can be more than one (e.g., observations on different polarizations). Then each observation might be considered as a separate sensor (Kovalenko et al., 2007b). However, the number of selected features should be kept to a minimum as an exhaustive search over all feature combinations would require an enormous amount of evaluations and is not feasible. As has been shown by Cremer et al. (2003b) for a polarimetric IR and GPR sensor suite, the following features can be considered: the level above the top-hat threshold, the area, the average Q value, the average U value and the contrast in I (for polarimetric infrared) and the variation in depth, the

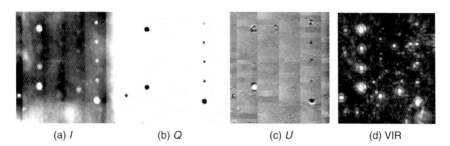

Figure 14.17 Polarimetric infrared (I, Q, and U) and ground penetrating radar (GPR) (VIR) images of the test ground (from Cremer et al., 2003b).

variation in energy, the average energy, the depth, and the variation of highest positive value (for GPR).

With the examples of these two sensors, it is possible to demonstrate the efficiency of the feature-fusion approach. The original images coming from both sensors are shown in Figure 14.17 (for the GPR sensor, the image is obtained by projection of a 3D-focused GPR image to the surface). The training set is small: only 21 landmines and $20\,\text{m}^2$ area. The number of potential false alarms is high for both sensors (1485 for the polarimetric IR and 2616 for the VIR). This joint dataset has been classified by two feature classification methods (LVQ-dist classifier and the Naive Bayes classifier). The results are shown in Figure 14.18. The fusion results for both classifiers at the training dataset are always better than the single-sensor results with the same classifier. Furthermore, a couple of sensor fusion receiver operating characteristic (ROC) points in the evaluation set were better (in terms of detectability by a fixed value of the FAR) than the single-sensor evaluation set results.

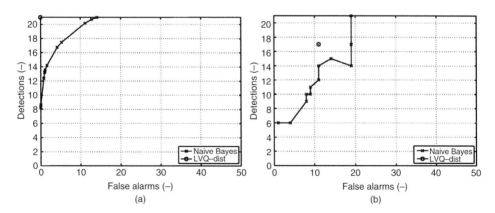

Figure 14.18 The sensor fusion results of the VIR and the polarimetric infrared system on the training set (a) and on an independent evaluation set (b) using leave-one-out as evaluation method. Both sets contain 21 landmines and cover $20\,\text{m}^2$ area (from Cremer et al., 2003b).

14.6. OVERALL PERFORMANCE OF GPR AS AN UXO/LANDMINE SENSOR

The overall performance of the GPR sensor for landmine detection heavily depends on the way in which it is used and what processing algorithms are implemented in it. Again, there are principal differences between handheld systems and vehicle-based systems.

The implementation of a GPR sensor in a handheld system is restricted by weight and size as the ground is surveyed manually. Typically a handheld system includes a relatively simple GPR with a single pair of transmit–receive antennas (single transmit and two receive antennas are used in HSTAMIDS) and relatively simple data processing. Often such a system detects objects based on B-scan processing. In ALIS, accurate-positioning information is available due to an additional sensor, allowing for computing 3D subsurface image and image-based object detection. In all handheld systems, classification of detected objects is left to the operator.

In the field trials of produced handheld mine detectors, it has been demonstrated that the GPR sensor improves the overall performance of the whole detection system. Statistics of the operational use of AN/PSS-14 from April 2006 to March 2007, during which 22 mine fields with a total area of 238 365 m^2 have been cleared, shows operational improvement from 7 to 17 times in comparison with a metal detector (Doheny, 2007). Improved performance results in the increase of the scanned area per day from 25 m^2 (performance of a conventional detector) to 275 m^2. Operational testing of MINEHOUND in 2005 has demonstrated reduction of the false alarms by a factor of better than 5:1 in Cambodia and by a factor of better than 7:1 in Bosnia and Angola (Daniels, 2006b). Similar performance of ALIS has been reported in Sato and Takahashi (2007).

Regarding the performance of GPR sensor alone, it has been reported in Daniels and Curtis (2003) that at a Bosnian test site of 200 cells of area 1×0.5 m^2, which consist of a gravel layer approximately 15 cm thick on top of a clay type soil, the GPR sensor of MINETECT (a prototype of MINEHOUND) has detected 100% of the mines, which were predominantly PMA2 and PMA3 AP mines. On a native Balkan topsoil test site with minimal grass vegetation (also 200 cells of area 1×0.5 m^2 consisted), the GPR sensor detected 94% of all the AP mines.

Vehicle-based GPR system in general allows for higher quality of the acquired data and much higher processing power than that in handheld systems. In these systems, complicated processing of the whole 3D volume of measured data is realized. Reports on field trials of such systems suggest that the GPR sensor does not suffer from major problems in detection of AT mines (Pressley et al., 2003) and under many circumstances can successfully detect all AP mines laid (Schavemaker et al., 2003). The main problem, however, is not an insufficient detection level but a high FAR. Reduction of the latter can be done via enhanced signal processing (especially the classification step of it) and improvement of the hardware in order to support extraction of features from the measured data necessary for classification.

14.7. CONCLUSION

Over the last two decades, GPR has surfaced as a promising technology for landmine and UXO detection. Under different environmental conditions, the GPR sensor demonstrated a high detection rate for UXO and antitank mines. The detection of AP mines is more difficult, but here very promising results are demonstrated as well (especially for shallowly buried mines). Furthermore, the GPR as close-in sensor has proven its ability to support classification of detected objects and drastic reduction of the FAR during demining operations. The application of GPR systems as standoff sensors for the detection of the former test ranges with UXO and landmine fields is not yet as successful as in close-in detection missions; however, this is the subject of continuing research.

The GPR sensor has a number of advantages over other landmine detection sensors. First, it is complementary to conventional metal detectors. Rather than detecting exclusively the presence of metal, it senses variations in the electromagnetic properties of the ground and therefore it can find mines with a wide variety of types of casing (not just those with metal). Second, it can often generate an image of the mine or another buried object based on dielectric constant variations because the size of the most mines is generally larger than the required radar wavelength at frequencies that still have a reasonable penetration depth. On the basis of the image or other features (which can be extracted from the image), the detected objects can be classified. Third, GPR scans at a rate comparable to that of an EMI system. Finally, GPR is a mature technology, with a long performance history.

However, GPR is not the ultimate solution for the landmine detection problem. GPR is not a specific sensor for explosives, and it detects only secondary signs of explosive devices (first of all, the dielectric contrast with environment). Because GPRs sense all electromagnetic inhomogeneities of the ground, all natural subsurface inhomogeneities (such as roots, rocks, and water pockets) will be a source of false alarms, which should be discriminated in later processing. Whether or not a GPR will detect a landmine/UXO highly depends on soil moisture, surface roughness, and mine location; such complex interplays make its performance highly variable and difficult to predict. Due to the above-mentioned reasons, a GPR sensor cannot perform the role of standalone sensor for landmine/UXO detection. However, it can play a crucial role within a multisensor suite. Ground penetrating radar and EMI sensors seem to be a very powerful combination, as together they can detect almost all landmines and classify them as landmines.

The design of a dedicated GPR sensor for landmine detection is a big challenge to a vast community of scientists and engineers active in the areas of system design, RF engineering, signal processing, and applied electromagnetics. After more than a decade of intensive R&D work, the three handheld systems with GPR sensor (AN/PSS-14, MINEHOUND, and ALIS) came into production. This gives hope that operational deminers worldwide will finally receive tools that can make a real difference in the field.

REFERENCES

Baum, C.E., 1971, On the singularity expansion method for solution of electromagnetic interaction problems, Albuquerque. Air Force Weapons Lab., EMP Interaction Note 88, 112p.

Baum, C.E., Rothwell, E.J. and Nyquist, D.P., 1991, The singularity expansion method and its application to target identification. Proceedings of IEEE, Vol. 79, pp. 1481–1491.

Berg, P.M. van den and Abubakar, A., 2003, Linear and non-linear inversion of GPR data. Proceedings, International Workshop on Advanced GPR, 2nd, Delft University of Technology, Delft, pp. 156–163.

Blagden, P.M., 1996, Kuwait: Mine clearing after Iraqi invasion. Army Quarterly & Defence Journal, Vol. 126, pp. 4–12.

Bloemenkamp, R. and Slob, E.C., 2003, The effect of the elevation of GPR antennas on data quality. Proceedings, International Workshop on Advanced GPR, 2nd, Delft University of Technology, Delft, pp. 201–206.

Brooks, J.W., 2000, Wavelet-based Shape-Feature Extraction for GPR Detection of Non-Metallic Anti-personnel Land Mines. Proceedings, Detection and Remediation Technologies for Mines and Minelike Targets V. Proceedings SPIE, Vol. 4038, Part II, pp. 1028–1036.

Carin, L., Geng, N., McClure, M., Sichina, J. and Nguen, L., 1999, Ultrawide-band synthetic-aperture radar for mine-field detection. IEEE Antennas and Propagation Magazine, Vol. 41, pp. 18–33.

Carin, L., Sichina, J. and Harvey, J.F., 2002, Microwave underground propagation and detection. IEEE Transactions on Microwave Theory and Techniques, Vol. 50, pp. 945–952.

Chan, L.C., Moffatt, D.L. and Peters, Jr., L., 1979, A characterization of subsurface radar targets. Proceedings IEEE, Vol. 67, pp. 991–1000.

Chen, C.-C. and Peters Jr, L., 1997, Buried unexploded ordnance identification via complex natural resonances. IEEE Transactions of Antennas and Propagation, Vol. 42, pp. 1645–1654.

Chen, C.-C., Nag, S., Burnside, W.D., Halman, J.I., Shubert, K.A. and Peters, Jr., L., 2000, A standoff, focused-beam land mine radar. IEEE Transactions of Geoscience and Remote Sensing, Vol. 38, pp. 507–514.

Chen, C.-C., Higgins, M.B., O'Neill, K. and Detsch, R., 2001, Ultrawide-bandwidth fully-polarimetric ground penetrating radar classification of subsurface unexploded ordnance. IEEE Transactions of Geoscience and Remote Sensing, Vol. 39, pp. 1221–1230.

Cherniakov, M. and Donskoi, L., 1999, Frequency band selection of radars for buried object detection. IEEE Transactions of Geoscience and Remote Sensing, Vol. 37, pp. 838–845.

Chignell, R.J., Dabis, H., Frost, N. and Wilson, S., 2000, The radar requirements for detecting anti-personnel mines. Proceedings International Conference on Ground Penetrating Radar, 8th, Proceedings SPIE, Vol. 4084, pp. 861–866.

Cloude, S.R., Milne, A., Thornhill, C. and Crisp, G., 1996, UWB SAR detection of dielectric targets. Proceedings, The Detection of Abandoned Land Mines: A Humanitarian Imperative Seeking a Technical Solution, EUREL, pp. 114–118.

Cosgrove, R.B., Milanfar, P. and Kositsky, J., 2004, Trained detection of buried mines in SAR images via the deflection-optimal criterion. IEEE Transactions on Geoscience and Remote Sensing, Vol. 42, pp. 2569–2575.

Collins, L.M., Zhang, Y., Li, J., Wang, H., Carin, L., Hart, S.J., Rose-Pehrsson, S. L., Nelson, H.H. and McDonald, J.R., 2001, A Comparison of the performance of statistical and fuzzy algorithms for unexploded ordnance detection. IEEE Transactions on Fuzzy Systems, Vol. 9, pp. 17–30.

Cremer, F., 2003a, Polarimetric Infrared and sensor fusion for the detection of landmines, PhD dissertation, Delft University of Technology, Delft, 2003.

Cremer, F., de Jong, W., Schutte, K., Yarovoy, A.G., Kovalenko, V. and Bloemenkamp, R.F., 2003b, Feature level fusion of polarimetric infrared and GPR data for landmine detection. Proceedings, EUDEM2-SCOT – 2003, International Conference on Requirements and Technologies for the Detection, Removal and Neutralization of Landmines and UXO, Vrije Universiteit Brussel, Brussels, Vol. 2, pp. 638–642.

Daniels, D.J., 1999, Resolution of ultra-wideband radar signals. IEE Proceedings, Radar, Sonar and Navigation, Vol. 146, pp. 189–194.

Daniels D.J. and Martel, C., 2001, Radar technology for mine detection. Proceedings, European Microwave Conference, 31st, pp. 1–10.

Daniels, D.J. and Curtis, P., 2003, MINETECT. International Workshop on Advanced GPR, 2nd, Delft University of Technology, Delft, pp. 110–114.

Daniels, D.J. (ed.), 2004, Ground Penetrating Radar, 2nd edition, The Institute of Electrical Engineers, London, 726p.

Daniels, D.J., Curtis, P., Amin, R. and Hunt, N., 2005, MINEHOUND™ production development. Proceedings, Detection and Remediation Technologies for Mines and Minelike Targets X. Proceedings SPIE, Vol. 5794, pp. 488–494.

Daniels, D.J., 2006a, A review of GPR for landmine detection: Sensing and imaging. An International Journal, Vol. 7, pp. 90–123.

Daniels, D.J., and Curtis, P., 2006, MINEHOUND™ trials in Cambodia, Bosnia and Angola, in Proceedings, Detection and Remediation Technologies for Mines and Minelike Targets X: Proceedings SPIE, Vol. 6217, p. 62172N.

Dogaru, T., Collins, L. and Carin, L., 2001, Optimal time-domain detection of a deterministic target buried under a randomly rough interface. IEEE Transactions on Antennas and Propagation, Vol. 49, pp. 313–325.

Doheny, R.C., 2007, Handheld Standoff Mine Detection System HSATMIDS: Program and Operations Update, presentation held at Mine Action National Directors and UN Advisors Meeting, 21 March 2007, at www.mineaction.org/.

Fukunaga, K., 1990, Introduction to Statistical Pattern Recognition, 2nd edition, Academic Press Professional, Inc., San Diego, 591p.

Gader, P.D., Mystowski, M. and Zhao, Y., 2001a, Landmine detection with Ground Penetrating Radar using hidden Markov models. IEEE Transactions Geoscience and Remote Sensing, Vol. 39, pp. 1231–1244.

Gader, P.D., Keller, J. and Nelson, B., 2001b, Recognition technology for the detection of buried land mines. IEEE Transactions on Fuzzy Systems, Vol. 9, pp. 31–43.

Gader, P.D., Lee, W.-H. and Wilson, J., 2004, Detecting landmines with ground penetrating radar using feature-based rules, order statistics, and adaptive whitening. IEEE Transactions of Geoscience and Remote Sensing, Vol. 42, pp. 2522–2534.

Gunatilaka, A. and Baertlein, B., 2001, Feature-level and decision-level fusion of noncoincidently sampled sensors for land mine detection. IEEE Transactions on Pattern Analysis and Machine Intelligence, Vol. 23, pp. 271–293.

Hatchard, C., 2003, A combined MD/GPR detector – The HSTAMIDS system, presentation at EUDEM2-SCOT – 2003, International Conference on Requirements and Technologies for the Detection, Removal and Neutralization of Landmines and UXO, 15-18 September 2003 – Vrije Universiteit Brussel, Brussels, Belgium.

Ho, K. and Gader, P., 2002, A linear prediction land mine detection algorithm for hand held ground penetrating radar. IEEE Transactions of Geoscience and Remote Sensing, Vol. 40, pp. 963–976.

Ho, K.C., Collins, L.M., Huettel, L.G. and Gader, P.D., 2004, Discrimination mode processing for EMI and GPR sensors for hand-held land mine detection. IEEE Transactions of Geoscience and Remote Sensing, Vol. 42, pp. 249–263.

Le-Tien, T., Talhami, H. and Nguyen, D.T., 1997, Target signature extraction based on the continuous wavelet transform in ultra-wideband radar. IEE Electronics Letters, Vol. 33, pp. 89–91.

Lijn, F. van der, Roth, F. and Verhaegen, M., 2003, Estimating the impulse response of buried objects from ground penetrating radar signals. Proceedings, Detection and Remediation Technologies for Mines and Minelike Targets VIII. Proceedings SPIE, Vol. 5089, pp. 387–394.

Ligthart, E.E., Yarovoy, A.G., Roth, F. and Ligthart, L.P., 2004, Landmine detection in high resolution 3D GPR images. Proceedings, MIKON'04, pp. 423–426.

Kerr, M., 1999, Multi-sensor ordnance detection and mapping system. Proceedings, UXO Forum, Atlanta.

Kositsky, J. and Milanfar, P., 1999, A forward-looking high-resolution GPR system. Proceedings, Detection and Remediation Technologies for Mines and Minelike Targets IV. Proceedings SPIE, Vol. 3710, pp. 1052–1062.

Kovalenko, V., Yarovoy, A., Roth, F. and Ligthart, L.P., 2003a, Full-polarimetric measurements over a mine field-like test site with the video impulse ground penetrating radar. Proceedings, Detection and Remediation Technologies for Mines and Minelike Targets VIII. Proceedings SPIE, Vol. 5089, Part I, pp. 448–456.

Kovalenko, V., Yarovoy, A. and Ligthart, L.P., 2003b, Application of Deconvolution and Pattern Search Techniques to High Resolution GPR Data. Proceedings, International Conference on Electromagnetics in Advanced Applications ICEAA'03, Torino, pp. 699–702.

Kovalenko, V. and Yarovoy, A., 2003c, Analysis of target responses and clutter based on measurements at test facility for landmine detection systems located at TNO-FEL. Proceedings, International Workshop on Advanced GPR, 2nd, Delft University of Technology, Delft, pp. 82–87.

Kovalenko, V., Yarovoy, A. and Ligthart, L.P., 2004, Object detection in 3D UWB subsurface images. Proceedings, European Microwave Conference, 34th, pp. 237–240.

Kovalenko, V., Yarovoy, A. and Ligthart, L.P., 2007a, A novel clutter suppression algorithm for landmine detection with GPR. IEEE Transactions on GeoScience and Remote Sensing, Vol. 45, pp. 3740–3751.

Kovalenko, V., Yarovoy, A. and Ligthart, L.P., 2007b, Polarimetric feature fusion in GPR for landmine detection. Proceedings, International Geoscience and Remote Sensing Symposium (IGARSS2007), pp. 30–33.

McDonald, J.R., Nelson, H.H., Robertson, R., Altshuler, T.W. and Andrews, A., 1997, Field demonstration of multi-sensor towed array detection system. Proceedings, UXO Forum, pp. 243–253.

McDonald, J., Lockwood, J.R., McFee, J., Altshuler, T., Broach, T., Carin, L., Harmon, R., Rappaport, C., Scott, W. and Weaver, R., 2003, Alternatives for landmine detection, Rand Corporation, 336p.

Milisavljevic, N. and Yarovoy, A., 2002, Position determination of a subsurface object by combined output of two receive GPR antennas,. Proceedings, International Conference on Digital Signal Processing (DSP 2002), 14th, Vol. 2, pp. 901–904.

Milisavljević, N., Bloch, I. and Acheroy, M., 2001, Application of the randomized Hough Transform to Humanitarian Mine Detection. Proceedings, IASTED International Conference on Signal and Image Processing (SIP2001), 7th, pp. 149–154.

Mine Action Equipment: Study of Global Operational Needs, 2002, Geneva: GICHD.

MineFacts, 1995, CD-ROM, v.1.2, The US Department of Defence.

Narayanan, R.M., Xu, Y., Hoffmeyer, P.D. and Curtis, J.O., 1995, Design and performance of a polarimetric random noise radar for detection of shallow buried targets. Proceedings, Detection and Remediation Technologies for Mines and Minelike Targets. Proceedings SPIE, Vol. 2496, pp. 20–30.

O'Neil, K., 2000, Broadband bistatic coherent and incoherent detection of buried objects beneath randomly rough surfaces. IEEE Transactions on Geoscience and Remote Sensing, Vol. 38, pp. 891–898.

O'Neil, K., 2001, Discrimination of UXO in soil using broadband polarimetric GPR backscatter. IEEE Transactions on Geoscience and Remote Sensing, Vol. 39, pp. 356–367.

O'Neil, K., Lussky Jr., R.F. and Paulsen, K.D., 1996, Scattering from a metallic object embedded near the randomly rough surface of a lossy dielectric. IEEE Transactions on Geoscience and Remote Sensing, Vol. 34, pp. 367–376.

Osumi, N. and Ueno, K., 1984, Microwave holographic imaging method with improved resolution. IEEE Transactions on Antennas and Propagation, Vol. 32, pp. 1018–1026.

Osumi, N. and Ueno, K., 1985, Microwave holographic imaging of underground objects. IEEE Transactions on Antennas and Propagation, Vol. 33, pp. 152–159.

Plett, G., Doi, T., and Torrieri, D., 1997, Mine detection using scattering parameters and an artificial neural network. IEEE Transactions on Neural Networks, Vol. 8, pp. 1456–1467.

Potin, D., Vanheeghe, P., Duflos, E. and Davy, M., 2006, An abrupt change detection algorithm for buried landmines localization. IEEE Transactions on Geoscience and Remote Sensing, Vol. 44, pp. 260–272.

Pressley, J.R.R., Page, L., Green, B., Schweitzer, T. and Howard, P., 2003, Ground standoff mine detection system (GSATMIDS). Block 0 contractor test results. Proceedings, Detection and Remediation Technologies for Mines and Minelike Targets VIII. Proceedings SPIE, Vol. 5089, Part II, pp. 1336–1344.

Ratcliffe, J.A., Sachs, J., Cloude, S., Grisp, G.N., Sahli, H., Peyerl, P., and De Pasquale, G., 2002, Cost effective surface penetrating radar device for humanitarian demining, in Smith, P.D. and Cloude, S.R. (eds), Ultra-wideband, Short-pulse Electromagnetics 5, Kluwer Academic/Plenum Publishers, New York, pp. 275–283.

Redman, J.D., Annan, A.P. and Das, Y., 2003, GPR for anti-personnel landmine detection: results of experimental and theoretical studies. Proceedings, Detection and Remediation Technologies for Mines and Minelike Targets VIII. Proceedings SPIE, Vol. 5089, Part I, pp. 358–374.

Roth, F., Genderen, P. van and Verhaegen, M., 2002, Radar response approximations for buried plastic landmines. Proceedings International Conference on Ground Penetrating Radar, 9th, Proceedings SPIE, Vol. 4758, pp. 234–239.

Roth, F., Genderen, P. van and Verhaegen, M., 2003, Processing and analysis of polarimetric Ground Penetrating Radar landmine signatures. Proceedings, International Workshop on Advanced GPR, 2nd, Delft University of Technology, Delft, pp. 70–75.

Sachs, J., Peyerl, P., Zetik, R. and Crabbe, S., 2003, M-sequence ultra-wideband-radar: State of development and applications. Proceedings of Radar 2003, Adelaide, Australia 6p.

Savelyev, T.G., van Kempen, L., Sahli, H., Sachs, J. and Sato, M., 2007, Investigation of time–frequency features for GPR landmine discrimination. IEEE Transactions on Geoscience and Remote Sensing, Vol. 45, pp. 118–129.

Sagues, L., Lopez-Sachez, J.M., Fortuny, J., Fabregas, X., Broquetas, A. and Sieber, A.J., 2001, Polarimetric radar interferometry for improved mine detection and surface clutter suppression. IEEE Transactions on Geoscience and Remote Sensing, Vol. 39, pp. 1271–1278.

Sato, M., 2003, A new bistatic GPR system using a passive optical sensor for landmine detection. Proceedings, International Workshop on Advanced GPR, 2nd, Delft University of Technology, Delft, pp. 164–167.

Sato, M. and Takahashi, K., 2007, The evaluation test of hand-held sensor ALIS in Croatia and Cambodia. Proceedings, Detection and Remediation Technologies for Mines and Minelike Targets IV. Proceedings SPIE, Vol. 6553, pp. 1–9.

Schavemaker, J.G.M., den Breejen, E., Cremer, F., Schutte, K. and Benoist, K.W., 2001, Depth fusion for antipersonnel landmine detection. Proceedings, Detection and Remediation Technologies for Mines and Minelike Targets VI. Proceedings SPIE, Vol. 4394, pp. 1071–1081.

Schavemaker, J., Breejen, E. de and Chignell, R., 2003, LOTUS field demonstration in Bosnia of an integrated multi-sensor mine-detection system for humanitarian demining. Proceedings, EUDEM2-SCOT – 2003, International Conference on Requirements and Technologies for the Detection, Removal and Neutralization of Landmines and UXO, Vrije Universiteit Brussel, Brussels, Vol. 2, pp. 613–616.

Slob, E.C., 2003, Toward true amplitude processing of GPR data. Proceedings, International Workshop on Advanced GPR, 2nd, Delft University of Technology, Delft, pp. 16–23.

Stiles, J.M., Parra-Bocaranda, P. and Apte, A., 1999, Detection of object symmetry using bistatic and polarimetric GPR observations. Proceedings, Detection and Remediation Technologies for Mines and Minelike Targets IV. Proceedings SPIE, Vol. 3710, pp. 992–1002.

Strifors, H.C., Brusmark, B., Gustafsson, A. and Gaunaurd, G.C., 1996, Analysis in the joint time–frequency domain of signatures extracted from targets buried underground. Proceedings, Automatic Object Recognition VI. Proceedings SPIE, Vol. 2756, pp. 152–163.

Sullivan, A., Geng, N., Carin, L., Nguyen, L. and Sichina, J., 1999, Performance analysis for radar detection of buried anti-tank and anti-personnel land mines. Proceedings, Detection and Remediation Technologies for Mines and Minelike Targets IV. Proceedings SPIE, Vol. 3710, pp. 1043–1050.

Torrione, P., Throckmorton, C. and Collins, L., 2006, Performance of an adaptive feature-based processor for a wideband ground penetrating radar system. IEEE Transactions on Aerospace and Electronic Systems, Vol. 42, pp. 644–658.

UXO Technology Demonstration Program at Jefferson Proving Ground, Phase III, 1997, U.S. Army Environmental Center Project Number SFIM-AEC-ET-CR-9701, 179p. (http://aec-www.apgea.army.mil:8080/).

Vines, A. and Thompson, H., 1999, Beyond the Mine Ban: Eradicating a Lethal Legacy, Research Institute for the Study of Conflict and Terrorism, London.

Vitebskiy, S. and Carin, L., 1996, Resonances of perfectly conducting wires and bodies of revolution buried in a lossy dispersive half-space. IEEE Transactions on Antennas and Propagation, Vol. 44, pp. 1575–1583.

Walton, E.K. and Cai, L., 1998, Signatures of surrogate mines using noise radar. Proceedings, Detection and Remediation Technologies for Mines and Minelike Targets III. Proceedings SPIE, Vol. 3392, Part 1, pp. 615–626.

Waltz, E. and Llinas, J., 1990, Multisensor data fusion, Norwood, Artech House, 464p.

Wilson, J.N., Gader, P., Lee, W.-H., Frigui, H. and Ho, K.C., 2007, A large-scale, systematic evaluation of algorithms using Ground Penetrating Radar for landmine detection and discrimination. IEEE Transactions on Geoscience and Remote Sensing, Vol. 45, pp. 2560–2572.

Witten, T.R. 1998, Present state-of-art in ground penetrating radars for mine detection. Proceedings, Detection and Remediation Technologies for Mines and Minelike Targets III. Proceedings SPIE, Vol. 3392, Part 1, pp. 576–585.

Won, I.J., Keiswetter, D.A. and Hanson, D.R., 1997, GEM-3: A monostatic broad electromagnetic induction sensor. Journal of Environmental Engineering Geophysics, Vol. 2, pp. 53–64.

Xu, L., Oja, E. and Kutltanen, P., 1990, A new curve detection method: randomized Hough transform (RHT). Pattern Recognition Letters, Vol. 11, pp. 331–338.

Xu, X., Miller, E., Rappaport, C. and Sower, G., 2002, Statistical method to detect subsurface objects using array ground-penetrating radar. IEEE Transactions on Geoscience and Remote Sensing, Vol. 40, pp. 963–976.

Yarovoy, A.G., Schukin, A.D., Kaploun, I.V. and Ligthart, L.P., 2001, Antenna system for UWB GPR for landmine detection. Proceedings, Detection and Remediation Technologies for Mines and Minelike Targets VI. Proceedings SPIE, Vol. 4394, pp. 692–699.

Yarovoy, A.G., Ligthart, L.P., Schukin, A.D. and Kaploun, I.V., 2002a, Polarimetric video impulse radar for landmine detection. Subsurface Sensing Technologies and Applications: An International Journal, Vol.3, pp. 271–293.

Yarovoy, A.G., Schukin, A.D., Kaploun, I.V. and Ligthart, L.P., 2002b, The dielectric wedge antenna. IEEE Transactions on Antennas and Propagation, Vol. 50, pp. 1460–1472.

Yarovoy, A., Kovalenko, V. and Fogar, A., 2003a, Impact of ground clutter on buried object detection by Ground Penetrating Radar. Proceedings, International Geoscience and Remote Sensing Symposium (IGARSS2003), pp. 755–757.

Yarovoy, A.G., Kovalenko, V., Bloemenkamp, R.F., Roth, F. and Ligthart, L.P., 2003b, Multi-waveform full-polarimetric GPR sensor for landmine detection: First experimental results. Proceedings, EUDEM2-SCOT – 2003, International Conference on Requirements and Technologies for the Detection, Removal and Neutralization of Landmines and UXO, Vrije Universiteit Brussel, Brussels, Vol. 2, pp. 554–560.

Yarovoy, A.G., Savelyev, T.G., Aubry, P.J., Lys, P.E. and Ligthart, L.P., 2007a, UWB array-based sensor for near-field imaging. IEEE Transactions on Microwave Theory and Techniques, Vol. 55, pp. 1288–1295.

Yarovoy, A.G., Roth, F., Kovalenko, V. and Ligthart, L.P., 2007b, Application of UWB near-field polarimetry to classification of GPR targets, in Sabath, F., Mokole, E.L., Schenk, U. and Nitsch, D. (eds), Ultra-wideband, Short-pulse Electromagnetics 7, , New York Springer Science + Business Media LLC, pp. 655–664.

Zhao, Y., Gader, P., Chen, P. and Zhang, Y., 2003, Training DHMMs of mine and clutter to minimize landmine detection errors. IEEE Transactions on Geoscience and Remote Sensing, Vol. 41, pp. 1016–1024.

Zoubir, A., Chant, I., Brown, C., Barkat, B. and Abeynayake, C., 2002, Signal processing techniques for landmine detection using impulse ground penetrating radar. IEEE Sensors Journal, Vol. 2, pp. 41–51.

CHAPTER 15

GPR Archaeometry

Dean Goodman, Salvatore Piro, Yasushi Nishimura, Kent Schneider, Hiromichi Hongo, Noriaki Higashi, John Steinberg *and* Brian Damiata

Contents

15.1. Introduction	479
15.2. Field Methods for Archaeological Acquisition	481
15.3. Imaging Techniques for Archaeology	482
15.4. Depth Determination	484
15.5. Case Histories	485
15.5.1. Case History No. 1: The Forum Novum, Tiber Valley, Italy	486
15.5.2. Case History No. 2: The Villa of Emperor Trajanus of Rome, Italy	488
15.5.3. Case History No. 3: Wroxeter Roman Town, England	494
15.5.4. Case History No. 4: Saitobaru Burial Mound No. 100, Japan	495
15.5.5. Case History No. 5: Saitobaru Burial Mound No. 111, Japan	498
15.5.6. Case History No. 6: Monks Mound, Cahokia, Illinois	501
15.5.7. Case History No. 7: Jena Choctaw Tribal Cemetery, Louisiana	502
15.5.8. Case History No. 8: Glaumbaer Viking Age, Iceland	505
Acknowledgments	507
References	507

15.1. Introduction

Solving subsurface problems in the field of archaeology without destructively intervening with the buried materials has become a prime focus of the archaeological community. The science to study, measure and quantify archaeological structures remotely has been designated as the field of Archaeometry. Remotely detecting archaeological structures is very important because excavation of a site can inadvertently destroy essential archaeological evidence which can then never be recovered. Because of the successful application of a variety of geophysical tools, and in particular GPR to probe beneath the ground, many archaeologists now regularly initiate geophysical surveys before studying or excavating potential sites. The application of GPR in archaeology has ranged from studying protected sites which can never be excavated, to using GPR to quickly and cost-effectively plan and carry out mitigation projects. GPR surveys at sites that are impacted by development fall under the category of rescue archaeology. This is the largest growing segment of the GPR applications in archaeology, and these kinds of surveys are expedited by a growing number of geotechnical consulting firms.

The first application of GPR in archaeology was initiated soon after the first commercial equipment became available in the 1970s. One of the earliest documented uses of GPR for archaeological prospection occurred in the mid-1970s when Bevan and Kenyon (1975) and Bevan (1977) used GPR to look for radar reflections from buried walls and variety of other historic structures; and Vickers and Dolphin (1975) used GPR to look for radar reflections from suspected buried walls associated with the native American Indian structures at Chaco Canyon. Dolphin et al. (1978) applied GPR in the successful search for caves on Victorio Peak in New Mexico.

A variety of GPR case histories were published in the 1980s and 1990s. Vaughn (1986) used GPR to discover a sixteenth century Basque whaling station and to locate the graves of fisherman. In Japan, Imai et al. (1987) applied GPR to discover pit house floors buried in volcanic soils with great precision. DeVore (1990) used GPR for investigations at the Fort Laramie National Historic Site. Other notable early GPR surveys for archaeological prospection includes studies by Bevan (1991), Sheets et al. (1985), Fischer et al. (1980) and Batey (1987).

These early GPR studies were primarily concerned with the discovery of buried features within known sites rather than imaging them. The early surveys used only paper records of the real-time GPR survey and never had the ability to perform any post processing on the data. Nonetheless, great efforts were used by the early investigators to create maps of subsurface anomalies by hand-contouring locations of continuous anomalies mapped in the field. These early crude maps have since been replaced by computer-generated time slice images. It is not known when the first computer-generated images that mapped horizontal changes in recorded reflections at constant time intervals were investigated. The authors have had access to unpublished reports (furnished by Bruce Bevan) showing as early as 1980, GPR surveys were done by researchers at Batelle National Laboratories in which a tractor-mounted digital-recording radar was used and computer-generated amplitude time slice maps were created from the recorded radargrams at an archaeological site. A crude form of time slice analysis was also available in 1986 from Geophysical Survey Systems Inc., the company that developed the first commercial GPR system. Nishimura and Kamei (1990) and Milligan and Atkin (1993) were among the first to employ a basic form of time slice analysis, in which radar reflections were mapped horizontally for archaeological applications.

The early time slice softwares created very pixilated maps where the profile spacing represented the width of the pixels displayed on the computer screen. These early maps were difficult to interpret and use because the line density remained fairly coarse. The early method has recently been explored again where a very fine line density recorded in the field has been applied in the data collection. Grasmueck and Weger (2002) have applied this older method to map marine sediments; however, their line density of 10 cm or less is rarely applied in archaeological environments. Goodman and Nishimura (1993) and Goodman et al. (1995) refined the GPR time slice method by binning data and applying interpolation procedures to estimate inter-profile locations. Interpolated time slice maps have helped to create more useful images for archaeological applications, particularly at sites where clutter is a problem that masks the continuity of features at a site. Several investigators have emulated the inter-line interpolation method for time slice analysis in radar, also with successful imaging results (e.g., Conyers and Goodman, 1997; Conyers and Cameron, 1998; Neubauer et al., 1999; Kvamme, 2001; Conyers and Connell, 2007; Piro and Goodman, 2008).

In this chapter we will briefly introduce the GPR field methods for archaeological investigation and look at a few GPR case histories that involve GPR-imaging techniques.

15.2. FIELD METHODS FOR ARCHAEOLOGICAL ACQUISITION

For most archaeological surveys with GPR, detection is the most important aspect of the investigation. When a sufficient density of profiles is recorded across a site, structural information can also be obtained regarding the buried features. Often investigators will choose a line density and profile direction that may yield the most successful imaging. In surveying grave sites, for example, where there is some initial knowledge that the graves are running in a roughly north–south direction, investigators will normally orient the line taking in an east–west direction. This will yield the best possible chance of traversing the graves. If a profile spacing of say 1 m were applied, there is a possibility of missing a grave if one were to profile parallel to the long axis, because the grave may be less than 1 m wide. Taking data perpendicular to the longer axis of the grave, or for that matter any archaeological structure, will aid detection. Recording of profiles in orthogonal directions will insure that a buried artifact will have been traversed and increase the probability of detection if the line density is at least half the smallest horizontal dimension of the buried targets.

Nonetheless, shallow and narrow features may require a very fine line density for detection. Shown in Figure 15.1 is a generalized description of the "unsampled" region versus the profile spacing. If the profile spacing is large, shallow areas can have no microwaves penetrating the region. At greater penetration depths however, the unsampled region decreases. If small targets are buried close to the ground surface, detection of these features may require denser line spacing than objects which are buried slightly deeper and within the penetration depth of the antenna. This thinking somewhat goes against what we believe about GPR: the shallower something is the better chance we can detect it. This is obviously not the case if buried materials are very shallow and a sufficient line density is not employed in the field survey.

In general, the smallest detectable size of archaeological materials is in general dependent on the frequency of the transmitting antenna. Typical frequencies for archaeological investigation range from about 200 to 800 MHz; however, even a low-frequency antenna on the order of 20 MHz might be used to discover structures buried below 15 m and 4 GHz antenna might be used to measure shallow features a few centimeters thick for instance on ancient mosaic floors (Utsi, 2006).

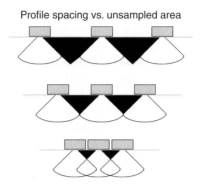

Figure 15.1 A depiction of the size of unsampled shallow areas of the ground versus line density is given. Shallowly buried targets that are smaller than half the aperture size of the prospecting antenna require slightly denser line spacing to be consistently detected. (The directional response function is assumed to have most of the energy transmitted within a beam that is about 120° wide).

Smaller high-frequency antennas can resolve smaller objects better than larger lower frequency antennas. Large antennas can detect very small objects, although the significance of the reflections from those small objects may not be realized by the naked eye on a single radargram profile. These reflections may be so minute that the changes may not be easily visualized. For instance, imaging a buried archaeological debris field, perhaps containing high concentrations of small fragments of ancient habitation (e.g., flint chips or other materials) may show some scattered energies on the radargrams which may only be one or two digital values different than in areas where no debris are located. Unless a fortunate color transform is used, it is unlikely that these small reflections will ever be noticed within the radargram – vertical slice – dataset. An alternative method for presenting GPR data from archaeological sites is needed to show these minute changes. Rather than showing the changes in reflected energies along the vertical radargram slices of the ground, it is more useful to map horizontal changes of reflected energy across a site. In this case, small, but consistent reflections above the background noise can be visualized by using time slice analysis.

15.3. IMAGING TECHNIQUES FOR ARCHAEOLOGY

There are a variety of imaging and analyses that can be implemented to solve the multitude of subsurface conditions in the archaeological field survey. Among a few of these imaging techniques and analyses are

Time slices – mapping of reflection anomalies horizontally across a site at various depths.

Isosurface rendering – showing equal amplitude surfaces in a 3D dataset by rendering the data, e.g. applying shading to the surface to give a 3D effect.

Static corrections – correcting and implementing the topography in the GPR dataset.
Overlay analysis – collecting (strong) reflectors from individual time slice maps and overlaying the results on a single, comprehensive 2D time slice.

The process of time slice analysis is not new for archaeological applications (Goodman and Nishimura, 1993). Goodman and Nishimura (1993) found from archaeological imaging, radargrams that were "binned" or spatially averaged in the vertical and horizontal windows to examine pulse energy, and then interpolated to create 3D volumes, provided more visually useful images for reading the continuity of subsurface reflection anomalies. Particularly at archaeological sites where subsurface structures may not be buried at equal depths (reflection times) and to reduce radar clutter, thicker slices of the radargrams often provide more useful images of the subsurface. Typically the radar signals are averaged over at least 1–2 wavelengths of the transmit pulse. Horizontal spatial averaging along the profile is set equal to or slightly less than the profile spacing. In the time slice analysis, the squared amplitude of the binned zone is averaged. A set of time slice datasets that are created by merging navigation data are then used to develop interpolated grid maps. Interpolation is usually done using a variety of algorithms; however, the authors have found the krigging algorithm (Isaaks and Srivastava, 1989) for finding interline estimation provides better resolution in developing time slice maps for archaeological applications.

One of the natural effects of time-slicing radargrams, and mapping variations in reflected amplitudes, is that this acts as a background filter. Often, GPR radargrams can have significant banding noise caused by equipment designs as well as reverberations from the ground surface. It is difficult to see reflective changes in the ground surface waves, or any part of the radargram where banding background noises are significant. By slicing the data, the banding noise is removed because only deviations from the averaged background noise are being detailed. GPR practitioners often will apply a series of filtering operations to remove background noise before beginning the time slice analysis. This filtering can have the unfortunate effect of removing features, which are more or less parallel to the radar profile. If a linear feature like a wall is being imaged, then a possibility of deleting this wall could occur if the structure is parallel to the profile direction. In data presentation for archaeological sites, radargrams that are *unfiltered* for background noise are often more useful for creating subsurface images. When line noise dominates the radargram dataset, using background filtered radargrams is the preferred dataset to create slices. For display purposes of 2D radargrams themselves, the background removed data are much more useful for identifying archaeological targets than are the raw radargrams.

Migrated data are often used at archaeological sites where the microwave velocities can be determined, subsurface features and stratigraphy are well defined, and where hyperbolic reflection patterns dominate a site. Migration is used to reduce hyperbolic diffraction tails, but often results in reduced relative signal amplitudes. Detection is usually the most important objective. The fact of a small target having a larger footprint because it has not been migrated is sometimes more useful when time slice images are made for archaeological sites. In un-migrated datasets, small features appear larger than they really are and can be detected. It is not unusual to use several sets of processed and unprocessed radargrams to extract different and essential information regarding an archaeological site.

Image processing is an important consideration in generating time slices. In some cases the time slices are presented as either linear, square root, or logarithm transforms to delineate various features that have a large dynamic range in recorded reflection intensity. Also, custom transforms are generated to amplify a particular reflection within a dataset, which may be the archaeological targets of interest. It is not always the case that the targets of interest are the strongest reflectors. Desired targets may be the mid-range to lowest reflection anomalies mapped across a site. In addition, minimum and maximum thresholding is often applied to the time slice maps to enhance or reduce background noise. Within each separate time window chosen, the full range of colors is generally applied between the relatively weakest and the strongest reflectors. The relative amplitude time slices can show the relative features within each time window clearly. However, the importance of these anomalies when compared with the strongest reflections in the entire dataset can be lost. For this reason, a series of time slices are sometimes displayed using an absolute reference amplitude in order to show the true recorded reflection amplitudes between different reflection times at a site. Finally, in order to convert time slices into actual depth slices, knowledge of the microwave velocities at the site is required.

15.4. Depth Determination

Determining the depth of a GPR survey can be accomplished by several methods:

1. making laboratory measurements of the dielectric and conductivity;
2. placement of a known object at depth to measure the two-way travel time;
3. wide angle measurement using separated transmitter–receiver antenna;
4. matching the shape of hyperbolas detected on the GPR radargram.

The first method involves collecting of samples of material at a site and then performing geophysical measurements back at the laboratory. One essential problem with performing laboratory measurements is that it is difficult not to disturb the material, and the in situ properties can be lost during extraction and transport. The second method of burying an object in the ground and then measuring the two-way travel time to the object provides reasonable estimates. However, usually the ground has been disturbed above the emplaced target and the measurement may be corrupted. Often a two-way travel time may be measured near a cut in the ground and an object can be inserted from the side so as not to disturb the upper material (Conyers and Lucius, 1996). This method provides a closer measurement to the microwave velocity at a site. In wide-angle measurements (sometimes referred to as "CMP") the travel–time slope of a reflecting surface is used to provide an estimate of the microwave velocity. This method requires a flat reflector at depth and relatively homogenous materials over the length of the field measurement.

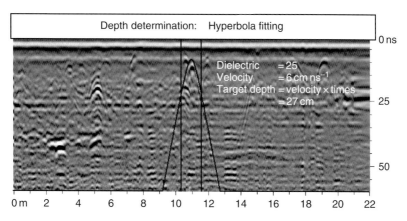

Figure 15.2 Example of matching a hyperbola to raw field data; the shape of the hyperbola gives a measure of the microwave velocity of the ground. Narrow hyperbolas indicate slow microwave velocities in the ground; broad hyperbolas indicate higher velocities. The depth to an object is the microwave velocity × travel time/2, where the one-half comes from the fact that travel time is the two-way travel time and thus the one-way travel time is half this.

The fourth method involves matching of hyperbola observed on recorded radargrams and is the best method for remotely measuring the electrical properties of material. The width of hyperbolas recorded off small (point source) objects is the "DNA" of the ground which provides an estimate of the microwave velocity. The narrower the hyperbola, the slower the microwave velocity is; the wider the hyperbola, the faster the microwave velocity of the ground surrounding the target is. Shown in Figure 15.2 is the match of hyperbola seen in data collected at a Native American burial site in Louisiana.

A common concern encountered by GPR practitioners in every archaeological survey is the depth of penetration to set the radar control unit for recording. In general, the depth of recording should be at least 1.5–2 times the depth of the targets. Also, the area to be surveyed should be much larger than the suspected target area. Just looking at target anomalies that fill up a grid is not very useful. It is necessary to see areas that have no anomalies to bring visual value to the surveyed areas that do.

15.5. Case Histories

In this section, several GPR case histories at archaeological sites are presented. The results show some very successful studies with radar to remotely sense and define subsurface archaeology – and all without excavation! In addition, other imaging techniques mentioned such as isosurface rendering, static corrections, overlay analysis, and synthetic radargrams, will be briefly introduced as they are applied to the selected studies.

15.5.1. Case History No. 1: The Forum Novum, Tiber Valley, Italy

The Forum Novum town and marketplace were constructed in the first century BC in Italy's Tiber Valley. The town flourished well into the fourth century AD. and fell into disarray after the Lombard invasions (Gaffney et al., 2004). At present, the ancient foundations of this town are below ground. In some locations, differences in tonal patterns on aerial photographs have provided indications of the presence of subsurface buildings. The tonal patterns that appear in agricultural fields are referred to as crop marks. These slight discolorations in the vegetation above buried walls are caused by various materials leaching out of the buried walls or changes in soil moisture above the walls, which eventually causes slight alteration in the color of the surface vegetation. Roman archaeologists can often map a site simply by reading aerial photographs; however, not all subsurface features are illuminated on the surface by crop marks. The Forum Novum is a good example of site where subsurface archaeology did not manifest itself on the surface vegetation.

A partial reconstruction of the marketplace at the Forum Novum is shown in Figure 15.3 in front of the church. Areas adjacent to this marketplace were surveyed with GPR in hopes of detecting continuation of buried structures. A GPR survey was completed using a 500-MHz antenna and collecting data along parallel lines spaced at 50 cm intervals across the site. Shown in Figure 15.4 (Goodman et al., 2004a) are two time slice images from an area located next to the market place. In this time slice map a very shallow image of near-surface reflections (0–7.8 ns) and a deeper image from about 50 cm (16.4–24.2 ns) depth are shown. In the shallower image various geometrical features can be seen, such as a circular

Figure 15.3 A photograph of the Forum Novum, Tiber Valley, in Vescovio, Italy is shown. A first century AD church in foreground with reconstructed ruin of a Roman marketplace is shown. Fields adjacent to the reconstructed ruins were surveyed during several field studies from 1998 to 2001.

GPR Archaeometry 487

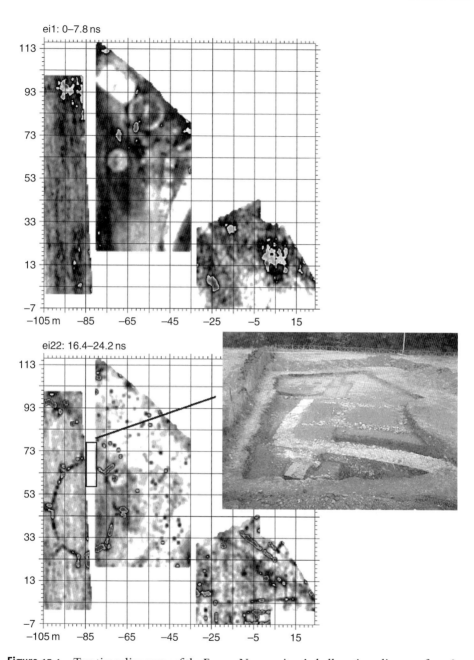

Figure 15.4 Two time slice maps of the Forum Novum site. A shallow time slice map from 0 to 7.8 ns shows numerous geometrical features, some of which can be seen by the naked eye as crop marks at the site. A deeper time slice map from 16.4 to 24.2 ns shows an oval reflection which is an amphitheater having eight entrances. Excavation of one entrance (photograph courtesy of the British School of Archaeology, Rome) shows a fallen coping stone which would have been a part of a base that supported wooden rafters and seating.

feature and a rounded rectangular feature. The rounded rectangular feature is the known foundation of a destroyed mausoleum. The time slice image at the deeper 50 cm level shows a comparatively large oval structure. The recorded reflections are from a buried wall defining the inner boundary of an undiscovered amphitheater. No crop marks on the surface of the site could be detected from aerial photographs indicating the presence of the destroyed urban amphitheater – that had gone undetected till the GPR survey illuminated this subsurface structure. Excavation at one of the eight entrances leading into the amphitheater shows a wall about 80 cm thick and buried at a depth of 50 cm (Figure 15.4). The total depth of the standing wall below ground is 1.5 m. The amphitheater could be dated to the late first century AD. based on shards of pottery that were found near the excavated wall of one of the entrances. From the GPR records, it could be determined that the amphitheater was probably a wood construction as no outer oval embankment wall was found. This interpretation was later corroborated by excavation results. From the subsequent discovery of the amphitheater, several of the geometric features imaged on the shallow time slice map, such as the round circle could be interpreted and are believed to be structures built for the training of gladiators.

In another area just next to the reconstructed ruins, time slice images reveal many rectangular forms that are believed to be living quarters (Goodman et al., 2004b) (Figure 15.5). Corridors and doorways into these rooms can be seen in the images. Increased levels of noise at greater depths (near 30–40 ns) suggest fallen wall materials. It is interesting to note that the very top time slice shows a Y-shaped anomaly, which corresponds to two dirt roads, void of vegetation, coming together on the site.

Several other important structures were discovered at the Forum Novum site over a period of 3 years of field surveys (Figure 15.6). In addition to the amphitheater, a large Roman villa containing an interior atrium and pond were imaged. Excavations at the villa revealed very shallow walls within 25 cm of the ground surface. The walls could not be seen clearly on the 2D radargrams because the ground wave interference was significant. Nonetheless, the time slice maps were able to easily distinguish the wall reflections from within the reverberating ground surface waves. The bathhouse at the Forum Novum is characterized by a large (flat) anomaly, which probably indicates a shallowly buried floor. A 7-m^2 rectangular feature in front of the church is believed to be a foundation, which would have supported a large monument. Several crypts flanking the angular garden walls were also discovered.

GPR imaging has been able to help archaeologists determine the major structural layout of the Forum Novum site. All was accomplished without any extensive excavation other than to verify structures and target potential sites with the most potential for recovering artifacts. The clarity in the GPR images on use in Roman sites from the Forum Novum site show the potential of this remote sensing tool.

15.5.2. Case History No. 2: The Villa of Emperor Trajanus of Rome, Italy

Roman Emperor Marcus Ulpius Trajanus (AD 52–117) enjoyed not only the lavish lifestyle that Rome offered, but he also enjoyed the ruggedness of the high alpines of the Affilani Mountains, located to the east of Rome (Figure 15.7). In this area a summer villa was erected for the Emperor on over 9 ha of land. Documents indicate

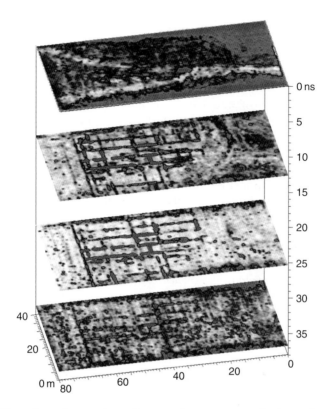

Figure 15.5 Time slices of an area just adjacent to the reconstructed ruins at the Forum Novum marketplace. Many rectangular rooms with entrances and corridors are clearly imaged in the time slices. The buildings are believed to be ancient living quarters for the people that ran the first and second century AD marketplace.

that Emperor Trajanus came to the villa often to engage in hunting expeditions (Fiore and Mari, 1999). The location of the villa was rediscovered after early eighteenth and nineteenth century excavations were made. The villa is believed to have been built over several terraces supported by thick walls with counterforts and niches. Walls from public buildings that are located at the entrance to the villa on the lowest terrace have undergone recent excavations and restorations (Figure 15.7). The entire villa is to be partially resurrected over the next several decades.

Less than 5% of the known buildings at the Villa were discovered before GPR surveys, which initially began in 1998 in conjunction with the Institute of Technologies Applied to Cultural Heritage (ITABC-CNR, Italy) and the Soprintendenza Archeologica per il Lazio (Italy). Extensive surveys of the villa grounds, made from 1998 to 2002, have covered approximately 2 ha. Some of the initial results from the GPR survey indicate that a beautiful geometric building once adorned the site. Shown in Figure 15.8 are two time slice maps for the Villa of Trajanus. In the upper map a large oval structure is imaged very close to the ground surface (Piro

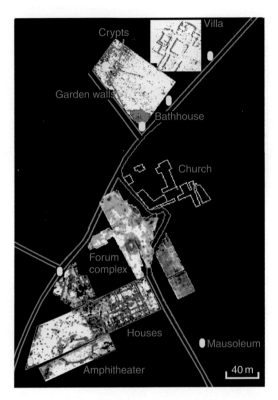

Figure 15.6 A complete map of GPR results from 1998 to 2001 overlaid on the Forum Novum site. The location of a Roman villa, amphitheater, housing, monument foundation, and angular garden walls were detected beneath the site.

Figure 15.7 Photograph of excavated walls that were once part of the public buildings located at the entrance to the Villa of Trajanus.

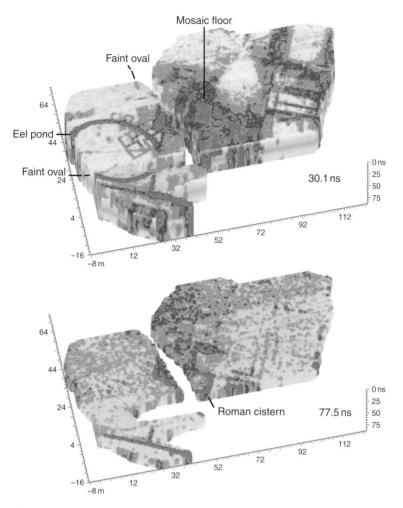

Figure 15.8 Two time slice images of the Villa of Trajanus at 30.1 ns and 77.5 ns are given. In the shallower time slice, an oval representing a buried eel pond, another portion of a faint oval from a previous and earlier construction, and many rectangular buildings are imaged. At the deeper depths at the site, a Roman cistern signified by a square and circular reflection anomaly is discovered.

et al., 2003). This prominent oval structure seen on the 30 ns time slice map is not a swimming pool but is believed to be an eel pond. Roman emperors ate eel fish for invigoration as these oily fishes are considered good for sustaining one in hot summer months.

Another interesting and subtle feature imaged shows part of what appears to be a part of a very faint oval, which is co-located with the stronger imaged oval. The faint oval may represent an earlier structure that pre-dates this main oval. This interpretation is consistent with an earlier oval having been robbed of its stonework and thus accounting for the lower reflections recorded here. Several rectangular features juxtaposed next to the eel pond, and slightly lower in reflection strength, suggest that these areas may have had the wall material removed as well. On the deeper time slice map (77.5 ns) a square-like feature appears. This reflection is thought to be a preserved mosaic floor. This inference is alluded to because at the deepest time slice map another structure that is square on its exterior and has a circular interior wall was discovered. This feature is more than likely a Roman cistern. A patio–mosaic floor would normally be adjacent to a cistern at a Roman villa. These time slice images, which detail the former Villa's structures at a variety of depths, were developed from 500 MHz GPR profiles collected along parallel transects that were spaced at 50 cm intervals (Piro et al., 2003).

Another possible method to display the 3D volume of the GPR amplitudes is to compute an isosurface rendering. One of the first archaeological applications of isosurface rendering was for imaging stone burials located beneath sixth century burial mounds in Japan (Goodman et al., 1997). Rendering is an analysis in which surfaces of equal amplitude within the 3D volume are illuminated. An equal amplitude surface within the volume to be rendered is called an "isosurface." In general, the calculation of reflected light from an isosurface is a very complicated and time-consuming process. A good approximation of the light reflected from an isosurface is to simply compute the tilt and rotation of an element on the isosurface and assign a color that will give the appearance of introducing shadowing along the isosurface. For instance, surfaces that are 90° away from the viewing angle will be assigned dark colors progressively becoming lighter in colors as the reflecting surface is tilted and rotated to 0°.

There are an infinite number of isosurfaces that can be displayed within a 3D volume. Any particular amplitude isosurface may show significant correspondence to real structures buried within the ground, but it is the choice of the archaeologist which surface has more relation to structures that may be present beneath the site. To show the many possible shapes and detail of subsurface structures, a large number of different isosurfaces within the 3D volume are computed. This analysis has been applied to the Villa of Trajanus. Shown in Figure 15.9 are three isosurface rendering. The isosurfaces represent the 20th, 80th, and 140th strongest reflectors out of a total of 256 reflecting surfaces chosen. The isosurface rendering can indicate the general shape of subsurface structures, and in this case the shape of buried walls comprising the eel pond. Other subtle structures that are known to exist at the site (Figure 15.8) are more difficult to ascertain with rendering images; nonetheless, where contrasts are good, as they are in the case of buried eel pond walls, a 3D visualization provides more lifelike renderings of the data. Viewing renderings in a computer animation that shows the 3D structures rotating in space provides the interpreter a very useful method to transcribe the geophysical data into a useful and more realistic presentation.

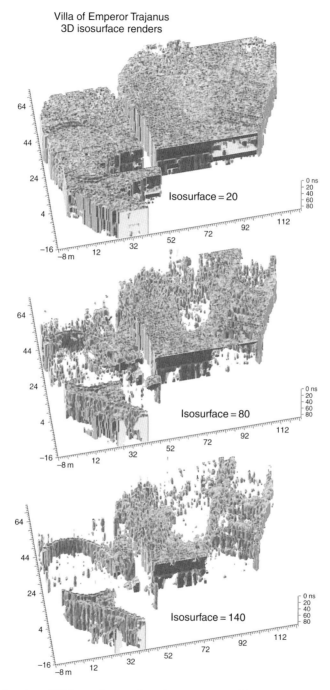

Figure 15.9 A series of 3D isosurface renders show a variety of different amplitudes, computed for the Villa of Trajanus. The rendered amplitude isosurface that is the 140th strongest amplitude of a possible 256 surfaces shows walls of the oval eel pond.

15.5.3. Case History No. 3: Wroxeter Roman Town, England

The Roman legions had built a fortress in the northern province of Brittania by 60 AD. (Renfrew and Bahn, 2000), about which the town of Wroxeter grew and prospered. Wroxeter was the fourth largest town in ancient England (180 acres), and its ruins, although mostly below ground, have been able to largely survive. Modern development in this part of Britain's hinterlands never culminated, and was fortunately diverted to the nearby and present day Shrewsbury. Today the site is marked by only a few standing walls of the bathhouse and garrison; the rest remains preserved underground. To begin the process of revealing this ancient town several GPR surveys (in addition to magnetometry and contact resistivity) were initiated in the 1990s in conjunction with the University of Birmingham and the University of Bradford (Nishimura and Goodman, 2000).

GPR surveys were expedited using a low-frequency 300 MHz GPR antenna with a relatively coarse profile spacing of 1 m. The soil was wet during data collection, which limited the depth of penetration and the quality of the data. Nonetheless, applying some filtering to the radargrams, relatively noise-free time slice maps were generated. Shown in Figure 15.10 is a shallow time slice map from 12 to 18 ns (approximately 36–48 cm depth). In this map a narrow rectangular feature was imaged. This feature is believed to be part of a Roman building that had all the stonework removed, leaving just a filled trench where the foundations once

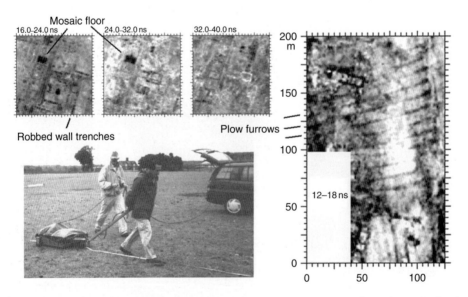

Figure 15.10 Photograph of the GPR survey at Wroxeter Roman Town, Shrewsbury, England. The linear features seen on one time slice image are believed to be medieval furrows to delineate agricultural ownership. In another area, a small rectangular feature was identified and interpreted as being an intact mosaic floor. Robbed wall trenches account for some of the linear reflections measured from destroyed building foundations.

existed. The radar detected differences in the material that formed and filled the trench. Other features on this time slice map include a series of linear striations, which are parallel to one another and have a spatial distance of about 10 m. The reflections are from the bottoms of trenches, or furrows, which were believed to have been dug in the medieval period to delineate agricultural ownership. It is remarkable that the direction of the present day field cropping is perpendicular to the medieval furrows and that GPR can differentiate these subtle signals in the depth record.

15.5.4. Case History No. 4: Saitobaru Burial Mound No. 100, Japan

Several centuries of mound building to entomb the dead in Japan were practiced during the Kofun period (300–700 AD). Many of these burial mounds are still intact today. One early sixth century burial mound in Saitobaru, Miyazaki, is a keyhole (zenpo koenbun) shaped mound (Figure 15.11). This mound is one of over 350 burial mounds preserved in this ancient cemetery. The length of this earthen work is over 70 m. During construction of this kind

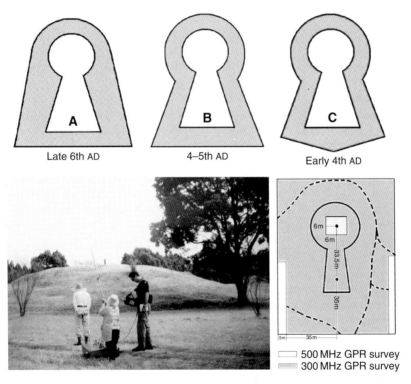

Figure 15.11 A photograph of Kofun No. 100 at the Saitobaru National Burial Mounds in Miyazaki Prefecture, Japan, is shown. The main objective of the survey was to delineate the shape of the outside moat wall. Three possible designs to be investigated could help archaeologists better determine the date of construction.

of burial mound, a moat was dug and the material from the moat used to construct the mound. The shape of the outside boundary of the moat can be used as an additional piece of evidence to help date the burial within few decades of its actual construction. Three possible designs for the shape of burial moat were to be investigated at the site (Figure 15.11). The earliest construction in the fourth century may have had an angled bottom part of the keyhole (model C), whereas the late sixth century would have had a flat bottom with straight embankments along the outside portion of the moat (model A). An out embankment that follows the designs of the mound itself (model B) can be dated to the mid-sixth century. The design of the ancient moats is rarely known because nearly 1500 years of natural deposition and weathering have completely buried these ancient mound features.

In the late 1990s, a GPR survey was initiated at Kofun Burial Mound No. 100 to discover the shape of the buried moat. Before the GPR survey, initial excavations were fruitless in finding the moat boundary. This is quite often the case when the materials that fill in the moat appear identical to the material below it and no texture or coloring changes in the soils can be seen. Two surveys were conducted at the mound. A high-resolution survey was completed on the top portion of the mound using a 500-MHz antenna, with a 25-cm profile separation and profiles collected in orthogonal north–south and east–west directions. This survey was designed to detect and map any possible burial remains in this location. The second survey was made with a lower frequency 300 MHz antenna with a profile line spacing of 1 m, in hopes of locating the burial moat surrounding the mound.

A series of time slice maps from the lower resolution study using the 300-MHz antenna over the site are shown in (Figure 15.12). Several modern features including a walking path can be seen on the shallowest time slice. At depth several flat terraces are imaged, suggesting the stonework that once completely adorned the entire mound at one time is still partially preserved on the flatter areas on the mound. The deeper time slice maps between 193 and 335 cm shows indications of portions of a burial moat. The moat structure does not appear completely on any individual slice made between these levels. To obtain a comprehensive map, which might better describe the shape of the moat, a special process called "overlay analysis" was applied.

In overlay analysis, the slicing levels, which contain structures of interest, are first identified. Next, the computer is programmed to collect only the strongest reflector at each location on the grid by tallying through the individual maps. A comprehensive map is then drawn which contains only the relative strongest reflectors from the series of time slice maps chosen. The word "relative" strongest reflector is used because each map has its own color transform applied to enhance anomalies. In implementation of overlay analysis, the strongest colors from the individual map are overlaid on the comprehensive image; thus each level has it own relative gaining. Absolute gaining can also be applied in overlay analysis; however, very often one level will have stronger overall reflected energies, which will overpower the comprehensive image and not allow the weaker levels to participate in the final image.

GPR Archaeometry

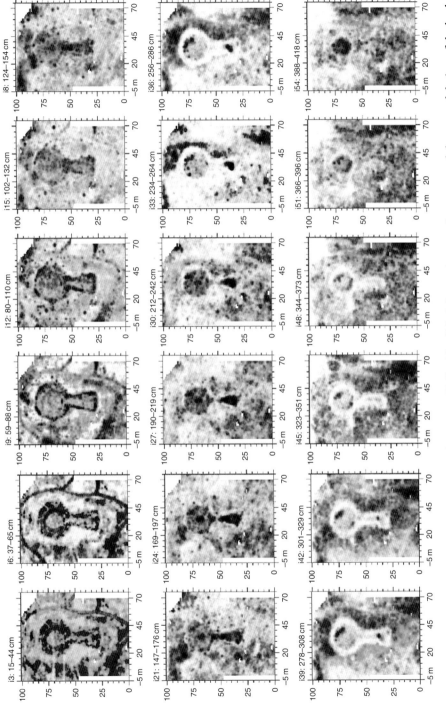

Figure 15.12 A complete set of time slice images taken across Kofun No. 100. A footpath, several terraces and the general design of the keyhole shaped mound are imaged on the shallower time/depth maps. At deeper depths greater than 190 cm, reflections from within the burial moat are recorded.

Overlay analysis is not limited to data from a single antenna or from a single source. This method can be applied for different data to create a comprehensive reflection map.

Some data from the 500 MHz survey in addition to the 300 MHz data are used in the final overlay image. The overlay process can be visualized using the data from Kofun Burial Mound No. 100 (Figure 15.13). One image from the high-resolution 500 MHz survey where the burial was detected very clearly, is shown along with six of the time slice levels from the 300-MHz survey that appears to have reflections from the moat of the burial mound. The 500-MHz time slice map showing the burial reflection is first completely drawn. Starting at 234 cm for the 300-MHz data, a time slice from this level is drawn next where only the pixels stronger than the previous map are shown. The next map, from 249 cm, shows only those areas that are stronger (in color) than the two previously drawn maps. The fourth map shows only those colors that are stronger than the previous three maps, and so on. At 278 cm depth map, reflections from the left side of the burial are beginning to be imaged as more strong reflection structures occur.

Making an overlay time slice where all these colors representing the relative-strongest-reflectors from each of the individual time slices is overlaid onto a single map is shown in the bottom image in Figure 15.13. From this overlay image, the shape of the moat can be partially discerned. Model A shown in Figure 15.11 can definitely be ruled out because the moat shape discovered does not have straight angled sides; it has corners that emulate the shape of the corners of the inside (keyhole) mound. There has been some debate whether model B or model C form was the ancient design for the mound. The data do not indicate the bottom of the keyhole has a straight, flat bottom, nor does it suggest a perfect symmetrical triangular bottom either. The eastern side of the keyhole bottom, however, does have some relative stronger reflectors that suggest half a triangular form is still present beneath the mound. Archaeologists at the Saitobaru Conservation and Archaeological Museum believe that model C is a more likely candidate given the radar data. Using GPR information, the mound moat was reconstructed during a restoration project. Several excavation trenches were placed across the moat after the radar survey, and no indication of a moat could be detected. For this site, GPR was superior to the human eye in helping to solve an archaeological problem.

15.5.5. Case History No. 5: Saitobaru Burial Mound No. 111, Japan

Often GPR data are collected at sites with significant topography that need to be studied. These situations require special needs to correct for the topography and to properly image subsurface structures. Inserting the topography, referred to as making "static corrections" to the 2D radargrams, normally requires that the 2D velocity across the site be known. Often, a single nominal velocity or a 1D velocity profile is used to adjust for the topography. Shown in Figure 15.14 is a radargram collected over a sixth century burial mound in Saitobaru, Miyazaki, Japan. Typical static corrections only adjust the radargram scans upwards or downward based on the two-way travel time of the topographic difference from some base level. When sites have significant topographic variations (i.e., on the order of the depth of

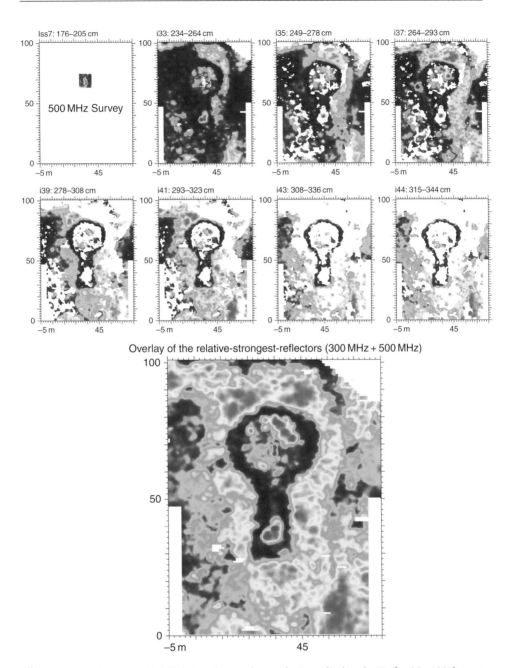

Figure 15.13 An image highlighting the overlay analysis applied to the Kofun No. 100 dataset. In the first step, a time slice image is drawn (from the 500 MHz survey), and then the portions from the next time slice (from the 300 MHz survey) that are stronger in reflection (color) are collected and overlaid. This continues until all the desired maps used in the overlay are included. The final overlay image detailing the shape of the burial moat is shown in the bottom figure and contains all the relative-strongest-reflectors chosen from time slice maps in the range from 176 to 344 cm. This analysis indicates that model C is the most likely candidate for describing the design for this early fourth century AD burial mound.

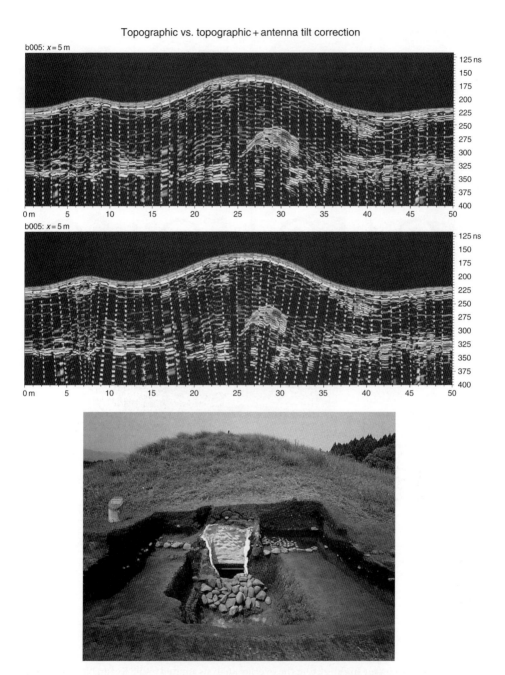

Figure 15.14 A radargram that has been corrected for topography (upper) and the same radargram corrected for the topography as well as the tilt of the antenna across a burial mound (Kofun No. 111, Saitobaru National Burial Mounds, Miyazaki Prefecture). The result indicates that in areas with significant topography that the antenna tilt must be accounted for.

penetration of the microwave signals recorded), the tilt of the antenna needs to be corrected (Goodman et al., 2006a).

In Figure 15.14 the lower radargram has been corrected for the antenna tilt. The tilt is computed from the slope of the topography. The steepest portions of the mound, which are about 25°, shifted the deepest reflectors almost 2 m horizontally considering the total depth of penetration of the microwaves. This amount of shift for a high-resolution archaeological survey can significantly alter interpretation of the images if it is not properly accounted for. In this particular example, the entrance over the subterranean chamber is shown on the radargram profile. The topography with antenna tilt makes the radargram footprint of the chamber slightly closer to its real size and shape.

15.5.6. Case History No. 6: Monks Mound, Cahokia, Illinois

The largest Native American mounds in North America are located just east of the confluences of the Mississippi, Illinois, and Missouri Rivers at Cahokia, Illinois. Monk's mound is by far the largest earthen-work mound in North America (Figure 15.15). The mound was built by thousands of workers near 1000 AD and was probably used as a temple site. The temple site was surrounded by a village

Figure 15.15 A complete set of time slice images processed from the GPR survey made on the top of Monk's Mound. Several features related to possible backfilled excavations, a footpath, and maybe gopher tunnels are detected.

sustaining over 25 000 inhabitants. Alluvial soils were loaded in baskets and laid down to create a temple mound that has a base larger than some pyramids. The site has probably been excavated numerous times, and the complete record of legal and illegal works on the site is not readily known. A GPR survey was conducted on the top of the Monk's Mound to determine whether an early twentieth century dwelling, since removed from the site, or evidence of a modern burial from the early twentieth century inhabitant of the site could be discovered.

A GPR survey was conducted using a 400-MHz antenna and a 50-cm profile spacing across the highest terrace on Monk's Mound. Shown in Figure 15.15 are a series of time slice maps from this area.

Several features are identified in the images. On the very top image, a strong linear reflection corresponds to a footpath across the upper terrace. The footpath is characterized by compacted soils that are free of vegetation. This feature reverberates into the deeper time slice maps. The "reverberation" of the footpath is caused by less energy being transmitted through this region to deeper areas of the time slice images. The top strong reflector (e.g., the smooth and compacted ground on the footpath) absorbs and reflects more energy away from deeper levels causing a gaining artifact from higher levels in the reflection history of the transmitted wave. Seismic reflection processes will often adjust deeper gains of return signals locally to account for shallower stratigraphic influences of deeper transmitted signals. This gaining correction procedure to date has not yet infiltrated the GPR field, but will be shortly.

On the second time slice from 5 to 12 ns, a rectangular feature can be identified. This feature may be related to previous excavation of the site; GPR is often able to detect the backfilled trenches even after many years since the excavation was completed and covered back. On a deeper map from 27 to 34 ns (~81–102 cm), several continuous and very narrow, linear features can be identified. These features may be gopher tunnels because several exit tunnels could be seen near these anomalies. A gopher tunnel could show up as a strong anomaly because void spaces are good GPR reflectors. Conclusive evidence for a modern burial cannot be unequivocally found on the site. Some strong reflections seen on the 9–16 ns and co-located on 36–43 ns time slice maps may be possible candidates for test excavation to determine whether these areas are in fact related or not to any modern burials.

15.5.7. Case History No. 7: Jena Choctaw Tribal Cemetery, Louisiana

An example of a GPR survey to rediscover unmarked graves from the 1960s and 1970s was recently investigated. The Jena Choctaw Tribal cemetery located in the Kisatchie National Forest, Louisiana, is a typical site in which headstones and other designations of gravesites were either poorly maintained or suffered the ill-effects of looters. The purpose of the survey at Kisatchie was to rediscover the graves without excavation, which was not a possibility in any event, for this sacred site. The kinds of targets to be located could be intact wood coffins, collapsed wood coffins, metal coffins, or just plain burial pits with no coffin.

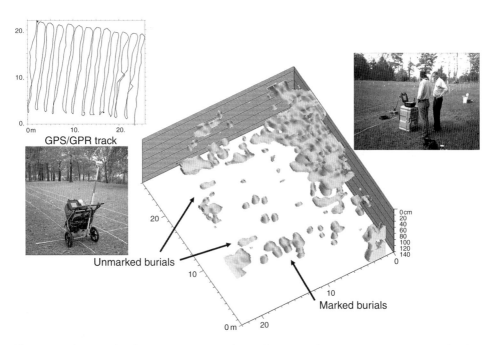

Figure 15.16 Results from GPR–GPS study to detect modern graves at a cemetery in the Kisatchie National Forest, Louisiana. Navigation relied solely on the GPS navigation; random GPR locations were monitored every 1 second in the field and used to create random grids which were interpolated to create complete 3D volumes. The location of marked and unmarked graves was identified on a 3D isosurface render for the Jena Chocktaw Whiterock Tribal Cemetery.

For creating GPR anomaly maps of gravesites, it is useful to have an absolute coordinate system. For this reason, a global positioning satellite system (GPS) was used to navigate the GPR antenna across the site (Figure 15.16). Using GPS to create maps in UTM coordinates would be useful for reoccupying the site by conservators to relocate the unmarked graves. A further advancement in the imaging for this study incorporated the (random) GPS coordinates to create 3D volumes of data that could be tied in with the UTM coordinates (Goodman et al., 2004b).

A total of 580 s of GPR data were recorded over the 20 × 18 m site. A single GPR radargram was collected for the entire survey which could be done quickly because the equipment was in a continuous recording mode. The data were processed to compute windowed averages of the squared reflection amplitude averaged every 0.5 s along the (random) profile track. Binning of the data in the vertical time domain was done at about 7 ns, with overlapped time windows. The results shown in a 3D isosurface indicate the location of several known intact graves at the site, as well as several anomalies that are interpreted as being from unmarked graves (Figure 15.16).

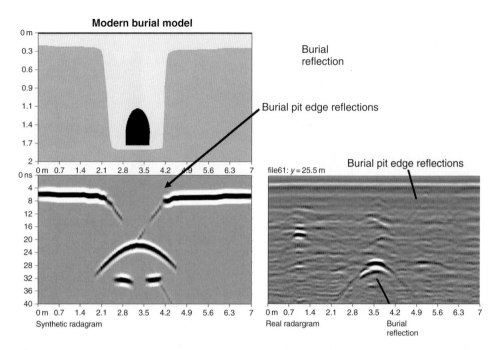

Figure 15.17 A synthetic radargram and a comparison with a real radargram taken across a coffin in a burial pit are shown. Reflections from the rounded edges of the top of the burial pit can create half hyperbolas which point downward toward the center of the burial.

There are other additional clues that can be used to identify potential gravesites. This involves studying the reflections from burial pit edges. A useful model to interpret the GPR dataset was to compute a synthetic radargram using GPRSIM© software (following Goodman, 1994) of a burial pit with an intact coffin (Figure 15.17). The synthetic radargram is also compared to a portion of a real radargram collected at the site. The profile is taken over a known gravesite at the Jena Choctaw Tribal cemetery in which a coffin is believed to be intact within the burial pit. At this particular site the coffin gives a strong reflected signature. The backfilled soil materials above the coffin also show some stronger scattered energy is recorded. The soils in this area do not re-compact themselves that quickly and the disturbed backfilled soils from the 30- to 40-year-old burials can be distinguished from unbroken ground at the site.

The telltale key that is often ignored regarding the signatures at burial pits, is that the edges of the pits can sometimes yield valuable information. Shown in the real data from the Jena Choctaw site is a half of a hyperbolic reflection that starts near the surface and points down toward the center of the suspected coffin (Figure 15.17). This half hyperbola is caused by a rough edge that exists at the pit boundary, which is still highly contrasted with the backfilled material at the shallower depths. The edges of pits in many kinds of soils can usually be detected, though they may represent only a very faint signature. The synthetic model has one side of the pit edge as a round digitized model and the other side is a perfect 90° corner. The rounded edge gives a half a hyperbola that points down toward the center of the pit. If a pit has both of its pit edges rough, then two half hyperbolas that may cross in the center of the pit can occur. Multiple reflections between an intact coffin and the walls of the pit can also create a complex radargram image (Goodman, 1994).

15.5.8. Case History No. 8: Glaumbaer Viking Age, Iceland

A Viking age site in northwest Iceland was the location for a GPR survey in 2005. The Glaumbaer site was identified as the location of a collapsed Viking age turf structure but the dimensions and layout were unknown. Turf structures in Iceland have an interior driftwood framework (Figure 15.18). In Iceland the Norse insulated their houses with thick (2 m) turf walls. The turf is primarily organic material, cut from the upper portions of bogs. In many cases, most large pieces of wood were removed and reused when the structures were abandoned. Over the last 1000 years, substantial aeolian deposition has completely obscured these structures. Once identified, Viking Age turf structures can, in many cases, be easily dated by tephrochronology, using the thin and fine variously colored volcanic tephra layers that fell from volcanic eruptions. The white/yellow tephra layer dated to 1104 AD (also seen in Figure 19) is the most distinct. Tephra layers are more commonly preserved in building turf than in general aeolian material that surrounds these buried structures. Therefore, other than the slightly elevated organic content, little else, other than soil color and mottling, is available to distinguish Viking Age turf from the surrounding aeolian deposition — in essence, the target contrasts are small.

The site was surveyed with a 400-MHz antenna in both x and y directions with a 0.5-m profile separation. Several time slice images were first tried before a special application for overlay analysis was implemented. Comprehensive time slice datasets from both combined x and y profiles were first generated independently. Using the overlay analysis described in Case History No. 5, the x profiles and the y profiles were graphically overlaid with the relative-strongest-reflectors from both surveys to generate comprehensive 2D time slice maps (after Goodman et al., 2006b). In this instance, the logic for applying time slice analysis was that target structures may not be completely on a level plane built in the ground, and that collecting the relative-strongest-reflectors over a large time slice range will "fill in" more of the subsurface reflections.

A comprehensive overlay image for the Glaumbaer Viking age site is shown in Figure 15.18. Some of the reflections identified from partial excavations at the site revealed fallen turf walls. An iron smithy could be inferred from small fragments of slag found near this southernmost reflection structure and a midden was also later verified by excavations.

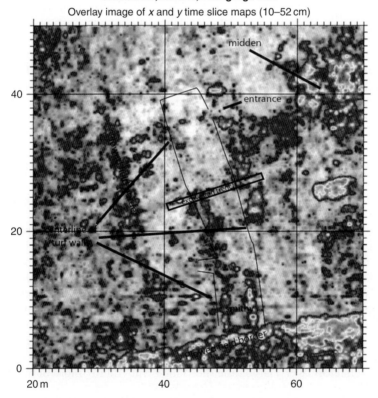

Figure 15.18 Overlay image comprising individual x and y time slice images, computed for the Viking age site in Glaumbaer, Iceland. Photograph at the top shows an excavation of a turf house wall; a stratigraphic cut shows a white tephra layer that is also found in the excavation. Several other features, including an iron smithy and a midden, are identified and verified by excavation (after Goodman et al., 2006b).

ACKNOWLEDGMENTS

The Icelandic surveys were sponsored in part by the US National Science Foundation under BCS grants 0107413, 0453892, and 0731371.

The support of Yuki, Lewis and Clark, Lazarus and Rachel Goodman made these writings possible. All images were constructed with GPR-SLICE© v5.0 Ground Penetrating Radar Imaging Software and GPRSIM© v3.0 Ground Penetrating Radar Simulation Software.

REFERENCES

Batey, R.A., 1987, Subsurface interface radar at Sepphoris, Israel, 1985. Journal of Field Archaeology, Vol. 14, pp. 1–8.

Bevan, B. and Kenyon, J., 1975, Ground-penetrating radar for historical archaeology. MASCA Newsletter, Vol. 11, No. 2, pp. 2–7.

Bevan, B.W., 1977, Ground-penetrating radar at Valley Forge, Geophysical Survey System, North Salem, New Hampshire.

Bevan, B.W., 1991, The search for graves. Geophysics, Vol. 56, pp. 1310–1319.

Conyers, L.B. and Connell, S., 2007, The applicability of using ground-penetrating radar to discover and map buried archaeological sites in Hawaii. Hawaiian Archaeology Journal, Vol. 11, pp. 62–77.

Conyers, L.B. and Cameron, C.M., 1998, Ground-penetrating radar techniques and three-dimensional computer mapping in the American Southwest. Journal of Field Archaeology, Vol. 25, pp. 417–430.

Conyers, L.B. and Goodman, D., 1997, GPR: An introduction for archaeologists: Walnut Creek, CA, Alta Mira Press, pp. 232p.

Conyers, L.B. and Lucius, J.E., 1996, Velocity analysis in archaeological ground penetrating radar studies. Archaeological Prospection, Vol. 3, pp. 25–38.

DeVore, S.L., 1990, Ground-penetrating radar as a survey tool in archaeological investigations: An example from Fort Laramie national historic site. The Wyoming Archaeologist, Vol. 33, pp. 23–28.

Dolphin, L.T., Tanzi, J.D. and Beatty, W.B., 1978, Radar probing of Victorio Peak, New Mexico. Geophysics, Vol. 43, pp. 1441–1448.

Fiore, M.G. and Mari, Z., 1999, La Villa di Traiano ad Arcinazzo Romano, Guida alla lettura del territorio, Soprintendenza Archeologica per il Lazio.

Fischer, P.M., Follin S.G. and Ulriksen, P., 1980, Subsurface interface radar survey at Hala Sultan Tekke, Cyprus, in Fischer, P.M. (ed.), Application of Technical Devices in Archaeology: Studies in Mediterranean Archaeology, Vol. 63, pp. 48–51.

Gaffney, V., Patterson, H. and Roberts, P., 2004, Forum Novum-Vescovio: Studying urbanism in the Tiber Valley. Journal of Roman Archaeology, Vol. 14, pp. 59–79.

Goodman, D., 1994, Ground penetrating radar simulation in engineering and archaeology. Geophysics, Vol. 59, pp. 224–232.

Goodman, D. and Nishimura, Y., 1993, Ground radar view of Japanese burial mounds. Antiquity, Vol. 67, pp. 349–354.

Goodman, D., Nishimura, Y., Hongo, H. and Noriaki, N., 2006a, Correcting for topography and the tilt of the GPR antenna. Archaeological Prospection, Vol. 13, pp. 157–161.

Goodman, D., Nishimura, Y., Okita, M. and Hongo, H., 1997, 3-D GPR amplitude rendering of Saitobaru Burial Mound #13, in Filtering, Optimization and Modelling of Geophysical Data in Archaeological Prospecting: Politecnico di Milano, Fondaazione ing. Carlo Maurilio Lerici, 50th Anniversary issue, pp. 93–101.

Goodman, D., Nishimura, Y. and Roger, J.D., 1995, GPR time slices in archaeological prospection. Archaeological Prospection, Vol. 2, pp. 85–89.

Goodman, D., Piro, S., Nishimura, Y., Patterson, H. and Gaffney, V., 2004a, Discovery of a 1st century AD Roman amphitheater and other structures at the Forum Novum by GPR. Journal of Environmental and Engineering Geophysics, Vol. 9, pp. 35–41.

Goodman, D., Schneider, K., Barner, M., Bergstrom, V., Piro, S. and Nishimura, Y., 2004b, Implementation of GPS navigation and 3D volume imaging of ground penetrating radar for identification of subsurface archaeology, in Proceedings, Symposium on the Application of Geophysics to Engineering and Environmental Problems, Environmental and Engineering Geophysical Society, Denver, pp. 806–810.

Goodman, D., Steinberg, J., Damiata, B., Nishimura, Y., Schneider, K., Hongo, H. and Higashi, N., 2006b, GPR overlay analysis for archaeological prospection, in Proceedings of the 11th International Conference on Ground Penetrating Radar, Columbus, June 2006, paper 141.

Grasmueck, M. and Weger, R., 2002, 3D GPR reveals complex internal structure of Pleistocene oolitic sandbar. The Leading Edge, July, pp. 634–639.

Imai, T., Sakayama, T. and Kanemori, T., 1987, Use of ground-probing radar and resistivity surveys for archaeological investigations. Geophysics, Vol. 52, pp. 127–150.

Isaaks, E.H. and Srivastava, R.M., 1989, An introduction to applied geostatistics: New York, Oxford University Press, 592 p.

Kvamme, K.L., 2001, Current practices in archaeogeophysics: Magnetics, resistivity, conductivity, and ground-penetrating radar, in Goldberg, P., Holliday, V.T. and Ferring, C. R., eds., Earth Sciences and Archaeology: New York, Kluwer/Plenum Publishers, pp. 353–382.

Milligan, R. and Atkin, M., 1993, The use of ground-probing radar within a digital environment on archaeological sites, in Andresen, J., Madsen, T. and Scollar, I., eds., Computing the Past: Computer Application and Quantitative methods in Archaeology: Aarhus, Denmark, Aarhus University Press, pp. 285–291.

Neubauer, W., Eder-Hinterleitner, A., Seren, S.S. and Melichar, P., 1999, Integrated geophysical prospection of Roman Villas in Austria, in Archaeological Prospection, 3rd International Conference on Archaeological Prospection, Munich, Sept., Bayerisches Landesamt fur Denkmalpflege, pp. 62–64.

Nishimura, Y. and Goodman, D., 2000, Ground penetrating radar survey at Wroxeter. Archaeological Prospection, Vol. 7, pp. 101–105.

Nishimura, Y. and Kamei, H., 1990, A study in the application of geophysical survey, in Pernifka, E. and Wagner, G., eds., Proceedings, Archaeometry '90 Conference, Heidelberg, April, pp. 757–765.

Piro, S. and Goodman, D., 2008, Integrated GPR data processing for archaeological surveys in urban areas: the case of Forum (Roma), in 12th International Conference on Ground Penetrating Radar, Birmingham, England (in press).

Piro, S. Goodman, D. and Nishimura, Y., 2003, The study and characterization of Emperor Traiano's villa using high-resolution integrated geophysical surveys. Archaeological Prospection, Vol. 10, pp. 1–25.

Renfrew, A.C. and Bahn, P., 2000, Archaeology: Theories, methods and practice: London, Thames and Hudson, 640 p.

Sheets, P.D., Loker, W.M., Spetzler, H.A.W. and Ware, R.W., 1985, Geophysical exploration for ancient Maya housing at Ceren, El Salvador, National Geographic Research Reports, Vol. 20, pp. 645–656.

Utsi, E., 2006, Improving definition: GPR investigations at Westminster Abbey, 11th.

International Conference on Ground Penetrating Radar, Columbus, Ohio, paper 83.

Vaughn, C.J., 1986, Ground-penetrating radar surveys used in archaeological investigations. Geophysics, Vol. 51, pp. 595–604.

Vickers, R. and Dolphin, L.T., 1975, A communication about an archaeological radar experiment at Chaco Canyon: Museum Applied Science Center for Archaeology, Univ. of Pennsylvania, Philadelphia, Newsletter, Vol. 11.

INDEX

Accessory equipment, 400
Accidental spill sites, 262
Active layers, 187–188
Adsorptive capacity, soil, 181
Advanced imaging and analysis tools, 166
Advective–dispersive transport behavior, 229–230
Aeolian dunes, 291
   bounding surfaces, 285
   composed of, 291
   preservation of, 288–289
Air-coupled antenna
   data processing, 407
Air-coupled systems, 398
Air-fall laminae, sand dunes, 276
Airfields, 433
Air pulse, 404
Alpine glaciers, 364
   Commonwealth Glacier, 365f
Amplitude *versus* offset (AVO) analysis, 256, 258, 265
   data for, 258
Analog-to-digital conversion (ADC) *see* Ground penetrating radar (GPR) systems
Ancient aeolian sandstones, 290
Antennas, 23
   classes, 100
   configurations, 399
   directivity, 24
      elevation in, 27f
   efficiency for, 106
   elevation, impact of, 450
   and GPR system testing, 399
   for ground penetrating radar, 112
      array antennas, 128
      coupling into dielectric, 113
      frequency domain antennas, 124
      time domain antennas, 115
   parameters, 102
      bandwidth, 106
      coupling energy into the ground, 105
      directivity, 105
      efficiency, 106
      energy transfer, 102
      footprint, 110
      gain, 104
      patterns, 108
      phase centre, 108
      polarisation, 107
      sidelobes and back lobes, 106
      time sidelobes and ring-down, 109
   shielding, 27
   system, 457
Archie's law variations, 8
Argillic horizons, 188
Array antennas, 128
   beam steering, 132
   32-channel, 130f
   16-channel radar system, 130f
   timed-array antenna, 131f
   see also Antennas
Attenuation and dispersion effects, 143–145
Attribute analysis, data processing, 167
Automated *vs.* user controlled systems, 411
AVO analysis *see* Amplitude *versus* offset (AVO) analysis

Back lobes, 100
Bagley Ice Field *see* Temperate firn
Band-pass filter in frequency domain, 153f
Bandwidth, 78
   antennas, 106
      fractional bandwidth, 106–107
Basic ray-tracing methods, 169
Bayesian techniques, 215–216
Beach ridges, 306
Biconical antennas, 120

Biodegradation effects of contaminant, 253
Biomonitoring, 193
  antenna and image analysis system, 193
  crosshole tomography, 195
  3D root mapping, 194–195
  GPR usage, 193
  high-frequency radar, 196
  postcollection processing, 194
Bituminous pavements:
  defect in, 420
  thickness and moisture, 420
Black Rapids Glacier *see* Temperate hydrology
Bonded water, 51
  dielectric phenomenon, 51–52
Borden aquifer, 262
Borehole:
  radar, 74
  transmission measurements, 234*f*
  moisture content estimation, 218
Bouncing test, 404
Bound charge effects, 48
Bounding surface facies, 282–283
Bound materials, 420
Bound structures, 420
Bow-tie antennas, 120
  see also Antennas
Brahmaputra (Jamuna), Bangladesh, 331
  cross strata, types of, 334–335
  depositional model for, 335*f*
  evolution of, 333*f*
Bridges:
  applications, 428
  deck surveys, 426
  general, 425
Bruggeman–Hanai–Sen (BHS)
  mixing model, 252
Butterworth filter, 153*f*, 155–157

Calamus, Nebraska, 329
Calcareous soils, 180–181
Capillary fringe, 228
Carbonate-fractured
  rock aquifer, 228–229
Carbon gas emissions, 232
Cation-exchange capacity (CEC), soil, 181

CEC *see* Cation-exchange capacity (CEC), soil
Center frequency and GPR, 19
Channel-belt deposits, 343–344
Channel geometry, 327, 329, 331, 340–342
  evolution of, 329
  model of, 332*f*
Characterizing system, 20
Chronostratigraphic gaps, 283
Clay particles in soils, 181
Clutter, 453
CMP *see* Common mid-point (CMP), surveys
Coastal environments and GPR:
  images, 305
  coastal erosion, signatures of, 307
  coastal paleochannels, 308
  coastal progradation, record of, 306
  deltas, 312
  lithological anomalies, signal response to, 310
  reservoir characterization, 313
  limitations in, 304
  methodology, 301
    seismic-reflection profiles, 301
    subsurface data, collection, 301–302
  strengths in, 303
  studies in, 305
Coastal erosion, signatures of, 307
  marker horizons, 308
  OSL dating, 308
Coastal paleochannels, 308
  channel-margin (contour) reflections, 309–310
  mixed-sediment barriers, 308–309
  storm–breach channels, 309
Coastal progradation, record of, 306
  GPR reflections, 306
  granulometric information, 306
Common mid-point (CMP), 31, 31*f*
  surveys, 158–159, 212, 258
Common-offset reflection survey, 30
  GPR frequency selection, 31
  parameters for, 30
Complex refractive index model (CRIM), 208, 251, 259

Concrete pavements, 422
Conductivity, 54
  imaginary component of, 55
  process of, 54f
Contaminant:
  case studies, 259
    accidental spill sites, 262
    controlled DNAPL injection, 260
    controlled LNAPL injection, 262
    leachate and waste disposal site
      characterization, 264
  data processing and interpretation, 257
    AVO analysis, 258
    data differencing, 257
    frequency-dependent properties,
      detection based on, 258
    NAPL, quantitative estimates
      of, 258
    trace attributes, 257
    visualization, 257
  distribution of, 254
    DNAPL, 254
    inorganics, 255
    LNAPL, 255
    saturated and unsaturated
      zone, 256
  electrical properties of, 249
    biodegradation effects, 253
    common organic, 250t
    inorganics, 253
    NAPLs, 249
    soil and rock with NAPL
      contamination, 250
  GPR methodology, 256
  plumes, 255–256
  residual zones of, 256
  types of, 248
Coupling energy into ground,
  antennas, 105
Coupling into dielectric, antennas, 113
  far-field power density, 114f
Crevasses, Ross Ice Shelf, 376, 377f
  GPR sensitivity of, 377
CRIM see Complex refractive index model
  (CRIM)
Crosshole tomography, biomonitoring, 195

Cross-strata:
  sand dunes, 278–279
    3D image of, 290
  small-scale, 324–326, 331, 334–335
CW, 75

Data collection:
  general, 401
  positioning, 404
  reference sampling, 405
  setups and files, 403
Data differencing and contaminant, 257
Data processing:
  developing practice, 145
  flow for GPR, 33–34
  and interpretation of contaminant, 257
  principles of, 143
  processing, imaging and visualisation,
    171
  steps for, 148
    advanced imaging and analysis
      tools, 166
    attribute analysis, 167
    data/trace editing and rubber-band
      interpolation, 148
    deconvolution, 158
    dewow filtering, 150
    elevation or topographic corrections,
      159
    filtering, 152
    gain functions, 161
    migration, 164
    numerical modelling, 168
    time-zero correction, 150
    velocity analysis and depth conversion,
      158
Data processing and interpretation:
  air-coupled antenna data processing, 407
  automated vs. user controlled systems,
    411
  dielectric values or signal velocities, 410
  general, 405
  GPR data preprocessing, 406
  ground-coupled data processing, 408
  interpretation of structures and other
    objects, 411

Data/trace editing and rubber-band
    interpolation, 148
  desaturation, editing tool, 149
  spline interpolation, 149
DDS source *see* Direct digital synthesis
    (DDS) source
Deconvolution, 34–35
  and data processing, 158
  spatial, 36
Deltas, 312
  GPR transect, 312*f*, 313*f*
  proglacial and glacial–marine, 312–313
Dense nonaqueous phase liquid (DNAPL),
    254, 260, 262
  pool development, 261–262
Desaturation:
  editing tool, 149
  function for saturated traces, 149*f*
Design criteria for GPR systems, 80
  total dynamic range (TDR), 81
Deterioration, 426
Dewow, 34
Dewow filtering, 150
  raw GPR trace, 151*f*
DFT *see* Discrete fourier transform (DFT)
DGPS *see* Differential global positioning
    systems (DGPS)
Dielectric constant, 7
Dielectric permittivity, 6, 251
Dielectric values or signal
    velocities, 410
Differential global positioning systems
    (DGPS), 129, 285
Diffractions, sand dunes, 293
  hyperbolae, 293
Digital IF, 95
Dipole, antennas, 116
  current and charge distribution, 116*f*
  electric field component, 117
  see also Antennas
Dipole moment density, 45–46
Direct digital synthesis (DDS) source, 86
Directional antennas, 100
  see also Antennas
Directivity, 105
Discrete fourier transform (DFT), 86

DNAPL *see* Dense nonaqueous phase liquid
    (DNAPL)
DOE *see* US Department of Energy (DOE)
Dry valleys *see* Alpine glaciers
Dune age and migration, 288
  OSL measurements, 288
  reflection terminations, 288
  relative chronology, 288
Dynamic hydrological processes for water
    resource research, 224
Dynamic range and radio frequency, 77
  stacking for, 77–78

East Antarctic Ice Sheet (EAIS), 362–364
Effective media modeling, 251–252
Electrical conductivity, 6, 8, 180
Electrical parameters of dielectrics, 44
Electrical resistivity tomography, 264
Electromagnetic material properties, 41
  electrical parameters of dielectrics, 44
    conductivity, 54
    permeability, 55
    permittivity, 45
  electromagnetic wave, relationship, 57
    complex effective permittivity
      expression, 59
    loss factor and skin depth, 59
  real materials characterising response, 62
    air borne radar, 42
    mixing models, 63
    universal dielectric response, 62
    volumetric and inclusion-based mixing
      models, 64
  real materials properties, 60
    time domain reflectometry (TDR),
      60–61
    vector network analyser method,
      60–61
  theory, 43
Electronic and atomic polarisation, 48
Elevation or topographic corrections for
    data processing, 159
  sampling interval and spatial accuracy
    requirements, 160*t*
Elevation static corrections, sand dunes,
    281

EM signal attenuation, zones of, 233–235
Energy transfer form antennas, 102
  components, 102
  field boundaries, 103
  Rayleigh distance, 104
Englacial stratigraphy *see* West Antarctica
Environmental noise, sand dunes, 291
  high-frequency electromagnetic waves, 291–293
Equiangular antennas, 127
Equivalent-time sampling (ETS), 83
  and real-time, 82*f*
  sampling interval, 83
ETS *see* Equivalent-time sampling (ETS)
Evaporites, sand dunes, 291

Faraday's law of induction, 43
Far Field/Fraunhofer zone, antennas, 104
FDTD technique *see* Finite-difference, time domain (FDTD) technique
Federal Highway Administration (FHWA), 361
Field cells, 169–171
Fifth wheel systems, 405
Filtering, data processing, 152
  filter types, 152
  frequency–wavenumber (FK) filters, 157–158
  two-dimensional filters, 157
Finite-difference, time domain (FDTD) technique, 169
  GPR bistatic reflection survey, 170*f*
Finite impulse response (FIR) filters, 155–157
FIR filters *see* Finite impulse response (FIR) filters
F–K migration, 164–165
Flicker noise, 89–90
Flow and transport parameters, distribution, 214
  direct mapping approaches, 216
  hydraulic conductivity, 214–215
  invert images of spatial variation, 215–216
  structural approach, 217
  time-lapse tomographic GPR data, 217

Fluvial deposits and GPR resolution, scales of, 324
Fluvial sediment:
  fluvial forms and stratasets, scales of, 325*f*
Fluvial sedimentology, use of GPR, 327
  antennae in, 325–326
  Brahmaputra (Jamuna), Bangladesh, 331
  Calamus, Nebraska, 329
  Fraser and Squamish Rivers, Canada, 349
  Mesozoic deposits of SW USA, 353
  Niobrara, Nebraska, 336
  Pleistocene outwash deposits in Europe, 350
  Sagavanirktok, northern Alaska, 343
  South Esk, Scotland, 327
  South Saskatchewan, Canada, 340
FMCW systems *see* Frequency modulation continuous wave (FMCW) systems
$1/f$ noise, 89–90
Footprint, antennas, 110
  low-attenuation media, 112
  plan resolution, 112
  radiated impulse, amplitude of, 110*f*
  radiation, 111*f*
  Ricker wavelet, 111*f*
Forum Novum, Tiber Valley, Italy, 486
FPG accumulation, 233–235
Fraser and Squamish Rivers, Canada, 349
Fraunhofer zone, 104
Free charge and interfacial polarisation, 53
  process, 53*f*
Free charge effects, 48
Free-phase gas (FPG)
  accumulation, 233–235
  content, 236*f*
Free space impedance, 11
Free water, 49
Frequency-dependent properties, 258
Frequency domain antennas, 124
  equiangular antennas, 127
  horn antennas, 127
  vivaldi, 126
Frequency modulation and GPR systems, 89
  sinc function, 89*f*

Frequency modulation continuous wave (FMCW) systems, 141–142, 362, 386–387
Frequency–wavenumber (FK) filters, 157–158
Fresnel reflection coefficients, 13
Fresnel zone, 104
 antennas, 104
Fusion with sensors, 469
FWD, 413

Gain, antennas, 104
Gain functions, 161
 AGC, 162
 filter function, 162–164
 SEC or energy decay, 162
 user-defined, constant, linear or exponential gains, 162
Gated, stepped-frequency, frequency-modulated continuous wave, 76
 factors, 76–77
Gating, 77, 90
Gaussian pulse, 115
Gauss' theorem, 43
Geophysical survey systems, 300f
Glaciers and ice sheets:
 Alaska, 379
  Bagley Ice Field, temperate firn, 384
  Black Rapids Glacier, temperate hydrology, 385
  Gulkana Glacier, temperate valley glacier, 382
  Matanuska Glacier, temperate valley glacier, 380
 Antarctica:
  dry valleys, alpine glaciers, 364–365
  glacial features of, 363
  ice velocity, 363
  McMurdo sound, ice shelf, 373
  radarsat composite image of, 364f
  Ross Ice Shelf, crevasses, 376
  West Antarctica, englacial stratigraphy, 371
  West Antarctica, polar firn, 367
  Whitmore Mountains, 372f
 Glaumbaer Viking Age, Iceland, 505

GPR applications on roads and streets:
 bituminous pavement thickness and moisture, 420
 bound structures, 420
 concrete pavements, 422
 defect in bituminous pavements, 420
 general, 416
 GPR in QC/QA, 423
 gravel road-wearing course, 423
 other subgrade applications, 418
 soil moisture and frost susceptibility, 417
 subgrade quality and presence of bedrock, 416
 subgrade surveys, site investigations, 416
 unbound pavement structures, 419
GPR archaeometry, 479
 case histories, 485
  Forum Novum, Tiber Valley, Italy, 486
  Villa of Emperor Trajanus of Rome, Italy, 488
 depth determination, 484
 field methods:
 imaging techniques, 482
  Glaumbaer Viking Age, Iceland, 505
  Jena Choctaw Tribal Cemetery, Louisiana, 502
  Monks Mound, Cahokia, Illinois, 501
  Saitobaru Burial Mound No. 100, Japan, 495
  Saitobaru Burial Mound No. 111, Japan, 498
  Wroxeter Roman Town, England, 494
GPR data analysis with other road survey data:
 and FWD, 413
 general, 413
 GPS, digital video and photos, 415
 other data, 416
 profilometer data, 414
GPR data preprocessing, 406
GPR hardware and accessorie:
 accessory equipment, 400
 air-coupled systems, 398

Index

antenna and GPR system testing, 399
antenna configurations, 399
general, 397
ground-coupled systems, 398
GPR in QC/QA, 423
GPS, digital video and photos, 415
Gravel road-wearing course, 423
Grid Area West (GAW), 376f
Ground-coupled data processing, 408
Ground-coupled systems, 398
Ground Penetrating Radar (GPR)
  and antennas, 112
  data processing and analysis steps, 147f
  methodology and contaminant, 256
    for monitoring, 256
  plateau, 11
  source near an interface, 11
    wavefronts from localized, 12f
Ground penetrating radar (GPR), electromagnetic principles of, 4–5
  components of, 18f
  constitutive equations, 6
  data analysis and interpretation, 33
    deconvolution, 35
    dewow, 34
    migration, 36
    time gain, 34
    topographic correction, 36
  electromagnetic fields, wave nature of, 8
    reflection, refraction, and transmission at interfaces, 13
    resolution and zone of influence, 14
    scattering attenuation, 16
    source near an interface, 11
    wave properties, 10
  material properties, 7
  Maxwell's equations, 6
  signal measurement, 17
    antennas, 23
    antenna directivity, 24
    antenna shielding, 27
    center frequency, 19
    characterizing system response, 20
    recording dynamic range, 22
    signal acquisition, 20
    time ranges and bandwidth, 18

survey methodology:
  common-offset reflection survey, 30
  multioffset common midpoint/wide-angle reflection and refraction velocity sounding design, 31
  sampling criteria, 29
  surveys, 30
  transillumination surveys, 31
Ground Penetrating Radar (GPR), water resource research:
  dynamic hydrological processes, monitoring, 224
  carbon gas emissions from soils, 232
  hyporheic corridor, 231
  moisture content in vadose zone, 225
  rhizosphere, 232
  solute transport in fractures, 229
  water table detection, 228
  flow and transport parameters, distribution/zonation of, 214
  hydrostratigraphic characterization, 209
  moisture content estimation, 217
  petrophysics, 206
Ground Penetrating Radar (GPR) and coastal environments:
  images, 305
    deltas, 312
    lithological anomalies, signal response to, 310
    paleochannels, 308
    progradation, record of, 306
    reservoir characterization, 313
    signatures of, 307
  limitations in, 304
  methodology, 301
    seismic-reflection profiles, 301
    subsurface data, collection, 301–302
  strengths in, 303
Ground Penetrating Radar (GPR) and sand dunes, 274
  aeolian bounding surfaces, 285
    interdune surfaces, 286
    reactivation surfaces, 285
    superposition surfaces, 285
  age and migration, 288
  ancient aeolian sandstones, 290

Ground Penetrating Radar (GPR) and sand dunes (*Continued*)
  imaging sedimentary structures and stratigraphy, 281
  pedogenic alteration and diagenesis, 291
    diffractions, 293
    environmental noise, 291
    evaporites, 291
    multiples, 293
    water table, 293
  radar facies, 282
  radar stratigraphy and bounding surfaces, 283
  stratigraphic analysis, 288
  survey design, 277
    direction, 278
    line spacing, 277
    orientation, 278
    step size, 277
    vertical resolution, 278
  three-dimensional images, 290
  topography, 279
    apparent dip, 281
    correction, 281
Ground Penetrating Radar (GPR) in data processin:
  developing practice, 145
  principles of, 143
  processing, imaging and visualisation, 171
  and seismic data, 143–145
  steps for, 148
    advanced imaging and analysis tools, 166
    attribute analysis, 167
    data/trace editing and rubber-band interpolation, 148
    deconvolution, 158
    dewow filtering, 150
    elevation or topographic corrections, 159
    filtering, 152
    gain functions, 161
    migration, 164
    numerical modelling, 168
    time-zero correction, 150
    velocity analysis and depth conversion, 158
Ground penetrating radar (GPR) systems:
  continuous-wave, 86
    frequency modulation, 89
    gating, 90
    stepped-frequency technique, 86
  design criteria for, 80
    system performance, 81
  high-speed data sampling, 73–74
  impulse ground penetrating radar:
    equivalent-time sampling, 83
    operation theory of, 81
    real-time sampling, 83
  impulse radar, design parameters, 84
    implementation, 85
  radio frequency, specifications and definitions, 77
    bandwidth, 78
    dynamic range, 77
    lateral resolution, 79
    unambiguous range, 79
  stepped-frequency radar, design parameters, 92
    implementation of, 93
  types of, 74
    gated, stepped-frequency, frequency-modulated continuous wave, 76
    impulse, 75
    stepped frequency, frequency-modulated continuous wave, 76
    swept frequency-modulated continuous wave, 75
Ground Penetrating Radar Soil Suitability Map of Wisconsin (GSSM-WI), 184
Ground wave measurements, moisture content estimation, 218
GSSM-WI *see* Ground Penetrating Radar Soil Suitability Map of Wisconsin (GSSM-WI)
Gulkana Glacier *see* Temperate valley glacier
Gypsiferous soils, 180–181
Gypsum, evaporites, 291

Half-power beamwidth (HPBW) of
    antennas, 105
Half width, 15
High-amplitude reflections on GPR
    images, 311
Home Free South (HF-S), 376f
Horn antennas, 127
HPBW see Half-power beamwidth
    (HPBW) of antennas
Hydrocarbon fuels, LNAPL contaminants,
    248–249
Hydrostratigraphic characterization, water
    resource research, 209
  common midpoint (CMP) survey, 212
  fractured bedrock characterization, 213f
  ground penetrating radar:
    for fractured bedrock, 213–214
    fractured rock, imaging, 213–214
    images, 211
    for mapping, 209–210
    northern peatland, 211
    northern peatland basin, image, 210f
    stratigraphy, 211–212
    structural boundaries, 211
  hydrogeological models, data for, 211
Hyperbolic velocity matching, 159f
Hyporheic corridor, 231

Ice bottom reflections, 376f
Ice shelf, McMurdo sound, 373–375, 375,
    376–377
IEE see Institution of Electrical Engineers
    (IEE)
IIR filters see Infinite impulse response
    (IIR) filters
Imaging techniques, 482
Impulse, 75
  antenna role, 75
  A-scope presentation, 75
  equivalent time sampling (ETS) in, 75
Impulse radar and GPR systems, 84
  implementation of, 85
    block diagram, 85f
    sampler and signal processing, 86
    timing source, 86
    transmitter, 85

Impulse radar technique, 82f
Induction term, 102
Infinite impulse response (IIR) filters,
    155–157
Initial detection, 463
Inorganic contaminant, 253
    Archie's Law, 253
Inorganic distribution and contaminant,
    255
Institution of Electrical Engineers
    (IEE), 100
Interdune surfaces, sand dunes, 286
Internal defects in forest, 196
International Roughness Index
    (IRI), 414
International Transantarctic Scientific
    expedition (ITASE), 367–369
Interpretation of structures, 411
I & Q data, 93
IRI see International Roughness Index
    (IRI)

Jena Choctaw Tribal Cemetery, Louisiana,
    502
Jurassic Navajo Sandstone, interdune
    surfaces, 286
  and GPR profile, 287f

Kerosene, LNAPL, 248–249, 253, 262
Kirchhoff migration, 164–165

Landmine and unexploded ordnance, 445
  detection in GPR, 462
  electromagnetic analysis, 446
  fusion with sensors, 469
  performance of GPR as, 472
  system design, 455
Lateral resolution for GPR
    systems, 79
Leachate and waste disposal site
    characterization, 264
Light Non-Aqueous Phase Liquids
    (LNAPL), 255–256, 262–264
  biodegradation of, 262–264
Linear reflectors, 188–189
Line spacing, sand dunes, 277

Lithological anomalies and coastal environments, 310
    individual reflections, significance of, 310–311
LNAPL see Light Non-Aqueous Phase Liquids (LNAPL)
Loaded antennas, 117
    efficiency, 118–119
    radiated field pattern, 117f
    resistivity, 119
Loss factor and skin depth, 59
Low-reflectivity zones, 233–235

Macrodispersion experiment (MADE) site, 211
MADE site see Macrodispersion experiment (MADE) site
Magnetic permeability, 6
Magnetic relaxation frequencies and permeability, 55
Mapping geology, 247–248
Marker horizons for coastal erosion, 308
Mars radar exploration programmes, 56
Matanuska Glacier see Temperate valley glacier
Maxwell's EM field equations, 43
Maxwell's equations, 5–6, 8
Maxwell's modified circuit Law, 43
Maxwell–Wagner polarisation effect, 48
McMurdo sound see Ice shelf
Mesozoic deposits of SW USA, 353
Methane, 232–233
    GPR for, dynamics, 233
Microtidal coastlines, 308–309
Migration, 36
Migration, data processing, 164
    hyperbola matching, 165–166
    methodological principle of, 165f
Mixing models, 63
MOG survey see Multioffset gather (MOG) survey
Moisture content estimation, 217
    Borehole transmission methods, 222
    GPR surveying techniques, 218
    GPR tomography, 222
    NMO velocity estimation, 220–221
    procedure for, schematic representation, 220f
    zero-offset profiling (ZOP) technique, 222
Monitoring remediation processes, 247–248
Monks Mound, Cahokia, Illinois, 501
Multioffset gather (MOG) survey, 32
Multioffset measurements for GPR surveys, 31
Multiple-offset reflection methods, moisture content estimation, 218
    GPR reflection:
        datasets, 220–221
        survey, 232f
        pumping test across a sandstone aquifer, 230f
Multiples, sand dunes, 293
    reflecting horizon, 293–294

Nanowires, 253
NAPLs see Non-aqueous phase liquids (NAPLs)
NDT transportation, 396
    bridges:
        applications, 428
        bridge deck surveys, 426
        general, 425
    data collection:
        data collection setups and files, 403
        general, 401
        positioning, 404
        reference sampling, 405
    data processing and interpretation:
        air-coupled antenna data processing, 407
        automated vs. user controlled systems, 411
        dielectric values or signal velocities, 410
        general, 405
        GPR data preprocessing, 406
        ground-coupled data processing, 408
        interpretation of structures and other objects, 411

GPR applications on roads and streets:
  bituminous pavement thickness and moisture, 420
  bound structures, 420
  concrete pavements, 422
  defect in bituminous pavements, 420
  general, 416
  GPR in QC/QA, 423
  gravel road-wearing course, 423
  other subgrade applications, 418
  soil moisture and frost susceptibility, 417
  subgrade quality and presence of bedrock, 416
  subgrade surveys, site investigations, 416
  unbound pavement structures, 419
GPR data analysis with other road survey data:
  and FWD, 413
  general, 413
  GPS, digital video and photos, 415
  other data, 416
  profilometer data, 414
GPR hardware and accessories:
  accessory equipment, 400
  air-coupled systems, 398
  antenna and GPR system testing, 399
  antenna configurations, 399
  general, 397
  ground-coupled systems, 398
  railways:
    ballast surveys, 431
    data collection from railway structures, 430
    general, 429
    subgrade surveys, site investigations, 432
Near field, 104
  antennas, 104
Negative peaks, 406–407
Niobrara, Nebraska, 336
  splay data, 337f
  splay model, 339f
NMO see Normal moveout corrections (NMO)

Non-aqueous phase liquids (NAPLs), 248–249
  and contaminant, 249
  dissolved phase components of, 250
  electrical properties of, 249
  quantitative estimates of, 258
  soil and rock properties with, contamination, 250
  solubility and toxicity of, 249
  Topp relationship, 259
  volumetric content, 252–253
Noninvasive geophysical monitoring, 224–225
Normal moveout corrections (NMO), 159–160
Notch filter, 153f
Numerical forward modelling, GPR analysis tools, 166–167
Numerical modelling for data processing, 168
  methods for, 169

Object classification, 467–468
Omni-directional antennas, 100
Operational bandwidth, 455
Optically stimulated luminescence (OSL) dating, 308
Organic rich soil horizons and luminescence (OSL) dating, 288–289
Organic soils and peatlands, 190
  lower-frequency antennas, 192
  soil classification, 191–192
Orientation, sand dunes, 278
Orientational polarisation and permittivity, 49
OSL dating see Organic rich soil horizons and luminescence (OSL) dating
Overdeepening, 381–382

Paraglacial coasts, 309–310
Patterns for antennas, 108
  far-field radiation, 108f
  linear plot of, 109f
Peak signal amplitude, 9
Permeability, 55
  magnetic relaxation frequencies, 55
  typical material attenuation values, 56f

Permittivity, 45
  dipolar or orientational polarisation, 49
    bonded water, 51
    complex polar materials, 52
    free water, 49
  dipole moment density, 45–46
  electronic and atomic polarisation, 48
  free, pure water spectrum, 51*f*
  free charge and interfacial polarisation, 53
  free space, 45
  real and imaginary components, 49–50
  relaxation phenomena, 48*f*
  static conductivity and relative, 46*t*
  time-dependent displacement mechanism, 46–47
Petrophysics, 206
  critical hydraulic parameters, 206
  dielectric mixing models, 208
  field, global and local scale, correlation, 209*f*
  geologic heterogeneity, 208–209
  Topp equation, 208
Phase centre for antennas, 108
Phase lock loop (PLL) circuitry, 93–94
Pleistocene outwash deposits in Europe, 350
  radar faciesmodel of, 352*f*
  scour-fill deposits, 351–352
  sedimentary and radar facies, 350–351
  trough-shaped cross strata, 352–353
Polar firn *see* West Antarctica
Polarimetry, 449–450, 461
Polarisation of antennas, 107
  circular, 107
Polar liquids, 49
Polar materials, 52
  broadening factor, 52–53
  Cole–Cole formulation, 53
Pomona soil, spodic and argillic horizons of, 188*f*
Positioning, 404
Positive peaks, 406–407
Positive reflection, 406–407
Postcollection processing, biomonitoring, 194

Power density patterns, 114*t*
PPL circuitry *see* Phase lock loop (PLL) circuitry
Practical evaluations of real materials, 60
PRI *see* Pulse repetition interval (PRI)
Principal plane cuts, 108
Processing, imaging and visualisation, data processing, 171
Pulse repetition interval (PRI), 79
Pumping test analyses, 228–229
  across a sandstone aquifer, 230*f*

Quasi-stationary term, 102
Quaternary gravelly deposit, 353

Radar facies, 327, 340–342, 349–353
  sand dunes, 282
    aeolian, 282–283
Radar stratigraphic analysis, 302
Radar stratigraphy and bounding surfaces of sand dunes, 283
Radial resolution length, 15
Radiation term, 102
Radiocarbon dating, sand dunes, 288–289
Railways:
  ballast surveys, 431
  data collection from railway structures, 430
  general, 429
  subgrade surveys, site investigations, 432
Range resolution and GPR systems, 77–78
  and dielectric constant, 79*f*
Rayleigh distance, 104
Rayleigh scattering, 17
Reactivation surfaces, sand dunes, 285
Real-time kinematic (RTK), 285
Real-time sampling, 83
  and equivalent-time, 82*f*
  impulse radar technique with, 84*f*
  sampling bridge, 83*f*
Recording dynamic range, 22
Reflection, refraction, and transmission at GPR interfaces, 13
Relative chronology for dune age, 288
Relaxation frequency, 46–47

Reservoir, coastal environments, 313
  hydrocarbon and hydrogeology
    applications, 313
Resolution:
  and penetration depth of, 180
  and zone of influence, 14
    components, 14
Reverse time migration, 164–165
Rhizosphere, 232
  sap flow measurements, 232
Ricker wavelet, 111f
Ring-down response of GPR, 21–22
Rising water table, sand dunes, 288–289
3D root mapping, biomonitoring, 194–195
Ross Ice Shelf, 363–364, 376
RTK see Real-time kinematic (RTK)
Rubber-band interpolation, 148

Sagavanirktok, northern Alaska, 343
  basal erosion, 343–344, 345f
  channel-belt deposits, 343–344, 345f
  compound braid bars, 343
  dune migration, 347–348
  gravelly braided river deposits, model,
    344–345
  open-framework gravels, 344–345
Saitobaru Burial Mound:
  No. 111, Japan, 498
  No. 100 Japan, 495
Sampling criteria for GPR, 29
Sand conductivity, dependence of, 251f
Sand dunes and GPR, 274
  aeolian bounding surfaces, 285
    interdune surfaces, 286
    reactivation surfaces, 285
    superposition surfaces, 285
  age and migration, 288
  ancient aeolian sandstones, 290
  classification, 274–276
  clay minerals in, 291
  imaging sedimentary structures and
    stratigraphy, 281
  morphology, 275f
  pedogenic alteration and diagenesis, 291
    diffractions, 293
    environmental noise, 291
    evaporites, 291
    multiples, 293
    water table, 293
  pedogenic modification of, 289
  radar facies, 282
  radar stratigraphy and bounding surfaces,
    283
  stratigraphic analysis, 288
  survey design, 277
    direction, 278
    line spacing, 277
    orientation, 278
    step size, 277
    vertical resolution, 278
  three-dimensional
    images, 290
  topography, 279
    apparent dip, 281
    correction, 281
    survey, 280
Sandy coasts, evolution of, 307
Sap flow measurements, 232
SAR see Synthetic aperture radar (SAR)
Saturated and unsaturated zone,
  contaminant, 256
Scaling, 426
Scattering attenuation, 16–17
SEC gain see Spherical exponential
  compensation (SEC) gain
Seismic data, 143–145
  digital processing of, 143–145
  and GPR, 143–145
Sensor fusion, 470
Shear zones, 376–377
  Radarsat image of, 376f
Shielded antennas, 29
Side lobes, 100
Sidelobes and back lobes,
  antennas, 106
Side-swipe, 380–381
Signal acquisition, 20
Signal amplitude of GPR, 5f
Signal measurement and GPR, 17
Signal scattering, 234f
Single-offset reflection methods, moisture
  content estimation, 218–219

Soil:
    attribute index values, 184
    carbon gas emissions from, 232
    clay minerals, 181
    GPR in organic soils and peatlands, uses of, 190
    grain dissolution process, 253
    ground penetrating data and surveys, 185
    moisture and frost susceptibility, 417
    moisture dynamics, 227
    properties, 180
        adsorptive capacity, 181
        electrical conductivity, 180
        surface conduction, 181
    and rock with NAPL contamination, 250
    suitability maps for GPR, 181
Soil Survey Geographic (SSURGO) database, 184
Soil surveys, data, 185
    antennas for, 186
    GPR usage in, 187
    horizons, thickness and depth of, 187–188
    radar interpretations for, 187
Solute transport in fractures, 229
South Esk, Scotland, 327
South Saskatchewan, Canada, 340
    basal erosion, 340–342, 341f
    compound-bar and channel-fill deposits, 340–342
    radar profiles, 341f
    unit bars, 340
Spatial filter, data processing, 154
Spherical exponential compensation (SEC) gain, 257
Spodic horizons, 188
Spreading and exponential correction (SEC) gains, 141–142
SSURGO database *see* Soil Survey Geographic (SSURGO) database
Star dunes, 274–276
State Soil Geographic (STATSGO) database, 182
STATSGO database *see* State Soil Geographic (STATSGO) database

Stepped frequency, frequency-modulated continuous wave, 76
    advantages of, 76
Stepped-frequency radar and GPR systems, 92
    implementation of, 93
        frequency-synthesized source, 93
        sampler and signal processing, 95
        transmitter and receiver, 94
    quadraphase modulation, 93
    technique for, 86
Stepped-frequency technique, 86
    direct digital synthesis (DDS) source, 86
    discrete fourier transform (DFT), 86
    gating, 90
        diagram of, 90f
        gate signals, timing sequence of, 90f
        time domain pulse response equivalent, 92
    sampled sine wave, 89f
Step size, sand dunes, 277
    Nyquist sampling interval, 277–278
    spatial aliasing, 277–278
Storeys, 324
Subgrade applications, 418
Subgrade quality and presence of bedrock, 416
Subgrade surveys, site investigations, 416
Subsurface data, collection, 301–302
Superposition surfaces, sand dunes, 285
Surface conduction, soil, 181
Surface Penetrating Radar, 100
Surface reflection methods, moisture content estimation, 218
Survey design for sand dunes, 277
Survey direction, sand dunes, 278
Swept frequency-modulated continuous wave, 75
    synthesized pulse, 75
Synthesized pulse, 75
Synthetic aperture radar (SAR), 74

Tapered impedance travelling wave antenna (TWIT), 122
Target separation, 18

TDR see Time domain reflectometry (TDR)
TEM horn antennas, 122
Temperate firn, Bagley Ice Field, 384
Temperate hydrology, Black Rapids Glacier, 385, 386f, 387f
 GPR profiles of, 386
Temperate valley glacier:
 Gulkana Glacier, 382, 383f
 Matanuska Glacier, 380
  ablation zone of, 381f
  location of, 374f
  off-axis reflections, 380–381
  profile elevation, 380–381
Temporal filter, data processing, 152
Temporal gains, 161
Tetrachloroethene plume, 231–232
Tetrachloroethylene, 248–249, 252–253, 260
Three-dimensional images, sand dunes, 290
Tidal inlets:
 channel-fill sequences, 308–309
Time domain antennas, 115
 biconical antennas, 120
 bow-tie antennas, 120
 dipole, 116
 loaded antennas, 117
 TEM horn antennas, 122
Time domain reflectometry (TDR), 208
 probes, 251, 259, 260–261
Time gain, 34–35
Time ranges and bandwidth for GPR, 18
Time sidelobes and ring-down, 109
Time-zero correction, 150
Topographic correction and GPR, 36
Topography for sand dunes, 279
 correction, 281
 dip angle and dip direction, 281
 effects, 281
 methods for measuring, 285
 surveys, 280
Topp equation, 208
Topp relationship, 259
Trace attributes, contaminant, 257
Tracer experiment in fractured rock, 231f

Tracer injection experiment and change in moisture content, 227f
Transillumination surveys, 31
Transition frequency, 10
Transmission–reflection techniques, 169
Transmit antenna, 457
Transmitter, 85
Transmitter blanking, 18
Transverse vector wave fields, EM waves, 13f
Transverse wave equation, 9
Trenches, 327, 334–335, 343
True dielectrics, 44
TWIT see Tapered impedance travelling wave antenna (TWIT)

Ultra-wideband (UWB), 74
Unambiguous range, 79
 and dielectric constant, 80f
 pulse repetition interval (PRI), 79
Unbound pavement structures, 419
Uranium-contaminated aquifer, 231–232
US Department of Energy (DOE), 215–216
UWB see Ultra-wideband (UWB)

Vadose zone, moisture content, 225
VCO see Voltage-controlled oscillator (VCO)
Velocity analysis and depth conversion, data processing, 158
Vertical radar profiling (VRP), 227
Vertical resolution, sand dunes, 278
 theoretical values for, 279t
Vibracores, 327, 331, 334–335
 channel-bar deposits reflect, 331
Villa of Emperor Trajanus of Rome, Italy, 488
Visualization, contaminant, 257
Vivaldi antenna, 126
 consists of, 126
 cutoff frequency, 126
 radiation patterns of, 127f
 see also Antennas
Voltage-controlled oscillator (VCO), 93–94

Voltage-standing wave ratio (VSWR), 106–107
Volumetric and inclusion-based mixing models, 64
    complex refractive index model (CRIM), 64
    Hanai–Bruggeman and Bruggeman–Hanai–Sen (BHS) models, 65–66
VPR *see* Vertical radar profiling (VRP)
VSWR *see* Voltage-standing wave ratio (VSWR)

Water resource research and GPR:
    dynamic hydrological processes, monitoring, 224
        carbon gas emissions from soils, 232
        hyporheic corridor, 231
        moisture content in vadose zone, 225
        rhizosphere, 232
        solute transport in fractures, 229
        water table detection, 228
    flow and transport parameters, distribution/zonation of, 214
    hydrostratigraphic characterization, 209
    moisture content estimation, 217
    petrophysics, 206
Water table, sand dunes, 293
    and GPR profiles, 293
Water table detection, 228
Wave properties, 10
West Antarctica:
    englacial stratigraphy, 371
    polar firn, 367
West Antarctic Ice Sheet (WAIS), 363
Wide-angle reflection and refraction (WARR) sounding mode, 31
Wind ripple laminae, sand dunes, 276
"WOW," 34
Wroxeter Roman Town, England, 494

Zero level, 408–409
Zero-offset profiling (ZOP), 32
    moisture content estimation, 222
    Borehole transmission, 234f
ZOP *see* Zero-offset profiling (ZOP)